# ENGINEERING ASPECTS OF THERMAL FOOD PROCESSING

# Contemporary Food Engineering

**Series Editor**

Professor Da-Wen Sun, Director

*Food Refrigeration & Computerized Food Technology*
*National University of Ireland, Dublin*
*(University College Dublin)*
*Dublin, Ireland*
*http://www.ucd.ie/sun/*

Engineering Aspects of Thermal Food Processing, *edited by Ricardo Simpson* (2009)

Ultraviolet Light in Food Technology: Principles and Applications, *Tatiana N. Koutchma, Larry J. Forney, and Carmen I. Moraru* (2009)

Advances in Deep-Fat Frying of Foods, *edited by Serpil Sahin and Servet Gülüm Sumnu* (2009)

Extracting Bioactive Compounds for Food Products: Theory and Applications, *edited by M. Angela A. Meireles* (2009)

Advances in Food Dehydration, *edited by Cristina Ratti* (2009)

Optimization in Food Engineering, *edited by Ferruh Erdoğdu* (2009)

Optical Monitoring of Fresh and Processed Agricultural Crops, *edited by Manuela Zude* (2009)

Food Engineering Aspects of Baking Sweet Goods, *edited by Servet Gülüm Sumnu and Serpil Sahin* (2008)

Computational Fluid Dynamics in Food Processing, *edited by Da-Wen Sun* (2007)

# ENGINEERING ASPECTS OF THERMAL FOOD PROCESSING

Edited by

RICARDO SIMPSON

CRC Press is an imprint of the
Taylor & Francis Group, an **informa** business

CRC Press
Taylor & Francis Group
6000 Broken Sound Parkway NW, Suite 300
Boca Raton, FL 33487-2742

First issued in paperback 2019



ISBN-13: 978-1-4200-5858-1 (hbk)
ISBN-13: 978-0-367-38554-5 (pbk)



**Library of Congress Cataloging-in-Publication Data**

Engineering aspects of thermal food processing / editor, Ricardo Simpson.
p. cm.
Includes bibliographical references and index.
ISBN 978-1-4200-5858-1 (hardcover : alk. paper)
1. Food--Effect of heat on. 2. Food industry and trade. I. Simpson, Ricardo.

TP371.E54 2009
664'.028--dc22 2009001576

**Visit the Taylor & Francis Web site at**
**http://www.taylorandfrancis.com**

**and the CRC Press Web site at**
**http://www.crcpress.com**

*This book is dedicated to my wife, Anita; family, José Ignacio, María Jesús, Enrique; my beloved mother, Carmen; and to the memory of my father, Jorge.*

# Contents

## PART II Modeling and Simulation



## PART III Optimization

# Series Preface

Food engineering is a multidisciplinary field of applied physical sciences combined with a knowledge of product properties. Food engineers provide technological knowledge essential to the cost-effective production and commercialization of food products and services. In particular, food engineers develop and design processes and equipment in order to convert raw agricultural materials and ingredients into safe, convenient, and nutritious consumer food products. However, food engineering topics are continuously undergoing changes to meet diverse consumer demands, and the subject is being rapidly developed to reflect the market needs.

For the development of the field of food engineering, one of the many challenges is to employ modern tools and knowledge, such as computational materials science and nanotechnology, to develop new products and processes. Simultaneously, improving food quality, safety, and security remain critical issues in food engineering study. New packaging materials and techniques are being developed to provide a higher level of protection to foods and novel preservation technologies are emerging to enhance food security and defense. Additionally, process control and automation are among the top priorities identified in food engineering. Advanced monitoring and control systems have been developed to facilitate automation and flexible food manufacturing. Furthermore, energy saving and minimization of environmental problems continue to be important food engineering issues and significant progress is being made in waste management, efficient utilization of energy, and the reduction of effluents and emissions in food production.

The *Contemporary Food Engineering* series consists of edited books and attempts to address some of the recent developments in food engineering. Advances in classical unit operations in engineering applied to food manufacturing are covered as well as such topics as progress in the transport and storage of liquid and solid foods; heating, chilling, and freezing of foods; mass transfer in foods; chemical and biochemical aspects of food engineering and the use of kinetic analysis; dehydration, thermal processing, nonthermal processing, extrusion, liquid food concentration, membrane processes, and applications of membranes in food processing; shelf life, electronic indicators in inventory management, and sustainable technologies in food processing; and packaging, cleaning, and sanitation. These books are intended for use by professional food scientists, academics researching food engineering problems, and graduate level students.

These books have been edited by leading engineers and scientists from many parts of the world. All the editors were asked to present their books so as to address market needs and pinpoint cutting-edge technologies in food engineering. Furthermore, all contributions have been written by internationally renowned experts who have both academic and professional credentials. All authors have attempted to provide critical, comprehensive, and readily accessible information on the art and science of a relevant topic in each chapter, with reference lists provided for further information. Therefore, each book can serve as an essential reference source to students and researchers at universities and research institutions.

**Da-Wen Sun**
**Series Editor**

# Preface

In the last 10 years, there has been a remarkable growth in research in the field of thermal processing, which indicates that the process is thriving and expanding all over the world. This book has been written with the intention of revising and updating the physical and engineering aspects of thermal processing of packaged foods.

Each chapter has been contributed by a renowned authority on a particular process and in this way the book covers all aspects of thermal processing. The book consists of four parts: (I) Fundamentals and New Processes, (II) Modeling and Simulation, (III) Optimization, and (IV) Online Control and Automation.

Part I consists of six chapters. Dr. Donald Holdsworth has written an outstanding introduction emphasizing the increased use of new packaging materials, including retortable pouches, and the use of containers made from other plastic composite materials. Dr. Silva and Dr. Gibbs have contributed the most complete and up-to-date chapter on pasteurization including a detailed account of the importance of *sous vide* processing. Chapter 3 has been written by top researchers from Unilever and deals with aseptic processing, a field which has expanded and developed in the last decade due to customer demand for better quality products. Chapter 4 is devoted to new and emerging technologies. This chapter is the result of collaboration among selected authors from academia and the industry. Traditional methods have been successful; however, limitations in heat transfer mean that this technology is not capable of providing convenient and high quality products. To overcome these limitations, methods using electromagnetism have been investigated and developed. The first part concludes with two excellent chapters on high-pressure processing by Dr. Gustavo Barbosa-Cánovas and coworkers. Chapter 5 discusses the principles behind four modeling approaches—analytical, numerical, macroscopic, and artificial neural networks—that can be used to predict temperatures in a high-pressure system. Chapter 6 highlights some applications of each modeling approach to high-pressure/low-temperature systems and high-pressure/high-temperature conditions reported in the literature.

Part II also consists of six chapters. Starting with this part, we have included two viewpoints on the crucial topic of thermal inactivation of microbial cells and bacterial spores. Due to the relevance of this subject in thermal food processing, we have asked the most prominent authors to collaborate on this work. Chapter 7 was written by Dr. Micha Peleg and coworkers and Chapter 8 was written by Dr. Arthur Teixeira and Dr. Alfredo Rodriguez. As the processing of heat-preserved foods in flexible pouches has gained considerable commercial relevance worldwide in recent years, Dr. Amézquita from Unilever and Dr. Almonacid from Chile cover the most important aspects of retortable pouch processing and mathematical modeling in Chapter 9. Although thermal processing, or canning, has proven to be one of the most effective methods of preserving foods while ensuring the product remains safe from harmful bacteria, it also has strong effects on the sensory characteristics of the product, such as color, texture, and nutritional value. In Chapter 10, Dr. Ramaswamy and Dr. Dwivedi discuss rotary processing and how it can be used to overcome this difficulty. The last two chapters of this part deal with mathematical modeling. Chapter 11 has been written by Dr. Michele Chiumenti and coworkers and focuses on the mathematical modeling of ohmic heating as an emerging food preservation technology currently used by the food industry. Chapter 12 includes a comprehensive review of computational fluid dynamics and has been written by the well-known professor Da-Wen Sun and coworkers.

Part III consists of four chapters. The whole concept is to understand that mathematical optimization is the key ingredient for computing optimal operating policies and building advanced decision support systems. Chapter 13 on optimization has been contributed by Dr. Julio Banga and his outstanding team. This chapter deals not only with global optimization in thermal processing, but several food processes such as thermal sterilization, contact cooking, and microwave processing that can also be analyzed to find optimal operating procedures computed via global optimization methods. Chapter 14 proposes a new economic evaluation procedure to optimize the system design and operation of multiple effect evaporators compared to the traditional chemical engineering approach based on total cost minimization. Chapter 15 describes the optimization of in-line aseptic processing and demonstrates that it is essential for successful commercial exploitation. Chapter 16 analyzes plant production productivity, although an important problem in food processing, it has received little attention in thermal processing. This type of optimization, scheduling to maximize efficiency of batch processing plants, has become well known and it is commonly practiced in many processing industries.

Part IV consists of two chapters. Chapter 17 describes a practical and efficient (nearly precise, yet safe) strategy for online correction of thermal process deviations during retort sterilization of canned foods. In Chapter 18, authors from academia (Dr. Osvaldo Campanella) and industry (Dr. Clara Rovedo, Dr. Jacques Bichier, and Dr. Frank Pandelaers) analyze and discuss manufacturers' businesses in today's competitive marketplace. For such purposes, manufacturers must face challenges of increasing productivity and product quality, while reducing operating costs and safety risks. Traditionally, plant automation has been the main tool to assist the manufacturer in meeting those challenges.

# Series Editor

**Professor Da-Wen Sun** was born in southern China and is a world authority on food engineering research and education. His main research activities include cooling, drying, and refrigeration processes and systems; quality and safety of food products; bioprocess simulation and optimization; and computer vision technology. His innovative studies on vacuum cooling of cooked meats, pizza quality inspection by computer vision, and edible films for shelf life extension of fruits and vegetables have been widely reported in national and international media. Results of his work have been published in over 180 peer-reviewed journal papers and in more than 200 conference papers.

Professor Sun received his first class BSc honors and MSc in mechanical engineering, and his PhD in chemical engineering in China before working in various universities in Europe. He became the first Chinese national to be permanently employed in an Irish university when he was appointed college lecturer at the National University of Ireland, Dublin (University College Dublin), Ireland, in 1995, and was then continuously promoted in the shortest possible time to senior lecturer, associate professor, and full professor. Sun is now Professor of Food and Biosystems Engineering and the director of the Food Refrigeration and Computerized Food Technology Research Group at University College Dublin.

As a leading educator in food engineering, Sun has contributed significantly to the field of food engineering. He has trained many PhD students, who have made their own contributions to the industry and academia. He has also given lectures on advances in food engineering on a regular basis at academic institutions internationally and delivered keynote speeches at international conferences. As a recognized authority in food engineering, he has been conferred adjunct/visiting/consulting professorships from 10 top universities in China including Zhejiang University, Shanghai Jiaotong University, Harbin Institute of Technology, China Agricultural University, South China University of Technology, and Jiangnan University. In recognition of his significant contribution to food engineering worldwide and for his outstanding leadership in the field, the International Commission of Agricultural Engineering (CIGR) awarded him the CIGR Merit Award in 2000 and again in 2006. The Institution of Mechanical Engineers based in the United Kingdom named him Food Engineer of the Year 2004. In 2008 he was awarded the CIGR Recognition Award in honor of his distinguished achievements in the top one percent of agricultural engineering scientists in the world.

Professor Sun is a fellow of the Institution of Agricultural Engineers and a fellow of Engineers Ireland. He has also received numerous awards for teaching and research excellence, including the President's Research Fellowship, and has received the President's Research Award from University College Dublin on two occasions. He is a member of the CIGR executive board and honorary vice president of CIGR; editor-in-chief of *Food and Bioprocess Technology*—an international journal (Springer); former editor of *Journal of Food Engineering* (Elsevier); series editor of the *Contemporary Food Engineering* book series (CRC Press/Taylor & Francis); and an editorial board member for the *Journal of Food Engineering* (Elsevier), *Journal of Food Process Engineering* (Blackwell), *Sensing and Instrumentation for Food Quality and Safety* (Springer), and the *Czech Journal of Food Sciences.* He is also a chartered engineer.

# Editor

**Ricardo Simpson** is currently working as a full professor at the Chemical and Environmental Engineering Department, Universidad Técnica Federico Santa María, Chile. He holds a biochemical engineering degree from the P. Universidad Católica de Valparaíso (PUCV, 1980), an MS in food science and technology (1990) and a doctorate in food science (1993) from Oregon State University, and a diploma in economics from the Universidad de Chile (1981). He lectured at PUCV from 1984 to 1999 and became a full professor in 1998. He was also a member of the Food Technology Study Group of CONICYT (equivalent to NSF).

Ever since Dr. Simpson obtained his PhD in 1993, he has been a prolific contributor to the food industry, not only in Chile, but also internationally (e.g., Unilever). His contributions have been summarized in more than 140 conference presentations and more than 50 refereed publications (as author or coauthor), thus advancing the understanding of many aspects of food engineering. He has also done extensive collaborative work with the Chilean food processing industry. He is one of the leading experts in the world in thermal processing of foods, having helped establish and improve food engineering programs at universities in Chile, Peru, and Argentina. He has presented short courses for the food industry in Costa Rica, Chile, Peru, and Argentina on energy conservation, thermal processing, and mathematical modeling applied to the food industry and also on project management. He has coplanned and codirected an international congress, the IV Ibero-American Congress in Food Engineering, and a national congress on food science and technology, both held in Valparaíso, Chile, in 1995 and 2003, respectively. He also planned and directed the national congress on mass and heat transfer held in Valparaíso, Chile, in 1996. He was vice president of the organizing committee of ICEF 10 (International Congress on Engineering and Food) held in Viña del Mar in April 2008. In recent years, he has published an average of eight refereed articles per year and has delivered several invited talks to international audiences. He has made outstanding contributions to engineering programs in education, research, development, consulting, and technology transfer that have resulted in improved food production, quality of life, and education for people living in Chile and Latin America. Recently, he completed a 4-month stay at Unilever's Food Research Center in Vlaardingen, and he was appreciated by the management for his work at its laboratory.

Dr. Simpson has consolidated his expertise as one of the leading experts in Latin America in thermal processing research (commercial sterilization of low-acid canned foods) in the last 3 years, and has been widely recognized in the international arena. Since 2002, he has published several manuscripts, patents, and book chapters on this field.

# Contributors

**Alik Abakarov**
Technical University Federico Santa María
Valparaíso, Chile

**Sergio Almonacid**
Technical University Federico Santa Maria
Valparaíso, Chile

**Antonio A. Alonso**
Process Engineering Group
Research Institute Marinas de Vigo
Vigo, Spain

**Alejandro Amézquita**
Unilever Safety and Environmental Assurance Centre
Bedfordshire, United Kingdom

**Eva Balsa-Canto**
Process Engineering Group
Research Institute Marinas de Vigo
Vigo, Spain

**Julio R. Banga**
Process Engineering Group
Research Institute Marinas de Vigo
Vigo, Spain

**Gustavo V. Barbosa-Cánovas**
Center for Nonthermal Processing of Food
Washington State University
Pullman, Washington

**Jacques Bichier**
JBT Technologies (former FMC Technologies)
Madera, California

**Peter M.M. Bongers**
Unilever Research and Development
Unilever Food & Health Research Institute
Vlaardingen, Netherlands

**Osvaldo Campanella**
Department of Agricultural & Biological Engineering
Purdue University
West Lafayette, Indiana

**Michele Chiumenti**
International Center for Numerical Methods in Engineering
Polytechnic University of Catalonia
Barcelona, Spain

**Pablo M. Coronel**
Unilever Research and Development
Unilever Food & Health Research Institute
Vlaardingen, Netherlands

**Maria G. Corradini**
Department of Food Science
Chenoweth Laboratory
University of Massachusetts
Amherst, Massachusetts

**Mritunjay Dwivedi**
Department of Food Science and Agricultural Chemistry
McGill University
Sainte Anne de Bellevue, Quebec, Canada

**Julio García**
Compass Engineering and Systems S.A.
Barcelona, Spain

**Paul Anthony Gibbs**
Leatherhead Food International
Surrey, United Kingdom

**S. Donald Holdsworth**
Withens
Moreton in Marsh, United Kingdom

**Pablo Juliano**
Center for Nonthermal Processing of Food
Washington State University
Pullman, Washington

**Soojin Jun**
Department of Human Nutrition, Food and Animal Sciences
University of Hawaii at Manoa
Honolulu, Hawaii

**Jasper D.H. Kelder**
Unilever Research and Development
Unilever Food & Health Research Institute
Vlaardingen, Netherlands

**Kai Knoerzer**
Innovative Foods Centre
Food Science Australia
Werribee, Victoria, Australia

**Danilo López**
Projects Department of Chemical Processes and Environmental Biotechnology
Technical University Federico Santa María
Valparaíso, Chile

**Cristian Maggiolo**
International Center for Numerical Methods in Engineering
Polytechnic University of Catalonia
Barcelona, Spain

**Mark D. Normand**
Department of Food Science
Chenoweth Laboratory
University of Massachusetts
Amherst, Massachusetts

**Tomás Norton**
Food Refrigeration and Computerised Food Technology Group
University College Dublin
National University of Ireland
Dublin, Ireland

**Frank Pandelaers**
JBT Technologies (former FMC Technologies)
Sint Niklaas, Belgium

**Micha Peleg**
Department of Food Science
Chenoweth Laboratory
University of Massachusetts
Amherst, Massachusetts

**Hosahalli S. Ramaswamy**
Department of Food Science and Agricultural Chemistry
McGill University
Ste. Anne de Bellevue, Quebec, Canada

**Alfredo C. Rodriguez**
National Center for Food Safety and Technology
Summit-Argo, Illinois

**Clara Rovedo**
JBT Technologies (former FMC Technologies)
Madera, California

**S. Salengke**
Department of Agricultural Technology
Hasanuddin University
Macassar, Indonesia

**Sudhir Sastry**
Department of Food, Agricultural and Biological Engineering
The Ohio State University
Columbus, Ohio

**Filipa Vinagre Marques da Silva**
Laboratory of Fonte Boa
National Institute of Biological Resources
Santarém, Portugal

**Ricardo Simpson**
Projects Department of Chemical Processes and Environmental Biotechnology
Technical University Federico Santa María
Valparaíso, Chile

**Josip Simunovic**
Department of Food, Bioprocessing and Nutrition Sciences
North Carolina State University
Raleigh, North Carolina

**Da-Wen Sun**
Food Refrigeration and Computerised Food Technology Group
University College Dublin
National University of Ireland
Dublin, Ireland

**Arthur Teixeira**
Food Science and Human Nutrition Department
University of Florida
Gainesville, Florida

# Part I

## Fundamentals and New Processes

# 1 Principles of Thermal Processing: Sterilization

*S. Donald Holdsworth*

## CONTENTS



## 1.1 INTRODUCTION

The objective of thermal processing of food products, which involves heating and cooling, is to produce a shelf-stable product, which is free from pathogenic organisms and will not produce food spoilage. The primary necessity is to destroy microorganisms capable of growing in the product and to prevent further spoilage by suitable packaging. In the conventional canning process, which uses a wide range of packaging materials, including tinplate, aluminum, glass, plastics, and composites, the filled and sealed containers are subjected to a heating and cooling regime. Alternatively, continuous flow heat exchangers can be used and the product packaged under aseptic conditions. The heating and cooling regime is known as the *process* and this chapter is concerned with the determination and validation of a process for a specific product, packed in a particular container size and heat processed in a given type of pressurized retort. The heating medium may involve steam, steam/air mixtures, or hot water, and the cooling medium is primarily water. The technology of canning is not discussed here but is detailed in numerous texts (see, e.g., Lopez, 1987; Fellows, 1990; Rees and Bettison, 1991; Brennan et al., 1992; Larousse and Brown, 1997; Ramaswamy and Singh, 1997). Current developments in the technology of in-container sterilization are fully discussed by Richardson (2001, 2004).

The operation of inactivating microorganism is generally referred to as *sterilization*, although it is not the same as the medical operation which involves the complete removal of microbial species. There is generally no need to remove thermophilic organisms, which have no public health significance and the process is described as *commercial sterilization*. The only requirement is that the products are not stored at a temperature in excess of 32°C, when the microorganisms will germinate causing product spoilage. This also requires products to be cooled rapidly after processing. If the ambient temperature of storage exceeds this temperature, e.g., hot climate countries, then it will be necessary to submit the product to a more severe process.

An important factor on deciding the severity of a process is the pH of the product, which may vary from neutrality pH 7 to acidic about pH 2.8. The food poisoning microorganism *Clostridium botulinum* and many other types of sporing and nonsporing bacteria are inhibited from growth at pH 4.5 (or slightly higher <4.7). Consequently this figure is often taken as the dividing line between the requirements of a mild process, e.g., pasteurization, 100°C and a more severe process, often known as a *botulinum process* involving temperatures of 118°C–125°C. The main classes of acidic products are fruit-based and preacidified products and these are discussed in Chapter 2.

It is possible to identify at least four groups of products:

Group 1: Low-acid products (pH 5.0 and above)—meat products, marine products, milk, some soups, and most vegetables

Group 2: Medium-acid products (pH 4.5–5.0)—meat and vegetable mixtures, specialty products, including pasta, soup, and pears

Group 3: Acid products (pH 3.7–4.5)—tomatoes, pears, figs, pineapple, and other fruits

Group 4: High-acid products (pH 3.7 and below)—pickles, grapefruit, citrus juices, and rhubarb

The bulk of food products are in the class requiring a sterilization process, e.g., meat, fish, and vegetables. This is a generalization and there are products which come on the dividing line, e.g., tomatoes and pears, and depend on the variety and maturity. For products in this pH region, it is necessary to conduct extensive trials to establish that food poisoning organisms are inhibited. Similarly for formulated products, it is necessary to examine the inhibitory effects of the ingredients.

Another factor that must be taken into account is the initial microbial loading of the product. This may be controlled by paying attention to handling and preparation procedures and hygiene conditions.

So far no consideration of the effect of heat processing on the food product has been mentioned; however, it is important to consider nutrient destruction, loss of vitamin potency, and overall quality deterioration. These will be affected but the duration and severity of the process, consequently, there is a need to determine an *optimum process* that delivers the necessary sterilization requirements and minimizes the quality degradation.

## 1.2 KINETICS OF THERMAL PROCESSING

### 1.2.1 Microbial Destruction

The engineering design of a process requires a quantitative measure of the effect of temperature and duration time on the destruction of microorganisms. It is usually considered that microbial death can be represented by a first-order kinetic equation, i.e., the destruction rate is proportional to the concentration $c$ of microorganisms (Equation 1.1)

$$-\mathrm{d}c/\mathrm{d}t = kc \tag{1.1}$$

where

$t$ is the time

$k$ is the reaction rate constant with units of reciprocal time

This can be integrated to give Equation 1.2 which expresses the concentration at any time $t$, where $c_0$ is the concentration at time zero.

$$c/c_0 = \mathrm{e}^{-kt} \tag{1.2}$$

The value of $k$ can be determined from the van't Hoff isochore equation (Equation 1.3).

$$k = A\mathrm{e}^{-E/RT} \tag{1.3}$$

This is usually known as the Arrhenius model for microbial inactivation, where $A$ is the pre-exponential factor, $E$ is the activation energy, and $R$ the universal gas constant. It is usual to specify a reference temperature and the corresponding $k$-value being $k_{\mathrm{ref}}$.

The traditional approach to this is slightly different and is based on the number, $N$, of microorganisms at time $t$. Thus, if the logarithm of the number of spores is plotted against time a semilinear plot is obtained with a negative slope. This has an intercept $N_0$ and a slope of $-1/D$, where $D$ is called the decimal reduction time of the microbial species, usually a highly heat-resistant spore. This can be represented by Equation 1.4.

$$\log N = \log N_0^{-t/D} \tag{1.4}$$

The two approaches are very similar in the temperature range around the figure of 120°C. The $D$-value is usually quoted in minutes and the $k$-value in seconds; hence $D = 2.3/60k$. Most spore survival curves are not linear but show *shoulders* and *tails* and many equations have been developed to deal with these curves and many theories discussed for their occurrence. A summary of some alternative models for microbial inactivation is given by Holdsworth and Simpson (2007), and also in Chapters 7 and 8.

The log $D$-value when plotted against temperature $T$ usually shows a linear relation and in order to compare differing organism it is necessary to use a reference temperature $T_{ref}$, e.g., 250°F or 121.1°C corresponding to $D_{ref}$. Using this to define the thermal death relationship results in Equation 1.5

$$\log(D/D_{ref}) = -(T - T_{ref})/z \qquad (1.5)$$

where $z$ is the temperature change necessary to change the $D$-value by 1 log-cycle, i.e., by a factor of 10.

Using the Arrhenius model outlined briefly above the $z$-value is given by Equation 1.6

$$z = 2.303RTT_{ref}/E \qquad (1.6)$$

Having established the necessary kinetic functions, either $k$ and $E$ or $D$ and $z$ these can be used for determining the times and temperatures for a satisfactory process (see Section 1.3). Specific values for these factors have been determined for a wide range of microorganisms in a variety of media and food products (see Holdsworth and Simpson, 2007).

For *C. botulinum* spores, $k$ can reliably be determined using $A = 2 \times 10^{60}$ s$^{-1}$ and $E = 310.11 \times 10^3$ J mol$^{-1}$ K$^{-1}$ using the Arrhenius approach and $z = 10$°C and $D_{121.1} = 0.3 \times 60$ s using the conventional canning approach.

More complex models to represent the thermal death of microorganisms, especially the effect of pH, have been established. These models which have a linear form are shown in Equation 1.7

$$\ln k = C_0 + C_1T^{-1} + C_2\text{pH} + C_3\text{pH}^2 \qquad (1.7)$$

For *C. botulinum*, in spaghetti/tomato sauce, the values of the constants were $C_0 = 105.23$, $C_1 = -3.704 \times 10^4$, $C_2 = -2.3967$, and $C_3 = 0.1695$ (Davey et al., 1995, 2001).

### 1.2.2 Kinetics of Food Quality Destruction

The effect of heat on the constituents of foods is generally deleterious to the overall quality. These include the degradation of vitamins, the softening of texture, loss of color, development of off-flavors, and destruction of enzymes. Some of these are desirable, e.g., enzyme inactivation fruits and vegetables and softening of texture in meat and fish products. All these reactions, chemical or physical, have different kinetics to microbial inactivation. Bacterial spores have $z$-values between 7°C and 12°C, whereas other constituents have values up to 50°C. This means that if high processing temperatures are used for short times (usually referred to as high temperature short time [HTST]-processes) there will be less destruction of thermolabile components and conversely for longer processes there will be greater loss. Milk is a typical example of a highly thermolabile product and consequently benefits from HTST processing. This is usually achieved in a continuous flow process followed by aseptic packaging.

## 1.3 PROCESS DETERMINATION

### 1.3.1 Heat Penetration *f*- and *j*-Factors

Establishing times and temperatures for processing of packaged foods depends on evaluating the amount of heat the product has received. The uniformity of heating will depend on the consistency of the product, for example, food products which are thick tend the heat by conduction whereas fluid products heat by convection. This means that in conduction-type packs the outer layers will heat more rapidly than the center of the container, whereas fluid products will heat more uniformly. The main objective is to ensure that the slowest heating part of the food in a container receives the minimum process necessary to achieve a sterilized product. It is for this reason that in-container heat penetration experiments are performed to establish the necessary process. This is achieved by placing a thermocouple at the point of slowest heating (often referred to as the critical point) and observing the temperature–time profile. This is plotted in the form of a log temperature/linear time plot and is known as a heat penetration curve. The slope of this curve gives the rate of heat penetration *f*-value ($f_h$ for heating and $f_c$ for cooling) and the intercept gives the *j*-value (with similar designations). These two values are the most important factors for describing the process characteristics. The *f*-value depends on the thermal diffusivity of the product and the container dimensions and can be determined by calculation (Ball and Olson, 1957; Holdsworth and Simpson, 2007). The *j*-value is known as the lag factor and its value depends on the position inside the container. While this applies to conduction packs, convection packs have much lower *f*-values, i.e., heat much more rapidly, and are related to the ratio of the can surface area to the volume of the container.

The simple division of heating regimes into conduction and convection is an idealized situation. In practice there are systems which heat by convection initially and as the product thickens conduction heating characteristics are shown and vice versa. The graphs from this type of behavior are known as broken-heating curves. Usually there is a distinct break between the two sections of the graph and the corresponding values for heating $f_1$ and $f_2$ can be determined.

### 1.3.2 Criteria for Adequacy of Processing *F*- and *C*-Values

The most important factor in food product sterilization is to be able to quantify the effect of the heating and cooling regime and establish that a given process is able to give a safe product. The universally agreed method of evaluating a process is based on the heat penetration curve and the use of lethal rates.

A measure of the lethal effect of heat on microorganism inactivation is the lethal rate *L* in minutes (Equation 1.8)

$$L = 10^{(T - T_{\mathrm{ref}})/z} \tag{1.8}$$

The basis of process evaluation is that lethalities are additive and the total lethality can be determined by converting the heat penetration curve into a lethality–time curve and integrating the area under the curve. The total lethality for the process is known as the *F*-value and can be determined from Equation 1.9

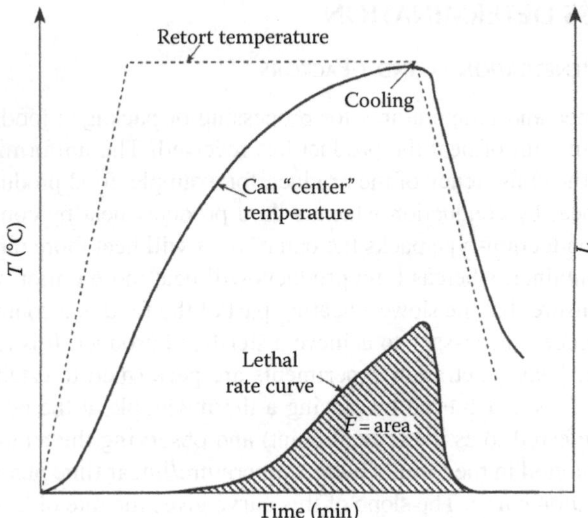


**FIGURE 1.1** Graph of temperature ($T$) and lethal rate ($L$) against time.

$$F_{T_{\mathrm{ref}}}^{z} = \int_0^t L\,\mathrm{d}t = 1 \tag{1.9}$$

This is illustrated in Figure 1.1.

When the reference temperature is 250°F or 121.1°C, the $F$-value is designated as $F_0$ (known as $F$ nought). The corresponding values of $z$ are 18°F or 10°C. However for processes not involving a botulinum process, e.g., heat destruction of yeast spores in beer the $F$-value would have adscripts indicating the $z$-value and the reference temperature, e.g., $F_{212}^{12.5}$ or $F_{100}^{7.0}$.

It is often more convenient for pasteurization studies to use the pasteurization unit (PU) which is defined in the same way as a lethal rate; however, there is no agreed reference temperature and consequently this must be stated in every case, along with the appropriate $z$-value.

For continuous flow sterilizers used in aseptic processing operations the microbial destruction $N/N_0$ can be estimated from Equation 1.10

$$\frac{N}{N_0}(2\pi/Q_{\mathrm{v}})\int_0^r rvx\,\mathrm{d}r \tag{1.10}$$

where

$x = 10^{-(l/v)D_{\mathrm{ref}}}$
$Q_{\mathrm{v}}$ is the volumetric flow rate
$r$ is the radius of the tube
$v$ is the fluid velocity
$l$ is the length of tube

Solutions of this equation for various systems are summarized in Holdsworth (1992) and Lewis and Heppel (2000).

While it is convenient to determine the *F*-value from the area under the curve or by addition of the lethalities at equal time intervals, it is possible to use the theoretical equations for the temperatures and times derived from analytical equations (see Holdsworth and Simpson, 2007). A number of computer programs are available for calculating process, the most recent having been developed by Simpson and now available on a computer disk (see Holdsworth and Simpson, 2007).

By analogy it is possible to define a *C*-value which will give a measure of the deterioration of any chemical or physical property of the food provides appropriate $z_c$-values are available (see Equation 1.11) where

$$C = \int_0^t 10^{T-T_{\text{ref}}/z} \mathrm{d}t \tag{1.11}$$

The original concept was developed by Mansfield (1962, 1974) and was first applied to determining the degree of cooking of a product.

While heat penetration studies are based at the point of slowest heating, *C*-values at this point are not relevant. This has led to the use of $C_s$-value which is a mass-average value for the whole of the container (see Equation 1.12)

$$C_s = D_{\text{ref}} \log(c/c_0) \tag{1.12}$$

where *c* is the concentration of the heat vulnerable component at times 0 and *t*. The reference temperature for cooking studies is usually taken as 100°C.

The value for $C_s$ may be obtained using Equation 1.13.

$$c/c_0 = \frac{1}{V}\int_0^V 10^{-C_c/D_{\text{ref}}} \mathrm{d}V \tag{1.13}$$

where

$$C_c = \frac{1}{V}\int_0^t 10^{(T-T_{\text{ref}})/z_c} \mathrm{d}t \tag{1.14}$$

For a more complete study of this subject, see Tucker and Holdsworth (1991) and Holdsworth and Simpson (2007).

## 1.4 OPTIMIZATION OF STERILIZATION AND COOKING

The fact that chemical and microbiological destruction kinetics differ leads to an important requirement for optimized processes (see Chapter 16). The current trend is to try to preserve the nutrients and flavors of food products by using techniques which reduce the heating load on the product and consequently increase the quality.

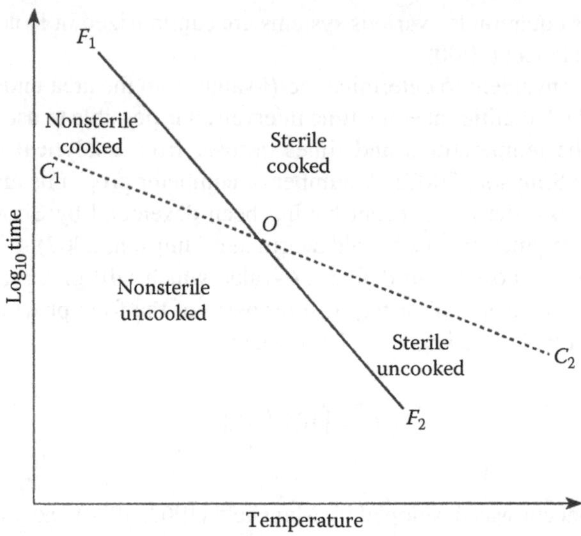


**FIGURE 1.2** Diagram of idealized $\log_{10}$ time versus temperature for microbial inactivation ($F_1OF_2$) and cooking ($C_1OC_2$) of food product, instantaneously heated.

The differing kinetic factors for nutrients and chemical species result in the need for optimization of heating conditions (see Section 1.2.2). The results of the incompatibility can be seen in Figure 1.2, where a log time versus temperature graph for an idealized situation of instantaneous heating of a food product to a given *F*-value is shown. Considering the sterilization line $F_1OF_2$, all time–temperatures on the right of the line will represent processing conditions which result in sterile product whereas all conditions to the left will be nonsterile. When the corresponding cooking line is plotted, the graph shows four product regions only two of which represent conditions which will result in a sterilized product. Processing at relatively high temperatures for a relatively short time will therefore result in maximum nutrient retention.

Various methods of calculating the process required to achieve optimization are discussed in Chapters 13, 15, and 16 (see also Tucker and Holdsworth, 1991; and Holdsworth, 2004).

## 1.5 ESTABLISHING SAFE CRITERIA HEAT-PROCESSED FOODS

For low-acid foods (pH $\geq$ 4.5) the most important criteria for ensuring the safety of heat-processed foods is that a minimum process must reduce the probability of survival of spores of *C. botulinum* to less than one spore in $10^{12}$ containers. A minimum process is usually taken as one which achieves an $F_0 = 3\,\text{min}$; in practice, processes are usually higher than this (6–10 being typical) either for controlling spoilage organisms or achieving the correct degree of texture softening.

The most definitive sources for assessing whether a process is suitable for a particular product are those produced by the National Food Processors' Association in the United States (NFPA, 1971, 1982), which apply to low-acid products. However, Brown

(1991) has discussed a number of other criteria arising from European countries, in particular the statutory requirements for the processing of milk and milk products. Special regulations apply to the canning of cured meats where a salt or sodium nitrite is added, and other products which contain added microbial inhibitors.

## REFERENCES


Ball, C. O. and Olson, F. C. W. 1957. *Sterilization in Food Technology—Theory, Practice and Calculations.* New York: McGraw-Hill.

Brennan, J. G., Butters, J. R., Cowell, N. D., and Lilly, A. E. V. 1992. *Food Engineering Operations.* London, UK: Applied Science Publishers.

Brown, K. L. Principles of heat preservation. 1991. In *The Processing and Packaging of Heat Preserved Foods*, eds. J. A. G. Rees and J. Bettison. Glasgow: Blackie & Sons.

Davey, K., Hall, R. F., and Thomas, C. J. 1995. Experimental and model studies of the combined effect of temperature and pH on the thermal sterilization of vegetative bacteria in liquid. *Food and Bioproducts Processing, Trans. IChemE* 73C3: 127–132.

Davey, K. R., Thomas, C. J., and Cerf, O. 2001. Thermal death of bacteria. *J. Appl. Microbiol.* 90(1): 148–150.

Fellows, P. 1990. *Food Processing Technology—Principles and Practice.* Chichester, UK: Ellis Horwood Ltd.

Holdsworth, S. D. 1992. *Aseptic Processing and Packaging of Food Products.* London: Applied Science Publishers.

Holdsworth, S. D. 2004. Optimizing the safety and quality of thermally processed packaged foods. In *Improving the Thermal Processing of Foods*, ed. P. Richardson, pp. 1–31. Cambridge: Woodhead Publishing.

Holdsworth, S. D. and Simpson, R. 2007. *Thermal Processing of Packaged Foods.* New York: Springer.

Larousse, J. and Brown, B. E. 1997. *Food Canning Technology.* New York: Wiley.

Lewis, M. J. and Heppel, N. J. 2000. *Continuous Thermal Processing of Foods—Pasteurization and UHT Sterilization.* Gaithersburg, MD: Aspen Publishers.

Lopez, A. 1987. *A Complete Course in Canning* (3 vols.). Baltimore, MD: The Canning Trade Inc.

Mansfield, T. 1962. High temperature/short time sterilization. *Proc. 1st Int. Congress Food Sci. Technol.* 4: 311–316.

Mansfield, T. 1974. *A Brief Study of Cooking.* San José, CA: Food and Machinery Corporation.

NFPA 1971. Processes for low-acid canned foods in glass containers. Bulletin 30-L. Washington, DC: National Food Processors' Association.

NFPA 1982. Processes for low-acid canned foods in metal containers. Bulletin 26-L. Washington, DC: National Food Processors' Association.

Ramaswamy, H. S. and Singh, R. P. 1997. Sterilization process engineering. In *Handbook of Food Engineering Practice*, eds. K. I. Valentos, E. Rotstein, and R. P. Singh, Chapter 2. Boca Raton, NY: CRC.

Rees, J. A. G. and Bettison, J. eds. 1991. *The Processing and Packaging of Heat Preserved Foods.* Glasgow: Blackie & Sons.

Richardson, P. ed. 2001. *Thermal Technologies in Food Processing.* Cambridge: Woodhead Publishing.

Richardson, P. ed. 2004. *Improving the Thermal Processing of Foods.* Cambridge: Woodhead Publishing.

Tucker, G. and Holdsworth, S. D. 1991. Mathematical modelling of sterilization and cooking for heat preserved foods—application of a new heat transfer model. *Food and Bioproducts Processing, Trans. IChemE.* 69C1: 5–12.

# 2 Principles of Thermal Processing: Pasteurization

*Filipa Vinagre Marques da Silva*
*and Paul Anthony Gibbs*

## CONTENTS

## 2.1 INTRODUCTION

Thermal pasteurization is a classical method of food preservation which extends the shelf life by inactivating vegetative cells of unwanted pathogenic and spoilage microorganisms with processing temperatures normally between 65°C and 95°C. This traditional physical process of food decontamination is still in common use today, being efficient, environmentally friendly, healthy, and inexpensive when compared with other technologies. As opposed to sterilization, the temperatures used are lower, allowing greater retention of the original properties of the raw food. A further step towards better quality can be achieved if pasteurization is used in combination with nonthermal food preservation methods such as the use of refrigerated distribution and storage (1°C–8°C), vacuum or modified atmosphere packaging, added preservatives, etc. This would allow the production of safe foods while minimizing the degradation of the "fresh" organoleptic and nutritive quality of the foods. Typical pasteurized foods include beverages such as milk, fruit juices, beer, low carbonated drinks, dairy products (e.g., cheese), meat and fish products (e.g., cured, cooked ham, hot-smoked fish), some sauces, pickles, and food ingredients. This chapter covers the pasteurization fundamentals, followed by a review focused on the heat resistance of relevant microbes in pasteurized foods, and finishes with a short section about the design of pasteurization processes for different types of foods.

## 2.2 FOOD PASTEURIZATION FUNDAMENTALS

This section presents the historical origin, definitions, and objectives of the pasteurization process. Also, the concept of quality and optimization of the process will be introduced, and finally, relevant equations used to model the inactivation of microbial and food-derived deteriorative enzymes, and the impact of the heat process on quality factors will be described and discussed.

### 2.2.1 Historical Background and Definitions of Food Pasteurization

The first investigations on pasteurization were carried out in 1765 by Spallanzani. He used a heat treatment to delay spoilage and preserve meat extract. From 1862 to 1864, Pasteur showed that temperatures of 50°C–60°C for a short time effectively eliminated spoilage microorganisms in wine. Pasteur (1876) also investigated beer spoilage. When milk producers adopted this process (Soxhlet, 1886; Davis, 1955; Westhoff, 1978), they were able to eliminate most of the foodborne illnesses. The main goal of pasteurization of low-acid chilled foods is the reduction of pathogens responsible for foodborne

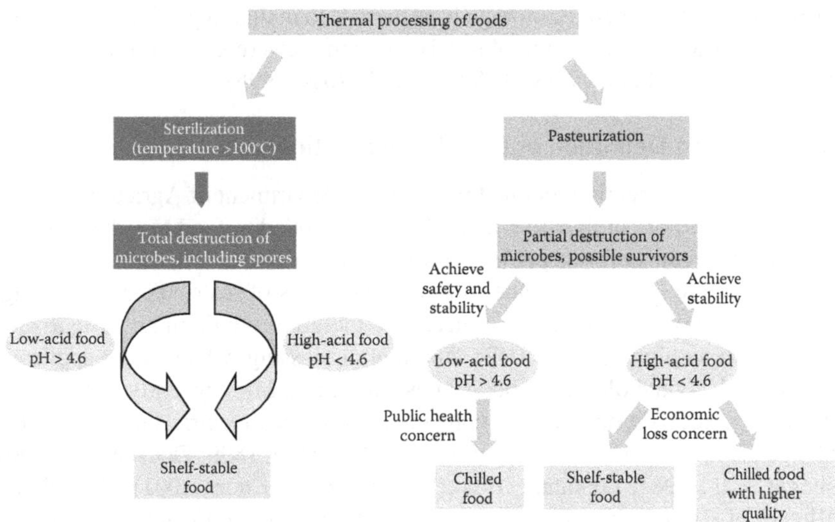


**FIGURE 2.1** Thermal processing of foods.

illness and human disease, whereas in the case of high-acid foods, pasteurization is intended to avoid spoilage and economic losses (Figure 2.1).

#### 2.2.1.1 Classical Definition of Food Pasteurization

The word "pasteurization" has its origin in the work of the French scientist Louis Pasteur, and refers to a mild heat treatment (50°C–90°C) used for food preservation, which aims to inactivate vegetative forms of pathogenic and spoilage microorganisms. Unlike sterilization, after pasteurization the food is not sterile since heat-resistant microbial spores are present (Lund, 1975a) (Figure 2.1). Therefore, other forms of preservation such as refrigeration (e.g., milk), atmosphere modification (e.g., vacuum packaging), addition of antimicrobial preservatives (e.g., salt, citric acid, benzoic acid, sorbic acid, sulphur dioxide, dimethyl dicarbonate, etc.), or combinations of the referred techniques are required for product stabilization and distribution. Exceptions are some processed foods that contain constituents or ingredients that are antimicrobial under certain conditions, not allowing microbial growth: fermented foods containing alcohol or acid (e.g., wine, beer, pickles), carbonated drinks (e.g., sodas), very sweet foods presenting $a_w < 0.65$ or soluble solids > 70°Brix (e.g., honey, jams, jellies, dried fruits, fruit concentrates), or salty foods (e.g. salted fish or meats). Other exceptions include the high-acid foods (pH < 4.6), which are stable at ambient conditions after a pasteurization process, because the acidic food environment is not conducive to the growth of harmful microorganisms and microbial spores in the pasteurized food. For these type of foods (pH < 4.6), a pasteurization process allows a long shelf-life (several months) at room temperature (Ramaswamy and Abbatemarco, 1996), and if refrigerated storage is used, a milder pasteurization may be applied and product quality is improved (Figure 2.1). If a food product has low acidity (pH > 4.6, e.g., milk), a shorter shelf-life (several days) is obtained after pasteurization, but refrigerated storage is necessary to maintain product safety during storage, by restricting the growth of surviving pathogens (e.g., sporeformers) in the foods (Potter, 1986; Fellows, 1988; Adams and Moss, 1995).

#### 2.2.1.2 Modern Definition of Food Pasteurization

Pasteurization was recently redefined by the U.S. Department of Agriculture as "any process, treatment, or combination thereof, that is applied to food to reduce the most resistant microorganism(s) of public health significance to a level that is not likely to present a public health risk under normal conditions of distribution and storage" (NACMCF, 2006). This definition therefore includes nonthermal pasteurization processes such as high pressure (HP). Thermal processing, which is the application of heat to foods, is the oldest method of pasteurization. More recently, the effects of nonthermal pasteurization, such as high intensity pulsed electric fields and HP on microorganisms and foods, have been investigated (Hite, 1899; Doevenspeck, 1961; Hülsheger et al., 1981; Lehmann, 1996; Hendrickx and Knorr, 2001). Nevertheless, the efficacy of HP in terms of spore (Lee et al., 2006) and enzyme inactivation is limited (Raso and Barbosa-Cánovas, 2003; Van Buggenhout et al., 2006). In fact, HP was responsible for stimulating germination of *Talaromyces macrosporus* mold spores (Dijksterhuis and Teunissen, 2003). Thus, HP treatments followed by thermal processing have been proposed in order to inactivate the spores (Heinz and Knorr, 2001; Raso and Barbosa-Cánovas, 2003). With respect to combinations of heat and pressure for the destruction of *Clostridium botulinum* spores, concerns were expressed by Margosch et al. (2006) who demonstrated higher spore survival when using temperature and HP treatments simultaneously in comparison to the exclusive use of temperature.

### 2.2.2 Desirable and Undesirable Changes in Foods with the Application of Heat: Quality Optimization of the Process

Thermal pasteurization, intended to inactivate pathogens and deteriorative microorganisms/enzymes, may also affect negatively the quality of foods (Lund, 1975b; Villota and Hawkes, 1986). Thus the heat application should be minimal and well balanced, being enough for food decontamination, while enabling maximum retention of the original food quality (Ramaswamy and Abbatemarco, 1996). The food color, aroma, flavor, and texture, readily perceived by food consumers, and nutritive/health value, are generally recognized as the major quality factors of foods, being used for quality optimization of the process. Knowing that various conditions of heating temperature ($T$) and time ($t$) lead to similar effects on microbial/enzymatic inactivation, a process that causes less impact on quality factors can be selected (Silva and Silva, 1997; Silva et al., 2003). Such optimization is possible because the thermal degradation kinetics of quality factors is much less temperature sensitive than the destruction of microorganisms (Teixeira et al., 1969; Holdsworth, 1985). The higher the pasteurization temperature applied, the shorter the time needed for the same microbial inactivation.

Most of the changes that occur in food as a result of pasteurization can be quantified with first-order kinetics (see Section 2.2.3.1), the $z$-value being a measure of the effect of temperature on the destruction rate of microbial/enzymatic/quality factors.

Approximate $z$-values of various thermally dependent factors were collected from the literature: vegetative bacteria, $z$ = 3.5°C–9.4°C (Table 2.3); fungal ascospores, $z$ = 5.2°C–9.2°C (Table 2.5); bacterial spores, $z$ = 4.2°C–15°C (Tables 2.1, 2.2, 2.4, and 2.6); enzymes, $z$ = 10°C–22°C (Table 2.7); flavor/odor, $z$ = 13°C–50°C (Ohlsson, 1980; Argáiz and López-Malo, 1995; Silva et al., 2000a); color, $z$ = 20°C–74°C (Sanchez et al, 1991; Silva and Silva, 1999); vitamin C, $z$ = 44°C–72°C (Silva and Silva, 1997); green olives texture, $z$ = 63°C (Sanchez et al., 1991). The $z$-values of quality factors (13°C–72°C) are, in general, higher than those found for microorganisms/enzymes (5°C–19°C). Thus the use of high temperature for short time causes a larger increase in the rate of microbial/enzyme inactivation than in the degradation rate of quality factors (Lund, 1975b). This is the basis for high-temperature short-time processing (HTST; 71.7°C for 15 s) commonly used for milk pasteurization.

### 2.2.3 Equations for Process Design and Assessment

Kinetic models are useful tools for the quantification of thermal inactivation of microorganisms or enzymes, and also quality changes. Classical equations to model isothermal microbial survivor curves will be presented in this section, although more recently some authors have demonstrated how thermal inactivation data can also be collected from nonisothermal experiments (Welt et al., 1997; Peleg and Penchina, 2000). Equations to assess process impact will also be presented. Finally, a brief overview of other models describing phenomena of "shoulders" and "tails" in microbial inactivation curves will be discussed.

#### 2.2.3.1 Bigelow or First-Order Kinetic Models

Chemical reaction kinetics is used to describe the microbial/cell thermal inactivation. The change/deterioration of most food factors with isothermal time exposure follows zero- (Equation 2.1) or first-order (Equations 2.2 through 2.4) reaction kinetics (Villota and Hawkes, 1992). Simple first order can be described either by Equation 2.2 or, when dealing with microorganisms which also exhibit log-linear spore inactivation kinetics, by the Bigelow model (Equation 2.3; Bigelow and Esty, 1920; Teixeira, 1992). First-order reversible (or fractional) reaction kinetics (Equation 2.4) has also been observed for some quality factors.

$$\frac{F}{F_0} = -k_T \times t \tag{2.1}$$

$$\frac{F}{F_0} = e^{-k_T \times t} \tag{2.2}$$

$$\frac{F}{F_0} = 10^{-(t/D_T)} \tag{2.3}$$

$$\frac{F - F_\infty}{F_0 - F_\infty} = e^{-k_T \times t} \tag{2.4}$$

where

$F$ is the microbial/enzyme/quality factor
$F_0$ and $F_\infty$ are the values of the factor at time zero and infinite time, respectively
$k_T$ is the reaction rate at temperature $T$ (min$^{-1}$)
$D_T$ is the decimal reduction time at temperature $T$ (min)
$t$ is the time (min)

The temperature effect on the reaction rate constant, $k_T$, is described by the Arrhenius equation (Equation 2.5). The Bigelow model (Equation 2.6) can also be used for first-order kinetics (Saguy and Karel, 1980; Wells and Singh, 1988).

$$k_T = k_{T_{ref}} \times e^{-\left[\frac{E_a}{R} \times \left(\frac{1}{T+273.15} - \frac{1}{T_{ref}+273.15}\right)\right]} \tag{2.5}$$

where

$k_{T_{ref}}$ is the reaction rate at reference temperature (min$^{-1}$)
$T_{ref}$ is the reference temperature (°C)
$E_a$ is the activation energy (J/mol)
$R$ is the universal gas constant (8.31434 J/mol/K)

$$D_T = D_{T_{ref}} \times 10^{\left(\frac{T_{ref}-T}{z}\right)} \tag{2.6}$$

where

$D_{T_{ref}}$ is the decimal reduction time at a reference temperature (min)
$z$ is the number of degrees Celsius required to reduce $D$ by a factor of 10 (°C)

#### 2.2.3.2 Nonlinear Survival Curves

Although log survivors vs time thermal inactivation kinetics is widely assumed to be linear, deviations from linearity (e.g., shoulder, tail, sigmoidal-like curves, biphasic curves, concave and convex curves) have been reported and remain unexplained, in particular with vegetative pathogens such as *Escherichia coli*, *Salmonella* spp., *Listeria monocytogenes*, *Staphylococcus aureus* (Chiruta et al., 1997; Juneja et al., 1997; Juneja and Marks, 2005; Valdramidis et al., 2006; Buzrul and Alpas, 2007). The observation of tails in the inactivation of *Mycobacterium avium* subsp. *paratuberculosis* (MAP) in milk was explained by cell clumps rather than the existence of a more heat-resistant cell fraction (Klijn et al., 2001). With respect to microbial spore thermal inactivation, a log-linear behavior is commonly observed. A few exceptions have been reported in the literature, *Bacillus sporothermodurans* spores exhibited an upper concavity (Periago et al., 2004). Tails were also observed in the heat inactivation of prions (Periago et al., 2003). The added complexity of new models might not be worth using in terms of microbial survival predictions (Halder et al., 2007), in particular spore-forming microbes.

#### 2.2.3.3 Process Design and Assessment

During the process, the integrated lethality at a single point within the food container, also known as pasteurization value ($P$), is the equivalent time of pasteurization at a certain temperature ($T_{ref}$) expressed in minutes (Equation 2.7) (Shapton, 1966):

$$P = \int_0^{PT} 10^{\left(\frac{T - T_{\text{ref}}}{z}\right)} \mathrm{d}t \tag{2.7}$$

where

$P$ is the pasteurization value (min)
$PT$ is the total process time (min)
$T_{\text{ref}}$ is the reference temperature for the pasteurization target (°C)
$z$ is the $z$-value for the pasteurization target (°C)

Knowing the time–temperature history at a point in the container, Equation 2.8 can be used to calculate the microbial reduction or retention of quality parameters in foods, which follow first-order kinetics of change, at that single point during the process time.

$$\frac{F}{F_0} = 10^{\left(\frac{-1}{D_{T_{\text{ref}}}} \int_0^{PT} 10^{\left(\frac{T - T_{\text{ref}}}{z}\right)} \mathrm{d}t\right)} \tag{2.8}$$

The Arrhenius model can be used to evaluate the temperature effect on the quality parameters with first-order reversible kinetics (fractional conversion model), such as flavor and aroma "cooked-notes" (Silva, 2000; Silva et al., 2000a). Equation 2.9 is used to calculate cooked-notes level at a single point at any time:

$$\frac{F}{F_0} = \frac{F_\infty}{F_0} + \frac{(F_0 - F_\infty)}{F_0} \times \exp\left[-k_{T_{\text{ref}}} \int_0^{PT} \mathrm{e}^{\left[-\frac{E_a}{R} \times \left(\frac{1}{T + 273.15} - \frac{1}{T_{\text{ref}+273.15}}\right)\right]} \mathrm{d}t\right] \tag{2.9}$$

The $P$-value and microbial/quality factor change can also be determined in terms of container volume average (see Holdsworth and Simpson, 2007).

## 2.3 HEAT RESISTANCE OF MICROBES TARGETING THE PRODUCTION OF SAFE AND STABLE FOODS

The global incidence of foodborne disease is difficult to estimate. However, 1.8 million people were reported to have died in 2005 from diarrheal diseases. A high proportion of these cases can be attributed to contamination of food and in particular drinking water (WHO, 2007). Low-acid foods have been the cause of human diseases such as gastroenteritis and listeriosis. Common symptoms of foodborne illness include diarrhea, stomach cramps, fever, headache, vomiting, dehydration, and exhaustion. Proper thermal processing of foods can eliminate most of the causative agents of foodborne diseases. Although microbial spoilage of thermally processed foods can be caused by incipient spoilage (growth of bacteria before processing) and recontamination after processing (leakage), we will focus our attention on the survival and growth of thermophilic microorganisms (e.g., sporeformers) due to insufficient heat processing. Furthermore, the increasing demand for minimally processed foods by consumers has resulted in recent outbreaks of foodborne illnesses and fatalities in the wake of underpasteurized foods. Food spoilage with economic losses has been also observed.

The highest incidence of rapid spoilage of processed foods is caused by bacteria, followed by yeasts and molds (Sinell, 1980). Parasites (protozoa and worms), natural toxins, viruses, and prions can also be a problem if industry uses contaminated raw materials (FDA, 1992).

This section starts with a short introduction about microbial spores, followed by a review of the heat resistance of thermally driven microbiological and biochemical (enzymes) criteria to be considered as targets in designing pasteurization processes for low-acid chilled foods and high-acid shelf-stable foods. Thermal resistances listed from higher to lower values will be presented for the most significant microorganisms able to grow in chilled low-acid foods and shelf-stable high-acid foods.

### 2.3.1 Spores: Heat-Resistant Microbial Forms

A spore is a highly resistant dehydrated form of a dormant cell produced under conditions of environmental stress and as a result of "quorum sensing." Molds and bacteria can produce spores, although mold spores are not as heat resistant as bacterial spores. Heat is the most efficient method for spore inactivation, and is presently the basis of a huge worldwide industry (Bigelow and Esty, 1920; Gould, 2006). Microbial spores are much more resistant to heat in comparison to their vegetative counterparts, generally being able to survive the pasteurization process. Spore heat resistance may also be affected by the food environment in which the organism is heated (Tables 2.1, 2.2, and 2.4 through 2.6). For instance, spores (and vegetative cells) become more heat resistant at low water activity (Murrel and Scott, 1966; Härnulv and Snygg, 1972; Corry, 1976; King and Whiteland, 1990; Tournas and Traxler, 1994; Silva et al., 1999). If after pasteurization the storage temperature as well as the food characteristics (pH, water activity, food constituents) are favorable for a sufficient amount of time, surviving spores can germinate and grow to attain high numbers (e.g., $10^7$ cells/g or mL). Subsequently, foodborne diseases and/or spoilage may occur.

The most dangerous sporeformers in low-acid chilled foods are the nonproteolytic strains of *Clostridium botulinum* (Gould, 1999; Carlin et al., 2000). Other human infections or intoxications from pasteurized (cooked) and chilled foods include spore-forming *Bacillus cereus* (Carlin et al., 2000). Unusual spoilage problems have been reported with *Alicyclobacillus acidoterrestris* in apple and orange juices (Brown, 2000) and other high-acid shelf-stable foods. There are a number of nonpathogenic sporeformers including facultative bacilli, butyric, thermophilic anaerobes, and molds that can cause significant economic losses to food producers. Control of spores during storage of pasteurized foods requires an understanding of both their heat resistance and outgrowth characteristics.

### 2.3.2 Microbial Heat Resistance in Low-Acid Pasteurized Chilled Foods (pH > 4.6)

Minimally heated chill-stored foods have been increasing by 10% each year in market volume, since they are convenient (ready-to-eat and with longer shelf-life than fresh) and can better retain the original properties of the foods. For reasons of public safety, low-acid pasteurized foods are stored, transported, and sold under refrigerated conditions and with a limited shelf-life (Figure 2.1), to minimize the outgrowth of pathogenic microbes in the foods during distribution. Beverages such as milk, certain fruit juices (e.g., tomato, pear, some tropical juices), dairy products (e.g., yoghurts, cheeses), poultry/meat/fish/vegetable products (e.g., cured, cooked ham), some shellfish (e.g., cockles), and some sauces are examples of low-acid pasteurized foods. Low-acid pasteurized and chilled foods also include refrigerated processed foods of extended durability (REPFED). Those are generally packaged under vacuum or modified atmospheres to ensure anaerobic conditions and submitted to mild heat treatments, being stored from a few days to several weeks depending on the food and severity of the heat process.

Following is a brief list of typical pathogens associated with foodborne diseases and outbreaks from improperly processed low-acid chilled foods. Nonproteolytic, psychrotrophic strains of *Clostridium botulinum* have been implicated in human botulism incidents caused by the following contaminated foods (Lindström et al., 2006): hot-smoked fish (Pace et al., 1967), canned tuna fish in oil (Mongiardo et al., 1985), canned truffle cream/canned asparagus (Therre, 1999), pasteurized vegetables in oil (Aureli et al., 1999), canned fish (Przybylska, 2003), and canned eggplant (Peredkov, 2004). Other examples of foodborne infections from raw and heated foods include *Bacillus cereus* (cooked rice and chilled foods containing vegetables), *Listeria monocytogenes* (milk, soft cheese, ice cream, cold-smoked fish, chilled processed meat products such as cooked poultry), *Escherichia coli* serotype O157:H7 (verotoxigenic *E. coli* VTEC; beef, cooked hamburgers, raw fruit juice, lettuce, game meat, cheese curd), *Salmonella enteritidis* (poultry and eggs), *Vibrio parahaemolyticus* (improperly cooked, or cooked, recontaminated fish and shellfish), *Vibrio cholerae* (water, ice, raw or underprocessed seafood), and foodborne trematodes from fish/seafood produced by aquaculture (FDA, 1992; Carlin et al., 2000; WHO, 2002; Keiser and Utzinger, 2005). Pasteurized milk and dairy products may also be contaminated with *Brucella*, thermophilic *Streptococcus* spp., and *Mycobacterium avium paratuberculosis* (MAP) (Grant, 2003), which can be infectious at low cell numbers, although they cannot grow at chill temperatures.

Psychrotrophic spoilage microbes such as lactic acid bacteria (LAB) (*Lactobacillus* spp., *Leuconostoc* spp., *Carnobacterium* spp.), molds (*Thamnidium* spp., *Penicillium* spp.), and yeasts (*Zygosaccharomyces* spp.) can occur in chilled low-acid foods during storage, in general due to postprocess contamination. These are very heat sensitive, for example, LAB $D_{63^\circ\mathrm{C}}$ is 14 s in meat sausages (Franz and vonHoly, 1996) and $D_{60^\circ\mathrm{C}}$ is 33 s in milk (De-Angelis et al., 2004).

#### 2.3.2.1 Psychrotrophic Strains of *Clostridium botulinum*

Anaerobic spore-forming *Clostridium* species can be a problem in REPFED foods which are increasingly selected by consumers. These include the nonproteolytic psychrotrophic strains of *C. botulinum* (toxin types B, E, F) and the food-poisoning pathogen *Clostridium perfringens*, although the latter is not psychrotrophic. The mild pasteurization process applied to REPFED foods, followed by extended storage at chill temperatures, favors the survival and growth of psychrotrophic strains (Group II) of *C. botulinum* (Lindström et al., 2006). In spite of the low incidence of this intoxication, the mortality rate is high, if not treated immediately and properly. *C. botulinum* is of greatest concern on account of its spore's heat resistance (Table 2.1), being able to

**TABLE 2.1**
**Heat Resistance of Psychrotrophic (Group II) Nonproteolytic Strains of *Clostridium botulinum* Spores**

| Food Product | Spore Inoculum, Botulinum Strains | *T* (°C) | *D*-Value (min) | *z*-Value (°C) | *T* Range (°C) | References |
|---|---|---|---|---|---|---|
| Crabmeat | Mixture of three strains: 'Ham', 'Kapchunka', 17B | 88.9 | 13 | 8.6 | 88.9–94.4 | Peterson et al. (1997) |
| | | 90.6 | 8.2 | | | |
| | | 92.2 | 5.3 | | | |
| | | 94.4 | 2.9 | | | |
| Cod homogenate | ATCC 25765, ATCC 9564 | 75.0 | 54 | 8.6 | 75.0–92.0 | Gaze and Brown (1990) |
| | | 80.0 | 18 | | | |
| | | 85.0 | 4.0 | | | |
| | | 90.0 | 1.1 | | | |
| | | 92.0 | 0.60 | | | |
| Turkey slurry | KAP B5 | 75.0 | 33 | 9.4 | 75.0–90.0 | Juneja et al. (1995) |
| | | 90.0 | 0.80 | | | |
| Carrot homogenate | ATCC 25765, ATCC 9564 | 75.0 | 19 | 9.8 | 75.0–92.0 | Gaze and Brown (1990) |
| | | 80.0 | 4.2 | | | |
| | | 85.0 | 1.6 | | | |
| | | 90.0 | 0.36 | | | |
| Turkey slurry | 'Alaska' | 70.0 | 52 | 9.9 | 70.0–85.0 | Juneja et al. (1995) |
| | | 85.0 | 1.2 | | | |
| Whitefish paste | 'Alaska', 'Beluga', 8E, 'Iwanai', 'Tenno' | 80.0 | 1.6–4.3 | 5.7–7.6 | 73.9–85.0 | Crisley et al. (1968) |
| Blue crab | 'Alaska', 'Beluga', crab G21-5, crab 25V-1, crab 25V-2 | 73.9 | 6.8–13 | 7.0–8.4 | 73.9–85.0 | Lynt et al. (1977) |
| | | 76.6 | 2.4–4.1 | | | |
| | | 79.4 | 1.1–1.7 | | | |
| | | 82.2 | 0.49–0.74 | | | |
| Oyster homogenate | 'Minnesota', 'Alaska', crab G21-5, crab 25V-1, crab 25V-2 | 73.9 | 2.0–9.0 | 4.2–7.1 | 73.9–82.2 | Chai and Liang (1992) |
| | | 82.2 | 0.080–0.43 | | | |

*T*, temperature (°C)

survive mild heat treatments, including pasteurization, and requiring special storage conditions (Peck, 2006). The use of refrigerated storage can reduce or at least retard toxin production, given that this organism needs much longer storage periods to produce the lethal toxin: within 31 days at 3.3°C in beef stew (Schmidt et al., 1961); within 22 days at 8.0°C (Betts and Gaze, 1995); ≥55 days at 4.4°C, 8 days at 10°C, and 2 days at 24°C in crabmeat homogenates (Cockey and Tatro, 1974). Thus, to control human botulism in low-acid pasteurized foods the use of refrigerated storage ($T < 8$°C) is required with a restricted shelf-life (Gould, 1999). Additional measures of safety with this risky class of foods include the use of added preservatives such as salt (>3.5%) and nitrites (>100 ppm) (e.g., cured meat products) (Graham et al., 1996). The unique use of such levels of salts is not sufficient to inhibit the mesophilic strains of *C. botulinum* (belonging to Group I, proteolytic, producing toxin types

A, B, and F) in pasteurized meat products, and these must be refrigerated (<8°C) (Peterson et al., 1997).

Psychrotrophic strains of *C. botulinum* present different thermal resistances depending first on the heating menstruum (the food), and in some cases on the spores of a particular strain. Similar results were obtained with toxin type E and type B strains in cod and carrot homogenates (Gaze and Brown, 1990). Table 2.1 presents thermal resistance data obtained from the literature in decreasing order of resistance to heat. In summary, $D_{90°C}$ varies from seconds to more than 8 min, $D_{85°C}$ from a few seconds to 37 min, and $D_{80°C}$ from a few minutes to 140 min. Crabmeat presented the highest *D*-value, 2.9 min at 94.4°C, using a mixture of 'Ham', 'Kapchunka', and 17B botulinum strains. Most of the authors published similar *z*-values, ranging between 7.2°C and 9.9°C. Oyster homogenate presented the lowest heat resistance for the five spore strains (two from outbreaks in 'Alaska' and 'Minnesota') of *C. botulinum* studied, the *D*- and *z*-values (4.2°C–7.1°C) being the lowest recorded (Chai and Liang, 1992). The *z*-values were also low in whitefish paste (Crisley et al., 1968).

#### 2.3.2.2 Other Pathogenic Spore-Forming Bacteria

Table 2.2 shows thermal resistance data of the spore-forming pathogens *Clostridium perfringens* and *Bacillus cereus*, which have been responsible for outbreaks in low-acid underpasteurized chilled foods. Studies with spores of six strains of *C. perfringens* demonstrated no growth at $T \leq 10°C$, although spore germination and extended survival at low temperatures occurred (de-Jong et al., 2004). However, since this is the most common foodborne illness from a sporeformer, it was also considered in this review (Table 2.2). As a result of temperature abuse during distribution or storage of heated/cooked foods (e.g., meat products), *C. perfringens* may grow, especially in establishments where large quantities of foods are prepared several hours before serving (e.g., school cafeterias, hospitals, nursing homes, prisons, etc.) with concomitant difficulties in rapid chilling to below 10°C. Spores of *C. perfringens* are more heat resistant than those of nonproteolytic *C. botulinum*, for example, $D_{99°C} = 23$ min in turkey, $D_{90°C} = 31$ min in pork roll, and $D_{90°C} = 14$ min in chicken breast. Some strains of *B. cereus* can grow at low temperatures ($T < 8°C$) and can also present problems in underpasteurized refrigerated foods (Dufrenne et al., 1994, 1995; García-Armesto and Sutherland, 1997; Carlin et al., 2000; Choma et al., 2000). Eleven strains of isolated psychrotrophic strains of *B. cereus* able to grow at ≤7°C in foods presented a 2.2 min $< D_{90°C} <$ 9.2 min in phosphate buffer (Dufrenne et al., 1995). Other published data gave $D_{90°C}$ of 10 min and 4 min in pork roll (Byrne et al., 2006) and water (Fernández et al., 2001), respectively. Psychrotrophic strains of *B. cereus* seem to be more heat resistant than psychrotrophic strains of *C. botulinum* (Figure 2.2).

#### 2.3.2.3 Non-Spore-Forming Psychrotrophic Pathogens

Table 2.3 shows the heat resistance of non-spore-forming pathogens able to grow at low temperatures ($T < 8°C$). The pathogenic vegetative bacteria, *Listeria monocytogenes*, *Yersinia enterocolitica*, and *Vibrio parahaemolyticus*, are also able to grow in foods with pH > 4.6 at temperatures lower than 6°C (Penfield and Campbell, 1990) but they are easily eliminated with a few seconds at 70°C or less (Table 2.3). Other non-spore-forming pathogens such as *Escherichia coli*, some strains of *Salmonella*, and *Aeromonas*

**TABLE 2.2**
**Heat Resistance of *Clostridium perfringens* and *Bacillus cereus* in Low-Acid Foods (pH > 4.6)**

| Bacteria | Spore Inoculum | Food Product | *T* (°C) | *D*-Value (min) | *z*-Value (°C) | *T* Range (°C) | References |
|---|---|---|---|---|---|---|---|
| *Clostridium perfringens* | Three strains: NCTC 8238/9, ATCC 10288 | Ground turkey | 99.0 | 23 | nr | nr | Juneja and Marmer (1996) |
| | Three strains: DSM 11784, NCTC 10614 (incidents), NCTC 08237 | Pork luncheon roll | 90.0<br>95.0<br>100.0 | 31<br>9.7<br>1.9 | 8.3 | 90.0–100.0 | Byrne et al. (2006) |
| | Three strains: NCTC 8238/9, ATCC 10288 | Marinated chicken breast | 90.0 | 14 | nr | nr | Juneja et al. (2006) |
| *Bacillus cereus* | Three strains: DSM 4313 (incident), DSM 626, NCTC 07464 | Pork luncheon roll | 85.0<br>90.0<br>95.0 | 30<br>10<br>2.0 | 8.6 | 85.0–95.0 | Byrne et al. (2006) |
| | Psychrotrophic strain INRA AVTZ415 | Distilled water | 85.0<br>90.0<br>95.0<br>100.0 | 16<br>3.9<br>1.0<br>0.24 | 8.2 | 85.0–100.0 | Fernández et al. (2001) |

*T*, temperature (°C); nr, not reported.

*hydrophila* (Palumbo et al., 1995; Schoeni et al., 1995; George, 2000; Papageorgiou et al., 2006) can also be a problem in chilled foods. However, pasteurization of a few seconds at 65°C or less should be sufficient to destroy these microorganisms (Table 2.3). In general, a few minutes at 60°C was enough to achieve one decimal reduction in *Listeria*, *E. coli*, *Salmonella*, and *Y. enterocolitica* populations in most of these foods (Figure 2.2). *A. hydrophila* presented the lowest heat resistance.

The potential for the growth of other vegetative bacterial pathogens such as *Campylobacter* spp., *V. cholerae*, *Shigella* spp., *Staphylococcus aureus*, *Enterococcus* spp. (FDA, 1992; Jay, 2000) in pasteurized/chilled foods is very low, since apart from being heat sensitive, food must be temperature abused during distribution to allow their growth.

Although the effect of MAP in humans is not known yet, this bacterium causes disease in cattle and as a precaution the design of pasteurization in milk and dairy products should consider this bacterium's thermal resistance. *D*-values in milk at 63°C, 66°C, and 72°C are 15, 5.9, and <2.03 s, respectively, and *z*-value is 8.6°C (Pearce et al., 2001). Keswani and Frank (1998) obtained much higher *D*-values in milk ($D_{63°C}$ = 1.6–2.5 min). *Coxiella burnetii* (the causative agent of 'Q-fever') has *D*-values of 4.14 min at 62.8°C and 2.21 sec at 71.7°C, *z*-value = 4.34°C in milk (Cerf and Condron, 2006).

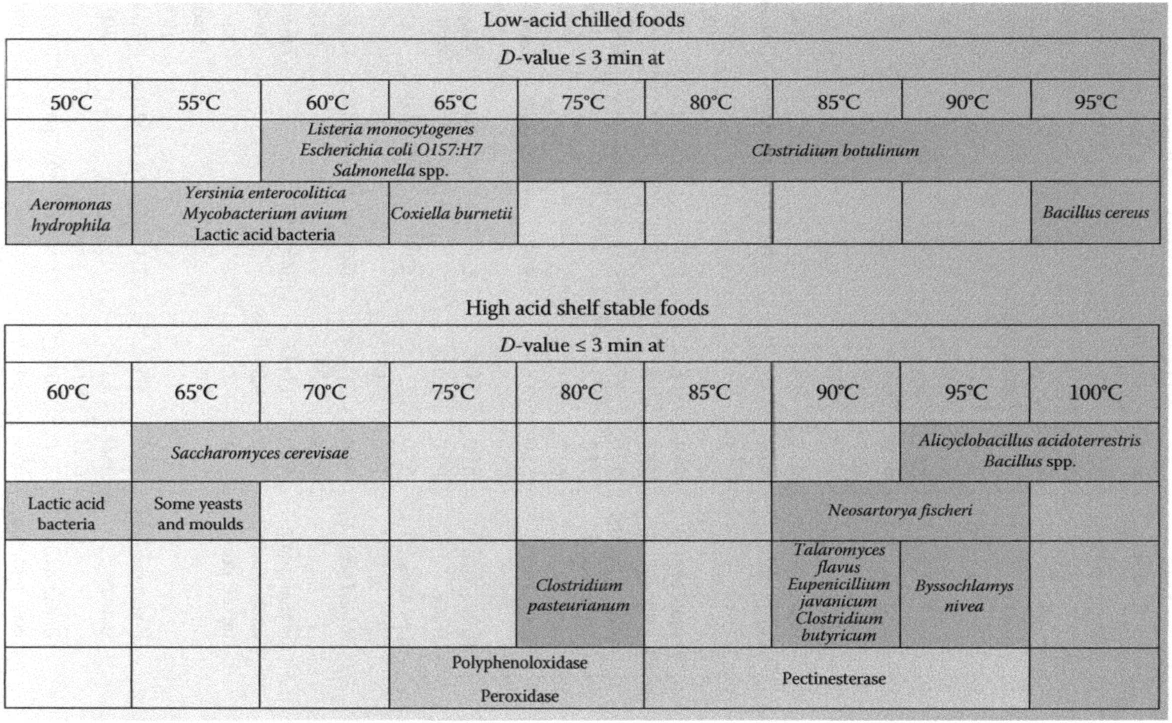

Low-acid chilled foods

| *D*-value ≤ 3 min at | | | | | | | | |
|---|---|---|---|---|---|---|---|---|
| 50°C | 55°C | 60°C | 65°C | 75°C | 80°C | 85°C | 90°C | 95°C |
| | | *Listeria monocytogenes*<br>*Escherichia coli O157:H7*<br>*Salmonella* spp. | | *Clostridium botulinum* | | | | |
| *Aeromonas hydrophila* | *Yersinia enterocolitica*<br>*Mycobacterium avium*<br>Lactic acid bacteria | | *Coxiella burnetii* | | | | | *Bacillus cereus* |

High acid shelf stable foods

| *D*-value ≤ 3 min at | | | | | | | | |
|---|---|---|---|---|---|---|---|---|
| 60°C | 65°C | 70°C | 75°C | 80°C | 85°C | 90°C | 95°C | 100°C |
| | *Saccharomyces cerevisae* | | | | | | *Alicyclobacillus acidoterrestris*<br>*Bacillus* spp. | |
| Lactic acid bacteria | Some yeasts and moulds | | | | | *Neosartorya fischeri* | | |
| | | | | *Clostridium pasteurianum* | | *Talaromyces flavus*<br>*Eupenicillium javanicum*<br>*Clostridium butyricum* | *Byssochlamys nivea* | |
| | | | Polyphenoloxidase<br>Peroxidase | | Pectinesterase | | | |

**FIGURE 2.2** Minimum pasteurization temperature to achieve 1 logarithmic microbial reduction (1*D*) in a few minutes.

**TABLE 2.3**
**Heat Resistance of Non-Spore-Forming Pathogenic Microbes in Low-Acid Foods (pH > 4.6)**

| Bacteria | Food Product | *T* (°C) | *D*-Value (min) | *z*-Value (°C) | *T* Range (°C) | References |
|---|---|---|---|---|---|---|
| *Listeria monocytogenes* | Ground pork | 55.0<br>70.0 | 47<br>0.085 | 5.9 | 55.0–70.0 | Murphy et al. (2004a) |
| | Chicken gravy | 50.0<br>65.0 | 119–195<br>0.19–0.48 | 5.2–6.1 | 50.0–65.0 | Huang et al. (1992) |
| | Cooked lobster | 51.6<br>54.4<br>57.2<br>60.0<br>62.7 | 97<br>55<br>8.3<br>2.4<br>1.1 | 5.0 | 51.6–62.7 | Buduamoako et al. (1992) |
| | Rainbow trout roe | 60.0<br>63.0 | 1.6<br>0.44 | 5.4 | 60.0–63.0 | Miettinen et al. (2005) |
| | Liquid egg yolk | 60.0<br>62.2 | 1.3<br>0.58 | 6.1 | 60.0–62.2 | Schuman and Sheldon (1997) |
| | Liquid egg white | 55.1<br>58.3 | 7.6<br>3.5 | 9.4 | 55.1–58.3 | |
| | Vacuum-packed minced beef | 50.0<br>55.0<br>60.0 | 36<br>3.2<br>0.15 | 4.2 | 50.0–60.0 | Bolton et al. (2000) |
| *Escherichia coli* O157:H7 | Ground pork | 55.0<br>70.0 | 33<br>0.048 | 4.9 | 55.0–70.0 | Murphy et al. (2004a) |
| | Fully cooked frank | 55.0<br>70.0 | 25<br>0.038 | 5.1 | 55.0–70.0 | Murphy et al. (2004b) |
| | Raw frank | 55.0<br>70.0 | 21<br>0.031 | 5.1 | 55.0–70.0 | |
| | Ground beef | 55.0<br>65.0 | 21<br>0.39 | 6.0 | 55.0–65.0<br>55.0–65.0 | Juneja et al. (1997) |
| | Ground meats: lamb, chicken, turkey, pork | 55.0<br>65.0 | 11–12<br>0.29–0.38 | 6.5–6.9 | | Juneja et al. (1997); Juneja and Marmer (1999) |
| | Ground morcilla sausage | 54.0<br>58.0<br>62.0 | 5.5<br>2.1<br>0.60 | 7.4 | 54.0–62.0 | Oteiza et al. (2003) |
| | Ground meat beef, pork sausage, chicken, turkey | 50.0<br>55.0<br>60.0 | 50–115<br>6.4–19<br>0.37–0.58 | 4.4–4.8 | 50.0–60.0 | Ahmed et al. (1995) |
| *Salmonella* spp. | Green pea soup | 60.0<br>65.6 | 10<br>1.0 | 5.7 | 60.0–71.1 | Thomas et al. (1966) |
| | Ground pork | 55.0<br>70.0 | 46<br>0.083 | 5.9 | 55.0–70.0 | Murphy et al. (2004a) |

**TABLE 2.3 (continued)**
**Heat Resistance of Non-Spore-Forming Pathogenic Microbes in Low-Acid Foods (pH > 4.6)**

| Bacteria | Food Product | *T* (°C) | *D*-Value (min) | *z*-Value (°C) | *T* Range (°C) | References |
|---|---|---|---|---|---|---|
| | Chicken Thigh meat | 55.0 | 12 | 6.9 | 55.0–62.5 | Juneja (2007) |
| | | 60.0 | 3.2 | | | |
| | | 62.5 | 0.84 | | | |
| | Chicken breast meat | 55.0 | 6.1 | 8.1 | 55.0–62.5 | |
| | | 60.0 | 3.00 | | | |
| | | 62.5 | 0.66 | | | |
| | Liquid egg yolk | 60.0 | 0.28 | 4.3 | 60.0–62.2 | Shuman and Sheldon (1997) |
| | | 62.2 | 0.087 | | | |
| | Liquid egg white | 55.1 | 8.0 | 3.5 | 55.1–58.3 | |
| | | 58.3 | 1.0 | | | |
| *Yersinia enterocolitica* | Vacuum-packed minced beef | 50.0 | 21 | nr | 50.0–60.0 | Bolton et al. (2000) |
| | | 55.0 | 1.1 | | | |
| | | 60.0 | 0.55 | | | |
| | Whole and skim milks | 62.8 | 0.17–0.18 | nr | nr | Toora et al. (1992) |
| *Aeromonas hydrophila* | Liquid whole egg | 48.0 | 3.6–9.4 | 5.0–5.6 | 48.0–60.0 | Schuman et al. (1997) |
| | | 60.0 | 0.026–0.040 | | | |

*T*, Temperature (°C); nr, not reported.

### 2.3.3 Microbial and Endogenous Enzymes Heat Resistance in High-Acid and Acidified Foods (pH < 4.6)

In high-acid and acidified foods, the main pasteurization goal is to avoid spoilage during distribution at room temperature, rather than avoiding outbreaks of public health concern (Figure 2.1). High-acid foods include most of the fruits, normally containing high levels of organic acids. The spoilage flora is mainly dependent on pH and soluble solids. The type of organic acids and other constituents of these foods such as polyphenols might also affect the potential spoilage microorganisms. Given the high acid content of this class of foods (pH < 4.6), the pathogens referred to in Section 2.3.2 (vegetative and spore cells) including the spore-forming *C. botulinum* are not able to grow. It is generally assumed that the higher the acidity of the food, the less probable is the germination and growth of bacterial spores, a pH < 4.6 being accepted as safe in terms of pathogenic sporeformers. However, various incidents in high-acid foods involving the spore-forming spoilage bacterium *Alicyclobacillus acidoterrestris* (Cerny et al., 1984; Jay, 2000) have been registered, since its optimum growth pH is between 3.5 and 4.5 for the type strain (Pinhatti et al., 1997) and optimum growth temperature is between 35°C and 53°C (Deinhard et al., 1987; Sinigaglia et al., 2003), depending on the strain.

Typical microbes associated with spoilage of high-acid and acidified shelf-stable foods are *A. acidoterrestris*, molds, yeasts, and some lactic acid bacteria (LAB).

Heat-resistant deteriorative enzymes such pectinesterase (PE), polyphenoloxidase (PPO), and peroxidase (PRO) may also degrade high-acid food quality during storage. Additionally, growth of spoilage spore-forming *Bacillus* and *Clostridium* has been registered in less acid foods (3.7 < pH < 4.6) such as tomato purée/juice, mango pulp/nectar, canned pear, and pear juice (Ikeyami et al., 1970; Shridar and Shankhapal, 1986). A review of the most thermally resistant pasteurization targets, such as microbial spores and enzymes, is presented.

#### 2.3.3.1 *Alicyclobacillus acidoterrestris* Spores

*A. acidoterrestris*, is a thermoacidophilic, nonpathogen, and spore-forming bacterium identified in the 1980s (Deinhard et al., 1987; Wisotzkey et al., 1992), which has been associated with various spoilage incidents in shelf-stable apple and orange juices. The presence of ω-alicyclic fatty acids as the major natural cell membrane lipid component gave the name *Alicyclobacillus* to this genus (Wisotzkey et al., 1992). Since this microbe does not produce gas, spoilage is only detected by the consumer at the end of the food chain, resulting in consumer complaints, product withdrawal, and subsequent economic loss. Spoilage aromas and taste are related to the production of a bromophenol and guaiacol. A relatively low level of $10^5$–$10^6$ cells/mL in apple and orange juices formed enough guaiacol (ppb) to produce sensory taint (Pettipher et al., 1997). Spoilage by *A. acidoterrestris* has been observed mainly in apple juice, but also in pear juice, orange juice, juice blends, and canned diced tomatoes (Cerny et al., 1984; Splittstoesser et al., 1994; Yamazaki et al., 1996; Pontius et al., 1998; Walls and Chuyate, 2000). Incidents were reported from all over the world (Germany, the United States, Japan, Australia, and the United Kingdom). A survey carried out by National Food Processors Association in the United States (Walls and Chuyate, 1998) in fifty seven companies, had shown that 35% of juice manufacturers had problems especially during warmer spring and summer seasons, possibly associated to *Alicyclobacillus*. Another incident with many complaints from consumers, referred to an iced tea (pH = 2.7) submitted to a thermal process of 95°C for 30 s, followed by hot-filling into cartons (Duong and Jensen, 2000). The slow cooling of the hot-filled tea or the high storage temperature may have allowed sufficient time for the spores to germinate and grow, causing taint problems. *A. acidoterrestris* spore germination and growth (to $10^6$ cfu/mL) under acidic conditions was reported in orange juice stored at 44°C for 24 h (Pettipher et al., 1997), and also in apple, orange, and grapefruit juices stored at 30°C (Komitopoulou et al., 1999). Spore germination and growth was observed after 1–2 weeks in apple juice, orange juice, white grape juice, tomato juice, and pear juice incubated at 35°C (Walls and Chuyate, 2000). Red grape juice did not support growth (Splittstoesser et al., 1994), possibly due to the polyphenols. The increase of soluble solids from 12.5°Brix ($a_w = 0.992$) to 38.7°Brix ($a_w = 0.96$) inhibited growth of *A. acidoterrestris* spores (Sinigaglia et al., 2003).

The spores of *A. acidoterrestris* are very resistant to heat compared to the major spoilage microbes and enzymes typical in high-acid shelf-stable foods (Tables 2.4 through 2.7), presenting 4 min < $D_{90°C}$ < 23 min, 1 min < $D_{95°C}$ < 5 min and 7°C < $z$-value < 13°C. Much lower $D$-values were recorded in wine ($D_{85°C} = 0.6$ min) (Splittstoesser et al., 1997), potentially due to the alcohol or other constituents created by fermentation. Further conclusions about *A. acidoterrestris* spore thermal resistance are

**TABLE 2.4**
**Heat Resistance of *Alicyclobacillus acidoterrestris* Spores in High-Acid Fruit Products (pH < 4.6)**

| Heating Medium | Spore Strain | pH | SS (°Brix) | *T* (°C) | *D*-Value (min) | *z*-Value (°C) | T Range (°C) | References |
|---|---|---|---|---|---|---|---|---|
| *Juices, nectars, fruit drinks, and wine* | | | | | | | | |
| Orange juice drink | nr | 4.1 | 5.3 | 95 | 5.3 | 9.5 | nr | Baumgart et al. (1997) |
| Fruit drink | nr | 3.5 | 4.8 | 95 | 5.2 | 10.8 | nr | |
| Fruit nectar | nr | 3.5 | 6.1 | 95 | 5.1 | 9.6 | nr | |
| Apple juice | VF | 3.5 | 11.4 | 85 | 56 | 7.7 | 85–95 | Splittstoesser et al. (1994) |
| | | | | 90 | 23 | | | |
| | | | | 95 | 2.8 | | | |
| Grape juice | WAC | 3.3 | 15.8 | 85 | 57 | 7.2 | 85–95 | |
| | | | | 90 | 16 | | | |
| | | | | 95 | 2.4 | | | |
| Orange juice | Type | 3.5 | 11.7 | 85 | 66 | 7.8 | 85–91 | Silva et al. (1999) |
| | | | | 91 | 12 | | | |
| Orange juice | DSM 2498; three isolated strains: 46; 70; 145. | 3.2 | 9.0 | 85 | 50–95 | 7.2–11.3 | 85–95 | Eiroa et al. (1999) |
| | | | | 90 | 10–21 | | | |
| | | | | 95 | 2.5–8.7 | | | |
| Orange juice | Z | 3.9 | nr | 80 | 54 | 12.9 | 80–95 | Komitopoulou et al. (1999) |
| | | | | 90 | 10 | | | |
| | | | | 95 | 3.6 | | | |
| Apple juice | Z(CRA 7182) | 3.5 | nr | 80 | 41 | 12.2 | 80–95 | |
| | | | | 90 | 7.4 | | | |
| | | | | 95 | 2.3 | | | |
| Cupuaçu extract | Type | 3.6 | 11.3 | 85 | 18 | 9.0 | 85–97 | Silva et al. (1999) |
| | | | | 91 | 5.4 | | | |
| | | | | 95 | 2.8 | | | |
| | | | | 97 | 0.57 | | | |
| Grapefruit juice | Z | 3.4 | nr | 80 | 38 | 11.6 | 80–95 | Komitopoulou et al. (1999) |
| | | | | 90 | 6.0 | | | |
| | | | | 95 | 1.9 | | | |
| Berry juice | nr | 3.5 | nr | 88 | 11 | 7.2 | 88–95 | Walls (1997) |
| | | | | 91 | 3.8 | | | |
| | | | | 95 | 1.0 | | | |
| Wine | nr | nr | nr | 75 | 33 | 10.5 | 75–85 | Splittstoesser et al. (1997) |
| | | | | 85 | 0.57 | | | |
| *Fruit concentrate* | | | | | | | | |
| Black currant concentrate | Type | 2.5 | 58.5 | 91 | 24 | nr | nr | Silva et al. (1999) |
| Light black currant concentrate | Type | 2.5 | 26.1 | 91 | 3.8 | nr | nr | |

SS, soluble solids (°Brix); *T*, temperature (°C); nr, not reported; *A. acidoterrestris* type strain, NCIMB 13137, GD3B, DSM 3922, ATCC 49025.

**TABLE 2.5**
**Heat Resistance of Spoilage Fungal Ascospores in High-Acid Fruit Products (pH < 4.6)**

| Fungal Ascospores | Fruit Product | pH | SS (°Brix) | *T* (°C) | *D*-Value (min) | *z*-Value (°C) | *T* Range (°C) | References |
|---|---|---|---|---|---|---|---|---|
| *Byssochlamys nivea* | Strawberry pulp | 3.0 | 15.0 | 80.0 | 193 | 6.4 | 80–93 | Aragão (1989) |
| | | | | 85.0 | 35 | | | |
| | | | | 90.0 | 6.3 | | | |
| | | | | 93.0 | 1.7 | | | |
| *Neosartorya fischeri* | Pineapple concentrate | 3.4 | 42.7 | 85.0 | 30 | 8.9 | 85–95 | Tournas and Traxler (1994) |
| | | | | 90.0 | 7.6 | | | |
| | | | | 95.0 | 2.3 | | | |
| | Pineapple juice | 3.4 | 12.6 | 85.0 | 20 | 9.2 | 85–95 | |
| | | | | 90.0 | 4.8 | | | |
| | | | | 95.0 | 1.7 | | | |
| *Neosartorya fischeri LT025* | Apple juice | 3.5 | 15.0 | 85.0 | 15 | 5.3 | 85–93 | Gumerato (1995) |
| | | | | 88.0 | 4.7 | | | |
| | | | | 90.0 | 2.6 | | | |
| | | | | 93.0 | 0.43 | | | |
| | Strawberry pulp | 3.0 | 15.0 | 80.0 | 60 | 6.4 | 80–93 | Aragão (1989) |
| | | | | 85.0 | 15 | | | |
| | | | | 90.0 | 2.6 | | | |
| | | | | 93.0 | 0.50 | | | |
| *Talaromyces flavus* | Apple juice | 3.7 | 11.6 | 87.8 | 7.8 | 5.2 | nr | Scott and Bernard (1987) |
| | | | | 90.6 | 2.2 | | | |
| | Strawberry pulp | 3.0 | 15.0 | 75.0 | 54 | 8.2 | 75–90 | Aragão (1989) |
| | | | | 80.0 | 18 | | | |
| | | | | 85.0 | 3.3 | | | |
| | | | | 90.0 | 0.90 | | | |
| *Eukenicillium javanicum* | Strawberry pulp | 3.0 | 15.0 | 80.0 | 15 | 7.9 | 80–90 | Aragão (1989) |
| | | | | 85.0 | 3.7 | | | |
| | | | | 90.0 | 0.80 | | | |

SS, soluble solids (°Brix); *T*, temperature (°C); nr, not reported.

dependent on the spore strain and/or fruit product. As expected, when increasing the soluble solids from 26.1°Brix to 58.5°Brix in black currant concentrate, the $D_{91°C}$-values increased from 3.8 to 24.1 min (Silva et al., 1999). However, the growth of *A. acidoterrestris* is inhibited at high soluble solids concentration, for example, no growth was observed in apple concentrate between 30°Brix and 50°Brix (Walls and Chuyate, 2000) and white grape juice with more than 18°Brix (Splittstoesser et al., 1997).

### 2.3.3.2 Fungal Ascospores

Fungal growth in pasteurized foods, raw materials, and food ingredients should be avoided, since some of them are able to produce mycotoxins. Spores and vegetative

**TABLE 2.6**
**Heat Resistance of Other Spore-Forming Bacteria in High-Acid Foods (pH < 4.6)**

| Bacterium Spores | Fruit Product | pH | *T* (°C) | *D*-Value (min) | *z*-Value (°C) | *T* Range (°C) | References |
|---|---|---|---|---|---|---|---|
| *Bacillus subtilis* | Tomato juice | 4.4 | 90 | 30 | 14.0 | 90–100 | Rodriguez et al. (1993) |
| | | | 95 | 16 | | | |
| | | | 100 | 5.7 | | | |
| *Bacillus licheniformis* | Tomato juice | 4.4 | 90 | 30 | 14.2 | 90–100 | Rodriguez et al. (1993) |
| | | | 95 | 12 | | | |
| | | | 100 | 5.9 | | | |
| | Tomato puree | 4.4 | 85 | 18 | 14.9 | 85–100 | Montville and Sapers (1981) |
| | | | 90 | 9.4 | | | |
| | | | 95 | 5.1 | | | |
| | | | 100 | 1.9 | | | |
| | Mango juice | 4.2 | 100 | 1.3 | nr | nr | Azizi and Ranganna (1993) |
| | Acidified papaya pulp | 4.2 | 100 | 2.2 | nr | nr | |
| *Bacillus megaterium* | Tomato juice | 4.2 | 100 | 1.6 | nr | nr | Gibriel and Abd-El (1973) |
| | Mango juice | 3.4 | 100 | 0.80 | nr | nr | |
| | Orange juice | 3.7 | 100 | 0.80 | nr | 100–110 | |
| | | | 110 | 0.025 | | | |
| *Bacillus coagulans* | Tomato concentrate (30.3° Brix) | 4.0 | 80 | 41 | 9.5 | 75–90 | Sandoval et al. (1992) |
| | | | 85 | 14 | | | |
| | | | 90 | 3.5 | | | |
| *Clostridium butyricum* | Peach | nr | 90 | 1.1 | 11.5 | 90–100 | Gaze et al. (1988) |
| | | | 95 | 0.39 | | | |
| | | | 100 | 0.15 | | | |
| *Clostridium pasteurianum* | Acidified papaya pulp | 3.8 | 75 | 9.7 | 8.8 | 75–80 | Magalhães (1993) |
| | | | 80 | 2.7 | | | |

*T*, temperature (°C); nr, not reported.

cells of most molds are inactivated upon exposure to 60°C for 5 min (Beuchat, 1998). Notable exceptions are the ascospores of certain strains of the molds *Byssochlamys nivea*, *Byssochlamys fulva*, *Neosartorya fischeri*, *Talaromyces flavus*, and *Eupenicillium javanicum* (King et al., 1969; Hatcher et al., 1979; Beuchat, 1986; Scott and Bernard, 1987; Aragão, 1989; Tournas and Traxler, 1994; Gumerato, 1995;

**TABLE 2.7**
**Heat Resistance of Endogenous Spoilage Enzymes in High-Acid and Acidified Fruit Products (pH < 4.6)**

| Enzyme | Fruit Product | pH | SS (°Brix) | *T* (°C) | *D*-Value (min) | *z*-Value (°C) | *T* Range (°C) | References |
|---|---|---|---|---|---|---|---|---|
| Pectinesterase | Papaya acidified nectar | 3.8 | 14.0 | 85 | 7.7 | 15.1 | 75–85 | Argáiz, 1994 |
| | Papaya acidified purée | 3.5<br>3.8<br>4.0 | 7.4–8.5 | 85<br>85<br>85 | 4.8<br>5.7<br>7.2 | 14.8<br>14.7<br>14.2 | 75–85 | |
| | Papaya acidified pulp | 4.0 | 10.5–12.0 | 85 | 3.9 | 15.0 | 82–102 | Nath and Ranganna 1981 |
| | Mandarin juice | 3.6<br>4.0 | 12.0<br>12.0 | 85<br>85 | 2.2<br>3.6 | 11.4<br>10.1 | 82–94<br>82–94 | Nath and Ranganna 1977 |
| Polyphenol-oxidase | Pineapple purée | 3.7 | 15.0 | 75<br>85<br>90 | 91<br>18<br>11 | 21.5 | 70–90 | Chutintrasri and Noomhorm 2006 |
| | Pears 'Bartlett' | 3.9 | nr | 75 | 6.5 | nr | 75–90 | Dimick et al. (1951) |
| | Apricots 'Royal' | 4.0 | nr | 75 | 1.3 | nr | 75–92 | |
| | Grapes 'Concord' | 3.3 | nr | 75 | 0.45 | nr | 65–80 | |
| | Peaches 'Elberta' | 3.5 | nr | 75 | 0.20 | nr | 70–75 | |
| | Apples 'Gravenstein' | 3.1 | nr | 75 | 0.13 | nr | 65–80 | |
| Peroxidase | Strawberry 'Selva' | nr | nr | 70 | 5.0 | 19.0 | 50–70 | Civello et al. (1995) |
| | Grapes 'Malvasia' | nr | nr | 70 | 4.8 | 35.4 | 65–85 | Sciancalepore et al. (1985) |

SS, soluble solids (°Brix); *T*, temperature (°C); nr, not reported.

Kotzekidou, 1997), where for high-acid fruit pulps/juices, a 0.8 min $< D_{90°C}$-value $<$ 6.3 min and 5.2°C $< z$-value $<$ 9.2°C were observed (Table 2.5). The ascospores of mold species have been associated with spoilage of canned tomato paste, surviving pasteurization of 85°C for 20 min, when initial numbers were near $10^5$ cfu/mL (Kotzekidou, 1997). Three strains of *B. fulva*, two strains of *B. nivea* and four strains of *N. fischeri* were identified. *T. flavus* has also been responsible for fruit juice spoilage incidents (King and Halbrook, 1987; Scott and Bernard, 1987). Thermal inactivation studies of *T. flavus*, *N. fischeri*, *Byssochlamys* spp. in five fruit-based concentrates (25.2°Brix −33.4°Brix) resulted in $D_{91°C}$ of 2.9–5.4 min, <2 min, and some seconds, respectively (Beuchat, 1986).

#### 2.3.3.3 Other Spoilage Bacteria and Fungi

Not so commonly, spoilage in less-acid foods or fruit pulps/drinks, such as tomato, pear, peach, mango, mandarin, and orange with pH between 3.7 and 4.5, may be caused by *Bacillus subtilis*, *B. coagulans*, *B. licheniformis*, *B. megaterium*, *B. polymyxa*, *B. macerans* (Vaughn et al., 1952; York et al., 1975; Montville and Sapers, 1981; Nakajyo and Ishizu, 1985; Shridhar and Shankhapal, 1986; Azizi and Ranganna, 1993; Rodriguez et al, 1993; Everis and Betts, 2001) and butyric anaerobes such as *Clostridium pasteurianum*, *C. butyricum*, *C. tyrobutyricum* (Ikeyami et al., 1970; Jacobsen and Jensen, 1975; De Jong, 1989; Everis and Betts, 2001). An accepted practice to avoid growth of these spore-forming bacteria is the acidification of the food with citric or ascorbic acids. The spores of *Bacillus* spp. have very high heat resistances, 3.5 min < $D_{90°C}$ < 30 min, 1 min < $D_{100°C}$ < 6 min, and 9°C < $z$-value < 15°C (Table 2.6). *Clostridium* spp. exhibited much lower temperatures/times for inactivation, $D_{90°C}$ in peach 1.1 min and $D_{80°C}$ in acidified papaya 2.7 min (Table 2.6).

*B. cereus* and two strains of *B. thuringiensis* isolated from spoiled mango pulp, exhibited a $D_{80°C}$ of 6 min in mango pulp (pH 4.0) (De-Carvalho et al., 2007). The heat resistance of the various fruit spoilage microorganisms (yeasts: *Saccharomyces cerevisiae*, *Rhodotorula mucilaginosa*, *Torulaspora delbrueckii*, and *Zygosaccharomyces rouxii*; molds: *Penicillium citrinum*, *P. roquefortii*, and *Aspergillus niger*; LAB: *Lactobacillus fermentum* and *L. plantarum*) was determined in acid juices, *S. cerevisae* being the most heat-resistant microorganism with $D_{57°C}$ ranging between 9.4 and 32 min (Shearer et al., 2002).

#### 2.3.3.4 Endogenous Enzymes

Major heat-resistant deteriorative enzymes in fruits are PE, PPO, and PRO. The enzyme activity is maximum at 30°C–40°C, being relevant for shelf-stable foods and not so important for refrigerated foods. PE alters the texture of fruits during storage. PPO causes enzymatic browning and alterations of flavor in fruits and vegetables. PRO can degrade nutritive (vitamin C), color, and flavor characteristics of foods, producing off-flavors in vegetables and browning in fruits, and is often used as an indicator for the effectiveness of vegetable blanching (Whitaker, 1972). Oxidases (PPO, PRO) require the presence of oxygen ($O_2$) for their activity. Depending on the food, these enzymes may have higher thermal resistance than certain microorganisms such as non-spore-forming bacteria, LAB, yeasts, and some molds (Table 3.7) (Shearer et al., 2002). Several studies have reported the use of these enzymes as targets of pasteurization (Dastur et al., 1968; Nanjundaswamy et al., 1973; Nath and Ranganna, 1977, 1980, 1981, 1983a,b). With orange juice, for example, the pasteurization objective can be the inactivation of PE. The dependence of enzyme inactivation on temperature can be expressed in the same way that has been used for microorganisms, with $D$-value and $z$-value. A review of thermal resistances of these enzymes in fruits has demonstrated that PE is more resistant to heat than

PPO (except for pineapple PPO) and PRO, and thermal inactivation varies considerably with their fruit origin (Table 2.7). For instance, mango extract submitted to 76.7°C for 1 min was sufficient for mango PE inactivation (Nanjundaswamy et al., 1973), whereas for PE found in papaya and mango purées, and papaya nectar, 1 min at 99°C was required (Argáiz and López-Malo, 1995).

The thermal resistance values for PE in various fruit products are 2.2 min < $D_{85°C}$-values < 7.2 min and 10°C < $z$-value < 15°C (Table 2.7). PPO of pineapple presented an exceptionally high resistance to heat ($D_{75°C}$ = 91 min, $D_{85°C}$ = 18 min) compared to five other fruits where $D_{75°C}$-values ranged between 0.13 and 6.5 min (Dimick et al., 1951). Also, the PPO of 'Anna' apples was stable at 60°C (Trejo-Gonzalez and Soto-Valdez, 1991) and PPO of Muscadine grapes ('Welder' and 'Noble' cultivars) was stable at 63°C (Lamikanra et al., 1992). The $D_{70°C}$ is 5 min for PRO of grapes and strawberries, but $z$-value was 35.4°C and 19°C for grapes and strawberries, respectively (Sciancalepore et al., 1985; Civello et al., 1995).

## 2.4 DESIGN OF PASTEURIZATION PROCESSES

As opposed to the design of sterilization processes where a "botulinum cook" must be applied (12*D* in mesophilic *C. botulinum*) to allow ambient storage of any food, the criteria for the design of pasteurization processes varies with the food product, and in low-acid foods, refrigerated storage is a necessary further "hurdle" to overcome the reduced severity of the heat treatment and assure food safety. The food industry designs pasteurization processes to obtain a safe and stable food product within a specified storage period, by using an empirical approach usually based on the experience with a particular food product, depending on the engineer, factory and processing plant characteristics, properties of the food, challenge tests, etc. Normally, cooked chilled low-acid foods have a short shelf-life, 1–2 weeks, and high-acid shelf-stable foods can have a longer shelf-life of several months. Also, after designing pasteurization processes based on *D*- and *z*-values (calculated from isothermal experiments), careful attention should be given to the microbial inactivation results obtained with real commercial processes, which deal with nonisothermal heating.

Before one can design a pasteurization process for a new food product, the following must be known or specified: if food is to be stored or distributed at ambient or if a cold chain is necessary; the desired shelf-life; if consumers belong to susceptible groups (e.g., babies, elderly, hospitalized, etc.). Then, the following steps must be taken to establish a pasteurization process for a new food product:

1. Identify the microbes (pathogens and spoilage) including spores able to germinate and grow in the food to be pasteurized (e.g., pH), as well as enzymes capable of causing food degradation, under the storage conditions (e.g., temperature, atmosphere).
2. Obtain from the literature heat resistance data (*D*- and *z*-values) regarding the microbes or enzymes previously identified (Section 2.3).
3. Select the most heat-resistant microbe or enzyme as the pasteurization target.
4. Always check experimentally for *C. botulinum* growth potential in the new food.

5. Experimentally determine *D*- and *z*-values of target microorganism in the new food.
6. Set a minimum P-value (pasteurization value) which delivers at least 6D in the most heat-resistant microbe (Betts and Gaze, 1992).
7. Perform challenge tests to validate the thermal process during storage under specific conditions (ambient for high-acid foods or chilled storage, $T \leq 10°C$, for low-acid foods).
8. Use preservatives for additional safety or longer shelf-life.

After carrying out an extensive review balance between the growth potential of microbes and their thermal resistance in the foods to be pasteurized, we recommend nonproteolytic *C. botulinum* spores and *A. acidoterrestris* spores as reference microorganisms to be used in the design of pasteurization processes in low-acid refrigerated foods and high-acid shelf-stable foods, respectively (Figure 2.2).

### 2.4.1 Low-Acid Cooked and Refrigerated Foods

With respect to low-acid cooked and chilled foods, food processors have to demonstrate that the processed food is safe, not being capable of supporting the growth and toxin production by *C. botulinum* within the specified storage life of the food. The following microbes are able to grow at pH > 4.6 and at low temperatures ≤8°C (Section 2.3.2): nonproteolytic *C. botulinum*, *Listeria*, *Y. enterocolitica*, *Aeromonas*, *V. parahaemolyticus*; a few strains of *B. cereus*, *E. coli* O157:H7, *Salmonella*; the spoilage LAB and yeasts. The review on microbial thermal resistance in chilled foods illustrates that spore-forming bacteria can be inactivated with temperatures 90°C–95°C, whereas temperatures of 65°C–70°C are enough for the destruction of vegetative forms of bacterial pathogens, and even lower temperatures of 55°C–60°C are enough for the destruction of psychrotrophic spoilage microbes such as LAB, molds, and yeasts (Figure 2.2):

1. *C. botulinum* (Table 2.1): $D_{94.4°C}$ (crabmeat) = 2.9 min, few seconds < $D_{90°C}$ < 10 min, few seconds < $D_{85°C}$ < 37 min, few minutes < $D_{80°C}$ < 140 min
2. *B. cereus* (Table 2.2): 1 min < $D_{95°C}$ < 2 min, 4 min < $D_{90°C}$ < 10 min, 16 min < $D_{85°C}$ < 30 min
3. Non-spore-forming pathogens (Table 2.3): $D_{65°C}$ few seconds, few seconds < $D_{60°C}$ < 10 min, few seconds < $D_{55°C}$ < 47 min
4. Spoilage microbes (LAB and yeasts): $D_{60°C}$ seconds

Based on nonproteolytic *C. botulinum* lethality to humans and its higher values of heat resistance compared to other psychrotrophic pathogens, we recommend the use of this microbe's heat resistance as the reference for the design of pasteurization processes for low-acid chilled foods. Being anaerobic, this is particularly important in vacuum-packed, and also semipreserved foods such as cured and cooked ham, cold-smoked fish, fermented marine foods, and dried fish (Peck, 2006). Then, the food processor should check if other psychrotrophic spore-forming pathogens and also spoilage microorganisms have a higher heat resistance than the nonproteolytic

*C. botulinum* in a particular food. Current recommendation is a process (*P*-value) of 10 min at 90°C or equivalent (Gould, 1999; Peck, 2006). If this process is not sufficient to achieve 6*D* inactivation of *C. botulinum*, the addition of preservatives is necessary for safety assurance. In these cases, the surviving *Clostridium* and *Bacillus* spores must be controlled with refrigeration ($T < 8$°C) and other hurdles such as salts (>3.5% salt-on-water, e.g., sodium chloride, sodium lactate) and nitrites (>100 ppm, e.g., sodium nitrite). It is known that a salt content ≥3.5% in the food stops botulinum growth during chill storage (Peck, 2006). For example, the addition of salt to levels of 2.5% and 4.3% increased the number of days required for growth of nonproteolytic *C. botulinum* strains in an anaerobic meat medium stored at temperatures from 5°C to 16°C (Graham et al., 1996).

In certain foods, milder thermal treatments than the recommended 10 min at 90°C are sufficient for safety. For example, crabmeat homogenate submitted to 1 min at 85°C was sufficient to inactivate $10^6$ cfu/g of type E *C. botulinum* spores (6*D*) and keep the food safe (nontoxic) for 6 months at 4.4°C (Cockey and Tatro, 1974). Peck et al. (1995) registered botulinum growth in a meat medium pasteurized at 80°C for 23 min after 12–40 days storage at 6°C–12°C, and recommended a minimum pasteurization of 19 min at 85°C, followed by storage at $T < 12$°C and a shelf life of not more than 28 days to reduce the risk of food botulism. This pasteurization (85°C, 19 min) retarded the growth at 6°C–12°C storage from 12 to 40 days to within 42–53 days. An even more intense pasteurization at 95°C for 15 min did not allow growth at $T \leq 12$°C for 60 days (Peck et al., 1995).

The survivors of mesophilic (optimum growth temperature, 10°C–50°C) and thermophilic (optimum growth temperature, 50°C–70°C) spores, including some strains of *C. botulinum* (Juneja and Marks, 1999) and *B. cereus*, *C. perfringens*, *B. coagulans*, *B. stearothermophilus*, *Desulfotomaculum nigrificans*, and *A. acidoterrestris*, will not be able to grow under refrigerated storage ($T < 8$°C). But the distribution chain should be carefully monitored and checked for temperature abuses. Following are examples of the pasteurization of two low-acid foods.

#### 2.4.1.1 Milk Pasteurization

Chilled milk is a particular low-acid food of high relevance and consumption. A study carried out with six strains of psychrotrophic nonproteolytic type E *C. botulinum*, could not detect their growth in milk stored at 4.4°C during 22 weeks. However, 'Beluga,' 'Tenno,' VH, and 'Alaska' strains could grow and produce toxin at 10°C after 21, 28, 42, and 56 days, respectively, while at 7.2°C only 'Tenno' could produce toxin after 70 days (Read et al., 1970). Normally, the objective of pasteurization for milk is mainly the destruction of pathogens *Mycobacterium tuberculosis*, *M. avium* subsp. *paratuberculosis* (MAP), *Coxiella burnetti*, *Brucella* spp., and *Streptococcus* spp. (Westhoff, 1978; Jay, 1992; Grant et al., 1996; Cerf and Condron, 2006). The HTST pasteurization standard, whereby milk is held at 71.7°C for at least 15 s (or equivalent process such 62.7°C for 30 min), was designed to achieve a 5-log reduction in the number of viable microorganisms in milk (European Economic Community, 1992; Stabel amd Lambertz, 2004). Pasteurizations of 72°C for 15 s, 75°C for 20 s and 78°C for 25 s resulted in 4 to >6 log reduction in MAP (McDonald et al., 2005).

#### 2.4.1.2 Surface Pasteurization of Eggs and Raw Meats

Hot water washes (72°C) of beef carcasses were not enough to control/eliminate pathogens during the storage at 4°C for 21 days (Dorsa et al., 1998). Recent studies on heat decontamination of food surfaces have been carried out with shell eggs (95°C < $T$ < 210°C, 2 s < time < 30 s; James et al., 2002), raw beef, raw pork, and raw chicken (steam for 60 s; McCann et al., 2006), but limitations regarding the changes on food surface quality have to be overcome.

### 2.4.2 Shelf-Stable High-Acid and Acidified Foods

A critical factor in high-acid and acidified foods is the pH. This factor should be controlled before and after the pasteurization and during storage since the thermal process and subsequent storage may increase the pH allowing growth of pathogens (Section 2.3.2) and some heat-resistant spore-forming *Bacillus* species (Table 2.6). The pH of less acid foods must be controlled at least below 4.5 by acidification. Another concern refers to the endogenous heat-resistant enzymes from foods. Those should be inactivated by pasteurization, since if they are active during storage, they can also modify the food pH allowing food-poisoning outbreaks and spoilage.

In low pH (<4.6) high-acid foods, no *C. botulinum* has been detected that will germinate, grow, and produce toxin (Blocher and Busta, 1983). The majority of bacterial spores would be inhibited by the low pH and only acid tolerant microorganisms such as *A. acidoterrestris*, molds, yeasts, LAB, and acetic acid bacteria might develop. This potential flora of high-acid products is less heat resistant than most of the thermophilic bacterial spores and therefore a milder thermal process of pasteurization (80°C–100°C) is sufficient for their inactivation and subsequent storage under ambient conditions. The collection of heat resistance data of microbes/enzymes able to grow/survive in high-acid foods (Section 2.3.3) demonstrated that a temperature of 95°C or more is necessary for the destruction of *A. acidoterrestris*, followed by 90°C–95°C for fungal ascospores, while 90°C is sufficient for PE inactivation and 70°C–80°C for PPO/PRO inactivation (Figure 2.2).

So far, there is no general criterion for the design of pasteurization of shelf-stable high-acid foods (pH < 4.6). Silva et al. (1999) conducted a systematic study to determine the effect of temperature, soluble solids, and pH on the $D$-values of *A. acidoterrestris* spores and to construct a mathematical model which was further validated with real fruit juices/concentrates. A new methodology to design pasteurization processes for shelf-stable high-acid and acidified food products based on the heat-resistant *A. acidoterrestris* spores was proposed and implemented with a fruit pulp (Silva et al., 2000b; Silva and Gibbs, 2001, 2004). The initial approach is to set up a minimum $P$ (pasteurization value) or exposure to heat (time/temperature) based on predictions of *A. acidoterrestris* heat resistance in the food to be pasteurized, followed by validation experiments. The following experiments should be carried out with the new food product: (1) potential for *A. acidoterrestris* spore germination and growth during food storage for at least 1 month at 25°C and 43°C; (2) prediction (Silva et al., 1999) and/or determination of $D$- and $z$-values of *A. acidoterrestris* spores in the food product to be pasteurized; and (3) monitoring product quality/stability during food storage following pasteurization treatments of differing intensities.

The use of *A. acidoterrestris* as a reference microorganism for the design of pasteurization in high-acid shelf-stable foods is appropriate given that this is a thermoacidophilic bacterium associated with various spoilage incidents registered worldwide (Section 2.3.3.1). However, knowing that *D*-values are very high compared to other microbes/enzymes in high-acid foods, if the storage experiments do not reveal *A. acidoterrestris* growth in the food, then a *P*-value of less than 6*D* may be applied for the food stabilization. This occurs, for example, with fruit concentrates, where the high content of soluble solids inhibits the spore growth. However, care must be taken when pasteurizing final products such as nectars or mixed fruit juices/drinks, since those diluted products may support *A. acidoterrestris* growth.

The use of other hurdles such as chemical preservatives to reduce the thermal treatment intensity was studied. Chemical control of *A. acidoterrestris* was first studied with nisin (Komitopoulou et al., 1999; Peña and Massaguer, 2006). Cerny et al. (2000) registered growth inhibition by adding ascorbic acid (≥15 mg/100 mL) or providing anaerobic conditions. The growth of *A. acidoterrestris* was prevented in drinks containing 1%–5% of orange juice or apple juice by carbonation, or by adding 300 mg/L of sorbic acid, 150 mg/L benzoic acid, or both (Pettipher and Osmundson, 2000).

#### 2.4.2.1 Beer Pasteurization

Analysis conducted with 461 beers obtained from 169 breweries in Germany, which included bottom- and top-fermented beers, and also light and dark colored beers, proved that beer pH can range between 3.96 and 4.74, with an average value of 4.4 (Lachenmeier, 2007). Given that beer is carbonated, contains alcohol and is bittered with hops, which are all natural antimicrobials, a mild pasteurization is effective for its stabilization at room temperature (e.g. 20 to 120 PU, or pasteurization units). 1 PU is equivalent to 1 min at 60°C, with $z = 7°C$ for beer spoilage microorganisms. However, safety concerns have been expressed, especially in alcohol-free beers and also in less bitter beers, the last being a trend in consumer preference. Lanthoen and Ingledew (1996) studied alcohol-free beers and reported that LAB presented a four- to sevenfold increase in *D*-value in alcohol-free beer compared to 5% v/v alcohol beer, and pathogens such *E. coli* and *S. typhimurium* were also more heat resistant by 3 to 17 times. Presently, the beer industry applies a more severe pasteurization process (e.g. 120 to 300 PU), to cope with these modifications in the traditional beer composition.

## REFERENCES

Adams, M.R. and Moss, M.O. 1995. *Food Microbiology*. Cambridge: The Royal Society of Chemistry.

Ahmed, N.M., Conner, D.E., and Huffman, D.L. 1995. Heat resistance of *Escherichia coli* O157:H7 in meat and poultry as affected by product composition. *Journal of Food Science* 60(3):606–610.

Aragão, G.M.F. 1989. Identificação e determinação da resistência térmica de fungos filamentosos termoresistentes isolados da polpa de morango, Master thesis, Universidade de Campinas, Brazil.

Argáiz, A. 1994. Thermal inactivation kinetics of pectinesterase in acidified papaya nectar and purees. *Revista Española de Ciencia y Tecnologia de Alimentos* 34(3):301–309.

Argáiz, A. and López-Malo, A. 1995. Kinetics of first change on flavour, cooked flavour development and pectinesterase inactivation on mango and papaya nectars and purees. *Revista Española de Ciencia y Tecnologia de Alimentos* 35(1):92–100.

Aureli, P., Fenicia, L., and Franciosa, G. 1999. Classic and emergent forms of botulism: The current status in Italy. *Eurosurveillance* 4:7–9.

Azizi, A. and Ranganna, S. 1993. Spoilage organisms of canned acidified mango pulp and their relevance to thermal processing of acid foods. *Journal of Food Science and Technology* 30(4):241–245.

Baumgart, J., Husemann, M., and Schmidt, C. 1997. *Alicyclobacillus acidoterrestris*: Occurrence, importance, and detection in beverages and raw materials for beverages. *Flussiges Obst* 64(4):178–180.

Betts, G.D. and Gaze, J.E. 1992. *Food Pasteurization Treatments*, Gloucestershire, United Kingdom: Campden Food and Drink Research Association, Technical Manual No. 27, Part II.

Betts, G.D. and Gaze, J.E. 1995. Growth and heat-resistance of psychrotrophic *Clostridium-botulinum* in relation to sous vide products. *Food Control* 6(1):57–63.

Beuchat, L.R. 1986. Extraordinary heat resistance of *Talaromyces flavus* and *Neosartorya fischeri* ascospores in fruit products. *Journal of Food Science* 51(6):1506–1510.

Beuchat, L.R. 1998. Spoilage of acid products by heat-resistant molds. *Dairy Food of Environmental Sanitation* 18(9):588–593.

Bigelow, W.D. and Esty, J.R. 1920. Thermal death point in relation to time of typical thermophilic organisms. *Journal of Infectious Diseases* 27:602–617.

Blocher, J.C. and Busta, F.F. 1983. Bacterial spore resistance to acid. *Food Technology* 37(11):87–99.

Bolton, D.J., McMahon, C.M., Doherty, A.M., Sheridan, J.J., McDowell, D.A., Blair, L.S., and Harrington, D. 2000. Thermal inactivation of *Listeria monocytogenes* and *Yersinia enterocolitica* in minced beef under laboratory conditions and in sous-vide prepared minced and solid beef cooked in a commercial retort. *Journal of Applied Microbiology* 88(4):626–632.

Brown, K.L. 2000. Control of bacterial spores. *British Medical Bulletin* 56(1):158–171.

Buduamoako, E., Toora, S., Walton, C., Ablett, R.F., and Smith, J. 1992. Thermal death times for *Listeria-monocytogenes* in lobster meat. *Journal of Food Protection* 55(3):211–213.

Buzrul, S. and Alpas, H. 2007. Modeling inactivation kinetics of food borne pathogens at a constant temperature. *LWT* 40:632–637.

Byrne, B., Dunne, G., and Bolton, D.J. 2006. Thermal inactivation of *Bacillus cereus* and *Clostridium perfringens* vegetative cells and spores in pork luncheon roll. *Food Microbiology* 23:803–808.

Carlin, F., Girardina, H., Peck, M.W., Stringer, S.C., Barker, G.C., Martinez, A., Fernandez, A., Fernandez, P., Waites, W.M., Movahedi, S., van-Leusden, F., Nauta, M., Moezelaar, R., del-Torre, M., and Litman, S. 2000. Research on factors allowing a risk assessment of spore-forming pathogenic bacteria in cooked chilled foods containing vegetables: FAIR collaborative project. *International Journal of Food Microbiology* 60:117–135.

Cerf, O. and Condron, R. 2006. *Coxiella burnetii* and milk pasteurization: An early application of the precautionary principle? *Epidemiology and Infection* 134(5):946–951.

Cerny, G., Hennlich, W., and Poralla, K. 1984. Fruchtsaftverderb durch bacillen: Isolierung und charakterisierung des verderbserregers. *Z. Lebensm. Unters. Forsch.* 179(3):224–227.

Cerny, G., Duong, H.A., Hennlich, W., and Miller, S. 2000. *Alicyclobacillus acidoterrestris*: Influence of oxygen content on growth in fruit juices. *Food Australia* 52(7):289–291.

Chai, T.J. and Liang, K.T. 1992. Thermal resistance of spores from 5 type E *Clostridium-botulinum* strains in eastern oyster homogenates. *Journal of Food Protection* 55(1):18–22.

Chiruta, J., Davey, K.R., and Thomas, C.J. 1997. Thermal inactivation kinetics of three vegetative bacteria as influenced by combined temperature and pH in a liquid medium. *Food and Bioproducts Processing* 75(C3):174–180.

Choma, C., Guinebretiere, M.H., Carlin, F., Schmitt, P., Velge, P., Granum, P.E., and Nguyen-The C. 2000. Prevalence, characterization and growth of *Bacillus cereus* in commercial cooked chilled foods containing vegetables. *Journal of Applied Microbiology* 88(4):617–625.

Chutintrasri, B. and Noomhorm, A. 2006. Thermal inactivation of polyphenoloxidase in pineapple puree. *LWT-Food Science and Technology* 39(5):492–495.

Civello, P.M., Martinez, G.A., Chaves, A.R., and Añón, M.C. 1995. Peroxidase from strawberry fruit (*Fragaria ananassa* Duch.): Partial purification and determination of some properties. *Journal of Agricultural and Food Chemistry* 43(10):2596–2601.

Cockey, R.R. and Tatro, M.C. 1974. Survival studies with spores of *Clostridium botulinum* type E in pasteurized meat of the blue crab *Callinectes sapidus. Applied Microbiology* 27(4):629–633.

Corry, J.E.L. 1976. The effects of sugars and polyols on the heat resistance and morphology of osmophilic yeasts. *Journal of Applied Bacteriology* 40(3):269–276.

Crisley, F.D., Peeler, J.T., Angelotti, R., and Hall, H.E. 1968. Thermal resistance of spores of five strains of *Clostridium botulinum* type E in ground whitefish chubs. *Journal of Food Science* 33(4):411–416.

Dastur, K., Weckel, K.G., and Elbe, J. 1968. Thermal processes for canned cherries. *Food Technology* 22:1176–1182.

Davis, J.G. *A Dictionary of Dairying*, 2nd ed., London: Leonard Hills Books, 1955, pp. 786–810.

De-Angelis, M., Di-Cagno, R., Huet, C., Crecchio, C., Fox, P.F., and Gobbetti, M. 2004. Heat shock response in *Lactobacillus plantarum. Applied and Environmental Microbiology* 70(3):1336–1346.

De-Carvalho, A.A.T., Costa, E.D., Mantovani, H.C., and Vanetti, M.C.D. 2007. Effect of bovicin HC5 on growth and spore germination of *Bacillus cereus* and *Bacillus thuringiensis* isolated from spoiled mango pulp. *Journal of Applied Microbiology* 102(4):1000–1009.

Deinhard, G., Blanz, P., Poralla, K., and Altan, E. 1987. *Bacillus acidoterrestris* sp. nov., a new thermotolerant acidophile isolated from different soils. *Systematic and Applied Microbiology* 10:47–53.

De-Jong, J. 1989. Spoilage of an acid food product by *Clostridium perfringens, C. barati* and *C. butyricum. International Journal of Food Microbiology* 8(2):121–132.

De-Jong, A.E.I., Rombouts, F.M., and Beumer, R.R. 2004. Behavior of *Clostridium perfringens* at low temperatures. *International Journal of Food Microbiology* 97:71–80.

Dijksterhuis, J. and Teunissen, P.G.M. 2003. Dormant ascospores of *Talaromyces macrosporus* are activated to germinate after treatment with ultra high pressure. *Journal of Applied Microbiology* 96(1):162–169.

Dimick, K.P., Ponting, J.D., and Makower, B. 1951. Heat inactivation of polyphenolase in fruit purées. *Food Technology* 237–241.

Doevenspeck, H. 1961. Influencing cells and cell walls by electrostatic impulses. *Fleischwirtschaft* 13:986–987.

Dorsa, W.J., Cutter, C.N., and Siragusa, G.R. 1998. Long-term profile of refrigerated ground beef made from carcass tissue, experimentally contaminated with pathogens and spoilage bacteria after hot water, alkaline, or organic acids washes. *Journal of Food Protection* 61(12):1615–1622.

Dufrenne, J., Soentoro, P., Tatini, S., Day, T., and Notermans, S. 1994. Characteristics of *Bacillus-cereus* related to safe food-production. *International Journal of Food Microbiology* 23(1):99–109.

Dufrenne, J., Bijwaard, M., te-Giffel, M., Beumer, R., and Notermans, S. 1995. Characteristics of some psychrotrophic *Bacillus cereus* isolates. *International Journal of Food Microbiology* 27:175–183.

Duong, H.A. and Jensen, N. 2000. Spoilage of iced tea by *Alicyclobacillus. Food Australia* 52(7):292.

Eiroa, M.N.U., Junqueira, V.C.A., and Schimdt, F.L. 1999. *Alicyclobacillus* in orange juice: Occurrence and heat resistance of spores. *Journal of Food Protection* 62(8):883–886.

European Economic Community. 1992. Requirements for the manufacture of heat-treated milk and milk-based products. Council directive 92/46/EEC, Annex C, No. L 268, p. 24. European Economic Community, Brussels.

Everis, L. and Betts, G. 2001. pH stress can cause cell elongation in *Bacillus* and *Clostridium* species: A research note. *Food Control* 12:53–56.

Fellows, P. 1988. *Food Processing Technology: Principles and Practice.* Chichester, UK: Ellis Horwood.

Fernández, A., Ocio, M.J., Fernández, P.S., and Martínez A. 2001. Effect of heat activation and inactivation conditions on germination and thermal resistance parameters of *Bacillus cereus* spores. *International Journal of Food Microbiology* 63:257–264.

FDA. 1992. Foodborne Pathogenic Microorganisms and Natural Toxins Handbook. US Food and Drug Administration. Center for Food Safety and Applied Nutrition (http://www.cfsan.fda.gov/~mow/badbug.zip).

Franz, C.M.A.P. and vonHoly, A. 1996. Thermotolerance of meat spoilage lactic acid bacteria and their inactivation in vacuum-packaged vienna sausages. *International Journal of Food Microbiology* 29(1):59–73.

García-Armesto, M.R. and Sutherland, A.D. 1997. Temperature characterization of psychrotrophic and mesophilic *Bacillus* species from milk. *Journal of Dairy Research* 64:261–270.

Gaze, J.E., Carter, J., Brown, G.D., and Thomas, J.D. 1988. Application of particle sterilisation under dynamic flow. Gloucestershire, United Kingdom: Campden Food and Drink Research Association, Technical Memorandum No. 508.

Gaze, J.E. and Brown, G.D. 1990. Determination of the heat resistance of a strain of non-proteolytic *Clostridium botulinum* type B and a strain of type E, heated in cod and carrot homogenate over the temperature range 70°C to 92°C. Gloucestershire, United Kingdom: Campden Food and Drink Research Association, Technical Memorandum Vol. 592, pp. 1–34.

George, M. 2000. Managing the cold chain for quality and safety. Technical manual of Flair-Flow Europe F-FE 378A/00, The National Food Centre, Dublin, Ireland.

Gibriel, A.Y. and Abd-El A1, A.T.H. 1973. Measurement of heat resistance parameters for spores isolated from canned products. *Journal of Applied Bacteriology* 36(2):321–327.

Gould, G.W. 1999. Sous vide foods: Conclusions of an ECFF botulinum working party. *Food Control* 10(1):47–51.

Gould, G.W. 2006. History of science—spores. Lewis B Perry Memorial Lecture 2005. *Journal of Applied Microbiology* 101(3):507–513.

Graham, A.F., Mason, D.R., and Peck, M.W. 1996. Inhibitory effect of combinations of heat treatment, pH, and sodium chloride on growth from spores of nonproteolytic *Clostridium botulinum* at refrigeration temperature. *Applied and Environmental Microbiology* 62(7):2664–2668.

Grant, I.R. 2003. *Mycobacterium paratuberculosis* and milk. *Acta Veterinaria Scandinavica* 44(3/4):261–266.

Grant, I.R., Ball, H.J., Neill, S.D., and Rowe, M.T. 1996. Inactivation of *Mycobacterium paratuberculosis* in cows' milk at pasteurization temperatures. *Applied and Environmental Microbiology* 62(2):631–636.

Gumerato, H.F. 1995. Desenvolvimento de um programa de computador para identificação de alguns fungos comuns em alimentos e determinação da resistência térmica de *Neosartorya fischeri* isolado de maçãs. Master thesis, Universidade de Campinas, Brazil.

Halder, A., Datta, A.K., and Geedipalli, S.S.R. 2007. Uncertainty in thermal process calculations due to variability in first-order and Weibull kinetic parameters. *Journal of Food Science* 72(4):E155–E161.

Härnulv, B.G. and Snygg, B.G. 1972. Heat resistance of *Bacillus subtilis* spores at various water activities. *Journal of Applied Bacteriology* 35:615–624.

Hatcher, W.S., Weihe, J.L., Murdock, D.I., Folinazzo, J.F., Hill, E.C., and Albrigo, L.G. 1979. Growth requirements and thermal resistance of fungi belonging to the genus *Byssochlamys*. *Journal of Food Science* 44(1):118–122.

Heinz, V. and Knorr, D. 2001. Effects of high pressure on spores. In *Ultra High Pressure Treatments of Foods*, Eds. Hendrickx, M.E.G. and Knorr, D. New York: Kluwer/Plenum, pp. 77–113.

Hendrickx, M.E.G. and Knorr, D. Eds. 2001. *Ultra High Pressure Treatments of Foods*. New York: Kluwer/Plenum.

Hite, B. 1899. The effect of pressure in the preservation of milk. *Bulletin of West Virginia University Agricultural Experimental Station* 58:15–35.

Holdsworth, S.D. 1985. Optimisation of thermal processing—a review. *Journal of Food Engineering* 4(2):89–116.

Holdsworth, S.D. and Simpson, R. 2007. *Thermal Processing of Packaged Foods*. New York: Springer.

Huang, I.P.D., Yousef, A.E., Marth, E.H., and Matthews, M.E. 1992. Thermal inactivation of *Listeria monocytogenes* in chicken gravy. *Journal of Food Protection* 55(7): 492–496.

Hülsheger, H., Potel, J., and Niemann, E.G. 1981. Killing of bacteria with electric pulses of high field strength. *Radiation and Environmental Biophysics* 20:53–65.

Ikeyami, Y., Okaya, C., Samayama, Z., Mori, D., and Oku, M. 1970. Gaseous spoilage by butyric anaerobes in canned fruits, I. In: canned mandarin orange. *Canners Journal* 49(11):993–996.

Jacobsen, M. and Jensen, H.C. 1975. Combined effect of water activity and pH on the growth of butyric anaerobes in canned pears. *Lebensmittel Wissenschaft und Technology* 8:158–160.

James, C., Lechevalier, V., and Ketteringham, L. 2002. Surface pasteurisation of shell eggs. *Journal of Food Engineering* 53:193–197.

Jay, J.M. 1992. High temperature food preservation and characteristics of thermophilic microorganisms. In *Modern Food Microbiology*, 4th ed., Chapman & Hall, New York, pp. 335–355.

Jay, J.M. 2000. Intrinsic and extrinsic parameters of foods that affect microbial growth. Chapter 3 in *Modern Food Microbiology*, 6th ed., Springer-Verlag, pp. 35–56.

Juneja, V.K. 2007. Thermal inactivation of *Salmonella* spp. in ground chicken breast or thigh meat. *International Journal of Food Science and Technology* 42(12):1443–1448.

Juneja, V.K., Eblen, B.S., Marmer, B.S., Williams, A.C., Palumbo, S.A., and Miller, A.J. 1995. Thermal resistance of nonproteolytic type B and type E *Clostridium botulinum* spores in phosphate buffer and turkey slurry. *Journal of Food Protection* 58(7):758–763.

Juneja, V.K. and Marmer, B.S. 1996. Growth of *Clostridium perfringens* from spore inocula in sous-vide turkey products. *International Journal of Food Microbiology* 32:115–123.

Juneja, V.K., Snyder, O.P., and Marmer, B.S. 1997. Thermal destruction of *Escherichia coli* O157:H7 in beef and chicken: Determination of *D*- and *z*-values. *International Journal of Food Microbiology* 35(3):231–237.

Juneja, V.K. and Marks, H.M. 1999. Proteolytic *Clostridium botulinum* growth at 12–48°C simulating the cooling of cooked meat: Development of a predictive model. *Food Microbiology* 16:583–592.

Juneja, V.K. and Marmer, B.S. 1999. Lethality of heat to *Escherichia coli* O157:H7: *D*- and *z*-value determinations in turkey, lamb and pork. *Food Research International* 32:23–28.

Juneja, V.K. and Marks, H.M. 2005. Heat resistance kinetics variation among various isolates of *Escherichia coli*. *Innovative Food Science and Emerging Technologies* 6:155–161.

Juneja, V.K., Fan, X.T., Pena-Ramos, A., Diaz-Cinco, M., and Pacheco-Aguilar, R. 2006. The effect of grapefruit extract and temperature abuse on growth of *Clostridium perfringens* from spore inocula in marinated, sous-vide chicken products. *Innovative Food Science & Emerging Technologies* 7(1–2):100–106.

Keiser, J. and Utzinger, J. 2005. Emerging foodborne trematodiasis. *Emerging Infectious Diseases* 11(10):1507–1514.

Keswani, J. and Frank, J.F. 1998. Thermal inactivation of *Mycobacterium paratuberculosis* in milk. *Journal of Food Protection* 61(8):974–978.

King, A.D., Michener, H.D., and Ito, K.A. 1969. Control of *Byssochlamys* and related heat resistant fungi in grape products. *Applied Microbiology* 18(2):166–173.

King, A.D. and Halbrook, W.U. 1987. Ascospore heat resistance and control measures for *Talaromyces flavus* isolated from fruit juice concentrate. *Journal Food Science* 52(5):1252–1254.

King, A.D. and Whiteland, L.C. 1990. Alteration of *Talaromyces flavus* heat resistance by growth conditions and heating medium composition. *Journal of Food Science* 55(3): 830–832.

Klijn, N., Herrewegh A.A.P.M., and de Jong, P. 2001. Heat inactivation data for *Mycobacterium avium* subsp *paratuberculosis*: Implications for interpretation. *Journal of Applied Microbiology* 91(4):697–704.

Komitopoulou, E., Boziaris, I.S., Davies, E.A., Delves-Broughton, J., and Adams, M.R. 1999. *Alicyclobacillus acidoterrestris* in fruit juices and its control by nisin. *International Journal of Food Science and Technology* 34:81–85.

Kotzekidou, P. 1997. Heat resistance of *Byssochlamys nivea*, *Byssochlamys fulva* and *Neosartorya fischeri* isolated from canned tomato paste. *Journal of Food Science* 62(2): 410–412, 437.

Lachenmeier, D.W. 2007. Rapid quality control of spirit drinks and beer using multivariate data analysis of Fourier transform infrared spectra. *Food Chemistry* 101:825–832.

Lamikanra, O., Kirby, S.D., and Musingo, M.N. 1992. Muscadine grape polyphenoloxidase: Partial purification by high pressure liquid chromatography and some properties. *Journal of Food Science* 57(3):686–689, 695.

Lanthoen, N.C. and Ingledew, W.M. 1996. Heat resistance of bacteria in alcohol-free beer. *Journal of the American Society of Brewing Chemists* 54(1):32–36.

Lee S.Y., Chung H.J., and Kang D.H. 2006. Combined treatment of high pressure and heat on killing spores of *Alicyclobacillus acidoterrestris* in apple juice concentrate. *Journal of Food Protection* 69(5):1056–1060.

Lehmann, G. 1996. High pressure treatment—a new food technology. *Fleischwirtschaft* 76(10):1004–1005.

Lindström, M., Kiviniemi, K., and Korkeala, H. 2006. Hazard and control of group II (non-proteolytic) *Clostridium botulinum* in modern food processing. *International Journal of Food Microbiology* 108:92–104.

Lund, D.B. 1975a. Heat processing. Chapter 3 in *Principles of Food Science. Part II. Physical Principles of Foods Preservation*, Ed. Fennema, O.R. New York: Marcel Dekker, pp. 32–92.

Lund, D.B. 1975b. Effects of blanching, pasteurization, and sterilization on nutrients. In *Nutritional Evaluation of Food Processing*, Eds. Harris, R.S. and Karmas, E. New York: AVI Publishing Co., pp. 205–239.

Lynt, R.K., Solomon, H.M., Lilly Jr., T., and Kautter, D.A. 1977. Thermal death time of *Clostridium botulinum* type E in meat of the blue crab. *Journal of Food Science* 42(4):1022–1025.

Magalhães, M.A. 1993. Estudo cinético da inativação térmica de enzimas termorresistentes, com ou sem adição de sacarose na polpa de mamão "Formosa" (*Carica papaya* L.) acidificada e o estabelecimento do processsamento térmico requerido. Ph.D. thesis FEA—UNICAMP, Campinas, Brazil.

Margosch, D., Ehrmann, M.A., Buckow, R., Heinz, V., Vogel, R.F., and Gänzle, M.G. 2006. High-pressure-mediated survival of *Clostridium botulinum* and *Bacillus amyloliquefaciens* endospores at high temperature. *Applied and Environmental Microbiology* 72(5):3476–3481.

McCann, M.S., Sheridan, J.J., McDowell, D.A., and Blair, I.S. 2006. Effects of steam pasteurisation on *Salmonella Typhimurium DT104* and *Escherichia coli* O157:H7 surface inoculated onto beef, pork and chicken. *Journal of Food Engineering* 76:32–40.

McDonald, W.L., O'Riley, K.J., Schroen, C.J., and Condron, R.J. 2005. Heat inactivation of *Mycobacterium avium* subsp. *paratuberculosis* in milk. *Applied and Environmental Microbiology* 71(4):1785–1789.

Miettinen, H., Arvola, A., and Wirtanen, G. 2005. Pasteurization of rainbow trout roe: *Listeria monocytogenes* and sensory analyses. *Journal of Food Protection* 68(8):1641–1647.

Mongiardo, N., Rienzo, B., Zanchetta, G., Pellegrino, F., Barbieri, G.C., Nannetti, A., and Squadrini, F., 1985. Descrizione di un caso di botulismo di tipo E in Italia. *Bollettino dell Istituto Sieroterapico Milanese* 64(3):244–246.

Montville, T.J. and Sapers, G.M. 1981. Thermal resistance of spores from pH elevating strains of *Bacillus licheniformis*. *Journal of Food Science* 46(6):1710–1712.

Murphy, R.Y., Beard, B.L., Martin, E.M., Duncan, L.K., and Marcy, J.A. 2004a. Comparative study of thermal inactivation of *Escherichia coli* O157:H7, *Salmonella*, and *Listeria monocytogenes* in ground pork. *Journal of Food Science* 69(4):97–101.

Murphy, R.Y., Davidson, M.A., and Marcy, J.A. 2004b. Process lethality prediction for *Escherichia coli* O157:H7 in raw franks during cooking and fully cooked franks during post-cook pasteurization. *Journal of Food Science* 69(4):112–116.

Murrel, W.G. and Scott, W.J. 1966. The heat resistance of bacterial spores at various water activities. *Journal of General Microbiology* 43:411–425.

Nakajyo, M. and Ishizu, Y. 1985. Heat-Resistance of *Bacillus-coagulans* spores isolated from spoiled canned low-acid foods. *Journal of the Japanese Society for Food Science and Technology-Nippon Shokuhin Kagaku Kogaku Kaishi* 32(10):725–730.

Nanjundaswamy, A.M., Saroja, S., and Ranganna, S. 1973. Determination of thermal process for canned mango products. *Indian Food Packer* 27(6):5–13.

Nath, N. and Ranganna, S. 1977. Time–temperature relationship for thermal inactivation of pectinesterase in 'Mandarin' orange (*Citrus reticulata blanco*) juice. *Journal of Food Technology* 12(4):411–419.

Nath, N. and Ranganna, S. 1980. Determination of thermal schedule for 'Totapuri' mango. *Journal of Food Technology* 15:251–264.

Nath, N. and Ranganna, S. 1981. Determination of thermal schedule for acidified papaya. *Journal of Food Technology* 46(1):201–206, 211.

Nath, N. and Ranganna, S. 1983a. Determination of a thermal process schedule for guava (*Psidium guajava* Linn.). *Food Technology* 18(3):301–316.

Nath, N. and Ranganna, S. 1983b. Heat transfer characteristics and process requirements of hot-filled guava pulp. *Journal of Food Technology* 18(3):317–326.

National Advisory Committee on Microbiological Criteria for Foods. 2006. Requisite scientific parameters for establishing the equivalence of alternative methods of pasteurization. *Journal Food Protection* 69(5):1190–1216.

Ohlsson, T. 1980. Temperature dependence of sensory quality changes during thermal processing. *Journal of Food Science* 45(4):836–839, 847.

Oteiza, J.M., Giannuzzi, L., and Califano, A.N. 2003. Thermal inactivation of *Escherichia coli* O157:H7 and *Escherichia coli* isolated from morcilla as affected by composition of the product. *Food Research International* 36:703–712.

Pace, P.J., Krumbiegel, E.R., Angelotti, R., and Wisniewski, H.J. 1967. Demonstration and isolation of *Clostridium botulinum* types from whitefish chubs collected at fish smoking plants of the Milwaukee area. *Applied Microbiology* 15, 877–884.

Palumbo, S.A., Call, J.E., Schultz, F.J., and Williams, A.C. 1995. Minimum and maximum temperatures for growth and verotoxin production by hemorrhagic strains of *Escherichia-coli. Journal of Food Protection* 58(4):352–356.

Papageorgiou, D.K., Melas, D.S., Abrahim, A., and Angelidis, A.S. 2006. Growth of *Aeromonas hydrophila* in the whey cheeses Myzithra, Anthotyros, and Manouri during storage at 4 and 12 degrees C. *Journal of Food Protection* 69(2):308–314.

Pasteur, L. 1876. Études sur la bière. Ses meladies, causes qui les provoquent, procédé pour la rendre inaltérable, avec une théorie nouvelle de la fermentation. Paris: Gauthier-Villars, p. 387.

Pearce, L.E., Truong, H.T., Crawford, R.A., Yates, G.F., Cavaignac, S., and de Lisle, G.W. 2001. Effect of turbulent-flow pasteurization on survival of *Mycobacterium avium* subsp. *paratuberculosis* added to raw milk. *Applied and Environmental Microbiology* 67(9):3964–3969.

Peck, M.W. 2006. *Clostridium botulinum* and the safety of minimally heated, chilled foods: An emerging issue? *Journal of Applied Microbiology* 101(3):556.

Peck, M.W., Lund, B.M., Fairbairn, D.A., Kaspersson, A.S., and Undeland, P.C. 1995. Effect of heat treatment on survival of, and growth from, spores of nonproteolytic *Clostridium botulinum* at refrigeration temperatures. *Applied and Environmental Microbiology* 61(5):1780–1785.

Peleg, M. and Penchina, C.M. 2000. Modeling microbial survival during exposure to a lethal agent with varying intensity. *Critical Reviews in Food Science and Nutrition* 40(2):159–172.

Peña, W.E.L. and Massaguer, P.R. 2006. Microbial modeling of *Alicyclobacillus acidoterrestris* CRA 7152 growth in orange juice with nisin added. *Journal of Food Protection* 69(8):1904–1912.

Penfield, M.P. and Campbell, A.M. 1990. *Food Preservation in "Experimental Food Science,"* 3rd ed., San Diego, CA, Academic Press, pp. 266–293.

Peredkov, A. 2004. Botulism, canned eggplant—Kyrgyzstan (Osh). ProMED Mail, 20041203.3225.

Periago, P.M., Fernandez, A., Collado, J., and Martinez, A. 2003. Note: Use of a distribution of frequencies model to interpret the tailed heat inactivation curves of prions. *Food Science and Technology International* 9(1):29–32.

Periago, P.M, van Zuijlen, A., Fernandez, P.S., Klapwijk, P.M., ter Steeg, P.F., Corradini, M.G., and Peleg, M. 2004. Estimation of the non-isothermal inactivation patterns of *Bacillus sporothermodurans IC4* spores in soups from their isothermal survival data. *International Journal of Food Microbiology* 95(2):205–218.

Peterson, M.E., Pelroy, G.A., Poysky, F.T., Paranjpye, R.N., Dong, F.M., Pigott, G.M., and Eklund, M.W. 1997. Heat-pasteurization process for inactivation of nonproteolytic types of *Clostridium botulinum* in picked Dungeness crabmeat. *Journal of Food Protection* 60(8):928–934.

Pettipher, G.L., Osmundson, M.E., and Murphy, J.M. 1997. Methods for the detection and enumeration of *Alicyclobacillus acidoterrestris* and investigation of growth and production of taint in fruit juice and fruit juice-containing drinks. *Letters in Applied Microbiology* 24(3):185–189.

Pettipher, G.L. and Osmundson, M.E. 2000. Methods for detection, enumeration and identification of *Alicyclobacillus acidoterrestris. Food Australia* 52(7):293–295.

Pinhatti, M.E.M.C., Variane, S., Eguchi, S.Y., and Manfio, G.P. 1997. Detection of acidothermophilic *Bacilli* in industrialized fruit juices. *Fruit Processing* 7(9):350–353.

Pontius, A.J., Rushing, J.E., and Foegeding, P.M. 1998. Heat resistance of *Alicyclobacillus acidoterrestris* spores as affected by various pH values and organic acids. *Journal of Food Protection* 61(1):41–46.

Potter, N.N. 1986. *Food Science*, 4th ed., New York: Van Nostrand Reinhold Company.

Przybylska, A. 2003. Botulism in Poland in 2001. *Przeglad Epidemiologiczny* 57:99–105.

Ramaswamy, H.S. and Abbatemarco, C. 1996. Thermal processing of fruits. In *Processing Fruits: Science and Technology*, Eds. Somogyi, L.P., Ramaswamy, H.S. and Hui, Y.H. Vol. I. Lancaster, PA: Technomic Publishing Co., pp. 25–65.

Raso, J. and Barbosa-Cánovas, G.V. 2003. Nonthermal preservation of foods using combined processing techniques. *Critical Reviews in Food Science and Nutrition* 43:265–285.

Read, R.B., Bradshaw, J.G., and Francis, D.W. 1970. Growth and toxin production of *Clostridium-botulinum* type-E in milk. *Journal of Dairy Science* 53(9):1183–1970.

Rodriguez, J.H., Cousin, M.A., and Nelson, P.E. 1993. Thermal resistance and growth of *Bacillus licheniformis* and *Bacillus subtilis* in tomato juice. *Journal of Food Protection* 56(2):165–168.

Saguy, I. and Karel, M. 1980. Modeling of quality deterioration during food processing and storage. *Food Technology* 34(2):78–85.

Sanchez, A.H., Rejano, L., and Montano, A. 1991. Kinetics of the destruction by heat of color and texture of pickled green olives. *Journal of the Science of Food and Agriculture* 54(3):379–385.

Sandoval, A.J., Barreiro, J.A. and Mendoza, S. 1992. Thermal resistance of *Bacillus coagulans* in double concentrated tomato paste. *Journal of Food Science* 57(6):1369–1370.

Schmidt, C.F., Lechowich, R.V., and Folinazzo, J.F. 1961. Growth and toxin production by type E *Clostridium botulinum* below 40°F. *Journal of Food Science* 26:626–630.

Schoeni, J.L., Glass, K.A., McDermott, J.L., and Wong, A.C.L. 1995. Growth and penetration of *Salmonella enteritidis*, *Salmonella heidelberg* and *Salmonella typhimurium* in eggs. *International Journal of Food Microbiology* 24(3):385–396.

Schuman, J.D. and Sheldon, B.W. 1997. Thermal resistance of *Salmonella* spp. and *Listeria monocytogenes* in liquid egg yolk and egg white. *Journal of Food Protection* 60(6):634–638.

Schuman, J.D., Sheldon, B.W., and Foegeding, P.M. 1997. Thermal resistance of *Aeromonas hydrophila* in liquid whole egg. *Journal of Food Protection* 60(3):231–236.

Sciancalepore, V., Longone, V., and Alviti, F.S. 1985. Partial purification and some properties of peroxidase from 'Malvasia' grapes. *American Journal of Enology and Viticulture* 36(2):105–110.

Scott, V.N. and Bernard, D.T. 1987. Heat resistance of *Talaromyces flavus* and *Neosartorya fischeri* isolated from commercial fruit juices. *Journal Food Protection* 50(1):18–20.

Shapton, D.A. 1966. Evaluating pasteurization processes. *Processes Biochemistry* 1:121–124.

Shearer, A.E., Mazzotta, A.S., Chuyate, R., and Gombas, D.E. 2002. Heat resistance of juice spoilage microorganisms. *Journal of Food Protection* 65(8):1271–1275.

Shridhar, P. and Shankhapal, K.V. 1986. Bacterial spoilage of canned mango pulp and green garden peas. *Indian Journal of Microbiology* 26(12):39–42.

Silva, F.V.M. 2000. Design and optimization of pasteurization conditions for cupuaçu (*Theobroma grandiflorum*) fruit pulp. PhD thesis, Escola Superior de Biotecnologia, Universidade Católica Portuguesa, Portugal.

Silva, F.V.M. and Silva, C.L.M. 1997. Quality optimisation of hot filled pasteurised fruit purées: Container characteristics and filling temperatures. *Journal of Food Engineering* 32(4):351–364.

Silva, F.M. and Silva, C.L.M. 1999. Colour changes in thermally processed cupuaçu (*Theobroma grandiflorum*) purée: Critical times and kinetics modelling. *International Journal of Food Science and Technology* 34(1):87–94.

Silva, F.M., Gibbs, P., Vieira, M.C., and Silva, C.L.M. 1999. Thermal inactivation of *Alicyclobacillus acidoterrestris* spores under different temperature, soluble solids and pH conditions for the design of fruit processes. *International Journal of Food Microbiology* 51(2/3):95–103.

Silva, F.M., Sims, C., Balaban, M.O., Silva, C.L.M, and O'Keefe, S. 2000a. Kinetics of flavour and aroma changes in thermally processed cupuaçu (*Theobroma grandiflorum*) pulp. *Journal of Science of Food and Agriculture* 80(6):783–787.

Silva, F.M., Gibbs, P., and Silva, C.L.M. 2000b. Establishing a new pasteurization criterion based on *Alicyclobacillus acidoterrestris* spores for shelf-stable high-acidic fruit products. *Fruit Processing* 10:138–141.

Silva, F.V.M. and Gibbs, P. 2001. *Alicyclobacillus acidoterrestris* spores in fruit products and design of pasteurization processes. *Trends in Food Science and Technology* 12:68–74.

Silva, F.V.M., Martins, R.C., and Silva, C.L.M. 2003. Design and optimization of hot-filling pasteurization conditions: cupuaçu (*Theobroma grandiflorum*) fruit pulp case study. *Biotechnology Progress* 19(4):1261–1268.

Silva, F.V.M. and Gibbs, P. 2004. Target selection in designing pasteurization processes for shelf-stable high-acid fruit products. *Critical Reviews in Food Science and Nutrition* 44(5):353–360.

Sinell, H.J. 1980. Interacting factors affecting mixed populations in microbial ecology of foods, Ed. Silliker J.H., Vol. 1. Academic Press, New York, pp. 215.

Sinigaglia, M., Corbo, M.R., Altieri, C., Campaniello, D., D'amato, D., and Bevilacqua, A. 2003. Combined effects of temperature, water activity, and pH on *Alicyclobacillus acidoterrestris* spores. *Journal of Food Protection* 66(12):2216–2221.

Splittstoesser, D.F., Churey, J.J., and Lee, C.Y. 1994. Growth characteristics of aciduric sporeforming bacilli isolated from fruit juices. *Journal of Food Protection* 57(12):1080–1083.

Splittstoesser, D.F., Lee, C.Y., and Churey, J.J. 1997. Control of *Alicyclobacillus* in the juice industry, Paper no. 3 presented at Session 36, Institute of Food Technologists Annual Meeting, Orlando, FL, June 14–18, 1997.

Stabel, J.R. and Lambertz, A. 2004. Efficacy of pasteurization conditions for the inactivation of *Mycobacterium avium* subsp. *paratuberculosis* in milk. *Journal of Food Protection* 67(12):2719–2726.

Teixeira, A.A. 1992. Thermal processing calculations. In *Handbook of Food Engineering*, Eds. Heldman, D.R. and Lund, D.B. Marcel Dekker, New York, 1992.

Teixeira, A.A., Dixon, J.R., Zahradnik, J.W., and Zinsmeister, G.E. 1969. Computer optimization of nutrient retention in the thermal processing of conduction-heated foods. *Food Technology* 23(6):137–142.

Therre, H. 1999. Botulism in the European Union. *Eurosurveillance* 4:2–7.

Thomas, C.T., White, J.C., and Longree, K. 1966. Thermal resistance of *Salmonellae* and *Staphylococci* in foods. *Applied microbiology* 14(5):815–820.

Toora, S., Buduamoako, E., Ablett, R.F., and Smith, J. 1992. Effect of high-temperature short-time pasteurization, freezing and thawing and constant freezing, on the survival of *Yersinia enterocolitica* in milk. *Journal of Food Protection* 55(10):803–805.

Tournas, V. and Traxler, R.W. 1994. Heat resistance of a *Neosartorya fischeri* strain isolated from pineapple juice frozen concentrate, *Journal of Food Protection* 57(9):814–816.

Trejo-Gonzalez, A. and Soto-Valdez, H. 1991. Partial characterization of polyphenoloxidase extracted from 'Anna' apple. *Journal of American Society for Horticultural Science* 116(4):672–675.

Valdramidis, V.P., Geeraerd, A.H., Bernaerts, K., and Van Impe, J.F. 2006. Microbial dynamics versus mathematical model dynamics: The case of microbial heat resistance induction. *Innovative Food Science and Emerging Technologies* 7(1–2):80–87.

Van Buggenhout, S., Messagie, I., Van der Plancken, I., and Hendrickx, M. 2006. Influence of high-pressure–low-temperature treatments on fruit and vegetable quality related enzymes. *Journal European Food Research and Technology* 223(4):475–485.

Vaughn, R.H., Irving, H.K., and Mercer, W.A. 1952. Spoilage of canned foods caused by the *Bacillus macerans-polymyxa* group of bacteria. *Food Research* 17:560–570.

Villota, R. and Hawkes, J.G. 1986. Kinetics of nutrients and organoleptic changes in foods during processing. Chapter 2 in *Physical and Chemical Properties of Foods*, Ed. M.R. Okos, Michigan, *American Society of Agricultural Engineering*, pp. 266–339.

Villota, R. and Hawkes, J.G. 1992. Reaction kinetics in food systems. In *Handbook of Food Engineering*, Eds. D.R. Heldman and D.B. Lund. Marcel Dekker, New York, pp. 39–144.

von Soxhlet, F. 1886. Ueber Kindermilch und Säuglings-Ernährung. *Münchener Medizinische Wochenschrift* 33:253–256, 276–278.

Walls, I. 1997. *Alicyclobacillus*—an overview, Paper no. 1 at Session 36, Institute of Food Technologists Annual Meeting, Orlando, FL, June 14–18, 1997.

Walls, I. and Chuyate, R. 1998. *Alicyclobacillus*-Historical perspective and preliminary characterization study. *Dairy, Food and Environmental Sanitation* 18(8):499–503.

Walls, I. and Chuyate, R. 2000. Spoilage of fruit juices by *Alicyclobacillus acidoterrestris*. *Food Australia* 52(7):286–288.

Wells, J.H. and Singh, R.P. 1988. A kinetic approach to food quality prediction using full history time–temperature indicators. *Journal of Food Science* 53(6):1866–1871, 1893.

Welt, B.A., Teixeira, A.A., Balaban, M.O., Semerage, G.H., and Sage, D.S. 1997. Iterative method for kinetic parameter estimation from dynamic thermal treatments. *Journal of Food Science* 62(1):8–14.

Westhoff, D.C. 1978. Heating milk for microbial destruction: A historical outline and update. *Journal of Food Protection* 41:122–130.

Whitaker, J.R. 1972. *Principles of Enzymology for the Food Sciences*, New York and Baser: Marcel Dekker.

Wisotzkey, J.D., Jurtshuk, P., Fox, G.E., Deinhard, G., and Poralla, K. 1992. Comparative sequence analyses on the 16S rRNA (rDNA) of *Bacillus acidocaldarius*, *Bacillus acidoterrestris*, and *Bacillus cycloheptanicus* and proposal for creation of a new genus, *Alicyclobacillus* gen. nov. *International Journal of Systematic Bacteriology* 42:263–269.

World Health Organization. 2002. Foodborne diseases, emerging. Fact Sheet no. 124 (http://www.who.int/mediacentre/factsheets/fs124/en/print.html).

World Health Organization. 2007. Food safety and foodborne illness. Fact Sheet no. 237 (http://www.who.int/mediacentre/factsheets/fs237/en/print.html).

Yamazaki, K., Teduka, H., and Shinano, H. 1996. Isolation and identification of *Alicyclobacillus acidoterrestris* from acidic beverages. *Bioscience Biotechnology and Biochemistry* 60(3):543–545.

York, G.K., Heil, J.R., Marsh, G.L., Ansar, A., Merson, R.L., Wolcott, M.T., and Leonard, S. 1975. Thermobacteriology of canned whole peeled tomatoes. *Journal Food Science* 40(4):764–769.

# 3 Aseptic Processing of Liquid Foods Containing Solid Particulates

*Jasper D.H. Kelder, Pablo M. Coronel, and Peter M.M. Bongers*

## CONTENTS



## 3.1 INTRODUCTION

The ultimate goal of food processing is to provide safe and ambient stable foods that have the nutritional and quality attributes preserved as close to freshly prepared food as possible. Besides the requirement of high-quality ingredients, processing has to be carried out in a manner that will preserve the quality attributes of the food materials, while minimizing the risk of microbial contamination and spoilage.

Aseptic processing is a combination of continuous commercial sterilization of the product and subsequent packaging in a material that has been separately sterilized. While aseptic processing is not a new concept (Nielsen, 1913), it has been expanded and developed in the last decades due to the customer demand for products with better quality attributes. Aseptic processing and packaging were introduced to commercial scale in the 1950s by Tetra Pak. From the mid-1970s, aseptic packaging of heterogeneous products (tomato pieces in juice) became possible when Scholle Corporation (Irvine, California) introduced the aseptic bag-in-box system. Low-acid heterogeneous products were first offered in retail packs in 1996, when Tetra Pak was granted a patent on an aseptic process for low-acid foods in the United States (Palaniappan and Sizer, 1997). The current applications of aseptic processing are vast, ranging from liquid dairy products and fruit juices, to whole fruit pieces and ready to eat soups. However, aseptic processing of low-acid particulate foods is still a major field of research in academia and development in industry.

Thermally processed foods are subject to several desirable and undesirable changes as a result of the exposure of the food to combined time–temperature, the extent which is driven by the kinetics of each process. Thermal processing inactivates microorganisms and spores that may cause spoilage or may pose consumer safety hazards. In addition, it inactivates enzymes that cause undesired color, texture, or flavor changes. Thermal processing also generates the desirable changes in texture and flavor generally known as cooking. On the other hand, some undesirable changes may occur such as change of flavor and color, modifications in texture, and reduction of the nutrient content. As microbial destruction generally occurs faster at higher temperatures than the destruction of quality factors, aseptic processing is carried out at high temperatures and short holding times. Typical process temperatures are between 120°C and 150°C, and holding times between 5 and 90 s (Reuter, 1987; Holdsworth, 1992; David et al., 1996).

Aseptic continuous processing presents several advantages over batch processing, such as smaller footprint, more appealing packaging, and the ability to produce foods with higher quality and nutrient retention (Potter and Hotchkiss, 1985). However, aseptic processes require tight process control to ensure that the food products receive sufficient thermal treatment to obtain safe products while preserving quality. Other disadvantages include a high expenditure in equipment, the need for trained people, and the specific demands made on packaging and filling operation.

One of the requirements of aseptic processing and packaging is to keep all of the components free of contamination: product, processing and filling lines, surge tanks, packaging equipment, and materials. Aseptic processing systems have to be sterilized prior to processing, generally using hot water or saturated steam. Packaging materials are sterilized separately using peroxides, or other chemicals together with sterile air. This process can be complicated and time consuming and requires good engineering of the lines.

Though thermal processing is the mainstay of the food-processing industry, and aseptic operations have steadily increased their share within this field, aseptically processed particulate foods are still rare in the marketplace. Heterogeneous foods present many challenges to the processor due to the complexities associated with the thermal treatment of large solid pieces in liquid media of varying viscosity. Heat transfer, residence time distribution (RTD), and interactions between particulates must be well understood to obtain safe products with good retention of quality. The challenges presented by thermal processing of heterogeneous products are the main subject of this chapter.

## 3.2 FUNDAMENTALS OF ASEPTIC PROCESSING

This section contains the fundamentals of heat transfer, quality, and safety. In conventional aseptic processing, microbial inactivation and development of quality factors are thermally driven, and heat transfer is therefore of key importance. The required duration and intensity of the heat treatment is derived from microbial kinetics, and the efficacy of the treatment can be verified using a number of validation tools.

### 3.2.1 Fundamentals of Heat Transfer

Safe thermal processing of heterogeneous food products requires a good understanding of underlying phenomena that influence the flow and heat transfer in such products. Heat transfer to and from food materials using conventional means occurs through a combination of conduction and convection. At the temperatures normally employed in liquid food processing, radiation heat transfer is insignificant. Volumetric technologies such as ohmic, radio frequent, and microwave heating are emerging solutions that offer faster and more uniform heat transfer, and these are treated elsewhere.

With the exception of volumetric technologies, heat transfer to a particulate liquid product can be visualized as an energy cascade (Figure 3.1). Firstly, the energy of the medium (e.g., water or steam) is transferred into the liquid part of the product, either by a process of conduction and convection (such as in a tubular heat exchanger), or by condensation, convection, and conduction (such as in steam injection). Secondly, the energy is carried from the liquid to the solid particles by convection and conduction, and thirdly throughout the particle by conduction. During cooling, this cascade is reversed.

Estimating heat transfer to liquid food products can be of considerable complexity. Most food products exhibit non-Newtonian flow properties that can be strong functions of both temperature and time (Holdsworth, 1993). The remaining thermal properties density, heat capacity, and conductivity are also correlated to temperature, albeit not

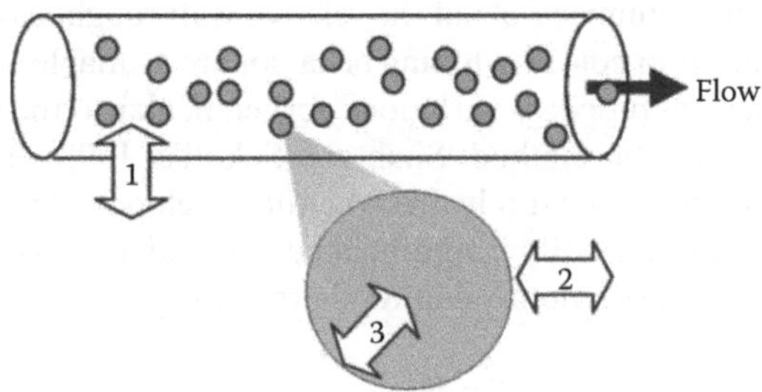


**FIGURE 3.1** Heat transfer cascade.

as strongly as viscosity. The thermal properties, the geometry of the heat exchangers, and the product flow rate determine whether the flow is laminar, transitional, or fully turbulent, and whether the contribution of free convection is significant. Obviously, turbulent mixing and free convection both allow for significantly higher heat transfer rates. In addition, internals (e.g., static mixers), bends, and auxiliary equipments (e.g., pumps) can greatly impact on heat transfer, especially for laminar flows. Finally, allowance should be made for fouling of the heat exchanger, both on the product and on the medium side.

Particles affect every step of the heat transfer cascade, and some of these effects have not been fully studied and understood. Particles may interact with the turbulent structure of the flow, or they may equilibrate existing temperature nonuniformities in laminar flows. In both cases, they affect the primary heat transfer from the medium into the carrier liquid. The second step in the cascade is energy transfer to the particles' surface, and this depends on particle shape, particle density, and the properties of the carrier liquid. In the final step, thermal energy is redistributed throughout the particles by conduction, which is governed by size, shape, and (inhomogeneous) thermal properties. At high mass fraction, particles may present a significant energy sink or source, especially when they undergo a phase transition (melting and crystallization).

Despite the complexities of heat transfer from particulate flows, an adequate estimate is required. Particles are regarded as crucial from the point of food safety, as their center often coincides with the coldest spot in the thermal process. In Section 3.2.2, a broad overview of heat transfer is given to particulate flows, with an emphasis on convective and conductive heating.

### 3.2.2 Flow and Heat Transfer to Liquids in Closed Channels

Closed channels are systems where a flow of product passes continuously through an arbitrarily shaped duct having a closed perimeter. The most common examples of closed channel flow in food processing are circular and annular tubes, or flow through the passages of a plate heat exchanger. Due to their practical importance, closed channel flow and heat transfer have been the subject of study for scientists and engineers for well over a century. A large body of analytical and experimental work is therefore available for friction factors and heat transfer, both in the laminar and turbulent flow regime (Ebadian and Dong, 1998).

Hydrodynamically developed friction factors, $f$, in tubes are customarily correlated to the Reynolds number $Re$ and the relative wall roughness $\varepsilon$ (Equation 3.1). Corrections resulting from non-Newtonian behavior are available, as well as for large temperature and thermal property variations perpendicular to the main direction of flow (e.g., the property ratio method; Shah and Sekuli'c, 1998). The dimensionless hydraulic length $z^{\approx}$ required for full hydrodynamic development depends on the flow Reynolds number, but is normally insignificant compared to the overall length of the continuous heat exchangers employed in food processing.

$$f = \Im\left(Re, \varepsilon, z^{\approx}\right) \tag{3.1}$$

Heat transfer in closed channels is expressed as the dimensionless Nusselt number *Nu* (Equation 3.2). For turbulent flows, the Nusselt number is correlated to the flow Reynolds number, the Prandtl number *Pr*, and the relative wall roughness. In laminar flow, thermal development can be very slow, and thermal correlations are therefore also a function of the dimensionless thermal length $z^*$.

$$Nu = \Im\left(Re, Pr, \varepsilon, z^*\right) \tag{3.2}$$

For plate heat exchangers no such general and comprehensive correlations are available. Due to the wide range of dimensions of the flow passages, and the structures created on the plate surface, each design needs to be characterized experimentally for flow and heat transfer behavior.

In laminar tube flow, heat transfer is also affected by curvature. Laminar flow and temperature fields in curved pipes deviate from those in straight tubes because of centrifugal forces acting perpendicular to the axial flow direction. As the axial velocity is nonuniform and centrifugal forces are proportional to the square of the axial velocity, fast-flowing fluid near the axis of the tube experiences a greater centrifugal force than slow fluid closer to the wall. This results in a secondary flow in the cross section, which consists of two symmetric vortices (Figure 3.2). These "Dean" vortices are named after W.R. Dean, who published the first mathematical treatise on curved flow (Dean, 1927). For strongly curved flows, the distribution and magnitude of the centrifugal forces are such that the maximum of the axial velocity is shifted from the tube centerline toward the outside of the tube. However, already at much lower centrifugal forces Dean vortices provide sufficient mixing to the main flow to significantly change the temperature distribution, effectively halving the thermal path of the flow. Thus, heat transfer in developed continuously coiled flow can be up to 500% higher (Manlapaz and Churchill, 1981; Kelder, 2003).

Dean vortices occur in bends of straight tubular heat exchangers as well. Temperature fields develop along a straight section, and heat transfer tends to decrease significantly. At every bend, however, the Dean effect mixes the flow and the temperature profile is restored to a less-developed state. Heat transfer in straight tubes having bends depends on the intensity of the Dean phenomenon in each bend, how far it propagates downstream from the bend and the length in between bends (Abdelmessih and Bell, 1999). When the straight sections are long compared to the flow path in the bends, it is recommended to consider the bends as a safety factor in terms of heat transfer performance.

When density is a function of temperature, buoyancy forces may give rise to secondary Morton (Morton, 1959) vortices (Figure 3.2). Morton vortices further

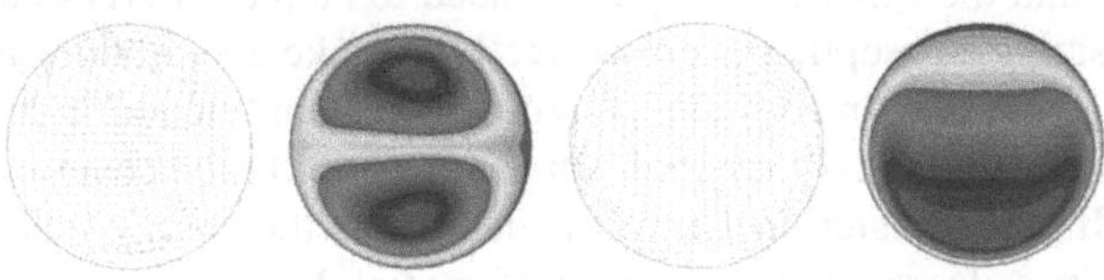

**FIGURE 3.2** Dean (left) and Morton (right) vortices and associated temperature profiles.

increase heat transfer in horizontal laminar flows (Morcos and Bergles, 1975). Though their strength diminishes at smaller temperature differentials between tube wall and product flow, in long heat exchanger tubes the impact of free convective currents on heat transfer can be considerable.

Whereas curvature and buoyancy can effectively increase heat transfer for medium viscosity laminar flows, this is less so in turbulent flow and particulate flows in very high-packing fraction (>50%). In turbulent flow, corrugated or dimpled tubes can be employed to increase the intensity of the turbulence, which can be especially effective in transitional flow (Bergles, 1998; Rozzi et al., 2007). For dense particulate flows, extremely viscous flow (paste-like) or flows which strongly foul, scraped surface heat exchangers may be employed (Rao and Hartel, 2006).

Particles interact with the flow structure of the liquid phase, especially at high mass fraction, thus impacting the wall-fluid heat transfer coefficients (Sannervik et al., 1996). However, data on particulate slurries are scarce, and aseptic process design would greatly benefit from reliable engineering correlations. One way to obtain such data would be to use a heat flux sensor (Barigou et al., 1998), which enables direct measurement of the heat transfer coefficient as a function of viscosity, flow rate, particle fraction, etc. under relevant food-processing conditions.

Fouling in food products can be severe and it is a major driver in terms of cost of production. Fouling increases energy costs due to higher heat transfer resistances, and in extreme cases deposits may spoil otherwise good product. It also limits the run-length of the process, thus increasing the downtime, production waste levels, and cost of cleaning. Fouling is especially relevant during the heating phase, though formation of deposits and sometimes biofilms in the cooling section should be considered as well.

Fouling may be caused by specific components in the product, for example, protein fouling associated with dairy ingredients. At higher temperatures, mineral salts may precipitate on the exchanger's surface, both on the product and on the medium side. Generally speaking, fouling is observed to increase when the heat exchanging process is intensified.

Thermal activation of thickeners (e.g., starch) inside the heat transfer passage may be regarded as an extreme case of fouling. The liquid close to the surface increases in viscosity, and the local flow slows down. As liquid close to the tube axis remains thin this can have a much higher velocity than anticipated and a wide distribution in residence times results. In that case, part of the liquid receives a minimal heat treatment, whereas part of the liquid may be severely overcooked (Kelder et al., 2004). Starch and other thickeners should therefore have been fully activated prior to any critical heating or holding stage.

To estimate the impact of fouling of heat transfer performance, the thickness of the deposits and their thermal properties need to be measured or calculated. It is paramount in sizing an aseptic system correctly to make assumptions about its components fouling behavior. In addition to sizing the system "clean," it should therefore also be sized at the end of its expected run-length, prior to the cleaning cycle.

The overall heat transfer coefficient is defined as the energy that is transferred per unit area per degree temperature differential between local bulk medium and local bulk product temperature. It can be calculated from the individual heat

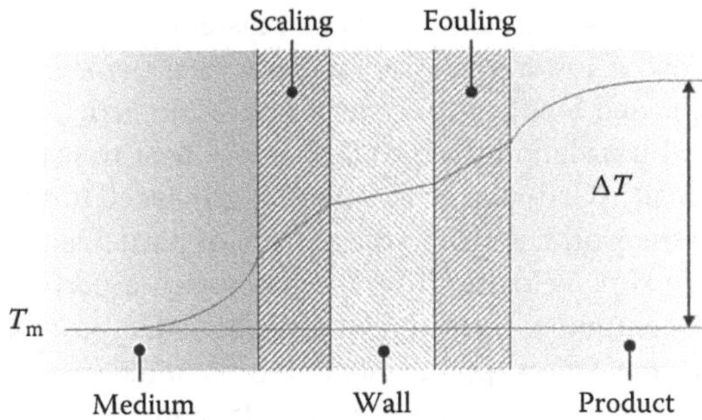


**FIGURE 3.3** Heat transfer resistance in a tubular heater.

transfer resistances in series. For example, in a tube in tube heat exchanger, an annular pressurized water flow heating a monotube containing product, as schematically drawn in Figure 3.3, the overall heat transfer coefficient $U$ is the inverse of the sum of all the heat transfer resistances in series (Equation 3.3).

$$\frac{1}{U}=\frac{1}{h_m}+\frac{t_s}{\lambda_s}+\frac{t_w}{\lambda_w}+\frac{t_f}{\lambda_f}+\frac{1}{h_p} \tag{3.3}$$

In Equation 3.3, $h_m$ and $h_p$ are the convective heat transfer coefficients on medium and product side, respectively. The thickness of the scaling, wall, and fouling layers $t_s$, $t_w$, and $t_f$, respectively, transfers heat by conduction only with a conductivity $\lambda_i$. Typical values for these contributions are given in Table 3.1.

### 3.2.3 Heat Transfer to Particulates

Fluid to particle heat transfer is the second step in the energy cascade. It strongly depends on the hydrodynamic conditions around each particle, and these in turn depend on the liquid properties, particle shape, and volume fraction. Several factors impact on the liquid particle heat transfer, and can be summarized as follows (Barigou et al., 1998):

**TABLE 3.1**
**Typical Values Heat Transfer Resistances in a 50 mm Tube**

| Resistance | Value (W/°K m²)⁻¹ | Description |
|---|---|---|
| $1/h_m$ | 0.0001–0.0005 | Heating or cooling medium |
| $t_s/\lambda_s$ | 0.002 | Scaling on medium side (0.5 mm) |
| $t_w/\lambda_w$ | 0.00005 | Stainless steel tube |
| $t_f/\lambda_f$ | 0.001 | Fouling layer on product side (0.5 mm) |
| $1/h_p$ | 0.001–0.01 | Product flow in tubes |

- *Liquid viscosity.* Viscosity governs heat transfer through the hydrodynamic state of the flow. A lower viscosity increases heat transfer in laminar flows (Chakrabandhu and Singh, 2002), since it permits particles greater freedom of rotation and translation. In turbulent flows, heat transfer also increases for lower viscosity (Gadonna et al., 1996), as enhanced turbulence promotes transfer from the wall but also to the surface of particulates.
- *Fluid–particle slip.* Velocity differences between a particle and the local liquid imply a continuous refreshment of the boundary layer surrounding the particle. Such convection appreciably improves liquid–particle heat transfer (Mankad et al., 1997). For particle Reynolds number $Re_p > 1$, flow around the particles becomes unsteady and this further increases heat transfer.
- *Particle rotation.* Generally, the relative velocity refers to the velocity difference between the particle centroid and the local fluid. Particle rotation provides an additional mechanism for local surface slip between fluid and particle. Especially near the wall of a tube velocity gradients exist and in consequence particles may rotate. Any rotational surface velocity thus generated is superimposed on the centroid velocity, resulting in local slip between liquid and particle surface which represents additional convective heat transport. Though evidence suggests that particle rotation indeed increases heat transfer (Krieth, 1963), particles tend to accumulate in the low shear center of the flow, where a sufficient particle density may actually limit rotation (Alhamdan and Sastry, 1998).
- *Particle–particle interactions.* When large volume fractions of particles are processed, the solids tends to displace the liquid and thus modify the local flow field. In some locations the particle density may be higher, in which case the particles may behave as a cluster with a thermally larger size. Conversely, in other locations the particle density may be correspondingly lower, implying channeling of the liquid. Such nonuniformities in particle density affect the local liquid–particle heat transfer rate unpredictably (Barigou et al., 1998). Also, turbulent motion or wake effects caused by one particle may alter the flow and temperature distribution for downstream particles (Kelly et al., 1995). However, this effect is complex with heat transfer increasing and decreasing, depending on the relative size and distance of the particles.
- *Particle–wall interaction.* When the particle size is an appreciable fraction of the tube diameter, particle wall interactions become important as the flow can no longer be regarded as uniform. Particles in that case neither see a uniform (slip) velocity, nor a uniform temperature at their boundary. These effects have not been studied in a systematic way, but their impact may (partly) explain the discrepancies between experimental data reported.

Despite the complexities associated with fluid–particle heat transfer, researchers have attempted to correlate experimental data on the fluid–particle heat transfer as correlations of the flow, thermophysical properties, and shape of the particles. The Ranz–Marshall correlation (Equation 3.4) is a well-known example (Ranz and Marshall, 1952a,b).

$$Nu = a + bRe_{\mathrm{p}}^{c}\, Pr^{d} \tag{3.4}$$

where

$Re_{\mathrm{p}}$ is the particle Reynolds number based on average slip velocity
$Pr$ is the Prandtl number
$a$–$d$ are the coefficients

For single spherical and infinite cylindrical particles in a quiescent liquid the value of $a$ equals 2, which represents the conductive lower limit of the particle Nusselt number. Equations of the shape of Ranz–Marshall do not apply to the slip velocity between the average particle and the average fluid velocity as derived from the average residence times of fluid and particle. Instead, the local fluid–particle slip-velocity (Balasubramaniam and Sastry, 1994) should be used to estimate heat transfer.

From Equation 3.4 and the discussion above it is clear that whereas the lower bound of the particle Nusselt number is 2, its actual value in closed channel flow is likely to be higher, thus resulting in over processed food when this conservative value is used in design calculations. Unfortunately, at the moment, it is not possible to reliably predict Nusselt numbers more realistically. For predictive purposes a conservative value of 2 has therefore to be assumed, unless data on the food processing line has been collected (e.g., using inoculated particles) that allows a more accurate estimate.

It is essential that general correlations become available accounting for at least some of the factors sketched above. A new experimental technique is "ice-ablation" (Tessneer et al., 2001), where weight loss of melting water particles is a measure of the heat transfer between liquid and particle (slurry). This technique is both simple and avoids many of the difficulties associated with existing methods as described by Barigou et al. (1998).

### 3.2.4 Heat Transfer Within Particulates

Intraparticle heat transfer is the third and final step in the heat transfer and occurs exclusively by conduction. Conductive heat transfer is well understood and for simple shapes (e.g., slabs, cubes, spheres, and cylinders) and constant thermal boundaries (temperature or flux), analytical solutions are readily available. In practical situations, such boundaries are rare and the resistance to heat transfer is divided between the liquid (boundary layer) and the inside of the particle. The relative importance of these is given by the particle Biot number $Bi$ (Equation 3.5). For low values of the Biot number ($Bi < 0.2$), the heat transfer resistance lies predominantly in the liquid phase, where the particle is approximately isothermal. For Biot values $Bi > 10$, heat transfer is limited by conduction inside the particle. Since the particle Nusselt number is at least 2 and the conductivities of liquid and solids are very similar, heat transfer to particles is mostly conduction limited (Equation 3.5).

$$Bi = \frac{h_{\mathrm{fp}} \cdot d_{\mathrm{p}}}{\lambda_{\mathrm{p}}} = \frac{\lambda_{\mathrm{f}}}{\lambda_{\mathrm{p}}} \cdot Nu_{\mathrm{p}} \tag{3.5}$$

The thermal equilibration time of a particle as a function of shape and Biot number can be calculated by using the Heisler–Gröber charts (Heisler, 1947). The degree of equilibration for a constant temperature boundary can be correlated to the dimensionless time or Fourier number *Fo*.

### 3.2.5 Residence Time Distribution in Particulate Flows

Heating and sterilization of particles depend not only on the heat transfer rates in the energy cascade (Figure 3.1), but also on the exposure times of the particles to the thermal boundary condition (local fluid temperature). Particles move through the system at different and varying speeds, and this RTD is therefore of key importance to establish the thermal time distribution and hence thermal impact of the process (Ramaswamy et al., 1995).

Different forces act on particles of arbitrary shape as they move with the fluid, of which drag forces, buoyancy forces, lift forces, and particle–particle forces are dominant in aseptic particulate food systems (Lareo et al., 1997). In curved (e.g., bends) and in turbulent flows, centrifugal and inertial forces may also be significant.

Generally drag forces between particle and liquid are large, and therefore the local slip velocity between fluid and particle is low. For low-particle Reynolds numbers ($Re_p < 1$), the drag force follows the Stokes law, whereas for higher Reynolds numbers, non-Newtonian fluid rheology and dense slurries, more involved correlations are available (Lareo et al., 1997).

Buoyant forces result from density differences between fluid and particle, which depend on material properties and local differences in temperature. Buoyant forces may cause sedimentation or flotation in the system, or separation in curved flows (e.g., bends). Though buoyant forces are generally small compared to viscous and inertial forces in food processing, for extended residence times (e.g., long heat exchanger tubes or large buffer vessels), they may produce significantly nonuniform particle distributions (Mankad and Fryer, 1997).

Lift forces are associated with nonuniform velocity fields around particles, where the local static pressure distribution exerts a net force. A well-known effect is the tendency of particles to migrate toward an equilibrium position at some radial distance in the tube (Lareo et al., 1997 and references therein). As a result, the particle concentration near walls is generally lower, with the particles accumulating in the core of the flow. On average particles, therefore, travel faster through the system than expected from the average fluid velocity.

Centrifugal forces result from flow curvature which is present in coils and bends. Due to the mixing effect of the Dean vortices (see Section 3.2.2), particles will change between streamlines of different velocity, and the RTD will generally be narrower (Sandeep and Zuritz, 1999; Sandeep et al., 2000).

Clearly, the interaction between flow and particles gives rise to a marked RTD for particles as they pass through any aseptic process. In addition, the average residence time and the shape of the distribution are also impacted by particle shape, size distribution, particle fraction, and possibly by particle deformability.

To satisfy food safety requirements, a worst-case approach is normally assumed of a critical particle traveling at the same speed as the fastest liquid in a laminar Newtonian (Poiseuille) flow. In that case the minimum particle residence time is half that of the average fluid residence time. However, many liquid foods are either low in viscosity and hence turbulent, or viscous and pseudoplastic. In those cases the velocity profile is more uniform with smaller differences between the fastest and the average liquid. An example is the velocity profile in Ostwald-de Waele or power-law flow (Equation 3.6), where $n$ is the power-law index (typically 0.5), and $v_m$ and $v$ are the maximum and the average fluid velocity, respectively (Hartnett and Cho, 1998).

$$v_m = v \cdot \frac{3 \cdot n + 1}{n + 1} \tag{3.6}$$

Using the pseudoplastic velocity profile to estimate the minimum particle residence time may still be overly conservative. As noted previously, particles have a tendency to move to the center of the flow, where their concentration may prevent relative motion and rotation. This may cause conditions for local plug-flow (Altobelli et al., 1991), which reduces the maximum velocity. In addition, particles may change position between different streamlines, as a result of mixing in bends. In practice, the residence time of the fastest particle is therefore longer than inferred from the maximum flow velocity (Fairhurst and Pain, 1999; Barigou et al., 2003). However, no reliable general correlations exist to account for these effects yet. Therefore, the fastest particle should be derived from the undisturbed (power law or otherwise) velocity profile, unless experimental data allow a more precise estimate.

In aseptic processing, liquid and particles are subject to a temperature profile and the particles therefore see a time-dependent thermal boundary. Simultaneous solution of both liquid and particle temperatures in aseptic systems—heating, holding, and cooling stage—therefore requires a numerical approach. Readers are referred to Chapter 15 for a more in-depth analysis.

## 3.3 FUNDAMENTALS OF SAFETY AND QUALITY

Safety and quality are at the heart of any successful aseptic operation. Kinetics of safety and quality factors strongly depend on temperature. Whether a process–product combination actually meets the specifications requires rigorous experimental validation. Process kinetics and process validation are therefore the topic of this section.

### 3.3.1 KINETICS OF SAFETY AND QUALITY FACTORS

Destruction of microorganisms and spores is achieved through the thermal process, though nutritional content and other quality factors are also affected. Traditionally, first-order kinetics have been used to describe changes in safety and quality factors such as the destruction of microbes or nutrients. Such kinetics can be used to fit kinetic data to the Bigelow or Arrhenius model. Both Bigelow and Arrhenius kinetics can be expressed as the relative rate of reaction $R_r$, which is the rate of change of a microbial or quality factor at an instantaneous temperature $T$, relative to that at reference temperature $T_r$ (Equation 3.7). Sensitivity of the process to temperature is accounted for through the $z$-value (the temperature difference required for a 10-fold change in reaction rate) and activation energy $Ea$ for Bigelow and Arrhenius kinetics, respectively. Though Arrhenius kinetics are considered to be more accurate in the high-temperature range (130°C–150°C) of ultrahigh temperature processing (Hallström et al., 1988; Datta, 1994), both models are extensively used.

$$R_r = 10^{\frac{T-T_r}{Z}}\bigg|_{\text{Bigelow}} \quad \vee \quad R_r = 10^{\frac{E_a}{R}\left(\frac{1}{T_r}-\frac{1}{T}\right)}\bigg|_{\text{Arrhenius}} \tag{3.7}$$

Integration of the relative reaction rate yields the equivalent process time $t_e$ of the heat treatment in seconds at the reference temperature $T_r$ (Equation 3.8).

$$t_e = \int_0^t R_r \, dt \tag{3.8}$$

Several common names exist for $t_e$ depending on the kinetic constants and the reference temperature chosen. For process safety, using a $z$-value of 10°C and a reference temperature $T_r$ = 121.1°C (250°F) yields the well-known $F_0$-value, or a "proteolytic *Clostridium botulinum*" cook. In pasteurization processes, the equivalent time is often referred to as the $P_{T_r}$-value, where a relevant reference temperature is chosen, normally in the 70°C–100°C range.

The time required at the reference temperature $T_r$ to obtain a 10-fold reduction (1 log) of the microorganism population is called the $D_{T_r}$-value. In safety calculations, dividing the equivalent process time by the $D_{T_r}$-value yields the number of logarithmic reductions in the original microbial load $N_0$ (Equation 3.9).

$$\frac{F_0}{D_{T_r}} = \log\frac{N_0}{N} \tag{3.9}$$

To enable relevant safety or spoilage calculations, a critical target organism must be identified and a number of log reductions chosen. Based on the $D_{T_r}$, $Z$, and $T_r$ for said microorganism, a minimum equivalent process time is then calculated.

Equations 3.7 and 3.8 can also be applied to quality factors such as color, texture, nutrient content, enzyme inactivation, or flavor development (Holdsworth, 1992). The equivalent process time is referred to as $C_{T_r}$- or cook-value, when pertaining to textural change and using a reference temperature of 100°C. In that case the $z$-value depends on both the textural aspect and ingredient. Application of the decimal reduction time $D_{T_r}$ to quality factors is also possible, and tables of the kinetic data are available in the literature.

It should be noted that Equations 3.7 through 3.9 apply to a liquid or solid element with a uniform temperature and no RTD. The intricacies of calculating safety and quality factors in nonuniform heterogeneous systems (both regarding temperature and residence time) are discussed in greater depth in Chapter 15.

### 3.3.2 Validation of Aseptic Lines

Thermal processes need to be validated in order to establish that sufficient thermal treatment has been applied. Legal requirements are different in each country: while the EU states that each company is responsible for the safety and spoilage, the United States has a very stringent control organization in the Food and Drug Administration (FDA). FDA rules separate acid and low-acid foods, the latter have been compiled in the 21CFR113.3 and constitute a complex body of regulation where scientific validation of the heat treatment is required. As product safety is obtained using a thermal process, validation focuses on the coldest part of the product either for liquids or for heterogeneous foods.

Traditional validation for retortable cans involves either the use of temperature sensors in the coldest spot or inoculation with viable thermally characterized microorganisms (bacterial endospores). This was first done in the 1920s by Bigelow and confirmed by Ball in the 1950s. Spores of surrogate microorganisms were also investigated, and one of the main concerns was to find a microorganism with similar death kinetics to the microorganisms of concern (*C. botulinum*). For the juice industry the FDA defined a surrogate microorganism as "any nonpathogenic microorganism that has acid tolerance, heat resistance, or other relevant characteristics similar to pathogenic microorganisms." *Clostridium sporogenes PA3679* is the recognized surrogate of *C. botulinum*. Pflug et al. (1980) demonstrated the use of *Geobacillus stearothermophilus* and *Bacillus subtilis* as surrogate microorganisms. Spores of these microorganisms are commercially available from several sources and have become the standard for validation of low-acid foods requiring sterilization.

In aseptic processing, products are processed separately from the final container and the methods of validation are different. When homogeneous products are processed, the line must be designed to properly sterilize the part of the product which receives the least thermal treatment, i.e., fastest and coldest throughout the process. This is normally the center of the holding tube, where the velocity is highest and the temperature is lowest. In this case, temperature data are required at the inlet and exit of the holding section in the center of the tube. However, the response time of the thermocouples must be short, and they should be sufficiently small not to act as heat sinks.

Heterogeneous flows will generally have the coldest spot in the center of the slowest heating particle. There are several methods to monitor the temperature inside the particulates in real time. Moving thermocouples have been used in research though this method is not practical in an industrial environment (Balasubramanian, 1994; Ahmad et al., 1999). Magnetic resonance imaging can also be used to monitor the temperatures of moving particulates, but the cost is high and industrial relevance is low.

More practical off-line solutions have therefore been developed to monitor the integrated thermal effect and these include either by using real or simulated particles that carry spores of a known microorganism. Such time–temperature integrators (TTIs) can be of any of the following types: immobilized spores, chemical or biochemical markers, thermomagnetic markers, and combinations of the above (Tucker, 1999; Tucker et al., 2002).

Immobilized spores in calcium alginate beads have been used to evaluate the ultra high temperature (UHT) processes for several decades, and viable spores of *G. stearothermophilus* and *B. subtilis* are inoculated into a sodium alginate solution to a known concentration (Holdsworth, 1992). The alginate is then formed into particulates of a certain size and shape and solidified by immersion in a calcium-salt solution. The alginate particulates can be made at any size or shape to simulate real food particulates or can be made as small alginate spheres (4 mm diameter), which are located into the center of existing food particulates (Tucker and Holdsworth, 1991).

Bioindicators are similar to the above, with the exception that the spores are inoculated in an indicator culture media contained inside small plastic pouches (Brinley et al., 2007), thus reducing the risk of contamination to the products. Residual viability following heat treatment is indicated by a change in color of the culture media or by further culturing of the surviving spores.

Chemical markers are substances that exhibit irreversible changes due to the exposure to time–temperature such as luminescence, color changes, or following known kinetics of destruction or generation. Most chemical markers cannot be reused and only few chemical reactions with similar kinetics to spores have been identified (Kim et al., 1996). Enzymes are probably the most promising of the chemical substances to be used as TTIs: inactivation kinetics are similar to those of microbes and food quality; they present low toxicity and can be obtained in high purity commercially (Tucker et al., 2002). Small quantities of enzyme can be located inside small tubes or pouches which are inside food or simulated particulates. A simple assay determines the remaining activity of the enzyme, which is related to the thermal treatment the enzyme received (Tucker and Holdsworth, 1991). However, the stability of most enzymes above 100°C is unsatisfactory and their use has been limited to temperatures in the pasteurization range (Tucker, 1999; Tucker et al., 2002). Recently, high-temperature enzyme isolates from *Pyrococcus furiosus* have been identified (Tucker et al., 2007) and also from *B. licheniformis* (Guiavarc'h et al., 2004). Applications of these are currently under development.

Thermomagnetic markers make use of changes in the magnetic field of two magnets after such magnets are allowed to combine. To do so, glue with a well-defined melting point (e.g., eutectic alloys) is placed between two magnets. When the melting point is reached, the glue melts and allows the magnets to join, thus generating disturbances in the magnetic field that can be monitored by sensors located outside of the tube. Placing sensors in different locations of the heating and holding units yields an indication of the thermal treatment of particles along the system. The trajectory and residence time of each particulate can also be reconstructed by monitoring the three-dimensional position of each particle in each sensing location. Since several magnet–glue pairs can be stacked inside one particle, a temperature history can be determined and used to optimize the thermal treatments (Simunovic et al., 2004). Recovery of magnetic particles at the end of the line is relatively simple: a metal detector can be used to recover only the particles containing the magnets.

Whatever the principle of TTIs, practical use is still far from straightforward in aseptic processing of heterogeneous food. A significant number of particles are required for statistically meaningful results (the FDA requires at least 300 particulates) and these must be injected and retrieved intact and in-full from the system. In addition, size, shape, buoyancy, and thermal conductivity must be representative of the target real particles. Though progress in this field is encouraging, the ideal would be miniature digital temperature loggers capable of surviving the aseptic process, which would be detectable throughout the line to establish particle flow trajectories.

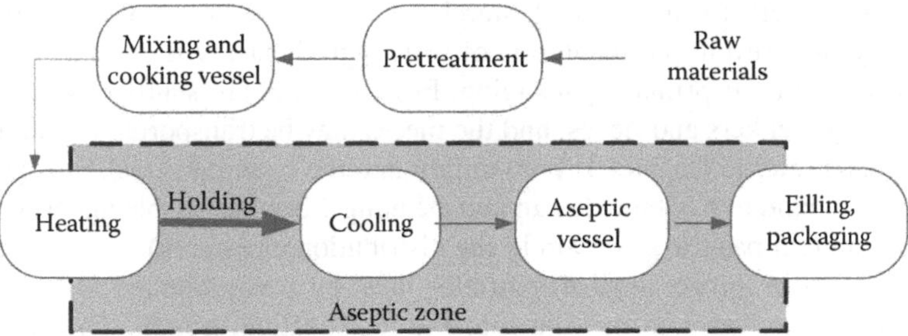


**FIGURE 3.4** Fundamental operations in aseptic processing.

## 3.4 ASEPTIC EQUIPMENT

From storage of raw materials to finished and stored product, an aseptic process consists of a sequence of essential steps shown in Figure 3.4.

A basic line consists of facilities to assemble and cook the raw materials, sterilize the product thermally, and fill this into the final product containers. To achieve ambient product stability, sterility must be maintained in the "aseptic zone" of the line.

### 3.4.1 Raw Materials Storage and Handling

Heterogeneous food can contain a myriad of ingredients that can be food products in their own right. Ingredients can be in the liquid (water, oils, and paste) or solid (salt, starch, and wet vegetable pieces) state. They may vary greatly in quantity, and depending on their shelf-life stability they are stored frozen, chilled, or at ambient temperature, where specific requirements on relative humidity or modified atmosphere may apply. In all cases, storage areas must prevent deterioration of both the microbial and the organoleptic properties, and the ingredients should only be removed from these areas immediately prior to use. Ingredients must be clearly marked and kept in separated areas when potential allergens are involved.

### 3.4.2 Cook Vessels

Cook vessels are operated in batch and their required size and auxiliary requirements depend on the batch cycle diagram (Figure 3.5). Normally, water is pumped into the cook vessel from the factory supply and powders (starch and salts) are premixed and

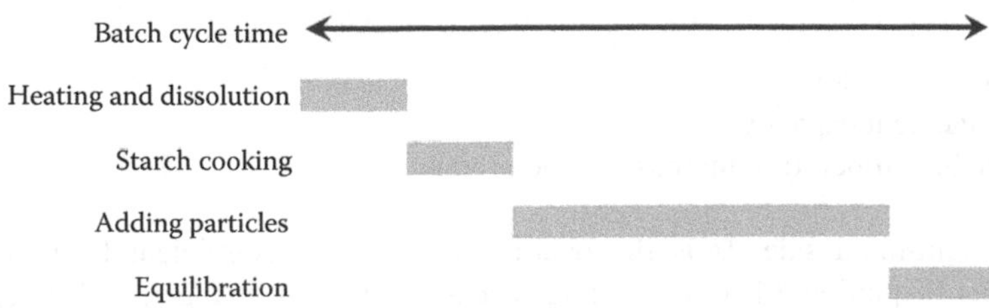


**FIGURE 3.5** Typical batch cycle diagram for cook vessel of particulate process.

added upon sufficient heating. Next, starch may be cooked to provide the required consistency and prevent sedimentation of solid ingredients. Oils and fats are generally added from their primary packaging. Frozen pieces are shuttled, separation is ensured using breakers and sieves, and the pieces may be transported to a tempering device (please refer to Chapter 15) or straight into the cooking vessel. Special attention should be paid to hygienic design and cleaning when perishable material is used on an intermittent basis (e.g., cream in the distribution pipe-work).

To ensure safe processing, all particles must have equilibrated to the average vessel temperature prior to passing into the first heating section. The worst-case scenario occurs when the largest particle, is added completely frozen at storage temperature (−20°C) at the end of the addition of the ingredients. The "final equilibration time" strongly depends on the thermal size of the particle: typically a doubling of the thermal size of the particle leads to a fourfold increase in equilibration time.

The size of the cook vessel and its operational duty depend on the batch cycle program of the cooking stage. Usually, multiple cook vessels are employed: while one is being drained, batch preparation is carried out in other vessels whereas the others may be cleaned.

Depending on the process chosen, the heating requirements for the cook vessel vary and different solutions for heating may be employed. The three main alternatives (in order of increasing heating duty) are a (scraped) jacketed vessel, a jacketed vessel with a heated stirrer, and a (scraped and jacketed) vessel using steam injection. For medium to low particulate mass fractions (<50%) agitation is required in the vessel to prevent sedimentation and separation of the particles. Alternatively, a thickener may be employed to ensure uniform particle distribution.

### 3.4.3 Heating Stage

Indirect heating is usually employed to raise the temperature of the liquid–particle mixture between the cook vessel and the holding tube; several heat exchangers exist for this purpose of which tubular, annular, and scraped surface heat exchangers are common examples. Sizing this equipment is straightforward and follows from product and medium flows and temperatures, and from estimates of the achievable overall heat transfer coefficient. An energy balance around a heat exchanger contains the factors governing the design of the heat exchanger (Equation 3.10).

$$m \cdot Cp \cdot \Delta T_\mathrm{p} = U \cdot A \cdot \Delta T_\mathrm{mp} \tag{3.10}$$

where

$m$ is the mass flow
$Cp$ is the heat capacity
$\Delta T_\mathrm{p}$ is the temperature increase of the product

On the right-hand side, $U$ is the overall heat transfer coefficient from medium to product (Equation 3.10), $A$ is the surface area of the exchanger, and $\Delta T_\mathrm{mp}$ is the temperature differential between product and medium.

To avoid excessive fouling, the heating medium is commonly pressurized water operated in countercurrent flow. To ensure a constant temperature differential between the product and the medium, the heat capacity flows ($m \cdot Cp$) should be equal. The heating medium may be heated in a secondary heat exchanger by high-pressure steam from the factory supplies.

Equation 3.10 is valid for low particle fractions (<5%) or for small particles (<3 mm), where the heat sink presented by the particulate phase is negligible and equilibration is (almost) instantaneous. For higher particle fractions and sizes >3 mm, the liquid may be successfully heated to the sterilizing temperature, but subsequent equilibration between particles and liquid lowers the equilibrated mixture temperature below the target. In this case, the heater should be sized to allow for equilibration.

Two points must be made regarding heat exchanger selection and sizing. First, it is paramount to minimize the thermal path over which the heat transfer is to be achieved. This is especially true for laminar flows, where the heat transfer rates are essentially conduction limited. In that case, halving the tube diameter approximately equates to halving the thermal path, and hence shortening the exchanger length by a factor of 4 and decreasing the product hold-up by a factor of 16. Though for particles the exchangers' geometry is dictated by the particle size, for liquids there is much more flexibility, also in terms of type of heat exchanger (annular and plate).

Second, the rate of heating of the product is proportional to the temperature differential with the medium, and to its temperature. However, the maximum medium temperature is limited: an excessive–medium temperature may exacerbate temperature nonuniformities in the flow, and hence in the final quality of the product. In addition, fouling rates often depend in a nonlinear fashion on wall temperatures and therefore medium temperature. In Chapter 15, it will also be shown that conductive limits to the heat transfer in particles pose another physical limit to the heating medium maximum temperature.

### 3.4.4 Holding Tube

For reasons of simplicity and its well-defined residence distribution, the aseptic holding section is generally a tube of circular cross-section. According to FDA regulations (FDA 21CFR113.3, 2006), lethality must be calculated using the fastest possible liquid element or slowest heating particle at the divert temperature. The divert temperature $T_d$ is the temperature measured at the exit of the holding tube, where product is diverted to drain, instead of being fed in the down stream cooler and aseptic tank.

To calculate lethality in the holding tube, a fastest particle or fluid element moving at twice the average flow velocity in the holding tube is evaluated, which assumes a developed Newtonian velocity profile. As mentioned in Section 3.2.5, this factor may be set less conservative (<2) when rheology and flow condition are more accurately known.

$$F_0 = \int_0^t 10^{\frac{T-T_r}{z}}\,dt = 10^{\frac{T_d-T_r}{z}} \cdot t_{min} \tag{3.11}$$

$$L_h \geq 2 \cdot v \cdot t_{\min_{F_0=3}} \tag{3.12}$$

where

$T_d$ is the divert temperature
$T_r$ is the reference temperature for *C. botulinum* (121.1°C)
$Z$ is the $z$-value of *C. botulinum* (10.8°C)
$t_{min}$ is the minimum residence time

The minimum length of the holding tube $L_h$ follows from Equation 3.12 for the average flow velocity $v$.

Temperature losses to the environment may be significant in long holding tubes, and evaluating Equation 3.12 based on the divert temperature produces an overestimation of the thermal treatment. Insulation around the holding tube may keep the exit temperature close to that of the inlet, and the inlet temperature can be set correspondingly lower.

### 3.4.5 Cooling Stage

For cooling both the liquid base and the particles, indirect heat transfer is the only feasible option. Flashing (evaporative cooling) generally impairs flavor and texture, and is limited to higher temperatures. Depending on liquid viscosity and cost constraints, tubular, annular, or scraped surface heat exchangers may be employed. Plate heat exchangers only are appropriate for inviscid (Newtonian) broths or for treating auxiliary culinary water.

Heat transfer in cooling is generally slower than in heating due to higher product viscosities. To maximize cooling rates the heat capacity flow of the cooling medium is usually set 5–10 times that of the product flow. To increase cooling rates, chilled water or glycol may be used as medium. The duty of the cooling medium follows from an energy balance, and the exchanger can be sized accordingly (Equation 3.10).

Liquid products without particles can be treated in any type of heat exchanger, but to minimize pressure drop and maximize heat transfer area a narrow gap annular heat exchanger is recommended. For extremely viscous liquids, an agitated heat exchanger (e.g., scraped surface heat exchanger [SSHE]) may be employed, in which case suitable precautions to ensure aseptic operation (steam barriers on seals) are required. For particulate flows, the preferred heat exchanger type strongly depends on the particle dimensions and their mass fraction. Both tubular and scraped surface heat exchangers are feasible solutions. Annular heat exchangers are not recommended for flows containing very large particles, as the gap width would become impractically large to allow the passage of particles without excessive shear damage.

During the cooling stage, a large flow resistance can be developed in the exchanger due to the increase in viscosity of the products. Whereas a certain pressure drop is required to provide the back-pressure to avoid boiling in the heater and holding tube, sometimes this pressure drop may become excessive. For such cases, the main pump cannot provide enough pressure, and the flow rate has to be reduced or an additional pump must be used.

### 3.4.6 Aseptic Tanks

Aseptic tanks are principally used to buffer product from the sterilizer line in case of unexpected filler interruption, or to feed the filler when the sterilizer line is briefly interrupted. Following start-up, the aseptic tank is filled up to the desired buffer level and is operated in steady state thereafter, since approximately equal flow rates of product enter into and are drained from the tank. The required size of the tank depends on the nominal product flow rate, and on assumptions regarding sterilizer and filler stoppage.

Sedimentation of particulates in the aseptic tank needs to be avoided as in the case of the cooking vessels. A uniform distribution of the particulates in the final pack again depends on a sufficient stirring rate, or on the use of thickeners of sufficient viscosity and yields stress.

### 3.4.7 Filler Operation and Pack Decontamination

Sterile operation of the filler and decontamination of the pack are essential for ambient stable products as both are part of the aseptic zone (Figure 3.5). Aseptic filling equipment must be chosen with the product range in mind, where a choice must be made between liquid and liquid/particulate fillers and also between single- and multiple-dose fillers. Sterilization and maintaining asepsis during production require integrated design of filler and pack system, examples being the Combibloc and Tetra Pak systems for retail size packages. Aseptic filling comprises three key steps: pack decontamination, filling, and sealing.

For best possible pack decontamination, the primary packaging material must be protected from contamination and moisture during transport and storage. Packaging material can be decontaminated using chemical means (e.g., liquid 35% hydrogen peroxide, peracetic acid, hydrogen peroxide vapor, or ozone) or using steam, and this must be performed prior to the entry of the packaging material into the aseptic zone. In aseptic filling, particle size and volume fraction may pose specific challenges. Particles require a sufficient dimension of the filler nozzle for undamaged passage, and this may impact negatively on filling accuracy (nozzle drip behavior) or soiling of the sealing area. Filler accuracy should be such that overfilling does not adversely affect seal integrity or stress the seals during transport. Filler-pack systems that are suitable for products containing large particulates are currently scarce.

Sealing is the final step of the filling operation, and it should consistently produce a hermetic seal with defined dimensions and strength which remains intact throughout the supply chain. Several sealing mechanisms are available such as heat, induction, and ultrasonic sealing. However, in all cases, the local heating of the material is the main factor affecting the seal quality.

### 3.4.8 Final Product Handling and Storage

Upon filling and sealing the packs are labeled, coded, and transported on conveyor belts to the secondary packaging area, to be subsequently palletized. Handling, storage, and transport systems must prevent pack damage and ensure high levels of pack integrity. Coding, secondary packaging, and palletization must meet trade distribution needs and minimize risks of primary pack damage.

### 3.4.9 Pumps and Back-Pressure

Flow through the process needs to be driven by a pressure differential, and a minimum absolute pressure level in the system is required to prevent boiling. Practically this pressure can be generated by three different mechanisms:

1. Gravity flow (through a height differential)
2. Gas pressure (e.g., pressurized nitrogen or air)
3. Mechanical displacement (e.g., lobe or piston pumps)

The pressure to drive the flow(s) through the system depends on the throughput, the equipment geometry, and the viscosity of the flows concerned. Its minimum value depends on the absolute pressure at the end of the line (before the back-pressure valve or other pressure reducing device) plus the flow resistance in the system. For viscous liquids, flow resistances greater than 5 bars are easily generated, and flow resistance in the cooling stage may restrict the throughput of the system. The back-pressure required to prevent boiling is determined by the maximum expected temperature in the system, plus a safety margin accounting for deviations in the heating medium temperature. Assuming a maximum operation product temperature of, for example, 140°C, back-pressure would approach 4 bars, of which 3 would have to be provided in the cooling section of the system.

Of the three alternative pressure delivery mechanisms, only mechanical displacement, such as lobe or piston pumps, is practical to drive the flow. When working in reverse these can also be used to create additional back-pressure prior to the filler.

An important consideration in pump selection is the damage inflicted to delicate particulates and any structures present in the liquid phase. This is especially relevant in the downstream section, where a considerable softening of the particles may have occurred due to the cooking treatment. All pumps should therefore be selected with emphasis on this requirement (see Chapter 15).

## 3.5 CONCLUDING REMARKS

High-quality aseptic processing and packaging of particulate liquid foods have opened up a new era in ambient stable food manufacturing. However, several issues need to be addressed and more fundamental knowledge needs to be amassed before this becomes a profitable reality.

In order to enable nonconservative designs and avoid overprocessing, there is a need for reliable engineering correlations for the wall–fluid heat transfer coefficients to particulate slurries as a function of liquid properties, and of particle loading and shape.

More understanding of fouling in heat exchangers is needed since at the moment modeling and predicting this are largely unchartered territory. Though this subject is vast given the range of products, equipment geometries, and processing conditions, control of fouling is essential to fully exploit the benefits of aseptic processing.

RTD of particles and heat transfer between liquid and particles require further research. Experimental tools to establish RTDs, particle trajectories, and ice-ablation techniques should be further developed to allow a reliable and routine determination of heat transfer coefficients.

Finally, practical means to validate thermal processes are required, particularly those applicable to particulate flows. New and emerging methods such as thermomagnetic switches and other TTIs may provide information and make the process easier to validate, control, and be accepted by regulatory agencies.

## ACKNOWLEDGMENTS

We would like to acknowledge the following persons for sharing their expertise and suggestions during the writing of this chapter: Pierre Debias, Gary Mycock, Hans Hoogland, David Dearden, Roger Jordan, and Annemarie Elberse.

## LIST OF SYMBOLS

| | | |
|---|---|---|
| $A$ | $m^2$ | surface area |
| $Bi = h \cdot d/\lambda$ | — | Biot number |
| $Cp$ | $J\ kg^{-1}\ °C^{-1}$ | specific heat capacity |
| $d$ | M | diamcter |
| $D$ | min | $D$-value |
| $Ea$ | $J\ mol^{-1}$ | activation energy |
| $f = 2 \cdot d \cdot \Delta p/L \cdot \rho \cdot v^2$ | — | friction factor |
| $Fo = Cp \cdot \rho \cdot t/\lambda \cdot d^2$ | — | Fourier number |
| $F_0$ | min | equivalent process time at 121.1°C |
| $H$ | $W\ m^{-2}\ °C^{-1}$ | heat transfer coefficient |
| $L_h$ | M | length of the holding tube |
| $M$ | $kg\ s^{-1}$ | mass flow rate |
| $N$ | — | power-law index |
| $Nu = h \cdot d/\lambda$ | — | Nusselt number |
| $Pr = Cp \cdot \mu/\lambda$ | — | Prandtl number |
| $R$ | $J\ mol^{-1}$ | universal gas constant |
| $R_r$ | — | relative reaction rate |
| $Re = \rho \cdot v \cdot d_h/\mu$ | — | Reynolds number |
| $T$ | M | thickness |
| $t_e$ | min | equivalent process time |
| $T$ | °C | temperature |
| $T_d$ | °C | divert temperature |
| $T_r$ | °C | reference temperature |
| $U$ | $W\ m^{-2}\ °C^{-1}$ | overall heat transfer coefficient |
| $V$ | $m\ s^{-1}$ | average velocity |
| $Z$ | M | streamwise distance |
| $Z$ | °C | $z$-value |
| $Z^{\approx} = \mu \cdot z/\rho \cdot d^2 \cdot v$ | — | dimensionless hydrodynamic length |
| $z^* = \lambda \cdot z/\rho \cdot Cp \cdot d^2 \cdot v$ | — | dimensionless thermal length |
| $\Delta T$ | °C | temperature increase or differential |
| $\varepsilon$ | M | absolute surface roughness |
| $\lambda$ | $W\ m^{-1}\ °C^{-1}$ | thermal conductivity |
| $\mu$ | Pa s | dynamic viscosity |

| | | |
|---|---|---|
| ρ | kg m$^{-3}$ | density |
| *Subscripts* | | |
| F | | of the fouling layer |
| M | | of the medium/max |
| P | | of the product/particle |
| W | | of the wall |

## REFERENCES


A.N. Abdelmessih and K.J. Bell. Effect of mixed convection and U-bends on the design of double pipe heat exchangers. *Heat Trans. Eng.* 20(3):25–36, 1999.

M.N. Ahmad, B.P. Kelly, and T.R.A. Magee. Measurement of heat transfer coefficients using stationary and moving particles in tube flow. *Trans. IChemE.* 77(C):213–222, 1999.

A. Alhamdan and S. Sastry. Bulk average heat transfer coefficient of multiple particles flowing in a holding tube. *Trans. IChemE.* 76(C):95–101, 1998.

S.A. Altobelli, R.C. Givler, and E. Fukushima. Velocity and concentration measurements of suspensions by nuclear magnetic resonance imaging. *J. Rheol.* 35(5):721–734, 1991.

V.M. Balasubramaniam and S.K. Sastry. Liquid-to-particle convective heat transfer in non-Newtonian carrier medium during continuous tube flow. *J. Food Eng.* 23:169–187, 1994.

M. Barigou, S. Mankad, and P.J. Fryer. Heat transfer in two-phase solid–liquid food flows: a review. *Trans. IChemE pt. C.* 76:3–29, 1998.

M. Barigou, P.G. Fairhurst, P.J. Fryer, and J.P. Pain. Concentric flow regime of solid–liquid food suspensions: theory and experiment. *Chem. Eng. Sci.* 58:1671–1686, 2003.

A.E. Bergles. Techniques to enhance heat transfer. In: W.M. Rohsenov, J.P. Hartnett, and Y.I. Cho, eds., *Handbook of Heat Transfer*, 3rd edition. pp. 11.1–11.76. McGraw-Hill, New York, 1998.

T.A. Brinley, C.N. Dock, V.-D. Truong, P. Coronel, P. Kumar, J. Simunovic, K.P. Sandeep, G.D. Cartwright, K.R. Swartzel, and L.-A. Jaykus. Feasibility of utilizing bioindicators for testing microbial inactivation in sweetpotato purees processed with a continuous-flow microwave system. *JFS.* 72(5):E235–E242, 2007.

K. Chakrabandhu and R.K. Singh. Fluid-to-particle heat transfer coefficients for continuous flow of suspensions in coiled tube and straight tube with bends. *Lebenm.-Wiss. U.-Technol.* 35:420–435, 2002.

A.K. Datta. Error-estimates for approximate kinetic-parameters used in food literature. *J. Food Eng.* 18(2):181–199, 1994.

J.R.D. David, R.H. Graves, and V.R. Carlson. *Aseptic Processing and Packaging of Food: A Food Industry Perspective.* CRC Press, Boca Raton, FL, 1996.

W.R. Dean. Note on the motion of a fluid in a curved pipe. *Philos. Mag. Ser.* 7(4):208–223, 1927.

M.A. Ebadian and Z.F. Dong. Forced convection, internal flows in ducts. In: W.M. Rohsenov, J.P. Hartnett, and Y.I. Cho, eds., *Handbook of Heat Transfer*, 3rd edition. pp. 5.1–5.137. McGraw-Hill, New York, 1998.

P.G. Fairhurst and J.P. Pain. Passage time distributions for high solid fraction solid–liquid food mixtures in horizontal flow: unimodal size particle distributions. *J. Food Eng.* 39:345–357, 1999.

J.P. Gadonna, J.P. Pain, and M. Barigou. Determination of the convective heat transfer coefficient between a free particle and a conveying fluid in a horizontal pipe. *Trans. IChemE.* 74(C):27–39, 1996.

Y. Guiavarc'h, A. vanLoey, F. Zuber, and M. Hendrickx. Development characterization and use of a high performance enzymatic time–temperature integrator for the control of sterilization process' impacts. *Biotechnol. Bioeng.* 88(1):15–25, 2004.

B. Hallström, C. Skjöldebrand, and C. Trägårdh. *Heat Transfer and Food Products*. Elsevier Applied Science, Amsterdam, The Netherlands, 1988.

J.P. Hartnett and Y.I Cho. Non Newtonian fluids. In: W.M. Rohsenov, J.P. Hartnett, and Y.I. Cho, eds., *Handbook of Heat Transfer*, 3rd edition. pp. 10.1–10.53. McGraw-Hill, New York, 1998.

M.P. Heisler. Temperature charts for induction and constant temperature heating. *Trans. ASME*. 69:227–236, 1947.

S.D. Holdsworth. *Aseptic Processing and Packaging of Food Products*. Elsevier Applied Science, Amsterdam, The Netherlands, 1992.

S.D. Holdsworth. Rheological models used for the prediction of the flow properties of food products: a literature review. *Trans. IChemE pt. C*. 71:139–179, 1993.

J.D.H. Kelder. Optimal design of spiral sterilisers. PhD thesis, Technische Universiteit Eindhoven, Unilever Research & Development, Vlaardingen, The Netherlands, 2003.

J.D.H. Kelder, K.J. Ptasinski, and P.J.A.M. Kerkhof. Starch gelatinisation in coiled heaters. *Biotechnol. Prog*. 20(3):921–929, 2004.

B.P. Kelly, T.R. Magee, and M.N. Ahmad. Convective heat transfer in open channel flow—effects of geometric shape and flow characteristics. *Trans. IChemE*. 73(C4):171–182, 1995.

H.J. Kim, Y.M. Choi, A.P.P. Yang, T.C.S. Yang, I.A. Taub, J. Giles, C. Ditusa, S. Chall, and P. Zoltai. Microbiological and chemical investigation of ohmic heating of particulate foods using a 5 kW ohmic system. *J. Food Process Pres*. 20(1):41–58, 1996.

C. Lareo, P.J. Fryer, and M. Barigou. The fluid mechanics of two-phase solid–liquid food flows: a review. *Trans. IChemE pt*. C. 75:73–105, 1997.

S. Mankad and P.J. Fryer. A heterogeneous flow model for the effect of slip and flow velocities on food steriliser design. *Chem. Eng Sci*. 52(12):1835–1843, 1997.

S. Mankad, K.M. Nixon, and P.J. Fryer. Measurement of particle–liquid heat transfer in systems of varied solids fraction. *J. Food Eng*. 31:9–33, 1997.

R.L. Manlapaz and S.W. Churchill. Fully developed laminar convection from a helical coil. *Chem. Eng. Commun*. 9:185–200, 1981.

S.M. Morcos and A.E. Bergles. Experimental investigation of combined forced and free laminar convection in horizontal tubes. *J. Heat Trans.-T. ASME*. 97(2):212–219, 1975.

B.R. Morton. Laminar convection in uniformly heated horizontal pipes at low Rayleigh number. *Q. J. Mech. Appl. Math*. 12:410–420, 1959.

S. Palaniappan and C.E. Sizer. Aseptic process validated for foods containing particulates. *Food Tech*. 51(8):60–68, 1997.

N.N. Potter and J.H. Hotchkiss. *Food Science*, 3rd edition. Aspen Publishers, New York, 1985.

I.J. Pflug, A.T. Jones, and R. Blanchett. Performance of bacterial spores in a carrier system in measuring the F0-value delivered to cans of food heated in a Steritort. *J. Food Sci*. 45:940–945, 1980.

H.S. Ramaswamy, K.A. Abdelrahim, B.K. Simpson, and J.P. Smith. Residence time distribution (RTD) in aseptic processing of particulate foods: a review. *Food Res. Int*. 28(3):291–310, 1995.

W.E. Ranz and W.R. Marshall. Evaporation from drops. *Chem. Eng. Prog*. 48(3):141–146, 1952a.

W.E. Ranz and W.R. Marshall. Evaporation from drops. *Chem. Eng. Prog*. 48(4):173–180, 1952b.

C.S. Rao and R.W. Hartel. Scraped surface heat exchangers. *CRC Cr. Rev. Food Sci*. 46:207–219, 2006.

H. Reuter, ed. *Aseptic Processing of Foods*. Technomic Publishing, Basel, Switzerland, 1987.

S. Rozzi, R. Massini, G. Paciello, G. Pagliarini, S. Rainieri, and A. Trifirò. Heat treatment of fluid food in a shell and tube heat exchanger: comparison between smooth and helically corrugated wall tubes. *J. Food Eng*. 79:249–254, 2007.

K.P. Sandeep and C.A. Zuritz. Secondary flow and residence time distribution in food processing holding tubes with bends. *J. Food Sci.* 64(6):941–945, 1999.

K.P. Sandeep, C.A. Zuritz, and V.M. Puri. Modelling non-Newtonian two-phase flow in conventional and helical-holding tubes. *Int. J. Food Sci. Technol.* 35:511–522, 2000.

J. Sannervik, U. Bolmstedt, and C. Trågårdh. Heat transfer in tubular heat-exchangers for particulate-containing liquid foods. *J. Food Eng.* 29(1):63–74, 1996.

R.K. Shah and D.P. Sekuli'c. Heat exchangers. In: W.M. Rohsenov, J.P. Hartnett, and Y.I. Cho, eds., *Handbook of Heat Transfer*, 3rd edition. pp. 17.1–17.169. McGraw-Hill, New York, 1998.

J. Simunovic, K.R. Swartzel, and E. Adles. Method and system for conservative evaluation, validation and monitoring of thermal processing. Patent US 6 776 523 B2; 2004.

W.A. Tessneer, B.E. Farkas, and K.P. Sandeep. Use of ablation to determine the convective heat transfer coefficient in two-phase flow. *J. Food Proc. Eng.* 24:315–330, 2001.

G.S. Tucker. A novel validation method: application of time–temperature integrators to food pasteurization treatments. *Trans. IChemE.* 77(C):223–231, 1999.

G.S. Tucker and S.D. Holdsworth. Mathematical modelling of sterilisation and cooking processes for heat preserved foods. *Food Bioprod. Process.* 69(C1):5–12, 1991.

G.S. Tucker, T. Lambourne, J.B. Adams, and A. Lach. Application of biochemical time–temperature integrators to estimate pasteurization values in continuous food processes. *Innovative Food Sci. Emerg. Technol.* 3:165–174, 2002.

G.S. Tucker, H.M. Brown, P.J. Fryer, P.W. Cox, F.L. Poole II, H.S. Lee, and M.W.M. Adams. A sterilization time–temperature indicator based on amylase from the hyperthermophilic organisms *Pyrococus furiosus. Innovative Food Sci. Emerg. Technol.* 63–72, 2007.

# 4 Ohmic and Microwave Heating

*Pablo M. Coronel, Sudhir Sastry, Soojin Jun, S. Salengke, and Josip Simunovic*

## CONTENTS



## 4.1 INTRODUCTION

Continuous flow thermal processing of food materials is the base for most aseptic and high temperature processes. Normally heat is transferred from a heating media into a product by either convection or conduction through walls (indirect heating) or by injecting the hot media into the products (direct heating).

There are, however, limitations to the heat transfer using such conventional means, such as low-conductivity and high-viscosity products, and products with pieces or particulates. To overcome these limitations, methods using electromagnetic energy have been explored and developed. These methods have, in principle, no temperature limitations, could be used for food with particulates, and due to their theoretical high speed and uniformity of heating they should have potential for the industry.

Two such methods that are available and have proven to be viable alternatives are ohmic and microwave heating. Ohmic works by heating the product using electrical current that is made to flow inside the product, the resistance of the food to this current flow generates heat. Microwave works by the movement of the molecules of the product in a rapidly changing electromagnetic field that generates heat.

Intensive research has been performed on these heating methods, and many scientific papers have been published, a review of this research can be found in the sections that follow, it is therefore no coincidence that ohmic and microwave heating seems to be making inroads into the industrial world and applications to continuous flow processing are starting to appear.

## 4.2 OHMIC HEATING

Ohmic heating involves the passage of an electric current directly through the material to be heated. Since an electric current involves the motion of charges—in the case of foods and other electrolytic materials, the current involves movement of ions—the collision of charges with molecules produces molecular agitation, and consequently, heat. Since the heat generation occurs directly within the food, heating is much more uniform than with conventional modes of heating.

Ohmic heating has been practiced since the nineteenth century, when a number of patents were filed for heating of flowable materials. In the twentieth century, ohmic heating was practiced on and off; first in the 1930s for electric pasteurization of milk, and later in the 1980s and 1990s for continuous flow sterilization and (presumably) aseptic packaging for solid–liquid food mixtures. After a few typical initial difficulties, the technology is now seeing significant use in Europe, Japan, and elsewhere for processing of solid–liquid mixtures, and for other specific food-related applications.

In practice, ohmic heating requires the food to be in contact with electrodes, across which a potential difference is applied. In order for heating to occur, the material must have some ability to conduct electricity. The rate of energy generation per unit volume ($q_{gen}$) may be expressed as shown in Equation 4.1.

$$q_{gen} = E^2\sigma \tag{4.1}$$

The electric field strength, $E$, may be modified either by physical design of the electrode gap, or more easily, by altering the applied voltage. The electrical conductivity $\sigma$ depends on the product, and may be altered by formulation to include various ingredients. If $\sigma$ is zero, as with pure fats and oils, heating is not possible; however, most foods and even water supplies have a nonzero value for $\sigma$, and may therefore be heated successfully.

*Differences from microwave heating.* As with microwave heating, ohmic heating involves the response of a material to an applied electric field. While ohmic heating may theoretically be done using direct current (DC); however it is not practical, since it results in electrolysis of aqueous media. Electrolysis (discussed in more detail later) may be mitigated either by operation at relatively high frequencies (1 kHz or greater), or by operating at lower frequencies (~50–60 Hz) but using special high-capacitance electrodes. By contrast, microwaves operate at much higher, but a restricted range of frequencies (typically 915 or 2450 MHz). Under these frequencies, water molecules respond to the applied field producing dipole rotation in addition to ionic motion.

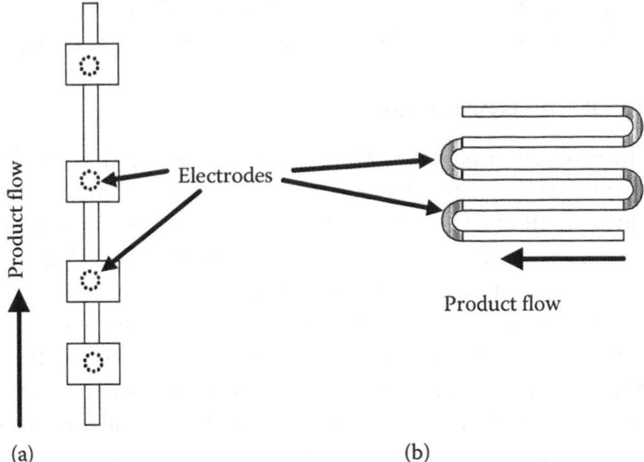


**FIGURE 4.1** Two designs of commonly used ohmic heating systems in industry practice (a) Front view (b) Lateral view.

### 4.2.1 Ohmic Heater Design

Ohmic heaters may be designed in many different ways, either for batch or continuous flow operation. The simplest arrangement involves a pair of parallel plates with the material either flowing (continuous) or being contained (batch) between them. However, in an industrial setting, most continuous flow devices fall into either the parallel plate, coaxial or in-line flow operation. Since many foods are of significant electrical conductivity, the two most common commercial embodiments involve the in-line electric field (Figure 4.1a and b). Other designs exist also, but it is not our intent to provide an exhaustive description here.

The design of ohmic heaters is still in its infancy, and many new and interesting design configurations are possible. Notably, most of the above designs are intended for continuous flow sterilization processing either of liquids, or more likely, solid–liquid food mixtures. Many future applications may be for other purposes, thus a plethora of design types may emerge as these develop.

### 4.2.2 Electrical Conductivity of Foods

Electrical conductivity of foods ($\sigma$) has been the subject of considerable research; principally because it shows sensitive dependence on the ingredient balance, and must be measured separately for each individual formulation. The major factors influencing $\sigma$ of a liquid food are temperature, frequency, and ingredients; for solids, the additional effect of microstructure must be considered; for solid–liquid mixtures, the relative proportions of each component and their individual variations also play significant roles. A detailed treatment of electrical conductivity is provided by Sastry (2005).

In general, it is noted that electrical conductivity increases with temperature; for liquids, the change tends to be linear, while for solids, the trend depends on cell structure. For solids with intact cells, breakdown of structure occurs as an electric field is applied; hence, electrical conductivity changes with temperature will depend on the presence or absence of this structure. For solids that have already been denatured by some prior thermal or electrical treatment, no further cell breakdown occurs, hence the electrical conductivity changes linearly with temperature, as for liquids.

### 4.2.3 Effect on Microorganisms

Historically, the efficacy of ohmic heating has been attributed to rapid and uniform heating alone, with no need for consideration of additional nonthermal effects of electricity. However, an emerging body of literature suggests that additional lethal nonthermal effects exist and are significant. If systematically verified, this approach affords the opportunity for significant process time reduction, adding to the advantages of ohmic heating. A review of some of this literature follows.

Palaniappan et al. (1990) reviewed the literature available at the time on the effect of electricity on microorganisms and concluded that much of the evidence was inconclusive. One difficulty in interpreting prior experimental data was the lack of adequate temperature control within experiments; thus it was not possible to conclude whether a given effect was due to thermal or electrical effects. Thereafter, it has been strongly recommended that studies purporting to study electrical effects ensure that control samples would be treated with the same thermal history as ohmic or moderate electric field (MEF) processed samples. This is a difficult task, but is possible to accomplish as shown in several papers. Thus in this review, we will specifically address whether or not a given study maintained equal temperatures.

Palaniappan et al. (1992) studied the death kinetics of yeast cells (*Zygo saccharomyces bailii*) under conventional and ohmic heating, and found no difference in the death rate. However, pretreatment by sublethal ohmic heating (now called MEF) caused a significant decrease in the subsequent death times of *Escherichia coli* vegetative cells at specific temperatures. This study was conducted with identical temperature histories for both ohmic and conventional treatments. However, with more sophisticated recent designs of kinetics equipment have resulted in more reliable data. Cho et al. (1999) studied the inactivation of *Bacillus subtilis* spores in buffer solution under conventional and ohmic heating, while maintaining identical temperature histories between ohmic and conventional cells. They found that thermal death times decreased significantly under ohmic heating conditions. Pereira et al. (2007) studied inactivation kinetics of *E. coli* in goat's milk, and *Bacillus licheniformis* ascospores in cloudberry jam during ohmic and conventional heating under identical temperature histories, and found that the thermal death times were shortened under ohmic heating. This shows that the results obtained in buffer media still hold within a food matrix.

A study on yeast cells by Yoon et al. (2002) showed that intracellular materials from *Saccharomyces cerevisiae* were exuded during ohmic heating with field strengths from 10 to 20 V/cm. Exudation was found to increase with field strength and frequency. It was concluded that ohmic heating induced electroporation, causing irreversible damage to cell membranes.

Sun et al. (2006) studied the death rates of *Streptococcus thermophilus* in milk, using conventional and combination (sublethal ohmic + conventional) treatments. Combination treatments were performed by ohmically heating samples from 10°C (50°F) to 42°C (108°F) for 12 successive cycles followed by the same conventional heating treatment as the purely conventionally treated samples. Inactivation of

*S. thermophilus* was significantly enhanced by the combination treatment, suggesting a nonthermal effect of electricity.

Leizerson and Shimoni (2005a) reported that ohmic heating resulted in high-quality orange juice, which was sensorially indistinguishable from fresh juice while completely (in their tests) inactivating bacteria, yeast, and molds, and reducing pectin esterase activity by 98%. Further studies by Leizerson and Shimoni (2005b) showed that the sensory shelf-life of ohmically processed juice was twice that of conventionally pasteurized juice. These results indicate the advantages of ohmic heating, but because of lack of identical temperature treatments, it is not possible to conclude if the stated advantages are due to rapid heating or to the electric field per se.

A study on mold (*Aspergillus niger*) in tomato by Yildiz and Baysal (2006) suggested some enhancement by ohmic heating. However, the lack of temperature control during experiments makes it difficult to separate thermal and nonthermal effects.

Even when ohmic heating is counteracted by cooling, electric fields as low as 1–2 V/cm have been observed to influence bacterial cells. Cho et al. (1996) studying fermentation of *Lactobacillus acidophilus*, found the lag phase to be significantly altered by MEF treatment. Subsequently, Loghavi et al. (2007, 2008) have not only confirmed these results, but also shown the sensitive influence of waveforms and frequency. While these studies are not specifically related to microbial inactivation, they serve to show that even mild electric fields have significant effects on cells.

## 4.2.4 Sterilization Applications

### 4.2.4.1 Continuous Sterilization

An attractive feature of ohmic heating is its capability of heating solid materials uniformly by internal generation, thus enabling the prospect of continuous flow sterilization of solid–liquid food mixtures, followed by aseptic packaging. The rate of heating of solids and liquids is dependent upon the electrical conductivity of the respective phases. Ideally, if the electrical conductivities of solid and liquid phases are equal, the heating rate is remarkably uniform, with the solid components often heating slightly faster than the liquid. This results in the interesting phenomenon of fluid temperatures rising during passage through the holding tube; something not encountered in conventional aseptic processing of particulate foods.

Often, many solid foods are of lower electrical conductivities than liquid phases, which may have dissolved salts which increase their electrical conductivity. Solids electrical conductivity may be increased by addition of salts or acids, thereby increasing ionic content and conductivity. This may be accomplished by marination (a slow process) or blanching in salt solutions. A key consideration is the optimization of product formulation such that product organoleptic properties are not compromised as a result of this method of formulation. This latter point has been addressed by Sarang et al. (2007), wherein the electrical conductivities of various vegetable and chicken components were increased by blanching within a soy sauce; followed by reincorporation of the blanching sauce into the final product to keep the composition the same as a conventional formulation. Sensory evaluation of the product revealed no loss of quality as a result of the formulation.

In some instances, solid components cannot easily be made more conductive; e.g., chicken. Under such conditions, it becomes necessary to consider and understand the rate of heating of each phase in the mixture. When electrical conductivities of solids are lower than that of the fluids, the relative heating rates depend on the electrical conductivity of the respective phases, the volume fraction of low-conductivity solids in the mixture, and the size of the nonconductive particle(s) in relation to overall system dimensions. It is important to understand these relationships, since the worst-case scenarios in ohmic heating differ considerably from conventional heating.

When only one low-conductivity particle is present in a high-conductivity fluid medium, and the particle dimensions are not of the same order as the system dimensions; the fluid typically heats faster than the particle. Temperature data from a sample experiment for such a case are plotted versus time in Figure 4.2 (the two top left curves) to demonstrate this point. The reason for this is that the conductive fluid provides a low-resistance parallel pathway enabling a large current density to pass through the fluid. While some of the current passes through the particle, it is of lower

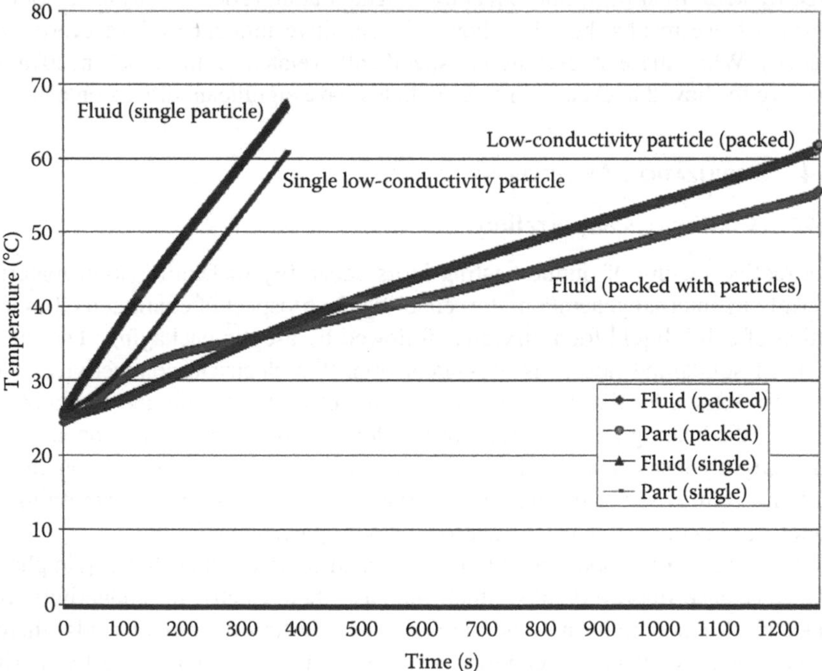


**FIGURE 4.2** Experimental data illustrating the effect of particle and fluid electrical conductivities and volume fractions on relative heating rates. Top left curves: Fluid with a single low-conductivity particle; the fluid heats faster than the particle, since it allows a low-resistance current pathway and thus carries a higher current density. Bottom right curves: High-conductivity fluid packed with low-conductivity particles. Note the slower heating rate, since the whole "circuit" is of high resistance. Note also that the particles now heat faster than the liquid, since the conduction path through the fluid has become tortuous, thus the current density in the fluid is now lower than through the particles.

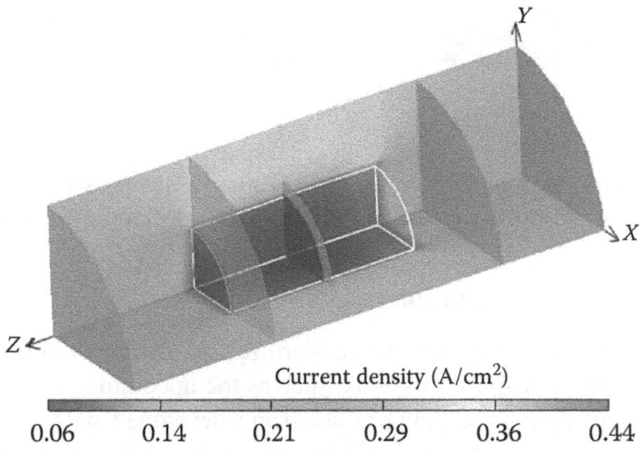


**FIGURE 4.3** Simulation results of current density in a particle of low-electrical conductivity suspended in a medium of high-electrical conductivity. (From Salengke, S. and Sastry, S.K., *J. Food Engr.*, 83, 324, 2007. With permission.) Simulation shows a one-quarter section of a tube within which the cylindrical particle is centered and aligned on the tube axis.

density and does not heat the particle to a significant extent. This is demonstrated by simulation results of Salengke (2000) shown in Figure 4.3.

However, when the fluid is packed with particles that are all of lower electrical conductivity, two notable changes occur, as shown in Figure 4.2 (two bottom right curves). Firstly, the heating rate of the entire fluid–solid mixture slows down, because now, the "circuit" is one of the high overall resistances, permitting lower current. It is possible to make the system heat faster, but will require a higher voltage. Secondly, the particles now begin to heat faster than the fluid. This is due to the now greater *tortuosity* of the current path through the fluid, which ensures that there is no longer a preferred pathway through either the fluid or the particles. This point has been shown previously by Sastry and Palaniappan (1992).

Another effect that is poorly understood, but nevertheless important is the role of fluid motion and mixing. When a particle is less conductive than a fluid, and is being heated by the fluid, its heating rate is actually *slower* when the fluid is *agitated* than if it is static. From the standpoint of conventional heat transfer, this may appear counterintuitive; nevertheless it is a prime example of how ohmic processing differs from conventional heating. The reason is that the electric field concentration near the particles creates hot zones around the particle (De Alwis and Fryer, 1990). These hot zones experience runaway heating in a static situation; however, an agitated system tends to dissipate the hot spots and actually slow down the heating of the particle. This effect has been discussed previously by Khalaf and Sastry (1996) and Salengke and Sastry (2007).

Other interesting effects occur if particles are of large dimensions, approaching those of the containing heater vessel under static conditions. Under these conditions, low-conductivity particles may heat rapidly, and high-conductivity particles may heat slowly. This has been demonstrated by the simulation studies of De Alwis and Fryer (1990).

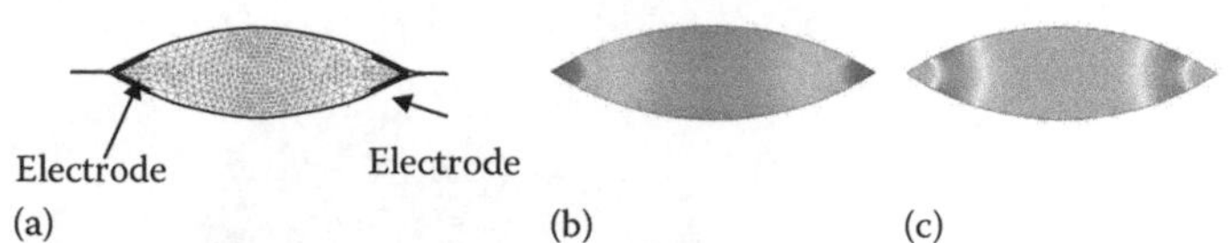


**FIGURE 4.4** Preliminary design of an ohmic heating pouch (a) showing electrodes and computational mesh, (b) electric field distribution, and (c) temperature distribution.

#### 4.2.4.2 In-Package Sterilization

An emerging application for in-package sterilization involves ohmic heating. In a project for long-duration space missions, such as the upcoming lunar base and Mars missions, ohmic heating food pouches are being developed with electrodes which permit reheating of foods contained within them. After food consumption, the packages may be reused to contain and sterilize waste. Packages for this purpose have been designed and optimized, and consist of electrodes on opposite ends of the package, with a means of electrical communication through the system walls. The optimization of these pouches has been the subject of detailed study (Jun and Sastry, 2005, 2007; Jun et al., 2005, 2007). In particular, it has been found that electrodes that conform to shapes of existing pouches result in significant nonuniformity in the field and temperature distributions. In particular, cold zones are found near the V-shaped pouch edges, and hot spots occur near electrode edges (Figure 4.4). This has necessitated a redesign of the pouch to a more rectangular shape, with flat parallel electrodes. Currently, the latest pouch design is undergoing testing to verify food sterilization capability. Capability of waste sterilization and food reheating has already been verified.

An important consideration for space missions is the equivalent system mass (ESM). At last estimate, it costs US$22,000 per kg that is placed in orbit. At such prices, it is essential not only that items be extremely lightweight, but also meet other restrictions, such as energy use and volume restrictions. Each of these factors is provided a weighting factor and the end calculation is the ESM. The ohmic system has been found in recent calculations to have a lower ESM than conventional modes of heat transfer (Pandit et al., 2008).

### 4.2.5 Electrolytic Effects

A question that has long been asked about ohmic heating is whether or not electrolytic effects exist that might impact the food product. It is well known from the electrochemistry literature that when DC power supplies are used, hydrogen and oxygen are produced at opposite electrodes, due to the electrolysis of water. Further, over time, metal components may dissolve within the product.

The key to prevention of electrolysis is the use of high-capacitance electrodes or frequencies that prevent Faradaic current from taking place. A number of electrode materials are now available that permit operation at 50 Hz (Amatore et al., 1998). Also, the advent of solid-state power supplies in the late 1990s has dramatically reduced the cost of supplying high-frequency power. Samaranayake et al. (2005) have shown that electrolytic effects can be controlled by careful selection of pulse waveforms in this manner.

### 4.2.6 Other Applications

Although the principal interest in ohmic heating has been in sterilization, there are a host of other potential emerging applications, particularly arising from the nonthermal effects of electricity. It is not within our scope to provide a comprehensive coverage of these applications, but we will mention these here.

Ohmic heating has been shown to increase exudation of intracellular liquids from cellular food materials, e.g., vegetables. This has led to the prospect that it could be used as a pretreatment for drying and juice expression (Wang and Sastry, 2000, 2002). Other potential uses include the detection of starch gelatinization (Wang and Sastry, 1997; Karapantsios et al., 2000; Chaiwanichsiri et al., 2001), blanching (Sensoy and Sastry, 2004a), and extraction (Sensoy and Sastry, 2004b). These are just a few examples of the many interesting applications of this emerging technology.

## 4.3 MICROWAVE HEATING OF FOODS

The use of microwaves to heat foods has become common in the American household. The microwave oven has become a normal appliance into all kitchens, being found in many variations, sizes, and powers. Microwave is also used in the industry, especially in tempering frozen foods and cooking of solid foods, and recently continuous flow microwave heating of fluids has emerged as a viable alternative for thermal processing.

Microwaves are a part of the electromagnetic spectrum that comprises frequencies between 300 and 3000 MHz. Microwaves find their most widespread use in communications, radar, and medical devices, thus in the United States the frequencies in which these microwave ovens operate have been regulated by the Federal Communication Commission. The allocated frequencies for industrial, medical, and household applications are 915 ± 13, 2,450 ± 50, 5,800 ± 75, and 24,125 ± 125 MHz for industrial, scientific, and medical applications (47CFR18.301, 2004). The 2450 MHz frequency is used mostly in household microwave ovens, while 915 MHz is used in industrial applications.

The heating of foods using microwaves relies on the generation of heat inside the food by the conversion of the electromagnetic energy from the microwaves into heat, and has several characteristics that differentiate it from conventional heating, such as

- Power can be turned on and off instantly
- It has very rapid dynamics
- It does not rely on contact with hot surfaces or a hot medium or electrodes
- It is selective, i.e., different materials, or portions of the same food material having different properties will heat at different rates
- It is volumetric, thus theoretically more uniform than conventional heating

With all those considerations, microwave heating is a promising technology for industrial heating of food materials that are difficult to heat conventionally such as viscous liquids and heterogeneous products. However, validation of processes using continuous flow microwave and observation supporting the claim of uniformity of heating still need to be performed in order to gain acceptance.

The development of industrial microwave heating of liquid foods has faced several challenges and setbacks over the years, mainly due to observed lack of uniformity of heating of food materials in household ovens. Drozd and Joines (1999, 2000, 2001, 2002) developed a continuous flow heating system, which has been manufactured and commercialized by Industrial Microwave Systems (Morrisville, North Carolina). Such system became the first commercial installation for shelf-stable low-acid pumpable food products processed using continuous flow microwave heating in 2008.

Heating using microwaves relies on the conversion of electromagnetic energy into heat, which differentiates it from conventional heating. Thus, the mechanisms of heating of foods when exposed to a microwave field need to be investigated in more detail (Metaxas and Meredith, 1988; Datta and Ramaswamy, 2001).

### 4.3.1 Heating Mechanisms of Foods

Foods are considered dielectric materials, which are materials that are neither conductors nor insulators. Dielectric materials heat when exposed to a microwave energy field; the rapidly changing electromagnetic field tries to polarize the charged molecules (ions, dipoles, quadrupole) present in the dielectric, however, the material molecules cannot follow this rapid polarization and thus some of that energy is lost in the form of heat (Metaxas and Meredith, 1988). It is generally accepted that water and ions (polar molecules) are responsible for the ohmic loss of microwave energy within a food, as stated in Equation 4.2 (Datta and Ramaswamy, 2001). This statement may also be related to the mobility of ions, which generates internal friction.

$$Q = \frac{\omega \varepsilon_0 \varepsilon''}{2} \left| E^2 \right| \tag{4.2}$$

where ε and ε'' are a representation of the dielectric properties of the material to be heated.

### 4.3.2 Dielectric Properties of Foods

The dielectric properties of the material, introduced in Equation 4.1 are the properties that relate the ability of the food materials to be heated using microwaves. Food materials belong to the group of dielectric materials, which are materials that are neither conductors, nor insulators. Electromagnetic waves propagate into dielectric materials but the amplitude of the waves decreases, and the energy that is lost by the waves is converted into heat inside the material. Dielectric properties of food materials can be written as shown in Equation 4.3, such that

$$\varepsilon_c = \varepsilon \left[ 1 - j\frac{\sigma}{\omega \varepsilon} \right] = \varepsilon' - j\varepsilon'' \tag{4.3}$$

The complex permittivity of a dielectric material consists of two parts, a real part and an imaginary part, which are generally expressed as factors of the permittivity of free space ($\varepsilon_0$). The real part, called dielectric constant ($\varepsilon'$), relates to the amount of energy that is reflected or transmitted by the material and to the ability of the material to store electromagnetic energy. The imaginary part, called loss factor ($\varepsilon''$), relates to the ability of the material to convert the electromagnetic energy into heat (Englender and Buffler, 1991). The conversion of the energy into heat is usually described as a function of frequency ($f$), electric field intensity (**E**), and loss factor, as shown in Equation 4.4 (Metaxas and Meredith, 1988; Buffler, 1993)

$$P = 2\pi f \mathbf{E}^2 \varepsilon_0 \varepsilon'' \tag{4.4}$$

It is, however, very difficult to predict the values of **E** in a practical manner. The amount of energy absorbed by food materials needs to be analyzed based on the value of the reflection and transmission coefficients. The value of these coefficients can only be analyzed if the values of dielectric properties can be predicted as a function of temperature, frequency, and composition of the food.

The prediction of dielectric properties based on the composition of the food materials is a subject that has attracted attention by academia and industry. The classic approach to the dielectric behavior of liquids in an electromagnetic field comes from the classic work of Debye (1929). Debye analyzed the behavior of a polar molecule in a nonpolar solvent as spheres suspended in a viscous liquid. From Debye's work Equation 4.5 was derived, where $\varepsilon_s$ and $\varepsilon_\infty$ are the dielectric constants at DC and very high frequencies, respectively, and $\tau$ is the relaxation time of the system that control the amount of energy that is converted into heat (Metaxas and Meredith, 1988).

$$\varepsilon = \varepsilon' - j\varepsilon'' = \varepsilon_\infty + \frac{\varepsilon_s - \varepsilon_\infty}{1 + j\omega\tau} \tag{4.5}$$

The dielectric constant at very high frequencies is independent of frequency, and a transition should occur at the relaxation frequency (Equation 4.6). At this frequency the loss factor should have a maximum. In the case of pure water, the static dielectric constant has a value of ~80, while the high-frequency dielectric constant has a value of 4.3, and the relaxation frequency is located in the vicinity of $10^{10}$ Hz (Kaatze, 1989).

$$f_{rel} = 1/(2\pi\tau) \tag{4.6}$$

The relaxation time is dependent on the viscosity of the system, and has been modeled using the Stokes equation. This model has been used extensively, even though it is based on simple interactions. Hasted (1948) modified the above equation to account for losses due to conductivity of the material ($\sigma$) which adjusted very well to aqueous solutions of ions, as observed by Stogryn (1971). Stogryn observed that the dielectric constant of salt water decreased and the loss factor increased with an increase in temperature. This observation was supported with a decrease in relaxation time and a decrease in the static dielectric constant with an increase in temperature. These observations were explained by Hasted (1948) as a result of the changes in internal energy and relaxation times of water with temperature. Since the main component responsible for microwave heating of foods is water, studies in aqueous dielectrics are relevant to food processing. Hill (1980) reviewed the behavior of dielectric materials, and concluded that a modified equation, like the Collie–Hasted–Ritson equation (Equation 4.7) in which correction parameters are added is a very good model for such materials. Debye parameters are found in reference books for many solvents, but generally for food products these parameters have to be estimated or generated.

$$\begin{aligned}\varepsilon' &= \varepsilon_{\infty} + \frac{\varepsilon_s - \varepsilon_{\infty}}{1+\omega^2\tau^2} \\ \varepsilon'' &= \omega\tau\frac{(\varepsilon_s - \varepsilon_{\infty})}{1+\omega^2\tau^2} + \frac{\sigma}{\omega\varepsilon_0}\end{aligned} \tag{4.7}$$

Composition is a very important factor in the dielectric behavior of any food material. However, foods have many complex components. Due to this complex nature of the components of a food material simple approaches like the Debye–Hasted equation are only valid for relatively simple food materials, such as milk or apple juice (Mudgett, 1974, 1986). Several researchers have tried to correlate the composition of foods to its dielectric properties. Water content, salt concentration, protein, carbohydrate, and fat contents have been used among other factors. Kudra et al. (1991) analyzed the characteristics of milk constituents in microwave heating. Funebo and Ohlsson (1999) made a prediction of the properties of fruits and vegetables as a function of temperature, bulk density, and water content. Their findings showed that other factors must be taken into account, and that the predictive equations failed to describe the products they targeted. Kent et al. (2000, 2001) measured the dielectric properties of meat, poultry, and fish products, and developed a method to detect the addition of water using dielectric properties. The method was able to determine liquid uptake, salt content, and protein, but had problems when more than one type of salt was added to the product.

Dielectric properties are a macroscopic effect of molecular interactions with the electromagnetic field, and temperature affects molecules by increasing the internal vibration of their bonds, elongating such bonds, increasing the mobility of molecules and their internal energy. Due to such changes in mobility of polar molecules dielectric properties have to be affected by temperature. Dependence on temperature of relaxation time follows an Arrhenius-type kinetics, which should allow the prediction of properties using equations. However, due to the complexity of the food matrixes experimental data are required.

Available data of dielectric properties of food materials (Nelson 1973, 1991, Nelson and Bartley 1994, Nelson et al. 2000; Kent 1987) showed that most liquid products follow the behavior observed in water dielectrics. The dielectric properties of solid foods were thought to behave in a similar manner, being water and minerals (ash) the main responsible for dielectric heating. However, it was observed that solid foods showed a different behavior than aqueous solutions. Datta et al. (1997) studied the dependence on temperature and composition of several food products, as a function of water and ash fraction. Their results confirm that water and ash contents are not enough to determine the dielectric properties of foods. It has been observed that foods containing hydrocolloids present differences in the dielectric properties behavior. Hydrocolloids are polymers, and have two characteristics that affect dielectric properties, water-binding capability, and that during heating phase changes occur. By binding water, the mobility of ions in the solution is restricted, thus the loss factor is affected. During changes of phase the dielectric properties of the whole material undergo changes, which are noticeable and have been observed by Rozzi and Singh (2000) when studying starch solution and by Mashimo et al. (1996). Mashimo et al. observed the dielectric properties of gellan gum solutions with temperature, when the solution goes through a phase transition the dielectric properties and relaxation time present a sudden change, quite similar to the one observed on the specific heat by differential scanning calorimetry (DSC) during such transitions.

Experimental data of dielectric properties are available in literature, with compilations like the ones by Kent (1987), Tinga and Nelson (1973), Nelson (1973, 1991, 1994, 2000), Datta and Ramaswamy (2001), and Funebo and Ohlsson (1999). It can be observed that food materials comprise a wide range of dielectric properties, thus making the development of a universal application apparatus a difficult task. More data are required, especially in the frequencies of industrial interest (915 MHz), which have been neglected in favor of the more widespread frequencies of household interest (2450 MHz).

### 4.3.3 Applications in the Food Industry

The food industry slowly accepts microwave heating, both as batch or continuous, as one of the methods available for processing. The applications that have initially succeeded in being adopted widely are shown in Table 4.1.

It is noticeable that most applications refer to solid foods, which can be conveyed by a belt or on a fluidized bed with no mention of fluid foods. Applications for continuous processing of fluid food material were researched by the North Carolina State University after Drozd and Joines (1999) invented a system especially designed for that purpose. The system relies on single-mode cavity applicators, such that the microwave energy is focused in one of the focal points of an elliptical applicator and a tube is placed so that it will be exposed to such focused field (Figure 4.5). Uniform heating was expected with this method, and Coronel (2005) investigated on this method to the point of making the technology commercially available.

**TABLE 4.1**
**Current Applications of Microwave Heating in the Food Industry**

| Product | Unit Operation |
|---|---|
| Pasta, onions, herbs | Drying |
| Bacon, meat patties, chicken, fish sticks | Cooking |
| Fruit juices | Vacuum drying |
| Frozen foods | Tempering |
| Surimi | Coagulation |
| Frozen vegetables | Blanching |

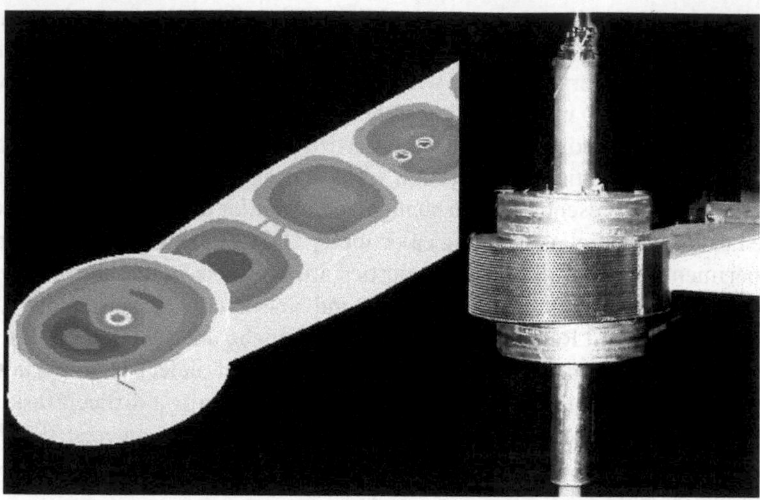

**FIGURE 4.5** Theoretical model of a single-mode cavity and actual continuous flow microwave heating system. (From Industrial Microwave Systems, Morrisville, NC. With permission.)

In the case of single-mode applicators, the focused field has a maximum in the center of the tube, and a minimum close to the walls of the tube, which is similar to the predicted laminar flow profile for Newtonian liquids. Theoretically, the streamlined flow closer to the walls would receive lower energy for longer times and the center would receive higher energy for shorter times, thus resulting in a uniform temperature profile. Experimental confirmation of this hypothesis was carried out using milk as test fluid in a 5 kW pilot scale microwave heating unit by Coronel et al. (2003). The resulting temperature field in the cross-sectional area of a 39 mm (1.5") tube showed that differences between temperatures in the cross-sectional area were relatively small, and the maximum temperature was always in the center of the tube as shown in Figure 4.6 (Coronel et al., 2003). Thus confirming the hypothesis of uniform temperature and opening the way for this technique in heat treatment of food products. Further research in the turbulent or plug flow regimes showed a need to use static or dynamic mixers to achieve perfect homogeneity due to the mismatch of the electromagnetic profile and flow profile as observed in Figure 4.7 (Coronel et al., 2005).

Due to the above-mentioned factors, low-acid foods that have high viscosity and which dielectric properties are within a range still under investigation are potential candidates to be processed using microwaves. This product category comprises fruit and vegetable purees, dressings, and sauces. Coronel et al. (2005) successfully processed and aseptically packaged sweet potato puree, obtaining a product that was shelf stable for more than 180 days and that showed increased retention of nutrients when compared to canned product. The process was later validated by Brinley et al. (2007) by utilizing pouches filled with bioindicators. The pouches were minimally heated by the microwave, thus a conservative validation was achieved, since the heat transfer was only convective heat transfer from the flow to the pouches. As a result

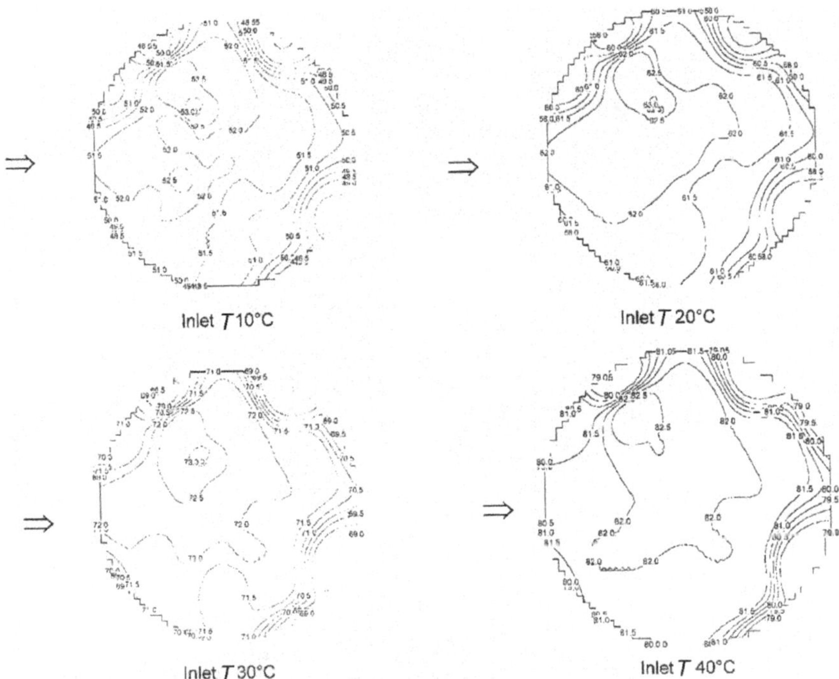


**FIGURE 4.6** Temperature profiles while heating milk in a continuous flow microwave system at different inlet temperatures. The arrow denotes the side of the tube that was directed to the waveguide.

of this research, YAMCO (Snow Hill, North Carolina) decided to invest in a plant to aseptically process nonacidified sweet potato puree without addition of water or preservatives. This facility started operating in 2008 and is the first of such operations known to the authors.

Kumar et al. (2007) presented a study of processing "Salsa con Queso" using continuous flow microwave and successfully processed the product to commercial sterility thus opening the field for particulate thermal processing using continuous flow microwave. Products containing larger particulates could also be processed using microwaves; the size and shape of the particulates may influence the process but this effect would be of little significance if the dielectric properties of the fluid and the particulates are close to one another. It would also be possible to process particulates with dielectric properties that allow for better absorption of microwaves than the liquid, since the particulates would heat more than the liquid, thus having higher thermal treatment. However, it still remains to be seen how the regulatory agencies would react to such process, and also experimentation confirmation would still be required.

Modeling of continuous flow microwave heating is a need for the future of this technology, since models could provide insight into the process that is almost impossible to attain experimentally. The modeling of electromagnetic fields is

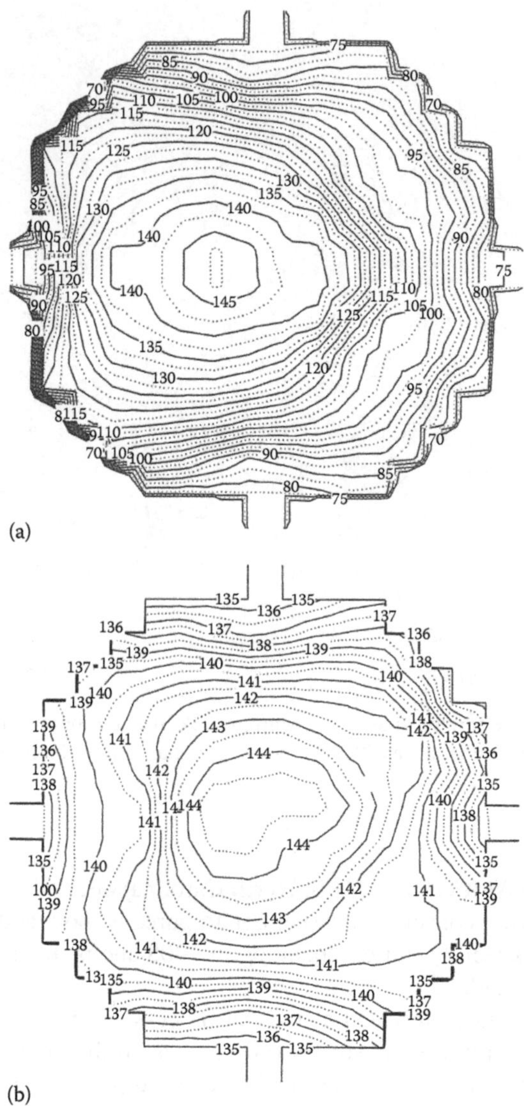


**FIGURE 4.7** Temperature profiles while heating sweet potato puree in a continuous flow microwave system, without mixers (a) and with mixers (b). The arrow denotes the side of the tube that was directed to the waveguide.

possible using a few commercial packages, even when full 3D simulation of the waves is required. However, combining heat transfer, fluid mechanics, and microwaves requires multiphysics programs that allow for the combined modeling, such programs are scarce and to our best knowledge only FLUENT (Ansys Inc., Cannonburg, Pennsylvania) and COMSOL (Comsol AB, Stockholm, Sweden) may provide opportunities for such modeling but still require further development (Datta, 2008).

### 4.3.4 Concluding Remarks

Continuous flow processing of foods using microwaves is a viable alternative for low-acid foods. Homogeneous viscous foods have been tested successfully and the first FDA compliant commercial operation started production in 2008.

Modeling and simulation of microwave is a requirement for further development of this technology and the current commercially available software is still developing to meet this demand. In order to further open the field to heterogeneous product, more research in products with particulates is needed.

## REFERENCES

47CFR18.301 Code of Federal Regulations (CFR) Title 47 Telecommunications, Part 18 Industrial scientific and medical equipment, Section 301 Operating frequencies. Office of the Federal Register, Washington, DC, 2004.

C. Amatore, M. Berthou, and S. Hebert. 1998. Fundamental principles of electrochemical ohmic heating of solutions. *J. Electroanal. Chem.* 457:191–203.

T. Brinley, C.N. Dock, V.D. Truong, P. Coronel, P. Kumar, J. Simunovic, K.P. Sandeep, G.D. Cartwright, K.R. Swartzel, and L.A. Jaykus. 2007. Feasibility of utilizing bioindicators for testing microbial inactivation in sweet potato purees processed with a continuous-flow microwave system. *J. Food Sci.* 72(5):E235–E242.

C. Buffler. *Microwave Cooking and Processing Engineering Fundamentals for the Food Scientist*, AVI Book, New York, 1993.

S. Chaiwanichsiri, S. Ohnishi, T. Suzuki, R. Takai, and O. Miyawaki. 2001. Measurement of electrical conductivity, differential scanning calorimetry and viscosity of starch and flour suspensions during gelatinisation process. *J. Sci. Food Agric.* 81:1586–1591.

H-Y. Cho, A.E. Yousef, and S.K. Sastry. 1996. Growth kinetics of *Lactobacillus acidophilus* under ohmic heating. *Biotechnol. Bioeng.* 49:334–340.

H-Y. Cho, A.E. Yousef, and S.K. Sastry. 1999. Kinetics of inactivation of *Bacillus subtilis* spores by continuous or intermittent ohmic or conventional heating. *Biotechnol. Bioeng.* 62(3):368–372.

P. Coronel. Continuous flow processing of foods using cylindrical applicator microwave systems operating at 915 MHz, Doctoral dissertation under supervision of K.P. Sandeep, North Carolina State University, Raleigh, NC, 2005.

P. Coronel, J. Simunovic, and K.P. Sandeep. 2003. Thermal profile of milk after heating in a continuous flow tubular microwave system operating at 915 MHz. *J. Food Sci.* 68(6):1976–1981.

P. Coronel, V.D. Truong, J. Simunovic, K.P. Sandeep, and G.D. Cartwright. 2005. Aseptic processing of sweetpotato purees using a continuous flow microwave system. *J. Food Sci.* 70(0):E531–E536.

A.K. Datta. 2008. Status of physics-based models in the design of food products, processes and equipment. *Comprehensive Reviews in Food Science and Food Safety* 7:121–129.

A.K. Datta and C.A. Ramaswamy (Eds.). *Handbook of Microwave Technology for Food Applications*, Marcel Dekker, New York, 2001.

A.K. Datta, S. Barringer, and M.T. Morgan. Effects of composition and temperature on dielectric properties of foods at 2450 and 27 MHz. In: *Conference on Food Engineering of the American Institute of Chemical Engineers 1997*, November 22–26, Los Angeles, CA, pp. 14–20, 1997.

A.A.P. De Alwis and P.J. Fryer. 1990. A finite element analysis of heat generation and transfer during ohmic heating of food. *Chem. Eng. Sci.* 45(6):1547–1559.

P.J.W. Debye. *Polar Molecules.* Chemical Catalog Company. New York. 1929.

M. Drozd and W.T. Joines. Electromagnetic exposure chamber for improved heating. U.S. Patent 5,998,774, 1999.

M. Drozd and W.T. Joines. Method and apparatus for rapid heating of fluids. U.S. Patent 6,121,594, 2000.

M. Drozd and W.T. Joines. Electromagnetic exposure chamber with a focal region. U.S. Patent 6,265,702, 2001.

M. Drozd and W.T. Joines. Cylindrical reactor with an extended focal region. U.S. Patent 6,797,929, 2002.

D.S. Englender and C.R. Buffler. 1991. Measuring dielectric properties of food products at microwave frequencies. *Microwave World* 12(2):6–14.

T. Funebo and T. Ohlsson. 1999. Dielectric properties of fruits and vegetables as a function of temperature and moisture content. *Journal Microwave Power Electromagnetic Energy* 34(1):42–54.

J.B. Hasted. *Aqueous Dielectrics*, John Wiley & Sons, New York, 1948.

N.E. Hill. 1980. The behavior of a dielectric obeying the Collie, Hasted and Ritson equation. *J. Phys. C Solid State Phys.* 13:6273–6277.

S. Jun and S.K. Sastry. 2005. Modeling and optimization of ohmic heating of foods inside a flexible package. *J. Food Proc. Eng.* 28(4):417–436.

S. Jun and S.K. Sastry. 2007. Reusable pouch development for long term space mission: 3D ohmic model for verification of sterility efficacy. *J. Food Eng.* 80:1199–1205.

S. Jun, B.F. Heskitt, S.K. Sastry, R. Mahna, J.E. Marcy, and M.H. Perchonok. 2005. Reheating and sterilization technology for food, waste and water: Design and development considerations for package and enclosure. Paper # 2005-01-2926, Proceedings of the 34th International Conference on Environmental Systems, Society of Automotive Engineers.

S. Jun, S.K. Sastry, and C. Samaranayake. 2007. Migration of electrode components during ohmic heating of foods in retortable pouches. *Innov. Food Sci. Emerg. Technol.* 8:237–243.

U. Kaatze. 1989. Complex permittivity of water as a function of frequency and temperature. *J. Chem. Eng. Data.* 34:371–374.

T.D. Karapantsios, E.P. Sakonidou, and S.N. Raphaelides. 2000. Electric conductance study of fluid motion and heat transport during starch gelatinization. *J. Food Sci.* 65:144–150.

M. Kent. *Electrical and Dielectric Properties of Food Materials: A Bibliography and Tabulated Data*, Science and Technology Publishers, London, 1987.

M. Kent, R. Knochel, F. Daschner, and U.K. Berger. 2000. Composition of foods using microwave dielectric spectra. *Eur. Food Res. Technol.* 210:359–366.

M. Kent, R. Knochel, F. Daschner, and U.K. Berger. 2001. Composition of foods including added water using microwave dielectric spectra. *Food Control* 12:467–482.

W.G. Khalaf and S.K. Sastry. 1996. Effect of fluid viscosity on the ohmic heating rates of liquid-particle mixtures. *J. Food Eng.* 27(2):145–158.

T. Kudra, F.R. Van de Voort, G.S.V. Raghavan, and H.S. Ramaswamy. 1991. Heating characteristics of milk constituents in a microwave pasteurization system. *J. Food Sci.* 56(4):931–937.

P. Kumar, P. Coronel, J. Simunovic, and K.P. Sandeep. 2007. Feasibility of aseptic processing of a low-acid multiphase food product (salsa con queso) using a continuous flow microwave system. *J. Food Sci.* 72(3):E121–E124.

S. Leizerson and E. Shimoni. 2005a. Effect of ultrahigh temperature continuous flow ohmic heating on fresh orange juice. *J. Agric. Food Chem.* 53:3519–3524.

S. Leizerson and E. Shimoni. 2005b. Stability and sensory shelf life of orange juice pasteurized by continuous ohmic heating. *J. Agric. Food Chem.* 53:4012–4018.

L. Loghavi, S.K. Sastry, and A.E. Yousef. 2007. Effect of moderate electric field on the metabolic activity and growth kinetics of *Lactobacillus acidophilus. Biotechnol. Bioeng.* 98(4):872–881.

L. Loghavi and S.K. Sastry. 2008. Effect of moderate electric field frequency on growth kinetics and metabolic activity of *Lactobacillus acidophilus*. Biotechnol. Progress 24(1):148–153.

S. Mashimo, N. Shinyashiki, and Y. Matsumura. 1996. Water structure in gellan gum-water system. *Carbohydr. Polym.* 30(2/3):141–144.

A.C. Metaxas and R.J. Meredith. *Industrial Microwave Heating*. Peter Peregrinus, London, 1988.

R.E. Mudgett. 1986. Microwave properties and heating characteristics of foods. *Food Technol.* 40(6):84–93.

R.E. Mudgett, A.C. Smith, D.I.C. Wang, and S.A. Goldblith. 1974. Prediction of dielectric properties in nonfat milk at frequencies of interest in microwave processing. *J. Food Sci.* 39(1):52–54.

S.O. Nelson. 1973. Electrical properties of agricultural products (a critical review). *Trans. ASAE.* 16(2):384–400.

S.O. Nelson. 1991. Dielectric properties of agricultural products. *IEEE Trans. Elect. Insul.* 26(5):845–869.

S.O. Nelson and P.G. Bartley. 2000. Measuring frequency- and temperature-dependent dielectric properties of food materials. *Trans. ASAE.* 43(6):1733–1736.

S.O. Nelson, W.R. Forbus, and K.C. Lawrence. 1994. Microwave permittivities of fresh fruits and vegetables from 0.2 to 20 GHz. *Trans. ASAE.* 37(1):183–189.

S. Palaniappan, S.K. Sastry, and E.R. Richter. 1990. Effect of electricity on microorganisms: A review. *J. Food Proc. Pres.* 14:393–414.

S. Palaniappan, S.K. Sastry, and E.R. Richter. 1992. Effects of electroconductive heat treatment and electrical pretreatment on thermal death kinetics of selected microorganisms. *Biotech. Bioeng.* 39:225–232.

R.B. Pandit, R. Somavat, S. Jun, B.F. Heskitt, and S.K. Sastry. 2008. Development of a lightweight ohmic food warming unit for a Mars Exploration Vehicle. World of Food Science. http://www.worldfoodscience.org/pdf/WFS_OSU_group_submission.pdf accessed February 21.

C. Pereira, J. Martins, C. Mateus, J.A. Teixeira, and A.A. Vicente. 2007. Death kinetics of *Escherichia coli* in goat milk and *Bacillus licheniformis* in cloudberry jam treated by ohmic heating. *Chem. Pap.* 61(2):121–126.

N.L. Rozzi and R.K. Singh. 2000. The effect of selected salts on the microwave heating of starch solutions. *J. Food Process. Preserv.* 24:265–273.

S. Salengke. 2000. Electrothermal effects of ohmic heating on biomaterials: Temperature monitoring, heating of solid–liquid mixtures, and pretreatment effects on drying rate and oil uptake. PhD dissertation. The Ohio State University, Columbus, OH.

S. Salengke and S.K. Sastry. 2007. Experimental investigation of ohmic heating of solid–liquid mixtures under worst-case heating scenarios. *J. Food Eng.* 83:324–336.

C.P. Samaranayake, S.K. Sastry, and Q.H. Zhang. 2005. Pulsed ohmic heating—a novel technique for minimization of electrochemical reactions during processing. *J. Food Sci.* 70 (8):E460–E465.

S. Sarang, S.K. Sastry, J. Gaines, T. Yang, and P. Dunne. 2007. Product formulation for ohmic heating: Blanching as a method to improve uniformity in heating of solid–liquid food mixtures. *J. Food Sci.* E227–E234.

S.K. Sastry. 2005. Electrical conductivity of foods. Chapter 10 in *Engineering Properties of Foods*. M.A. Rao, S.S.H. Rizvi, and A.K. Datta, Eds. Marcel Dekker Inc., pp. 589–639.

S.K. Sastry and J.T. Barach. 2000. Ohmic and inductive heating. *J. Food Sci.* 65(8):42s–46s.

S.K. Sastry and S. Palaniappan. 1992. Mathematical modeling and experimental studies on ohmic heating of liquid–particle mixtures in a static heater. *J. Food Proc. Eng.* 15(4):241–261.

I. Sensoy and S.K. Sastry. 2004a. Ohmic blanching of mushrooms. *J. Food Proc. Eng.* 27(1):1–15.

I. Sensoy and S.K. Sastry. 2004b. Extraction using moderate electric fields. *J. Food Sci.* 69(1):FEP7–FEP13.

A. Stogryn. 1971. Equations for calculating the dielectric constant of saline water. *IEEE Trans. Microwave Theory Techniq.* 1971:733–736.

H. Sun, S. Kawamura, J.-I. Himoto, and T. Wada. 2006. Ohmic heating for milk pasteurization; effect of electric current on nonthermal injury to *Streptococcus thermophilus.* ASABE Paper No. 066025. American Society of Agricultural and Biological Engineers, St. Joseph, MI.

W.R. Tinga and S.O. Nelson. 1973. Dielectric properties of materials for microwave processing—tabulated. *J. Microwave Power* 8(1):23–65.

W.-C. Wang and S.K. Sastry. 1997. Starch gelatinization in ohmic heating. *J. Food Eng.* 34(3):225–242.

W.-C. Wang and S.K. Sastry. 2000. Effects of thermal and electrothermal pretreatments on hot air drying rate of vegetable tissue. *J. Food Proc. Eng.* 23(4):299–319.

W.-C. Wang and S.K. Sastry. 2002. Effects of moderate electrothermal treatments on juice yield from cellular tissue. *Innov. Food Sci. Emerg. Technol.* 3(4):371–377.

H. Yildiz and T. Baysal. 2006. Effects of alternative current heating treatment on *Aspergillus niger*, pectin methylesterase and pectin content in tomato. *J. Food Eng.* 75:327–332.

S.W. Yoon, C.Y.J. Lee, K.M. Kim, and C.H. Lee. 2002. Leakage of cellular materials from *Saccharomyces cerevisiae* by ohmic heating. *J. Microbiol. Biotechnol.* 12(2):183–188.

# 5 High-Pressure Thermal Processes: Thermal and Fluid Dynamic Modeling Principles

*Pablo Juliano, Kai Knoerzer, and Gustavo V. Barbosa-Cánovas*

**CONTENTS**

## 5.1 INTRODUCTION

High-pressure high-temperature (HPHT) treatment, also known as pressure-assisted thermal processing (PATP), is an emerging preservation method for the development of shelf-stable low-acid food products. HPHT involves combining pressures of 600–800 MPa with moderate initial chamber temperatures of 60°C–90°C, and takes advantage of the increasing process temperature during pressurization to eliminate spore-forming bacteria (Matser et al., 2004; Margosch, 2005). For instance, pressurization temperatures of 90°C–116°C combined with pressures of 500–700 MPa have been used to inactivate a number of strains of *Clostridium botulinum* spores (Farkas and Hoover, 2000; Margosch et al., 2004). Other researchers showed that certain bacterial endospores (*C. sporogenes*, *Bacillus stearothermophilus*, *B. licheniformis*, *B. cereus*, and *B. subtilis*) in selected matrices like phosphate buffer, beef, vegetable cream, and tomato puree (Gola et al., 1996; Raso et al., 1998; Rovere et al., 1998; Meyer et al., 2000; Balasubramanian and Balasubramaniam, 2003; Krebbers et al., 2003) can be eliminated after short-time exposure to temperatures and pressures above 100°C and 700 MPa, respectively.

Several HPHT combinations have been proposed for spore inactivation (de Heij et al., 2003; Leadley, 2005; Barbosa-Cánovas and Juliano, 2008), among which, the application of pressures >600 MPa with short holding times (5 min or less) seems to be the most economical for industrial purposes, as well as a safer choice. Shorter processing times would increase productivity, equipment lifetime, and reduce maintenance costs. Furthermore, a shorter thermal pressure process would provide a product with increased quality retention, making the process potentially more advantageous than conventional in-container sterilization (i.e., canning), especially in the manufacture of high-quality foods in large containers. In fact in other studies, HPHT treatment has been shown to produce foods with higher pigment, flavor, and nutrient retention among other improved sensory attributes (Krebbers et al., 2002; Matser et al., 2004; Juliano et al., 2006a,b; Juliano et al., 2007a; Barbosa-Cánovas and Juliano, 2008).

Design of thermal process operations requires models able to predict temperature distribution during treatment inside the processing unit. In particular, heat transfer models can be used to express the amount of heat received by the product during processing. These models can assist in a number of tasks, as listed in the following (Nicolaï et al., 2001; Peleg et al., 2005):

1. Prediction of temperature uniformity to identify potential "cold spots" in a food product or "cold" regions in the high-pressure chamber during treatment.
2. Identification of an expression or a process performance parameter (similar to the thermal death time or *F*-value; or cook value) that represents temperature distribution and magnitude related to a target temperature during HPHT processing:
    - To account for "commercial sterility" achieved*; or to account for microbiological stability state attained throughout the volume of prepackaged food contained in vessel
    - To predict quality degradation as a result of temperature (pressure) use

* Commercial sterility in this case refers to the inactivation of spoilage spore-forming bacteria to achieve extended shelf-life at room temperature storage. In thermal processing, spoilage spore-forming bacteria are more heat resistant than *C. botulinum*. During HPHT, target spoilage spores such as *B. stearothermophilus* have been shown to be less resistant than *C. botulinum* (Margosch et al., 2004).

3. Other HPHT process design aspects: thermal process evaluation based on container size and shape, food composition, equipment modification and optimization, scale-up studies, energy use, modification and optimization of process conditions, etc.
4. Temperature control and monitoring (e.g., at pressure fluid inlet, inside packages, and other critical locations inside the vessel)

For evaluation of conventional in-container sterilization processes, a heat penetration model is developed using a temperature profile of the product during heating, holding, and cooling phases at the "coldest" region in a retort and, specifically, at the "coldest" point inside the container located in that region (Larousse and Brown, 1997). Minimum temperature requirements for commercially sterilizing low-acid foods have been specified by the U.S. Food and Drug Administration (21 CFR 113) based on process validation data determined for *C. botulinum* after retort processing. However, microbial validation of an HPHT process remains a challenge, not only due to the unavailability of *C. botulinum* inactivation data for several strains at HPHT conditions (Margosch et al., 2004), but also because a process performance expression that accounts for the temperature and pressure history during the process needs to be identified.

Strategies for HPHT process microbial validation are currently under development by a few research groups in Australia, Europe, and the United States. However, few have clearly reported about these strategies in the public literature (Sizer et al., 2002). Table 5.1 shows a number of steps that can be used for validation of an HPHT process.

The HPHT process consists of (a) preheating food packages in a carrier outside the vessel, (b) transferring the preheated carrier into the vessel and equilibrating up to an initial temperature, (c) pressurizing and holding at a target pressure, (d) releasing pressure, (e) removing carrier from vessel, and (f) cooling down products in the carrier and removing the products. Therefore, the temperature history inside an HPHT-processed food is determined by six main process time intervals (Juliano, 2006; Barbosa-Cánovas and Juliano, 2008) (Figure 5.1): (a) product preheating to a target temperature $T_h$, (b) product equilibration to initial temperature $T_s$, (c) product temperature increase to $T_{p1}$ due to compression heating, (d) product cooling down to $T_{p2}$ due to heat removal through the chamber, (e) product temperature decrease to $T_f$ during decompression, and (f) product cooling to $T_c$.

It is well known that hydrostatic compression of a food above room temperature increases temperature in both the compression fluid and food (Ting et al., 2002). However, compression heating in the steel walls of the pressure vessel is almost zero (Ting et al., 2002; de Heij et al., 2003). Therefore, a thermal gradient is developed during pressurization (Denys et al., 2000b; Otero and Sanz, 2003), leading to heat loss toward the chamber wall and subsequent cooling down of the fluid/sample system. Consequently, the chamber wall becomes the "coldest" region in the chamber (de Heij et al., 2003), even if pressure is uniformly distributed at all points, as indicated by Pascal's law.*

* Pascal's law states that when an increase in pressure occurs at any point in a confined fluid, there is an equal increase at every other point in the container.

**TABLE 5.1**
**Suggested Steps for Validation of an HPHT Process**

| Step | Purpose |
|---|---|
| 1. Mapping and quantification of temperature uniformity during pressure come-up and holding times. | Identification of "cold" zones in vessel and process modifications to maximize the temperature uniformity. |
| 2. Establishing the assumptions and deriving a model according to temperature distribution data obtained. | A number of assumptions can be made to derive 1- to 3D heat and fluid dynamic transfer models. |
| 3. Adjusting the empirical temperature data to selected model. | Verifying the modeling assumptions based on vertical and radial profiles. As a first approach, trials should be done in a vessel filled with water. Once predictive models are established for the simplified system, later studies should include packages inside vessel with products of different composition. |
| 4. Establishing the parameters for characterizing the temperature profiles. | (a) To represent 2- or 3D temperature uniformity inside the vessel.<br>(b) To represent and compare time–temperature history of an HPHT process in "coldest region" of system considered. |
| 5. Identifying the accurate inactivation kinetic models for *C. botulinum* at selected HPHT conditions and food media. | (a) To compare the inactivation efficiency determined in small scale (0.1–1 L) with pilot (2–35 L) or industrial scale (>100 L) vessel.<br>(b) To establish the surrogate microorganisms and enzymes that can represent *C. botulinum* inactivation. |
| 6. Mapping the *C. botulinum* inactivation distribution inside the vessel by placing microorganisms at different points in vessel and evaluating the inactivation. | (a) To validate the system from an industrial perspective.<br>(b) To compare the inactivation distribution data of *C. botulinum* with other microbial surrogates, by predicting the inactivation distribution. |
| 7. Correlating the inactivation distribution map with temperature distribution map. | To couple the heat transfer models with microbial inactivation kinetic models. |
| 8. Establishing the performance criterion for microbial inactivation in HPHT process. | Used in process design and control as criterion for food safety assurance. |

Figure 5.1 shows a typical temperature profile during HPHT treatment, indicating the cooling down experienced in the holding process. Loss of heat is reflected in the difference between initial and final temperature during holding time ($T_{p2} < T_{p1}$), and temperature at the beginning and end of the pressurization–depressurization period ($T_f < T_s$).

To achieve sterilization during pressurization, all parts of the treated food must at least reach $T_{p1}$ at pressure $P_1$ (Figure 5.1), which is the maximum temperature

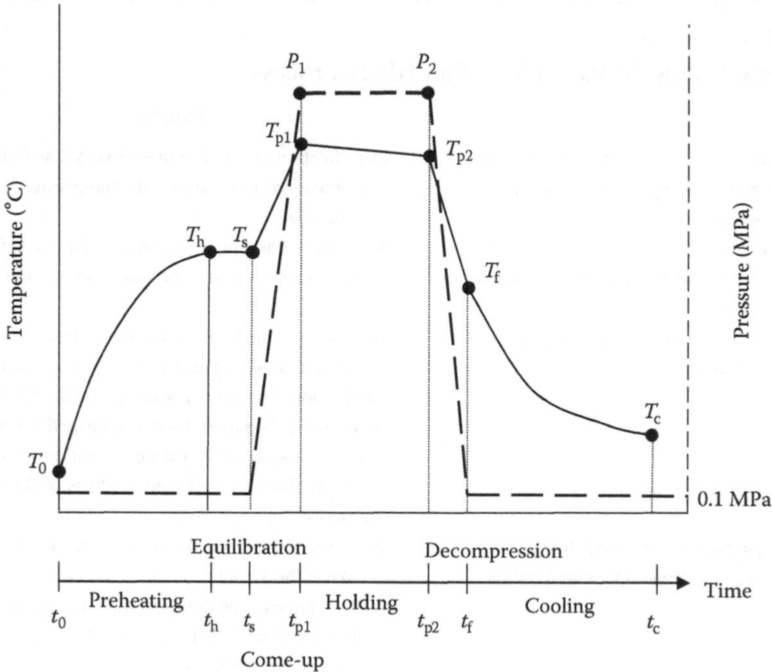


**FIGURE 5.1** Typical temperature profile of a pressure-assisted thermal process. (Adapted from Balasubramaniam, V., Ting, E., Stewart, C., and Robbins, J., *Innovat. Food Sci. Emerg. Technol.*, 5, 299, 2004.)

targeted for the maximum pressure holding time. To achieve this goal a number of variables must be controlled from the start (Balasubramaniam et al., 2004). Thus, understanding the effects of combined temperature and pressure on microbial inactivation distribution will require knowing the temperature within the pressure vessel at specific locations, and at all times during a high-pressure process (Carroll et al., 2003).

As mentioned before, a microbial performance model that includes *C. botulinum* inactivation kinetic parameters and HPHT process temperature and time needs to be established. Barbosa-Cánovas and Juliano (2008) investigated related criteria proposed in the literature and identified three main performance approaches that could be used to evaluate the HPHT process: (a) a traditional approach using conventional thermal processing parameters (Koutchma et al., 2005), (b) a Weibullian approach using adjusted inactivation distribution parameters (Peleg et al., 2005), and (c) a volumetric temperature distribution and spore inactivation distribution approach.

Koutchma et al. (2005) showed that the HPHT process could be validated by applying concepts traditionally used in conventional thermal processing of low-acid foods. Based on evidence reported on the linearity of microbial inactivation curves (semilog scale) of classical surrogates, namely *B. stearothermophilus* and *C. sporogenes* PA 3679, the authors determined their thermal sensitivity ($Z_P$-value) and pressure sensitivity ($Z_T$-value) at pressures of 600–800 MPa and temperatures

91°C–108°C. The $Z_P$- and $Z_T$-values of *C. sporogenes* PA 3679 spores did not to vary with pressure or temperature, respectively, at the selected conditions.* The authors adapted a concept established by Pflug (1987) to calculate process lethality and *F*-values for the HPHT sterilization process, and included the initial microbial load.

Peleg et al. (2005) showed that the first-order microbial inactivation kinetics assumed in the determination of *D*-values for *C. botulinum* strains, the building blocks for determining traditional thermal processing $F_0$ values, do not necessarily apply when spores are subjected to a lethal environment (e.g., high temperature, high pressure, antimicrobials, etc.). Alternatively, it was suggested that *C. botulinum* inactivation kinetics in a nonisothermal process can be fitted to a power law or "Weibullian" model with temperature-dependent parameters:

$$\log\left[\frac{N(t)}{N_0}\right] = -b(T)\,t^{n(T)} \tag{5.1}$$

where

$N_0$ is the initial number of microorganisms

$N(t)$ is the number of surviving microorganisms at time $t$

The parameters $b(T)$ and $n(T)$ are functions of the process temperature history

A similar model can be used to describe the HPHT process using model parameters $b$ and $n$ as functions of both pressure and temperature (Campanella and Peleg, 2001). Since the increase in temperature inside the product is a result of compression heating, temperature is a function of pressure and time. Hence, the combined effects of pressure and temperature can be accounted for in the following expression:

$$\log\left[\frac{N(t)}{N_0}\right] = -b(P,T)\,t^{n(P,T)} \tag{5.2}$$

To understand the effects of pressure and temperature on the parameters $b$ and $n$, the temperature $T(t)$ and pressure $P(t)$ histories must be known. Temperature can be represented as a function of pressure by using the compression heating expression, as shown later. Peleg et al. (2003, 2005) and Peleg (2006) proposed a method to estimate the survival parameters $b$ and $n$ for conditions under variable pressure and temperature. Once the parameters $b(t)$ and $n(t)$ are known, inactivation of microorganisms under variable pressure and temperature can be predicted. These parameters could eventually be used to measure the performance of an established HPHT sterilization process by using actual temperature–pressure records.

To date, little has been published on the variation of Weibullian power-law parameters with temperature and pressure. Individual Weibull parameters have been determined for a number of aerobic and anaerobic spores at specific pressure–temperature combinations within the ranges 95°C–121°C and 500–700 MPa, but have not been expressed as function of temperature and/or pressure (Rajan et al., 2006a,b; Ahn et al., 2007).

* In this case, $Z_P$ and $Z_T$ were determined at nonisobaric and nonisothermal conditions, respectively.

On the other hand, inactivation of a selected thermobaroresistant *C. botulinum* strain has been expressed as a function of temperature and pressure in an $n$th order model (Margosch et al., 2006). This was the first attempt to provide a model describing *C. botulinum* kinetics as a function of temperature and pressure.

$$\frac{dN(t)}{dt} = -k_i(T,P) \cdot N^n \tag{5.3}$$

A reaction order of 1.35 was found by fitting $k$ to curves obtained at different pressure–temperature combinations of 70°C–120°C and 600–1400 MPa and ambient pressure. Single values for each constant for each combination were condensed as follows:

$$k_i^{'}(T,P) = e^{A_0 + A_1 \cdot P + A_2 \cdot T + A_3 \cdot P^2 + A_4 \cdot T^2 + A_5 \cdot P \cdot T + A_6 \cdot P \cdot T^2} \tag{5.4}$$

where $k_i^{'} = k_i^{'}(T,P) = k_i \cdot N_0^{n-1}$ (Margosch et al., 2006).

As indicated in the HPHT validation steps (Table 5.1), temperature distribution of the process must be identified, modeled, and validated first. Then the traditional $F_0$, the Weibullian parameters, or other parameters can be applied to evaluate the microbial inactivation achieved during the process. Furthermore, understanding the extent of temperature uniformity (or temperature distribution) inside the vessel in an HPHT process is key to predict the extent of inactivation of target microorganisms and enzymes. If small amounts of heat are lost during the HPHT process, this will affect the inactivation rate of microbial spores significantly (de Heij et al., 2002, 2003; Ardia et al., 2004). Moreover, thermal gradients established during compression and holding periods (as shown later) can cause inhomogeneities in the inactivation of target microorganisms and enzymes due to over- and underprocessed fractions inside the vessel (Denys et al., 2000a,b; Hartmann and Delgado, 2003a; Carroll et al., 2003; Otero and Sanz, 2003; Hartmann et al., 2004). Therefore, the third approach (c), using thermal and fluid dynamic models to predict temperature distribution at specific locations inside the vessel, by combining (or coupling) these with selected microbial inactivation kinetics models in the form of a differential equation, could help predict the homogeneity in microbial spore inactivation (Knoerzer et al., 2007; Juliano et al., 2008). Chapter 6 will provide examples of coupling inactivation with conductive and convective numerical models.

The objective of this chapter was to introduce the different and most recent modeling approaches that have been proposed to express and represent the extent of temperature uniformity during a high-pressure process. The revision of existing HPHT systems, the system variables involved, as well as the governing equations inside the vessel, will help establish the assumptions needed to apply these models.

## 5.2 MODELING SYSTEM AND SOURCES OF VARIABILITY

Before explaining the different modeling methods applicable to HPHT processing, it is important to identify the components and materials interacting in a high-pressure system during the process, and how this interaction can affect the loss of compression heat during pressurization at different stages, and consequently, lead to a lower target process temperature ($T_{p1}$). Moreover, different approaches that have been proposed to eliminate these sources of system variability and heat loss will be outlined.

### 5.2.1 Components and Materials in a Typical High-Pressure Sterilization System

A high-pressure system designed for sterilization conditions must be able to withstand high pressures ranging from 600 to 800 MPa and chamber temperatures up to at least 90°C. This can be accomplished by building a pressure chamber of appropriate thickness and a pumping system with fast pressure come-up. Today, sterilization systems range from laboratory scale (0.02–1.5 L), pilot scale (2–50 L), and industrial scale (150 L). Existing types of equipment have varied configurations (Balasubramaniam et al., 2004) that offer different levels of heat retention and efficiency. A more efficient use of compression heating would be to lower the process temperature required to achieve sterilization (de Heij et al., 2002).

The following paragraphs will outline the different considerations in designing a system, particularly the vessel layout, pressure transmitting fluid selection, use of packaging materials, and the food contained within (to understand the heat transfer phenomena). These considerations will be revisited in Section 5.4.1 to establish the design assumptions for developing an analytical or a numerical (computational fluid dynamic [CFD]) model.

#### 5.2.1.1 Components of a High-Pressure Vessel

A typical (batch) high-pressure machine system, applicable to all temperatures, consists of (a) a thick cylindrical steel vessel with two end closures, (b) a means of restraining end closures (e.g., yoke, threads, and pin), (c) a low-pressure pump, (d) an intensifier for system compression, using liquid from the low-pressure pump to generate high-pressure process fluid, and (e) necessary system controls and instrumentation (Farkas and Hoover, 2000). For HPHT treatment at pressures over 400 MPa, pressure vessels can be built with two or more concentric cylinders made of high-tensile strength steel. An outer cylinder confines the inner cylinders such that the wall of the pressure chamber is always under some residual compression at the design operating pressure. In some cases the cylinders and vessel frame are prestressed by winding wire under tension layer upon layer. The tension of the wire compresses the vessel cylinders, reducing the diameter of the cylinders (Hjelmqwist, 2005). This special arrangement allows for an equipment lifetime of over 100,000 cycles at pressures 680 MPa or higher. Preferred practice in the design of high-pressure chambers is to make sure the contained food is in contact with parts made of stainless steel; that way the filtered (potable) water can be used as isostatic compression fluid (Farkas and Hoover, 2000).

For improved insulation, to prevent heat loss through the steel wall, a material with low-thermal conductivity (less than 1 W/m/K) could be used as part of the pressure vessel design (de Heij et al., 2003). Vessel materials suggested for this application are polyoxymethylene (POM), polyetheretherketone (PEEK), polytetrafluoroethylene (PTFE), polypropylene (PP), or ultrahigh-molecular-weight polyethylene (UHMWPE).

Today, high-pressure vessels are regularly insulated with these types of materials in the interior as a liner. A product container (wall thickness 5 mm or more) constructed from these materials can also be used (de Heij et al., 2003). Thermophysical properties (e.g., thermal diffusivity) of these insulation materials under high-pressure conditions have not been determined yet, however, empirical data on temperature inside the chamber have proven beneficial in preventing heat loss at holding times less than 5 min. Size of pressure vessel also plays an important role in compression heating retention (Ting et al., 2002), as larger vessels have been shown to retain more heat during holding times (Hartmann and Delgado, 2003b).

#### 5.2.1.2 Pressure Transmission Fluid

The selected compression fluid can contribute to compression heating of the food and possible heat gain or loss in the food. Farkas and Hoover (2000) mention food-approved oil or water-containing FDA- and USDA-approved lubricants, anticorrosion agents, and antimicrobial compounds as possible compression fluids. Water solutions of castor oil, silicone oil, propylene glycol, and sodium benzoate are sometimes used as pressure-transmitting fluids in laboratory-scale machines (Ting et al., 2002). For HPHT treatment, the typical fluids used in pressure vessels include water with glycerol, edible oils, and water/edible oil emulsions (Meyer et al., 2000). However, for commercial purposes, the use of potable water is the most recommended compression medium for maintaining the cleanliness of product packages (Farkas and Hoover, 2000).

The ideal scenario would be to find a compression fluid with minimal differences in compression heating behavior comparable to the food sample (Meyer et al., 2000; Balasubramanian and Balasubramaniam, 2003; de Heij et al., 2003). However, this would be impractical in processing foods of varied composition. For instance, foods high in lipid content show higher compression heating than water-based foods (Balasubramanian and Balasubramaniam, 2003). If water is used as the fluid medium, an increased temperature gradient is created within the lipidic portions of the sample. This phenomenon of temperature increase does not impede achieving the target process temperature ($T_{\mathrm{pl}}$) in the nonlipidic portions of the sample. Thus, even though a temperature gradient is developed due to the presence of lipids in the food, the process conditions could be maintained when using water as the pressure medium.

#### 5.2.1.3 Flexible Container

Packaging used for high-pressure treated foods must accommodate more than a 12% reduction in volume, and be able to return to its original volume, without loss of seal integrity and barrier properties (Farkas and Hoover, 2000; Caner et al., 2004). Until now, identification of suitable packages that can survive pressure sterilization conditions, i.e., that retain seal and overall integrity as well as adequate barrier properties against oxygen and water vapor, remains a challenge. For example, retort pouches, designed to withstand temperatures of 121°C, may delaminate and blister after thermal pressure processing; particularly at chamber temperatures >90°C and pressures >200 MPa (Schauwecker et al., 2002).

For packaged foods, the food-containing material constitutes an intermediate barrier to the transfer of heat from the heating medium into the product or, after thermal processing, from the product to the cooling medium. Few research studies (discussed in Chapter 6) have considered the influence of heat transfer between the heating medium, container material, and the product (Larousse and Brown, 1997; Hartmann et al., 2003; Hartmann and Delgado, 2003b). The packaging material selected can affect the preheating rate needed to achieve the initial temperature ($T_i$), depending on the composition and thickness of the material.

It is not yet known how the combined application of pressure and internal pressure inside the package can affect heat conductivity during processing of food products. The pressure increase inside the package is not only a result of mass augmentation from the pressure medium entering the chamber, but also due to shrinkage of the package under pressure, i.e., its structural response to pressure application and resistance to deformation. In most cases, polymeric laminates provide enough flexibility as to transfer the pressure inside the package, resulting in pressure equilibrium throughout the vessel.

Hartmann and Delgado (2003b) discussed the mathematical description of pressurization of a liquid (model enzyme solution) inside a package (see Chapter 6), mentioning that pressure increase in the pressure medium and its transmission from the packaging material to the contained enzyme solution is "a highly complex physical process." This mathematical expression describing the compression of pressure medium represents a fluid–structure interaction problem. To solve this problem, conservation equations of fluid dynamics must be applied using expressions from structure mechanics that describe the interface between the fluids (outside/inside packaging material) and structure (packaging material).

At present, there have been no data reported for the compression heating rates of packaging material layers forming a film. Assuming compression heating of the material gives a higher temperature rise than in both the food and compression medium, there would be a "warmer" layer surrounding the food. This phenomenon is an inherent part of the process and significant if 5% of the vessel is full of packaging material (assuming 5% in volume of packaging material per food package). However, testing the compression heating of different types of films would help identify their influence on temperature homogeneity inside the package. This is especially important when maximizing the use of vessel space and a large number of packages are included.

#### 5.2.1.4 Food Properties

Food characteristics (e.g., product composition, viscosity, phase state, density, and porosity) determine the pathways of heat transfer in a food during processing (Barbosa-Cánovas and Juliano, 2008). When foods are compressed, their volume is reduced as a function of the imposed pressure. During compression, foods are generally more dense and compacted, and those of a porous semisolid nature have lower porosity up to a certain degree. Thus, foods may experience increased homogeneity and improved temperature distribution during pressurization. However, a common problem is the release of air located within the food to the package, which adds to the initial air headspace. This headspace, located at certain points in the package, can alter temperature and pressure uniformity, as well as shape because of these two reasons: (a) compressed residual air may not transfer hydrostatic pressure to the food in the same way as the compression medium (Ting et al., 2002), (b) residual air reduces its volume and heats up at higher pressure. Heating up of retained air, as indicated by Boyle's law of gases, creates an extra temperature gradient in the food package system. The incidence of this effect on product temperature will mainly depend on the product size and amount of headspace in the package. Farkas and Hoover (2000) mentioned that the amount of air in the system is not a critical process factor since it has no effect on high-pressure inactivation kinetics. However, Balasubramaniam et al. (2004) indicated that the total headspace, oxygen in particular, should be minimized.

As discussed earlier, compression heating differences between semisolid food and the vessel wall can be significant, thereby causing heat diffusion out of the sample. Consequently, "cold spots" can be located on the surfaces of packages near the vessel wall (Matser et al., 2004). The situation becomes more complex when the food is nonhomogenous in composition or structure. In this case, although transient heat will transfer in different directions according to the location of different components, the main heat flux will probably be directed toward the vessel wall. The number of packages in the vessel and their location will determine the extent of temperature reduction inside the total food mass being processed.

### 5.2.2 Critical Process Variables

As already mentioned, the temperature evolution of the process is marked by changes taking place in several steps (Figure 5.1). Critical process variables (Tables 5.2 and 5.3) must be controlled in different portions of the system previously described (Section 5.2.1) to (a) assure reaching target process temperatures and pressures in all food packages, (b) provide the amount of heat required for spore inactivation, and (c) maintain heat retention efficiency. Sections below show how process target temperatures and pressures can be achieved, and variations reduced, by controlling factors outlined in Tables 5.2 and 5.3 during preheating time, equilibration, pressure come-up time, pressure holding time, decompression time, and cooling time.

#### 5.2.2.1 Initial Temperature Distribution in the Sample, Fluid, and Vessel before Pressurization

A uniform initial target temperature of the food sample ($T_s$) is desirable to achieve a uniform temperature increase in a homogenous system during compression (Farkas and Hoover, 2000; Meyer et al., 2000). If cold spots are present within the food, part of the product will not achieve the target process temperature ($T_{p1}$) during pressurization (Meyer et al., 2000). Several preheating methods can affect initial temperature distribution in the food package and thereby provide a nonhomogenous kill. Use of still water baths ($T_s$ or greater), steam, steam injection in water, or dielectric heating has been suggested as preheating alternatives. Faster preheating methods provide less uniformity, and thus require a longer time for equilibration to achieve temperature homogeneity. In practice, and to save high-pressure machine operating time, the sample may undergo two equilibration steps: (1) equilibration in a still water bath

**TABLE 5.2**
**Factors Affecting Heat Transfer during Preheating of Packaged Foods**

| Process Variable | System Element | Process Factors | Parameters |
|---|---|---|---|
| Fluid temperature | Preheating system | • Type of system<br>• Ratio of fluid mass: product mass (number of packages)<br>• Racking system (separation between container, circulation between layers, package restraint to specified thickness)<br>• Heat transfer aids (steam, steam/vapor, microwaves, radio frequencies, circulation pumps) | • Heat transfer coefficient<br>• Heating rate |
| Product initial temperature | Container geometry | • Packaging material (composition, thickness)<br>• Package thickness<br>• Fill weight<br>• Sample confining system (racks, trays, cassettes)<br>• Container headspace (amount of air in the package)<br>• Container shape<br>• Distribution of food particulates | • Package thermal diffusivity[a]<br>• Time to reach target temperature<br>• Temperature equilibration time |
| Target preheating temperature | Product characteristics | • Composition of ingredients<br>• Particulate size<br>• Soluble solids<br>• Physical state (fresh/cooked, liquid, semisolid, frozen)<br>• Food structure (homogeneity)<br>• Occluded gases<br>• Viscosity | • Product thermal diffusivity[a] |

*Source*: Juliano, P., Toldra, M., Koutchma, T., Balasubramaniam, V.M., Clark, S., Mathews, J.W., Dunne, C.P., Sadler, G., and Barbosa-Cánovas, G.V., *J. Food Sci.*, 71, E52, 2006b. With permission.

[a] Thermal diffusivity is determined by specific capacity, density, and thermal conductivity.

at $T_s$, and (2) equilibration inside the chamber after inserting the product and closing the chamber at $T_s$. An alternative approach is to preheat the food up to temperature $T_h$ greater than target temperature $T_s$. This will shorten equilibration time, saving step (1), and ensure all parts of the food reach initial temperature $T_s$. Juliano (2006) summarized the factors affecting heat transfer in a preheating system (Table 5.2).

Balasubramaniam et al. (2004) mentioned that the dimensions and thermal properties of the test sample and vessel determine equilibrium time inside the chamber (step 2).

**TABLE 5.3**
**Critical Variables and Factors to Control and Assure Achievement of Target Temperatures and Pressure during HPHT Treatment**

| System Elements | Variables | Process Factors | Parameters |
|---|---|---|---|
| Chamber | Temperature ($T_s$) | • Chamber composition and structure<br>• Chamber thickness<br>• Chamber volume<br>• Thermoregulation system<br>• Pressure pump system<br>• External and internal intensifier<br>• Insulation with liner/ use of polymeric carrier; type of polymer<br>• Heating power supplied | • Thermal diffusivity[b]<br>• Heat retention efficiency<br>• Compression heating rate of selected insulation polymer in carrier |
| Preheating system | Temperature ($T_h$) | • Type of system<br>• Sample size | • Heat transfer coefficient<br>• Heating rate |
| Compression medium | Temperature[a] ($T_s$)<br>Pressure ($P_1$) | • Composition<br>• Chamber thermoregulation system<br>• Ratio of fluid mass: product mass in chamber<br>• Incoming fluid during pressurization<br>• Overflow prefill fluid in carrier | • Heat transfer coefficient<br>• Compression heating rate<br>• Final pressurization temperature ($T_{p2}$)<br>• Final decompression temperature ($T_f$)<br>• Time needed to reach the thermal equilibration during pressure holding ($t/T_p = T_s$) |
| Package | Final preheating temperature ($T_h$)<br>Initial temperature ($T_s$)<br>Pressure ($P_1$)<br>Pressurization temperature ($T_p$) | • Food composition<br>• Food structure (porosity, phase state)<br>• Sample shape<br>• Sample size<br>• Film type (thickness, composition)<br>• Headspace<br>• Sample mobility<br>• Number of packages | • Thermal diffusivity[b]<br>• Compression heating rate of food components<br>• Compression heating rate of packaging film<br>• Thermal expansivity<br>• Final pressurization temperature ($T_{p2}$)<br>• Final decompression ($T_f$)<br>• Time needed to reach thermal equilibration during pressure holding ($t/T_p = T_s$) |
| Cooling system | Cooling temperature ($T_c$) | • Type of system<br>• Cooling power supplied | • Heat transfer coefficient<br>• Cooling rate |

[a] Initial temperature of fluid contained in machine and fluid pumped for pressurization should be the same.
[b] Thermal diffusivity is determined by specific capacity, density, and thermal conductivity.

They recommended that all locations in the pressure vessel and product be equilibrated within ±0.5°C of the target initial temperature $T_s$. Furthermore, prior to introducing a prepackaged food, the correct homogenous temperature of the high-pressure vessel chamber must be ensured (Meyer et al., 2000). To achieve in-chamber temperature homogeneity, fluctuations from the thermoregulation system should be minimized.

Initial temperature of the sample ($T_0$) is not critical from a food safety standpoint; however, it will influence overall processing time. Throughput could be improved, especially in larger packages, by setting $T_0$ to above room temperature (e.g., 50°C–60°C) using a water bath, thus shortening preheating time. This will depend on product composition and its sensitivity to long-time exposure to moderate temperatures.

#### 5.2.2.2 Temperature Distribution during Pressurization

As already mentioned, "cold spots" located close to the vessel wall develop during pressure come-up time. Consequently, heat will flow from the product to the vessel wall, and the product fraction closest to the vessel wall will cool down and not reach the final temperature achieved at the center of vessel (de Heij et al., 2002; Ting et al., 2002; Otero and Sanz, 2003; Ardia et al., 2004). Moreover, a certain amount of (preheated) liquid enters the vessel (2 mL to 1.7 L) depending on the vessel's dimensions. This can lead to the cooling down of the lower part of the vessel (Knoerzer et al., 2007). To decrease the cooling caused by the inflowing pressurizing fluid, the following system modifications have been suggested (de Heij et al., 2003; Balasubramaniam et al., 2004; Knoerzer et al., 2007): (a) an internal pressure intensifier can be incorporated to avoid liquid entering the vessel, (b) the pressurizing fluid, high-pressure pipes, and external intensifier system in the high-pressure pump can be preheated to desired initial temperature $T_s$, and (c) insulation with a special liner can be installed to avoid contact between the product and the entering fluid. These modifications should be implemented according to pressure vessel dimensions and initial temperature $T_s$, as specified for different operating parts of the equipment in Table 5.3. Option (c) can be further enhanced by inserting a free-moving piston at the bottom of the liner (Denys et al., 2000a,b) or leaving a small opening (at opposite end of liner away from pressure fluid inlet) to allow transfer of pressure to the compression medium into the liner.

Modern 35 L high-pressure machines (at least three exist today; developed by Avure Technologies, Kent, Washington) contain a metallic cylindrical "furnace" (metallic laminate heated by resistors) located inside the chamber near the vessel wall. The furnace surrounds a preheated polymeric liner for carrying the food packages. Furnace temperature is set higher than target initial temperature ($T_s$), eliminating the problem of heat loss into the vessel walls. Furthermore, a series of controls are set to automatically recirculate water from a fill tank, previous to starting the pressure pump, to ensure initial target temperature is reached in the compression fluid, the polymeric carrier, and the samples after closing the chamber.

Once initial variables (discussed in Section 5.2.2.1; Tables 5.2 and 5.3) are controlled, a uniform temperature rise will occur during come-up time throughout the vessel, assuming the product is of uniform structure and composition (Matser et al., 2004). Farkas and Hoover (2000) indicated that compression temperature increase will be uniform if the food is homogenous, assuming the food has less than 25% fat content (seen in lipid-based substances, due to higher compression heating).

#### 5.2.2.3 Compression Rate

During the pressure rise period (i.e., pressure come-up time), vegetative bacteria can be inactivated (Rodríguez, 2003), as well as bacterial spores, but only when increase in pressure is combined with temperature (Koutchma et al., 2005). The come-up time depends on the compression rate of the sample volume and the pressure-transmitting fluid and it is proportional to the power of the low-pressure pump (driving the intensifier) and the target process pressure (Farkas and Hoover, 2000; Balasubramaniam et al., 2004).

In general, compression/decompression curves are nearly linear (Figure 5.1) and, therefore, the compression rate can be assumed constant. At a given food package/compression fluid initial temperature, a constant compression rate should provide constant compression heating inside the package. Compression heating of the transmitting fluid and food will be higher at initial temperatures between 75°C and 90°C. At pressure of 100 MPa, near room temperature (25°C), water typically changes by 3°C, whereas at an initial temperature (80°C), compression heating changes by 4.4°C. (Farkas and Hoover, 2000; Ting et al., 2002).

Both the temperature of the product and compression fluid rise between 20°C and 40°C during high-pressure treatment, however as said before, the metal vessel used in the process will not be subjected to significant compression heating (de Heij et al., 2002; Ting et al., 2002). It can be assumed that different compression rates provide the same compression heating, and that the highest compression heating rate prevents heat loss through the vessel walls, decreasing overall processing time. However, lack of adequate insulation may affect the compression heating rate, as heat would be lost as pressure increases.

#### 5.2.2.4 Selecting Pressure and Holding Time

More recent high-pressure units have additional features such as pressure intensifiers and automatic pressure controls, which allow maintaining an almost constant pressure during holding time (Figure 5.1), i.e.

$$P_1 = P_2 \tag{5.5}$$

High-pressure equipment allows controlling pressure within ±0.5% (e.g., ±3.4 MPa at 680 MPa) and recording it at the same level of accuracy (Farkas and Hoover, 2000). A constant pressure during holding time is necessary to maintain a constant temperature profile and thereby retain compression heat. When all heat retention aids mentioned before are simultaneously applied, there is a minimum temperature loss during holding time. Controlling all the factors and use of proper insulation could even make the system adiabatic during a short-holding period. Nevertheless, there is no existing equipment that can guarantee a heat retention efficiency of 100% (i.e., $T_{p1} = T_{p2}$).

With heat retention efficiency lower than 100%, the observed change in sample temperature would be a result of the heat transfer process inside the high-pressure vessel. In this case, as shown later, the heat transfer not only depends on the thermal properties of the food and fluid, but also on those of the vessel/liner walls, the heat exchange possibilities inside the vessel/liner (convection, sample mobility, number of samples, etc.), the heating power supplied (Otero et al., 2002a), and other factors mentioned in Tables 5.2 and 5.3.

#### 5.2.2.5 Decompression Rate

Once the holding time is complete, the pressure relief valve is automatically opened and the water used for compression is allowed to expand, returning to atmospheric pressure within a short time (less than 30 s, depending on the target pressure). The decompression time has been defined as the recorded time to bring a mass of food from process pressure to 37% of said pressure (Farkas and Hoover, 2000). It is not clear if this interval of decompression is critical for spore inactivation. However, control of decompression rate is recommended. For example, single stroke intensifiers may be used to control the decompression rate of a system.

A high-decompression rate will cause the product temperature to decrease uniformly during decompression, irrespective of the value of $T_{p2}$. As the fluid mass and packaging expand (assuming there is no chamber insulation), heat is removed while the compression medium is released through the outlet valve. As mentioned before the difference between the system temperature (chamber, transmitting fluid, and samples), before and after high-pressure processing ($T_s$ and $T_f$), can also indicate the extent of heat loss during processing (Ting et al., 2002).

#### 5.2.2.6 Cooling Time after Decompression

The start of the cooling period after HPHT processing is not critical from a food safety standpoint, except to minimize unnecessary quality and nutritional damage due to heat dwelling, unlike in retort processing. After the heating period in a retort process, especially when dealing with larger cans or packages, there is a temperature distribution within the container, suggesting a compromise exists between the onset of cooling and end of heating (Holdsworth, 1997). While it is important to know the temperature distribution during preheating and pressurization, to establish an adequate process, on this basis, it is also important to know what contribution the cooling phase makes in preventing overheating and optimizing the process for maximum quality retention. In this case, convective cooling methods, such as the use of turbulent low-temperature water baths after removal of the sample from the pressure chamber, can be used for accelerated temperature reduction.

## 5.3 GOVERNING EQUATIONS INSIDE THE VESSEL

### 5.3.1 Fundamental Physics during a High-Pressure Process

Before focusing on the energy balance of a high-pressure process inside the chamber, it is important to determine the component of heat generation due to compression. Barbosa-Cánovas and Rodríguez (2005) derived an expression from the total derivative of the entropy of the pressurized system. Assuming that the variation in entropy, d$s$, in the pressurization system is a function of both temperature and pressure, i.e., $s = f[T, P]$, then

$$ds = \left(\frac{\partial s}{\partial T}\right)_P dT + \left(\frac{\partial s}{\partial P}\right)_T dP \tag{5.6}$$

If the process is assumed reversible, the total entropy change would be zero ($ds = 0$), then rearranging, the compression heating can be expressed as

$$\frac{dT}{dP} = -\frac{(\partial s/\partial P)_T}{(\partial s/\partial T)_P} \tag{5.7}$$

Maxwell's relation of states is expressed as

$$\left(\frac{\partial s}{\partial P}\right)_T = -\left(\frac{\partial V}{\partial T}\right)_P \tag{5.8}$$

Substituting Equation 5.8 into Equation 5.7, and multiplying and dividing the numerator by $V$, and the denominator by $T$, yields:

$$\frac{dT}{dP} = \frac{V\left(\frac{1}{V}\left(\frac{\partial s}{\partial P}\right)_T\right)}{\frac{1}{T}\left(T\left(\frac{\partial s}{\partial T}\right)_P\right)} \tag{5.9}$$

Thus, the compression heating expression includes the expansion coefficient $\alpha_p$ (1/K) and the specific heat capacity $C_p$ (J/kg K), and is given by

$$C_p = T\left(\frac{\partial s}{\partial T}\right)_P \tag{5.10}$$

$$\alpha_p = \frac{1}{V}\left(\frac{\partial V}{\partial T}\right)_P \tag{5.11}$$

Substituting Equations 5.10 and 5.11 into Equation 5.9, and expressing the specific volume $V$ ($m^3$/kg), as the inverse of density $\rho$, result in Equation 5.12, which expresses the compression heating and has been utilized by several authors (Carroll et al., 2003; Otero and Sanz, 2003). However, as indicated by Equation 5.12 for the compression heating equation, the temperature increase will depend on how the expansion coefficient $\alpha_p$, density $\rho$ (kg/$m^3$), and specific heat capacity $C_p$ of both the food and liquid change during pressurization. It follows that

$$\frac{dT}{dP} = \frac{T\alpha_p}{\rho C_p} \tag{5.12}$$

### 5.3.2 High-Pressure Process Equations

Barbosa-Cánovas and Rodríguez (2005) applied a typical energy balance to a pressure system, obtaining the following expression:

$$E_{\mathrm{st}} = E_{\mathrm{in}} - E_{\mathrm{out}} + E_{\mathrm{g}} \tag{5.13}$$

where

$E_{\mathrm{in}}$ is the energy entering the system or work of compression
$E_{\mathrm{out}}$ is the energy leaving the system (i.e., heat loss through vessel walls during compression, holding time, and decompression)
$E_{\mathrm{g}}$ is the energy of compression inside the system
$E_{\mathrm{st}}$ is the energy accumulated or stored inside the system

It can be assumed that no energy is produced from a chemical reaction (i.e., $E_{\mathrm{g}} = 0$). If the system is adiabatic (i.e., $E_{\mathrm{out}} = 0$), then Equation 5.13 can be reduced to a single-term equation, or, $E_{\mathrm{in}} = E_{\mathrm{st}}$. In this case, energy is generated due to compression heating during come-up time.

Following this criterion (i.e., assuming the system is adiabatic), Equation 5.12 can be rearranged into Equation 5.14 and differentiated as a function of time $t$ as Equation 5.15, to express the conversion of mechanical energy into thermal (internal) energy due to compression during a reversible adiabatic change (Denys et al., 2000a,b; Rodríguez, 2003).

$$\rho C_p \mathrm{d}T = T\alpha_p \mathrm{d}P \tag{5.14}$$

$$\rho C_p \frac{\partial T}{\partial t} = T\alpha_p \frac{\partial P}{\partial t} \tag{5.15}$$

In other words, Equation 5.15 indicates that the rate of energy accumulation, given by the rate of temperature increase, is equivalent to the heat produced due to pressurization at a given temperature in an adiabatic (isolated) system. When the system is adiabatic, Equation 5.15 is also valid for the holding (i.e., $\partial P/\partial t = 0$) and decompression steps.

#### 5.3.2.1 Heat Transfer Balance Using Fourier's Law

In a high-pressure system (insulated or not), heat can be diffused throughout the vessel boundaries. Then, the balance in Equation 5.13 can be expressed using terms for unsteady-state heat conduction during the pressure come-up step by using Fourier's law of heat diffusion, with generation term $Q$ (Davies et al., 1999; Datta, 2001).

$$\underbrace{\frac{\partial}{\partial t}\left(\rho C_p T\right)}_{\text{Rate of accumulation}} = \underbrace{\nabla \cdot \left(k \nabla T\right)}_{\text{Fourier's law of unsteady heat conduction}} + \underbrace{Q}_{\text{Compression heating}} \tag{5.16}$$

where $k$ is the overall thermal conductivity of the fluid/food system inside the chamber. In this case, convective currents within the pressurizing fluid are assumed negligible and are not included in this balance. Based on Equation 5.15, $Q$ represents the compression heating rate term expressed as

$$Q = T\alpha_p \frac{\partial P}{\partial t} \tag{5.17}$$

where

$Q > 0$ at $\forall t > t_s$ (pressure come-up step)
$Q = 0$ at $\forall t$, $t_{p1} < t < t_{p2}$ (pressure holding step)
$Q < 0$ at $\forall t$, $t_{p2} < t < t_f$ (pressure release step)

Hence, Equation 5.16 also applies to the representation of all high-pressure processing steps.

In conventional heat transfer calculations, thermal conductivity is generally assumed to be independent of temperature (Holdsworth, 1997). However, this assumption is questionable for most thermal properties in food, as their dependency on temperature and pressure is unknown (Otero and Sanz, 2003). The influence of high pressure and temperature on thermophysical properties is discussed in further detail in Section 5.3.2.4.

#### 5.3.2.2 Contribution of Fluid Motion to Heat Transfer

The fluid movement inside the chamber during pressurization and holding time can also influence temperature distribution (Hartmann and Delgado, 2003b). The thermodynamic and fluid-dynamic behavior of the pressure medium is described by conservation equations of mass, momentum, and energy (Hartmann et al., 2003). These equations account for the convective currents inside the vessel and are expressed as follows:

- Mass conservation or equation of continuity (Chen, 2006):

$$\frac{\partial \rho}{\partial t} + \nabla \cdot \left(\rho \vec{V}\right) = 0 \tag{5.18}$$

where $\vec{V}$ is the velocity vector. This equation assumes that the fluid is initially at rest. Density increases with increasing pressure and temperature, therefore the first term in Equation 5.18 becomes nonzero. Since the left side must equal zero, the fluid $\vec{V}$ must adopt nonzero values at pressure increments, thus enforcing a fluid motion.

- Energy conservation (Chen, 2006):

$$\frac{\partial \left(\rho C_p T\right)}{\partial t} + \nabla \cdot \left(\rho \vec{V} C_p T\right) = \frac{\partial P}{\partial t} + \nabla \cdot (k \nabla T) \tag{5.19}$$

- Momentum conservation or the Navier–Stokes equation of Motion (Chen, 2006):

$$\rho\left[\frac{\partial \vec{V}}{\partial t}+\left(\vec{V}\cdot\nabla\right)\vec{V}\right]=-\nabla P+\nabla\cdot\left(\eta\cdot\nabla\vec{V}\right)+\rho g \tag{5.20}$$

where

$\eta$ represents the viscosity of the compressed fluid

$g$ represents the gravity constant

These partial differential equations are solved simultaneously, or as referred to in the literature, "coupled" by using CFDs software (see Section 5.4.3). These equations can also be coupled with other differential equations representing inactivation or degradation kinetics, or turbulence. The coupling of governing thermal fluid dynamic equations with microbial inactivation equations will be covered in Section 5.4.3.

#### 5.3.2.3 Possible Initial and Boundary Conditions of the Computational Domain

The complete computational domain is defined by the sample, media, and pressure vessel setup (also called subdomains) forming the high-pressure system. The initial conditions of a system can be defined in its subdomains. Provided the system is initially at thermal equilibrium (i.e., uniform temperature distribution), the heat retention during pressure holding is limited by the sample's mass, thermal diffusivity, and heat transfer coefficient at the sample boundary. Boundary conditions represent the thermal and/or flow behavior at the system's external or internal boundaries. Descriptions of the most commonly applied boundary conditions in high-pressure modeling are described below.

##### *5.3.2.3.1 Symmetric Boundary*

Given the cylindrical geometry of a vertical high-pressure vessel, it is very common to assume that the system is axis-symmetric, for example, by assuming the compression fluid enters from the geometrical center of the bottom part of the vessel. Thus, the generic three-dimensional (3D) problem is reduced to a two-dimensional (2D) axis-symmetric problem through application of symmetric boundary conditions ($r = 0$):

$$-k_S \left.\frac{\partial T}{\partial r}\right|_{r=0} = 0 \quad \forall z, \forall t > 0 \tag{5.21}$$

and

$$-k_S \left.\frac{\partial \vec{V}}{\partial r}\right|_{r=0} = 0 \quad \forall z, \forall t > 0 \tag{5.22}$$

The advantage of a pseudo-3D model is that it can reduce the computational time required for a 3D model (discussed in Section 5.5). However, a 3D model would be needed once the packages are heterogeneously (asymmetrically) distributed throughout the vessel or when a horizontal system is modeled.

*5.3.2.3.2 Inflow Velocity Boundary*

In an indirect pressure system, i.e., a system using an external high-pressure intensifier pump, it is necessary to define an inflow velocity boundary condition for the inlet tube (pressure fluid entrance) for come-up and holding times. Thus

$$v_z\big|_r = v_{\text{in}} \quad \forall r,\ 0 < r < R,\ 0 < t < t_{\text{p1}} \tag{5.23}$$

where $v_{\text{in}} = 0\ \forall t, t > t_{\text{p1}}$.

*5.3.2.3.3 Pressure Boundary*

To account for pressure increase in the water and package subdomains, the following pressure conditions can be imposed:

$$\frac{\partial P}{\partial t} = p_{\text{rate}} \forall z,\ \forall r,\ \forall t > 0 \tag{5.24}$$

By assuming a linear pressure increase or a constant pressure rate $p_{\text{rate}}$, pressure come-up time is

$$p_{\text{rate}} = P_{\text{target}}/t_{\text{p1}}\ \ \forall t, 0 < t < t_{\text{p1}}$$

and pressure holding time can be assumed constant:

$$p_{\text{rate}} = 0 \quad \text{and} \quad P = P_1\ \forall t, t_{\text{p1}} < t < t_{\text{p2}}$$

For decompression, a normal flow pressure condition corresponding to the release of pressure fluid through the outlet can be imposed, for example:

$$p_{\text{rate}} = -\frac{P_{\text{target}}}{t_{\text{f}} - t_{\text{p2}}} = \forall z,\ \forall r,\ \forall t,\ t_{\text{r}} < t < t_{\text{p2}} \tag{5.25}$$

*5.3.2.3.4 Boundaries between the Different Subdomains*

A boundary condition describing the heat transfer between the liquid and the solid domains (i.e., food, carrier, and steel vessel walls) can be expressed as (Figure 5.2a):

$$-k_{\text{S}}\frac{\partial T}{\partial n} = h_{\text{S-F}}(T - T_{\text{L}}) \tag{5.26}$$

where

$T$ is the bulk system temperature
$T_{\text{L}}$ is the liquid temperature
$k_{\text{S}}$ is the thermal conductivity of a solid domain
$h_{\text{S-F}}$ is the convective heat transfer coefficient at the solid–liquid surface
$n$ represents a "normal" vector at the surface wall

In order to make the model more accurate the packaging material containing the food can be taken into account. In this case, the heat transfer coefficient, $h_{S-F}$, has to be modified as follows:

$$-k_v \frac{\partial T}{\partial n} = h_{S-Feq}(T - T_{S-P}) \tag{5.27}$$

with

$$\frac{1}{h_{S-Feq}} = \frac{1}{h_{S-P}} + \frac{d_P}{k_P} + \frac{1}{h_{P-F}} \tag{5.27a}$$

where

$h_{S-Feq}$ is the equivalent heat transfer coefficient between the food packages and compression fluid
$T_{S-P}$ is the temperature of the food package
$d_p$ is the thickness of packaging film
$k_P$ is the thermal conductivity of the packaging film
$h_{P-F}$ is the heat transfer coefficient between the packaging and the fluid

Unless a function $k_P$ in terms of pressure or temperature is known, $k_P$ and $d_P$ may be assumed constant, and may still be able to predict good results. Heat transfer coefficients can be extracted from empirical relations between dimensionless Nusselt, Prandtl, and Reynolds numbers according to laminar or turbulent conditions (Perry, 1997).

The boundary condition for heat transfer located at the vessel wall can be represented by the interaction between the compression fluid and the wall in the following expression (Figure 5.2b):

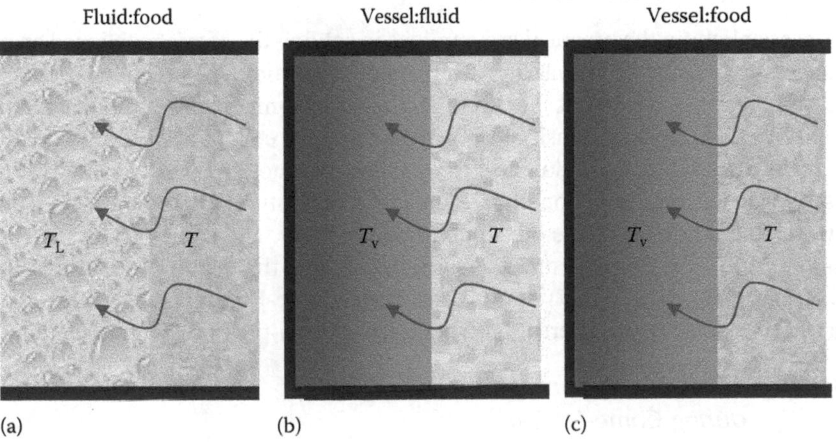


**FIGURE 5.2** Possible boundary conditions for transfer of heat from food package: (a) transfer from package to fluid, (b) transfer from fluid to vessel wall (or carrier), and (c) transfer from package to steel vessel wall (or carrier).

$$-k_V \frac{\partial T}{\partial n} = h_{V\text{-}F}(T - T_L) \quad (5.28)$$

where

$T_L$ is the fluid temperature
$k_V$ is the thermal conductivity of steel
$h_{V\text{-}F}$ is the convective heat transfer coefficient at the vessel wall surface
$n$ represents a "normal" vector at the surface wall

Another boundary condition is the continuity of heat flux at the vessel wall between all fluid–solid and solid–solid boundaries (Figure 5.2c), which can be considered, for example, when looking into the vessel and assuming the chamber is filled to capacity with food packages, and expressed as

$$-k_V \frac{\partial T}{\partial n} = k_{S/L} \frac{\partial T}{\partial n} \quad (5.29)$$

where $k_{S/L}$ is the thermal conductivity of the semisolid food contained in the compression fluid.

In case laminar flow conditions exist, fluid velocity at the chamber walls can be assumed zero due to fluid adhesion to the walls (no slip condition).

Boundary conditions applied to specific models, such as the use of time-dependent vessel walls (due to heat transferred into the steel mass) (Hartmann et al., 2004) and the description of boundaries for turbulent flow conditions inside the vessel (i.e., logarithmic wall function) (Juliano et al., 2007b; Knoerzer et al., 2007), will be described in Chapter 6.

#### 5.3.2.4 Thermal Properties of Foods at High-Temperature High-Pressure Conditions

Before establishing the practical assumptions in solving the above-mentioned energy balances, knowledge of the variation in thermal properties as a function of temperature and pressure is needed. For food and other polymeric materials the thermal expansivity, specific heat, density, viscosity, and thermal conductivity at high-pressure and high-temperature conditions are for the most part not known. It is assumed that these properties are a function of pressure and temperature (Otero and Sanz, 2003). However, the properties have been published for water, and some oils and alcohols (Harvey et al., 1996). Furthermore, as discussed later in this section, Zhu et al. (2007) and Ramaswamy et al. (2007) used a Dual-Needle Line-Heat-Source to measure the thermal conductivity of different food materials at high-pressure conditions.

##### *5.3.2.4.1 Changes in Thermophysical Properties during Come-Up Time*

It was previously shown that the magnitude of the compression heating rate depends on the thermal expansivity of the substance (i.e., fluid or food product), and its specific heat and density. This magnitude also depends on the initial product temperature (Hoogland et al., 2001; Matser et al., 2004). In fact, for most foods, the compression heating rate varies within ranges similar to water (Ting et al., 2002; Balasubramanian and Balasubramaniam, 2003), depending on the nature of the product. Foods notably different are oils and alcohol (Rasanayagam et al., 2003). Few researchers have reported data on compression heating rates for food products at HPHT conditions. Balasubramaniam et al. (2004) reported compression heating of water as 4.0°C, 4.6°C, and 5.3°C/100 MPa at initial temperatures ($T_s$), 60°C, 75°C, and 90°C, respectively.

Based on the experimental compression heating of 4.6°C/100 MPa at an initial temperature of 75°C, thermal and transport properties and their variation with pressure were obtained (Figure 5.3) from the NIST/ASME database (Harvey et al., 1996). This database software uses water property "formulations" that are developed and maintained by the International Association for the Properties of Water and Steam (IAPWS), as described in a software manual (Rodríguez, 2003); the manual gives a more detailed explanation on the derivation of these formulations and used the software to study thermal properties of water during high-pressure processing applications at isothermal conditions.

These formulations are based on a fundamental equation for Helmholtz's energy, expressed in dimensionless terms, as a function of temperature and density. The formulations are recommended for temperatures between 0.15°C and 1000°C and for all positive pressures up to 1000 MPa, except for thermal conductivity and viscosity, where pressures are valid up to 400 and 500 MPa, respectively. Uncertainties in the pressure/temperature ranges considered for sterilization are ±0.02% for density, with no estimates mentioned for $C_p$ or other properties. When considering the conduction balance in Equation 5.16, properties appear as multipliers and denominators of each term; therefore, variations in each property can be compared by their actual values in either decimal scale (Figure 5.3a) or in relative percent scale by comparing changes with respect to their values at 0.1 MPa (Figure 5.3b).

Otero et al. (2002b) also compiled a number of references reporting values for the thermodynamic properties of liquid water and ice in a wide range of temperatures and pressures. They developed a number of routines using MATLAB® (The MathWorks Inc., Natick, Massachusetts) to determine the variations in these properties at different pressure/temperature combinations valid for liquid water, ranging from 0.1 to 500 MPa and −40°C to 120°C. These values were used for the prediction of compression heating of water during pressure come-up time.

#### *5.3.2.4.1.1 Expansion Coefficient*

During pressure/temperature come-up time, the expansion coefficient $\alpha$ decreased gradually, reaching a 28% reduction at 700 MPa and 105°C (Figure 5.3b). By increasing temperature from 75°C to 99°C at atmospheric pressure (0.1 MPa), $\alpha$ increases approximately 22% in liquid water. Otero et al. (2002b) determined that $\alpha$ is a function of temperature and pressure, using a numerical model determined by Ter Minassian et al. (1981). Although the established working limit is 500 MPa, a similar value of $\alpha$ was obtained at 700 MPa and 105°C, and corresponded with an approximate reduction of 28%. The 3D representation of $\alpha_p$ as a function of temperature and pressure in Figure 5.3c shows that this property is more sensitive to temperature than pressure.

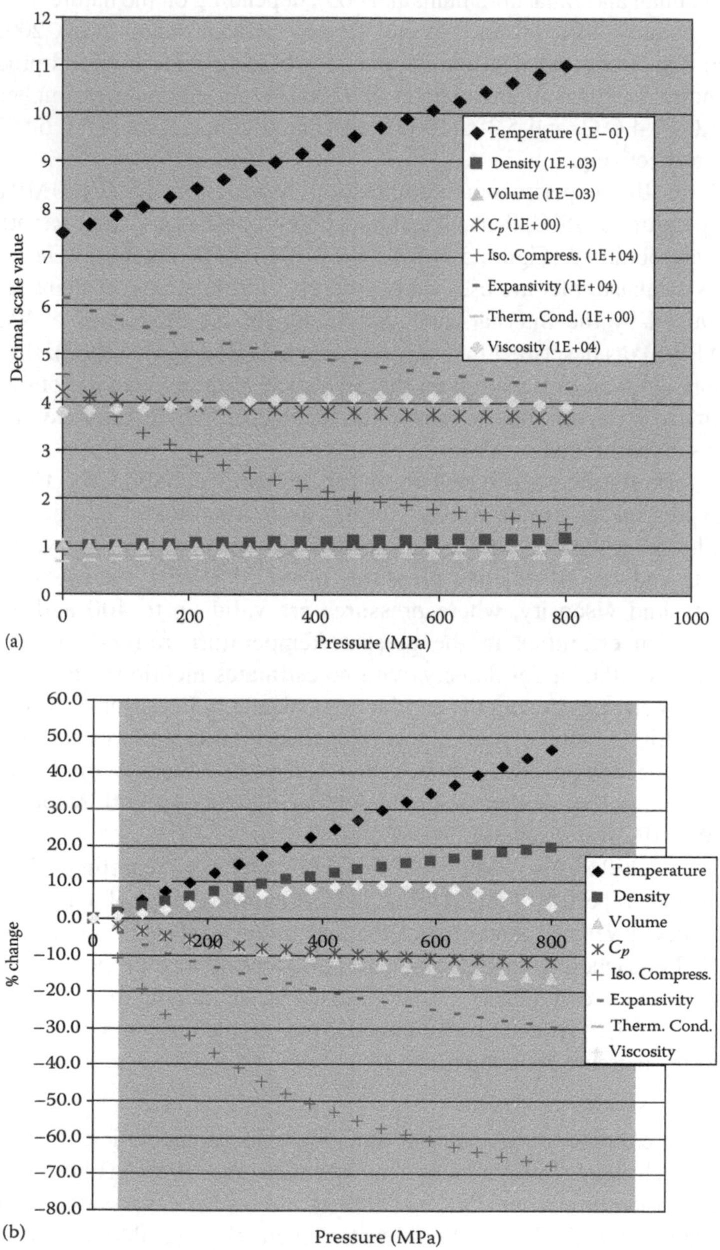


**FIGURE 5.3** Variation in thermophysical properties of liquid water at increasing pressure and temperature, according to compression heating rate of 4.5°C/100 MPa. Properties were calculated from NIST/ASME properties database: (a) values plotted in decimal scale, (b) values expressing change (%) from initial value at 75°C and 0.1 MPa, and (c) response surface plots of selected properties as function of temperature and pressure. Plots prepared by the authors of this chapter. Temperature, °C; density, kg/m³; specific volume, m³/kg; $C_p$, kJ/K kg; compressibility, 1/MPa; expansion coefficient, 1/K; thermal conductivity, W/m K; viscosity, Pa s.

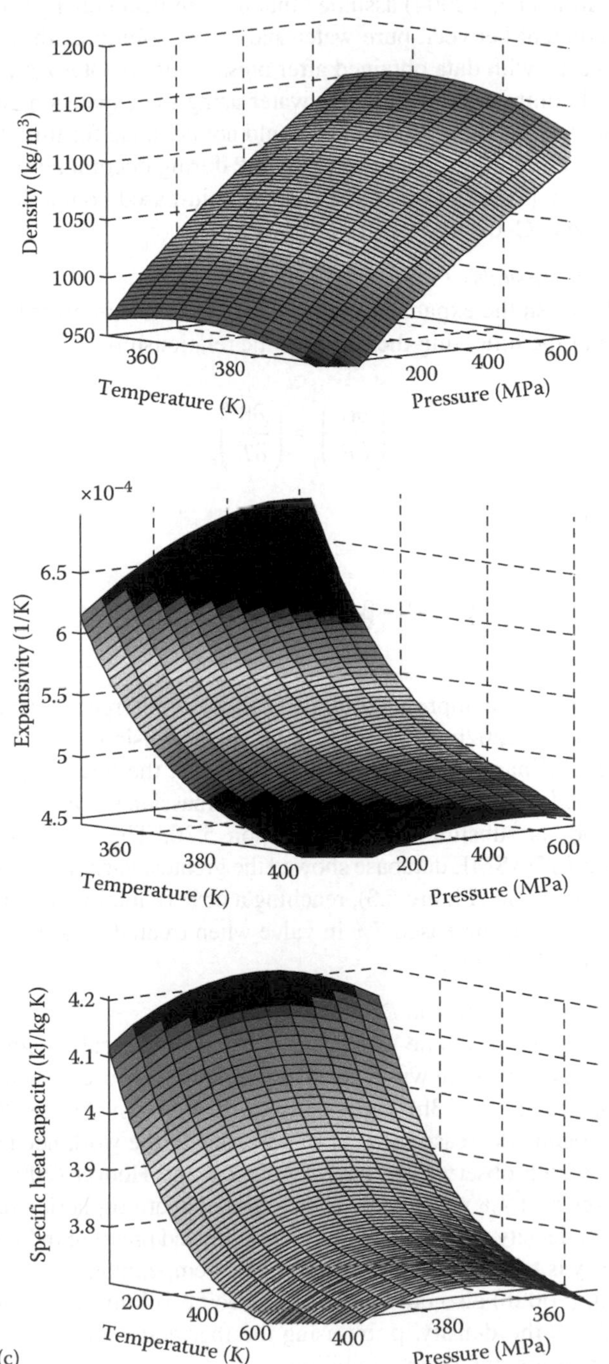


**FIGURE 5.3 (continued).**

When modeling temperature rose due to compression in sucrose solutions and orange juice, Ardia et al. (2004) assumed that no significant changes occurred in the expansion coefficient between pure water and mixed solutions; they were still able to fit Equation 5.12 with data obtained after pressurization. Otero et al. (2000) accurately predicted compression heating in water using the expansion coefficient from literature, however, the same predictions could not be made for milk fat, cream, and ethylene glycol. Regarding variations in $\Delta T/\Delta P$ during decompression, it was mentioned that $\alpha$ has a positive sign in the case of lipid-based compounds between 0.1 and 400 MPa and 0°C–62°C.

#### 5.3.2.4.1.2 *Compressibility Coefficient*

The relation between the expansion coefficient $\alpha$ and the compressibility coefficient $\alpha$ is described in the following thermodynamic relationship:

$$\left(\frac{\partial \alpha}{\partial P}\right)_T = -\left(\frac{\partial \beta}{\partial T}\right)_P \tag{5.30}$$

and can be rewritten as

$$\beta(T,P) - \beta(T_0,P) = -\int_{T_0}^{T}\left(\frac{\partial \alpha}{\partial P}\right)_T \mathrm{d}T \tag{5.31}$$

where $\beta(T_0, P)$ is the compressibility throughout a referenced isothermal ($T_0$ = 323.16). Otero et al. (2002b) determined $\beta$ of water by using a polynomial equation according to Ter Minassian et al. (1981). In this case, the integral part in Equation 5.31 was obtained from the derivative of Ter Minassian's numerical model for $\beta$ (not shown) and its substitution into Equation 5.30. The coefficient $\beta$ of water obtained from NIST/ASME database showed the greatest variation with pressure and temperature application (Figure 5.3), reaching a 65% reduction in value at 700 MPa and 105°C; however, it increased 7% in value when treated at atmospheric pressure and temperatures 75°C–99°C.

#### 5.3.2.4.1.3 *Specific Volume and Density*

When observing specific volume $V$ and density $\rho$ in decimal scale (Figure 5.3a), no significant variations can be seen with increasing pressure. However, when expressing the percentage change (Figure 5.3b), variations were found from 15% to 18% at 700 MPa and 105°C. A volume decrease in water of 15%, due to the work needed to compress water, coincided with observations made by Farkas and Hoover (2000). The specific volume increases by 1.7% when water is only heated at atmospheric pressure. The 3D representation of density as a function of temperature and pressure in Figure 5.3c shows that this property is more sensitive to pressure than temperature.

Otero et al. (2002b) also determined the specific volume $V$ of liquid water, i.e., the inverse of specific density, $\rho$, by using the thermodynamic equation based on concepts of expansion and compressibility:

$$\mathrm{d}V = -V\beta \cdot \mathrm{d}P + V\alpha \cdot \mathrm{d}T \tag{5.32}$$

which gives the following expression for specific volume:

$$V = V_0 \cdot \exp\left\{ -\int_{P_0}^{P} \beta\left(P, T_{K_1}\right) \cdot \mathrm{d}P + \int_{T_{K_1}}^{T_K} \alpha\left(P_0, T_K\right) \cdot \mathrm{d}T_K \right\} \tag{5.33}$$

where $V_0 = 110^{-3}$ m$^3$/kg corresponds to the specific volume of liquid water at atmospheric pressure and temperature of 0°C.

Other authors used an equation of state (Saul and Wagner, 1989) to account for the compressibility or rate of density change of pure water under high pressures up to 2500 MPa. They developed a numerical code to express water compressibility as a function of temperature and pressure.

*5.3.2.4.1.4 Specific Heat Capacity*

Observed in decimal scale, the change in $C_p$ with pressure application at elevated temperature seems insignificant with respect to other properties. However, $C_p$ showed an approximate 12% reduction in specific heat capacity compared to its value at atmospheric pressure and temperature 75°C. The reduction in specific heat capacity of water at high pressures implies that less heat is required to increase temperature of water at higher pressures than at atmospheric pressure (Rodríguez, 2003). Increase in water's $C_p$ at temperatures ranging from 75°C to 99°C at 0.1 MPa is almost zero percent (0.5%). The 3D representation of $C_p$ as a function of temperature and pressure in Figure 5.3c shows that this property is more sensitive to temperature than pressure.

The specific heat capacity ($C_p$, kJ/kg·K) of liquid water can be expressed in the following thermodynamic equation (Otero et al., 2002b):

$$C_p(P,T) = C_p(P_0,T) - T_K \int_{P_0}^{P} V\left(\frac{\partial \alpha}{\partial T} + \alpha^2\right) \mathrm{d}P \tag{5.34}$$

where $C_p(P_0,T)$ is the specific heat capacity at atmospheric pressure. Sato et al. (1988) determined an empirical polynomial of $C_p$ at atmospheric pressure as a function of temperature, which can be incorporated into Equation 5.31. Similar to the method used to determine the expansion coefficient, Ter Minassian's model on compressibility and Equation 5.30 can be substituted to numerically determine $C_p$ at a given temperature and pressure. Otero et al. (2002b) found agreement between Equation 5.34 and experimental data with a maximum relative error of 1%.

Ardia et al. (2004) modeled the adiabatic temperature rise of sucrose solutions for different concentrations, based on Equation 5.12, using mixing rules for pure substances (Knoerzer et al., 2004) at ambient pressure to determine the density of the mix ($\rho_{\mathrm{mixture}}$) and the specific heat ($C_{p\mathrm{mixture}}$) of this model:

$$\rho_{\mathrm{mixture}} = \left[\frac{[\mathrm{W}]_{\mathrm{M}}}{\rho_{\mathrm{water}}} + \frac{[\mathrm{S}]_{\mathrm{M}}}{\rho_{\mathrm{solid}}}\right]^{-1} = \left[\frac{[\mathrm{W}]_{\mathrm{M}}}{\rho_{\mathrm{water}}} + \frac{[\mathrm{S}]_{\mathrm{M}}}{\left[\dfrac{1587.9}{1+0.000107(T-15)}\right]}\right]^{-1} \tag{5.35}$$

$$C_{p\,\text{mixture}} = [\text{W}]_\text{V} \cdot C_{p\,\text{water}} + [\text{S}]_\text{V} \cdot C_{p\,\text{solid}} = [\text{W}]_\text{V} \cdot C_{p\,\text{water}} + [\text{S}]_\text{V} \cdot \xi \cdot (1622 + 7.125T) \quad (5.36)$$

where [W] and [S] are the relative amounts of water and solid in mass (M) and volume (V) percentage, respectively. Both $\rho_{\text{mixture}}$ and $C_{p\,\text{mixture}}$ were determined as a function of temperature. An empirical correction factor was used to correctly fit the experimental results for $C_{p\,\text{mixture}}$:

$$\xi = \frac{C_{p\,\text{water}}(T, 0.1\,\text{MPa})}{C_{p\,\text{water}}(T, p)^{0.75}} \quad (5.37)$$

where the numerator is represented by $C_p$ of water calculated at atmospheric pressure and the denominator denotes $C_p$ of water at higher pressure conditions.

*5.3.2.4.1.5 Thermal Conductivity*

Thermal conductivity $k$ increased by 44% at 700 MPa and 105°C, at a similar rate as temperature (Figure 5.3b). However, when increasing temperature from 75°C to 99°C at atmospheric pressure, thermal conductivity in water only changes by 1.8%. It has been reported that "during compression, thermal conductivity is greatly accelerated, since the molecules are in closer proximity, facilitating vibrational energy transfer" (Sizer et al., 2002). Thus, from the values plotted above, water at high-pressure sterilization conditions will transfer heat 1.4 times faster than water at atmospheric pressure and temperature of 75°C.

Denys and Hendrickx (1999) utilized the line heat source probe method to determine thermal conductivity. They compared thermal conductivity of water, apple pulp, and tomato paste at 30°C and 65°C in combination with pressures between 0.1 and 400 MPa. They found a linear increase of thermal conductivity due to the combined effect of temperature and pressure at initial temperatures 30°C and 65°C (Table 5.4). However, Zhu et al. (2007), utilizing their Dual-Needle Line-Heat Source method, did not find the same linear behavior even though they found good agreement with the NIST data (Harvey et al., 1996).

**TABLE 5.4**
**Thermal Conductivity Values of Water, Tomato Paste, and Apple Pulp**

| | | Thermal Conductivity *k* (W/m°C) | | |
|---|---|---|---|---|
| ***P*** **(MPa)** | ***T*** **(°C)** | **Water** | **Tomato Paste** | **Apple Pulp** |
| 0.1 | 30 | 0.53 | 0.53 | 0.55 |
| 0.1 | 65 | 0.58 | 0.60 | 0.61 |
| 400 | 30 | 0.75 | 0.67 | 0.72 |
| 400 | 65 | 0.79 | 0.73 | 0.76 |

*Source:* Denys, S. and Hendrickx, M.E., *J. Food Sci.*, 64, 709, 1999. With permission.

Kowalczyk et al. (2005) determined the density, thermal conductivity, and specific heat of potatoes, pork, and cod at 200 MPa and temperatures around freezing. Properties were found by curve fitting the time, pressure, and temperature values into a model based on a heat balance (which included a term for phase change), and by assuming the properties are independent of temperature. Similarly to water, increased pressure increased both the density and thermal conductivity, whereas the specific heat capacity decreased.

#### 5.3.2.4.1.6 Viscosity

Even though viscosity is not a thermal property, it is important to consider the role of a compressed fluid in predicting temperature evolution inside the pressure chamber. In decimal scale (Figure 5.3a), viscosity does not seem to vary with pressure at high-temperature conditions. Regarding the percent changed, an increase was observed with respect to its initial value at atmospheric pressure (Figure 5.3b), reaching a peak point around 500 MPa, and coming down to a relative difference of 6% at 700 MPa and 105°C. Since the NIST database was specified up to 500 MPa for viscosity, this decrease in relative difference cannot be assured. Viscosity decreases by 25% when atmospheric temperature is increased from 75°C to 99°C. Thus, pressurized water at 700 MPa and high temperature behaves more like a fluid at atmospheric conditions and temperature of 75°C as viscosity values are closer to atmospheric values when pressure is higher. Furthermore, if the validity of the model remains for pressures higher than 500 MPa, then the pressure fluid or pressurized liquid food would not be as viscous as expected in high-pressure application.

When numerically simulating the convective and diffusive transport effect on a high pressure-induced inactivation process, Hartmann and Delgado (2002a) implemented a model of viscosity as a function of pressure and temperature using a numerical code adapted from Watson et al. (1980) and Forst et al. (2000). The model was valid up to 600 MPa and 150°C.

#### 5.3.2.4.2 Compression Heating Prediction

Equation 5.12 has proven useful to predict compression heating of water (Otero et al., 2000; Ardia et al., 2004). Otero et al. (2000) measured temperature drop in water during the decompression step (instead of compression heating step) after holding time at selected pressures ranging from 0.1 to 350 MPa and initial temperatures between 22°C and 62°C (Otero et al., 2000). Temperature readings were successfully predicted in Equation 5.12 using small pressure intervals of $\Delta P$ (around 10 MPa) and experimental data for values $\rho$, $\alpha$, and $C_p$ taken from literature (Ter Minassian et al., 1981). Moreover, Ardia et al. (2004) predicted the temperature rise of water, sugar solutions, and orange juice in a high-pressure machine. For this purpose, Equation 5.12 was rewritten as

$$\Delta T = \int_{P_0}^{P_1} \frac{\alpha}{\rho \cdot C_p} \cdot T \cdot \mathrm{d}P \tag{5.38}$$

where $C_p$ and $\rho$ of both the sugar solutions and orange juice were substituted by the mixture formulations expressed in Equations 5.35 and 5.36, and $\alpha$ in all solutions was assumed to be the same as in water. A regressive calculation was done to express each property as a function of temperature and pressure, implementing NIST formulations for water (Harvey et al., 1996). They found no significant deviations between properties predicted by NIST and experimental results for water obtained from a Multivessel Model U111 (Unipress, Warsaw, Poland), even at sterilization conditions (initial liquid temperature 80°C and 600 MPa).

For sugar solutions, no significant deviations were detected in the range 0.1–600 MPa, while good continuity was found in extrapolating properties provided by NIST between 600 and 1400 MPa. When Equation 5.38 was applied to simulate the heat of compression in orange juice, using a sucrose solution with 9% solids content, no obvious deviations were evident from measured temperature. In summary, this work showed that the model was able to reproduce compression temperature rise even at sterilization conditions. They also verified that $\alpha$, $\rho$, and $C_p$ values of pure water, as determined from the NIST database, can be substituted into the model to predict the increase in temperature of some liquid foods such as sugar solutions and juices.

##### 5.3.2.4.3 *Water Thermal Diffusivity and Other Relations during Come-Up Time*

Thermal properties in the energy balance shown in Equation 5.16 can be rearranged so that the thermal diffusivity $\gamma$ is part of the unsteady heat conduction term:

$$\frac{\partial T}{\partial t}=\frac{T\alpha_p}{\rho C_p}\frac{\partial P}{\partial t}+\gamma\nabla^2 T \tag{5.39}$$

The thermal diffusivity is directly proportional to the thermal conductivity at a given density and specific heat, i.e.,

$$\gamma=\frac{k}{\rho\cdot C_p} \tag{5.40}$$

Thus, it measures the ability of the material to conduct heat relative to its ability to store heat. Furthermore, thermal diffusivity determines the speed of heat during 3D propagation or diffusion through the material. As shown in Figure 5.4, thermal diffusivity increases as water is pressurized, resulting from an increase in thermal conductivity. When comparing Figures 5.3b and 5.4b, it can be seen that the relative increase of both diffusivity and conductivity follows the same trend as the increase in temperature is due to compression. The increase in water thermal diffusivity with pressure indicates that the relative ability of water to conduct energy, with respect to the ability to store energy, is favored with high pressure (Rodríguez, 2003). Zhu et al. (2007) also found an increase in thermal diffusivity with increased pressure, not only in water, but also in potato and cheese.

Carroll et al. (2003) determined an analytical solution (described later), which was used to extract values of thermal diffusivity and thermal expansivity in pure water from an experimental cooling curve obtained during holding time at 600 MPa and 51°C initial temperature. Thermal diffusivity was found to be 20% higher than the value from NIST/ASME database (Harvey et al., 1996) at the corresponding temperature/pressure combination, whereas the thermal expansivity was 20% to 25% lower. Differences were attributed to the model, which assumed conduction to be solely occurring and neglected the effect due to convection. However, the theoretical cooling curve generated, including the calculated heat transfer properties, closely

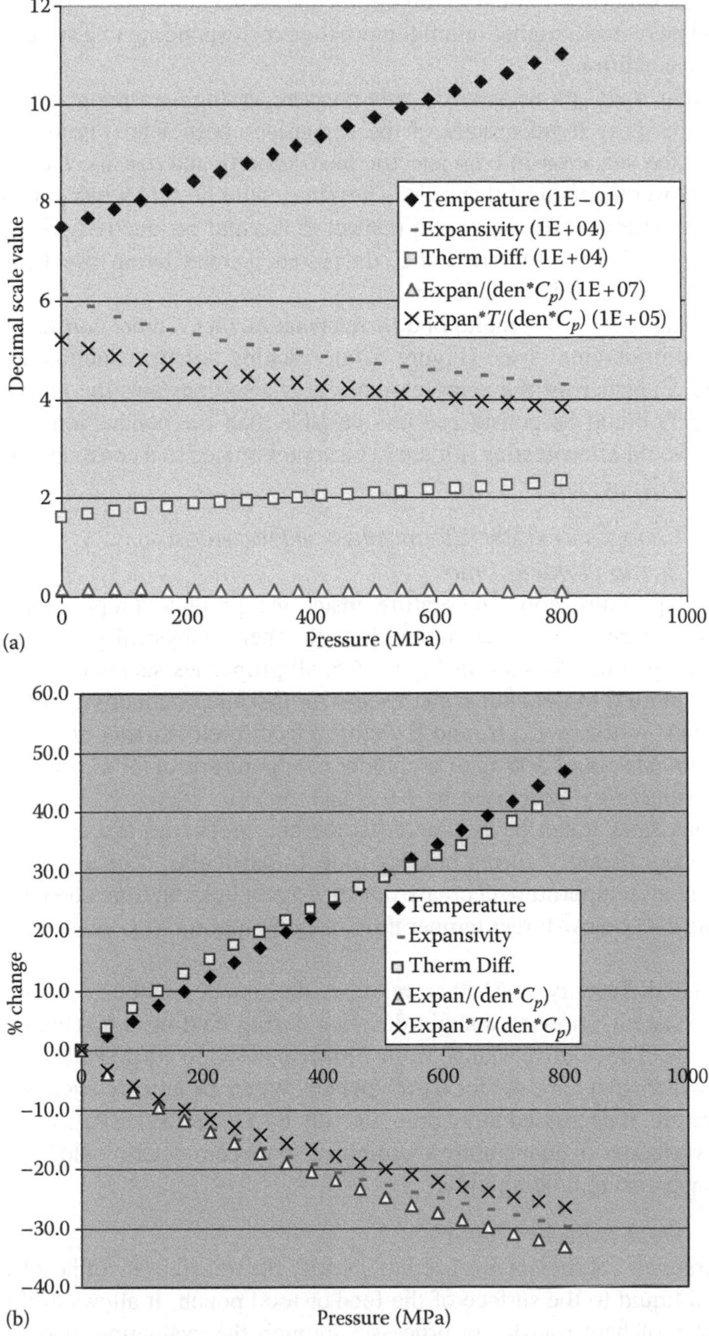


**FIGURE 5.4** Variation in property ratios corresponding to compression heating and rate of heat conduction terms in Equation 5.44, for liquid water at increasing pressure and temperature, according to compression heating rate of 4.5°C/100 MPa. Values shown: (a) values in decimal scale and (b) values expressing change (%) from initial value at 75°C and 0.1 MPa. Thermal diffusivity, $m^2/s$; expan/(den*$C_p$), $m^3/kJ$; expan T/(den $C_p$), $m^3$ K/kJ.

predicted the experimental curve, except for the higher initial process temperature. The model predicted a higher initial temperature corresponding to a system working in adiabatic conditions.

The ratio $\alpha_p/(C_p\rho)$ decreased with pressure at high-temperature conditions, following the same trend exactly of the expansion coefficient (Figure 5.4), which proved that the variation in both specific heat capacity and density was not significant for the overall trend of this ratio. The same could be said when comparing the behavior of thermal diffusivity with respect to thermal conductivity. In fact, since density increased with pressure and $C_p$ decreased, a cancellation effect can be seen in their variation at HPHT conditions.

If the ratio $\alpha_p/(C_p\ \rho)$ is multiplied by temperature (K), a more constant trend can be observed in decimal scale (Figure 5.4a), reaching a 21% reduction at 700 MPa and 105°C. Considering the compression rate is also constant, the term shown in Equation 5.39 could be considered less variable than the conduction term, which eventually would allow testing if it could be approximated to a constant value for the derivation of the model.

#### *5.3.2.4.4 Changes in Water Thermophysical Properties during Holding Time*

Once target pressure and temperature inside the product/compression fluid are reached, it is necessary to understand how the thermophysical properties change during holding time. As seen in Figure 5.5, all properties seemed to remain constant when plotted in decimal scale, except for thermal conductivity and viscosity (Figure 5.6a). While $\rho$, $C_p$, $\alpha$, and $\beta$ changed from their starting pressure holding values at 700 MPa and 105°C to a chamber temperature of 75°C, by 1.4%–2.5%, thermal conductivity decreased by 6.6% and viscosity increased by 39% (Figure 5.6b). In this case, it can be assumed that thermal properties (except viscosity) do not change significantly during holding time. In particular, if insulation is present in the chamber, temperature decrease during a 5 min holding time could be reduced to less than 2°C below target temperature, thereby maintaining most properties as constant.

Thermal diffusivity and other relations discussed previously also remained relatively constant, showing a slight decrease during holding time (Figure 5.5a) in decimal scale. The relative change from 700 MPa/105°C to 700 MPa/75°C was 8.8% for thermal diffusivity and 4% for $\alpha_p/(C_p\ \rho)$ ratio. When the latter relation is multiplied by temperature, it decreased more pronouncedly by 12% at 700 MPa/75°C. However, assuming variations in temperature are minimal, these parameters could be considered to not change during holding time.

#### *5.3.2.4.5 Heat Transfer Coefficient*

The heat transfer coefficient $h$ is used to quantify the transfer rate of heat by convection from a liquid to the surface of the food or food pouch. It allows evaluating the effectiveness of heat transfer in processes through the evaluation of overall resistances participating in the system of study. Particularly in high-pressure processing, for sterilization purposes especially, this property is used to calculate the amount of heat transfer between the food package and the surrounding fluid. Some authors (Carroll et al., 2003) assume that the pressurizing fluid is typically undisturbed

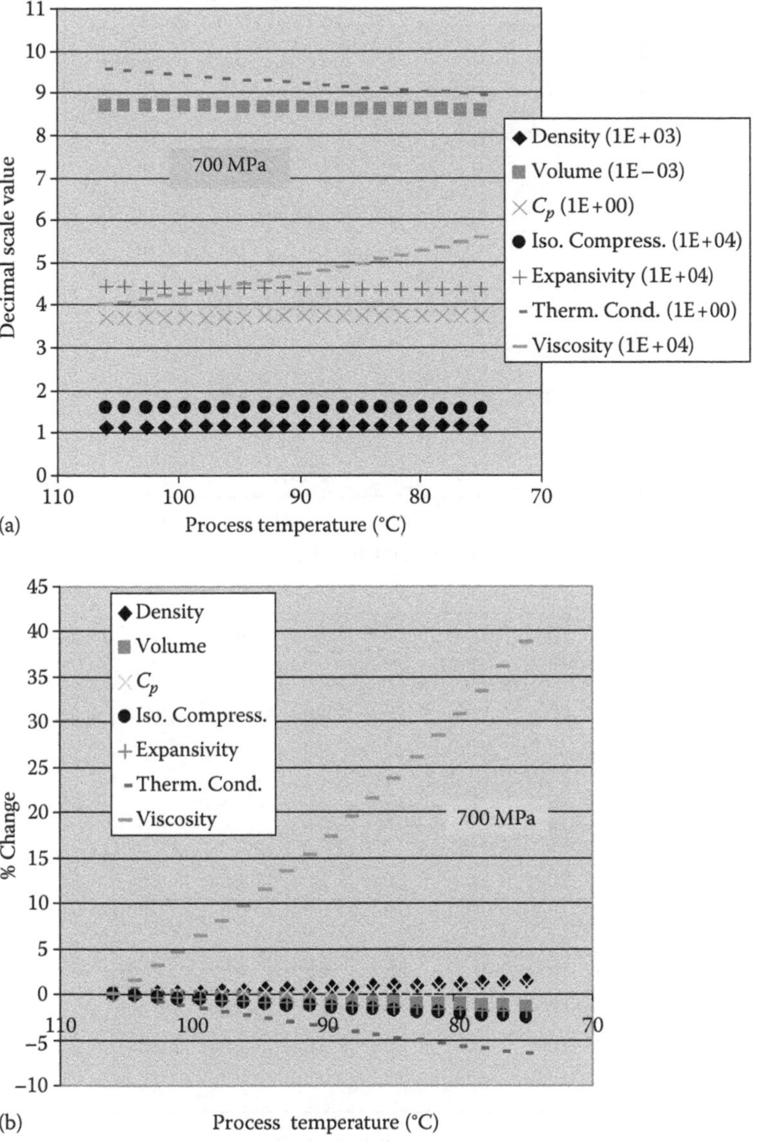


**FIGURE 5.5** Variation in thermophysical properties of liquid water during holding time at 700 MPa and decreasing process temperature, according to cooling rate of 1.63°C/time unit. Properties were calculated from NIST/ASME properties database. Values shown: (a) in decimal scale and (b) values expressing change (%) from initial value at 105°C and 700 MPa during pressure holding. Temperature, °C; density, kg/m$^3$; specific volume, m$^3$/kg; $C_p$, kJ/K kg; compressibility, 1/MPa; expansion coefficient, 1/K; thermal conductivity, W/m K; viscosity, Pa· s.

and convection within the pressurizing fluid is not considered. Carroll et al. (2003) considered the heat transfer coefficient from the chamber inner wall to the heating jacket (referenced to inner area); where the jacket temperature was the reference

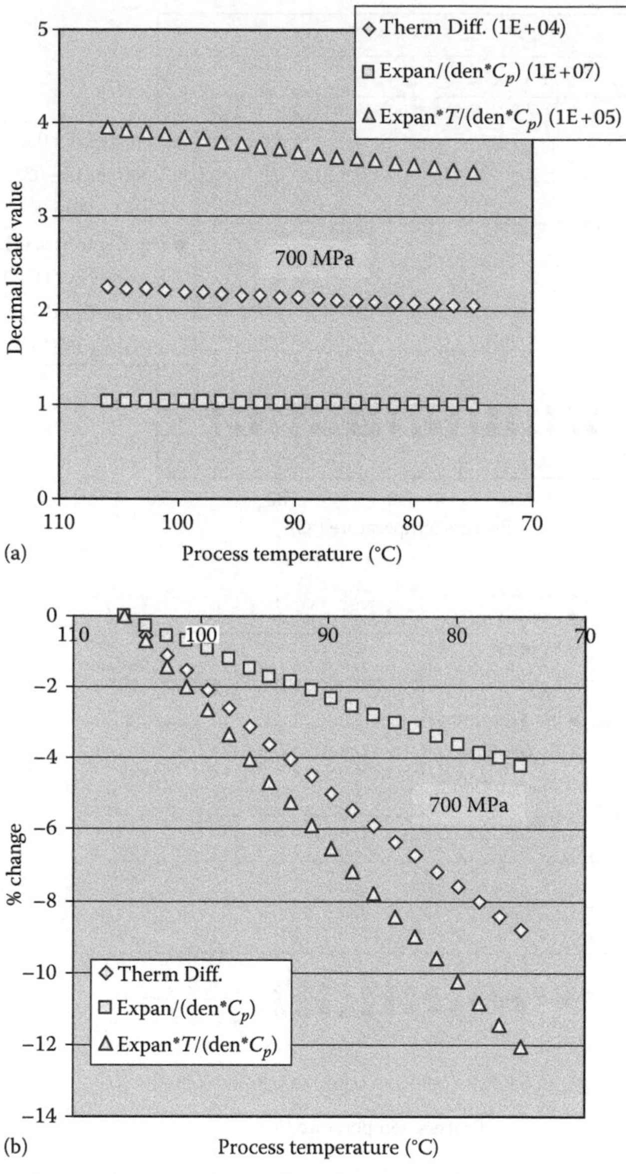


**FIGURE 5.6** Variation in property ratios of liquid water during holding time at 700 MPa and decreasing process temperature, according to cooling rate of 1.63°C/time unit. Values shown: (a) values in decimal scale and (b) values expressing change (%) from initial value at 105°C and 700 MPa during pressure holding. Thermal diffusivity, $m^2/s$; expan/(den*$C_p$), $m^3/kJ$; expan T/(den $C_p$), $m^3$ K/kJ.

initial temperature used to determine a heat transfer model. As mentioned before, heat transfer coefficients can be extracted from empirical relations between dimensionless Nusselt, Prandtl, and Reynolds numbers according to laminar or turbulent conditions (Perry, 1997). However, not much information has been published on the validity of empirical equations, as applied to extracting heat transfer coefficients at high-pressure conditions. Different regions inside the vessel may require a particular equation (including dimensionless numbers) at the boundary to represent accurate heat transfer coefficients.

## 5.4 MODELING APPROACHES FOR PREDICTING TEMPERATURE UNIFORMITY

Factors discussed in the above sections affect the heat transfer in the pressure vessel in a substantial way. For a model to accurately predict the evolution of temperature in packages located at different points in the high-pressure vessel, the model should account for all factors (Otero and Sanz, 2003). However, the identification of reliable modeling approaches representative of the extent of temperature uniformity and its evolution during high-pressure processing is work in progress. Four types of modeling methods have been proposed by different research groups: analytical, numerical, macroscopic, and artificial neural networks (ANN). The sections below will review the proposed models and assumptions, providing insight into their advantages and limitations.

### 5.4.1 Important Design Assumptions in Developing a Model

In order to design a model that allows accurate prediction of chamber temperature profiles during pressurization, a number of assumptions need to be made. The assumptions made for a particular vessel system should provide the closest prediction of the process as influenced by factors such as the chamber system, insulation conditions, compression medium composition, etc. (see Table 5.3).

#### 5.4.1.1 Assumptions for the Chamber System

Of the two main compression systems existing for high-pressure processing, the indirect system using an external intensifier design is most commonly selected for sterilization purposes. Therefore, an indirect pressure chamber, rather than a plunger press design, should be the modeling system of choice.

As mentioned in Section 5.2.2.2, modern high-pressure sterilization systems include a series of design features and controls to ensure initial temperature homogeneity inside the sample carrier before pressure is applied. Thus, initial temperatures in the vessel furnace (if any), the fluid surrounding the polymeric carrier, the polymeric carrier inner fluid, and food packages forming the system ought to be assumed homogenous and constant (at thermal equilibrium). In particular, the temperatures of the fluid and food packages inside the polymeric carrier should be the same ($T_s$). Farkas and Hoover (2000) mentioned that the initial temperature $T_s$ must not be less than 0.5°C below the specified value in all food locations.

In addition, pressure applied to the food sample and compression fluid must be assumed equal at all points. Most pressure pump systems increase the pressure inside the vessel at a constant rate, therefore the pressurization rate as well as pressure release rate can be assumed constant (Otero et al., 2000; Carroll et al., 2003).

The compression heating rate of most polymers chosen to build a liner is still unknown in the public domain. Some of these polymers may have a much higher compression heating rate than steel (rate is ~0), even matching the compression heating of water (3°C–5°C/100 MPa). Furthermore, it is reasonable to assume that the thermal conductivity of these polymers will remain much lower than stainless steel under pressure and therefore retain compression heat. Thermal conductivity and diffusivity of these materials at high-pressure conditions are also unknown, especially as function of temperature and pressure. Therefore, based on the current knowledge, an additional safety factor would be to assume the material has zero compression heating, but still maintains insulating properties.

It is also assumed that the steel vessel has a predefined constant volume, thus neglecting the expansion of the pressure vessel due to associated piping during compression. Vessel expansion may add several percentage points in fluid to the vessel's volume (Otero and Sanz, 2003). A filled 100 L vessel will require an additional volume of water (15–20 L) for it to reach a pressure of 680 MPa, and some of this energy exchange might not only translate into a pressure or viscosity increase, but also into a vessel volume change.

The vessel walls should be considered, as discussed before, to have variable temperature over the wall surface during pressure application (Hartmann et al., 2004), rather than an "infinite reservoir where heat is rapidly distributed and temperature gradients in the steel vanish immediately" (Hartmann and Delgado, 2002b). Furthermore, it should be assumed the chamber is completely filled with water at the beginning. The thermostat adapted to the heating system (including a water jacket or heat resistors) might also show variations during holding time, thereby giving a fluctuating boundary condition. These fluctuations should also be ignored when developing a model. Radiation heating from the vessel walls can also be assumed as negligible.

As mentioned in Section 5.3.2.3, different approaches to boundary conditions for temperature and velocity prediction on the liquid–solid interfaces can be applied. Furthermore, in vertical systems, it may be assumed that the entrance of compression fluid is at the geometrical center, allowing consideration of a 2D axis-symmetric system. This assumption is not possible for horizontal vessels since convective motion due to gravity is not axis symmetric; therefore, a full 3D model is required.

#### 5.4.1.2 Assumptions for Compression Fluid

Based on previous discussion in Section 5.2.1.2, it can be assumed the compression liquid is pure water, since the compression fluid used in industrial-type vessels for pasteurization and sterilization purposes is actually drinking water with no additives. Therefore, a model could include thermophysical properties for water only (plus properties of food and carrier material).

In theory, the heat generated by compression is dissipated by a combination of conduction and convection within the pressurizing fluid in the chamber, and by transfer across the chamber wall into the surroundings (Hartmann and Delgado, 2002a,b; Carroll et al., 2003). Assuming an external intensifier is used, a certain volume of fluid is forced into the chamber to reach the target pressure, thereby creating convective currents. Carroll et al. (2003) and Kowalczyk et al. (2004) assumed the convection effect within the pressurizing fluid is negligible based on the fact that during the pressure holding time, the pressurizing fluid and chamber volume are typically undisturbed. However, the model developed by Knoerzer et al. (2007) has shown that buoyancy forces due to temperature and thus density gradients lead to a fairly strong convection.

Furthermore, Carroll et al. (2003) calculated the resultant error for a laboratory plunger press system (17 mm in diameter) and attributed a 20% error caused by neglecting the contribution of convection. Hartmann and Delgado (2003b) have shown that even though the incoming pressurizing fluid is tempered before entering the high-pressure vessel, it is not heated further due to the effect of pressure (in reaching temperature of existing fluid in chamber), and it may thus "cool down" the bottom section of the vessel as shown in a 6.3 L vessel initially at 40°C. Knoerzer et al. (2007) also established that convective currents inside the vessel, especially in the absence of a carrier, play a major role in the temperature value achieved as well as uniformity.

In a comparative study between an indirect system without packages and a direct system with packages filled with water, Hartmann and Delgado (2003b) assumed that all liquids are Newtonian, compressible and chemically inert. The fluid was assumed to be initially at rest and in thermal equilibrium. Hartmann (2002) estimated a Reynolds number during pressurization of approximately 10. Based on this, the flow of the liquid was assumed as laminar. In addition, once target pressure is reached, it should be assumed the inflow of compression fluid stops.

During "instant" pressure transfer during come-up time, increase of mass in the pressure medium leads to a volume reduction in the samples. This phenomenon can be neglected by assuming the volume as constant and density as a function of pressure.

#### 5.4.1.3 Assumptions for Food Packages

For food packages, the first possible assumption is that the thickness of the packaging material allows for sufficient heat penetration and does not affect the temperature distribution at the end of preheating period. Furthermore, Hartmann and Delgado (2003b) pointed out that it should be assumed that deformation of packages obeys a prescribed kinematics, to maintain its integrity and to not provide significant resistance to compression. In addition, the compression heating effect of the packaging material layer surrounding the food should be assumed as similar to water.

The pressure of compression fluid should be assumed as almost identical to that of contained food in the package. This assumption has the following implications:

- No air headspace should be assumed to remain inside the package.
- Gas released from food structure due to temperature rise at preheating step or due to mechanical hydrostatic shrinking of food should be negligible.
- Shrinkage due to compression of contained air should be negligible.

A further assumption is that the food does not change its composition, structure, and phase state between initial and termination points of the process (Farkas and Hoover, 2000). However, the HPHT used might produce compositional and structural changes (e.g., changes in porosity or density) in selected food composites. The most common occurrence in high-protein foods is the pressure-induced protein denaturation (Otero and Sanz, 2003). Thus, chemical/physicochemical reactions and changes in food volume or compressibility that might affect mass and heat transfer pathways can be neglected.

The shape, amount, and position of pouches in the vessel can also affect the way heat is transferred during preheating and pressurization (Table 5.3). An initial approach should consider several options: (a) no pouch inside, (b) one pouch fixed at the geometric center of the vessel, and (c) several pouches geometrically distributed and fixed within the vessel (undisturbed by flow of incoming pressure transmitting fluid). The location, size, and shape of the sample should be fixed to establish boundary conditions. An initial consideration due to the cylindrical shape of the vessel could be to have a cylindrical type of package for an axis-symmetric model (Hartmann and Delgado, 2003b).

Thermal gradients due to differences in compression heating properties between sample and sample holder can also be neglected. Due to lack of knowledge on the thermophysical properties of different food components, these can be assumed similar to water. Depending on whether a solid, liquid, or solid:liquid mixture, or porous material is contained in a flexible container, mass transfer considerations can be considered negligible as an initial approach.

Gravity influence, however, cannot be assumed as negligible. Especially during holding stage, natural convection is mainly influenced by gravitational forces (Knoerzer et al., 2007). Carroll et al. (2003) assumed the food and fluid properties to be isotropic but still dependent on temperature and pressure. Nevertheless, temperature differences throughout the vessel lead to nonisotropic material properties, which need to be accounted for.

#### 5.4.1.4 Assumptions for Processing Conditions

In this case, a single pressurization step may be considered and the processing times (including for preheating and pressurization) may be assumed constant based on a constant fluid inlet velocity. Depending on inlet geometry and velocity, the governing differential equations for either laminar or turbulent conditions have to be selected. Input variables for the model may be the initial temperature inside the product and vessel ($T_s$), target pressure $P_1$, pressure rate, holding time $[t_{p1} - t_{p2}]$, and decompression rate (Figure 5.1). CFD software packages allow starting from an initial temperature distribution from a previous model (e.g., by considering vessel interactions with all components or by accounting for the product and carrier preheating step) or by using different initial temperatures in different components. For example, Juliano et al. (2008) studied the effect of having a colder top vessel lid at the start of the pressurization process (discussed in Chapter 6).

### 5.4.2 Analytical Modeling

Analytical modeling involves the fitting of experimental temperature data from a heating or cooling curve obtained from pressure come-up, pressure come-down, or pressure holding periods into an equation, and then determining the model parameters needed to represent the thermal effects occurring during HPHT processing. This type of modeling enables calculating the thermal expansivity, thermal diffusivity, and heat transfer coefficient (for any vessel) from an analytical solution by means of adjusting experimental temperature data (Carroll et al., 2003). In particular, the model should at least be able to predict the final temperature during compression $T_{p2}$, the final temperature after pressure release $T_f$, and the overall process time during $t_f$ at specified vessel locations. As a rule, if an energy balance can be solved analytically, its solutions will be more general, straightforward, and easier to apply than a numerical solution. These models are one-dimensional (1D) in nature since they do not consider flow occurring inside the vessel. In order to account for flow, equations described in Section 5.3.2.2 must be accounted for and also require numerical methods to be solved. Chapter 6 describes an example of an analytical model applied for the prediction of temperature at a single vessel location after compression and holding time.

### 5.4.3 Numerical Modeling

For calculation of heat and mass transfer in more complicated situations (e.g., when flow or heat transfer coefficients are considered, or variables are used as functions of pressure and temperature) numerical models are needed. As already mentioned, analytical models are not able to contemplate several scenarios in a single expression, especially when modeling the process in two or three dimensions. For this purpose the problem is reduced significantly by requiring a solution for a "discrete" number of points (or grid), rather than for each point of the space–time continuum in which the governing equations of mass, energy, and momentum are applied. These partial differential equations describing the entire system are transformed into a system of equations and solved numerically to approximate the exact solution (Nicolaï et al., 2001).

#### 5.4.3.1 Discretization Methods

Various discretization methods can be used for the numerical solution of high-pressure problems. In all cases, computational grids are tailored to provide a "mesh" independent solution for the numerical approximation of the governing equations. The most commonly used are the finite volume method, the finite element method, and the finite difference method in CFD.

##### *5.4.3.1.1 Finite Volume Method*

The finite volume method of discretization is most widely used in CFD software packages at the moment. It obeys the clear physical principle of conservation of incoming and outgoing mass, energy, and momentum from each volume element. First, the given computational domain is subdivided into finite volume elements. Second, the system of general conservation equations is written in every volume element, independently of the coordinates. Third, the system is integrated over finite volume $V$ with surface $A$. Volume integrals in the governing partial differential equations for all transported quantities in the system (i.e., energy, mass, and momentum), containing a divergence term, are converted to surface integrals using the divergence (or Gauss) theorem.* By integrating, these terms are then evaluated as fluxes at the surface of each finite volume. Given that

* The divergence theorem states that outward flux of a vector field through a surface is equal to the triple integral of the divergence on the region inside the surface, i.e., it states that *the sum of all sources minus the sum of all sinks gives the net flow out of a finite volume element.*

the flux entering a given volume is identical to that leaving the adjacent volume, these methods are conservative. More information about the theory behind the finite volume method is covered elsewhere (Versteeg and Malalasekera, 1995).

The finite volume method has been used by several authors (Hartmann, 2002; Hartmann and Delgado, 2002a,b, 2003a,b; Hartmann et al., 2003, 2004; Ghani and Farid, 2007) to numerically solve conservation laws and thus simulate temperature and flow distributions inside high-pressure vessels with different configurations (described in Chapter 6).

###### *5.4.3.1.2 Finite Element Method*

This method provides an overall solution in the entire computational domain and consists of five steps:

1. The given computational domain is subdivided into a collection as a number of finite elements and subdomains of variable size and shape. These are interconnected in a discrete number of nodes. A large number of element shapes have been suggested in literature and are provided in most finite element software packages. A number of 2D and 3D element shapes are shown in Figure 5.7.
2. The solution of the partial differential equation is approximated for each element by a low-order polynomial in such a way that it is defined uniquely (i.e., coefficients are determined in terms of the "approximate" solution) at the nodes (Nicolaï et al., 2001).
3. The global approximate solution, composed of all solutions at the nodes, can then be written as a series of low-order discrete polynomials.
4. Residuals (or errors) are obtained by substituting the approximate solution into the differential equations.
5. The unknown coefficients of the polynomials are found by orthogonalization of these errors with respect to the polynomials. After the coefficients are determined, a system of equations (algebraic or ordinary differential) is solved by certain techniques to obtain predicted temperature fields.

The theory of this method escapes the scope of this chapter. More detailed information can be extracted from Nicolaï et al. (2001) and Zienkiewicz (1977). Otero et al. (2007) and Knoerzer et al. (2007) used COMSOL Multiphysics (COMSOL AB, Stockholm, Sweden), which incorporates the finite element method, to predict flow and temperature distributions in laboratory and pilot scale high-pressure systems.

###### *5.4.3.1.3 Finite Difference Method*

The finite difference method is based on the approximation of the derivatives in the governing equations by the ratio of two differences (i.e., temperature or velocity over distance or time). For example, the first time derivative of the temperature as a function of time $T(t)$ at time $t_i$ can be approximated by

$$\left.\frac{\mathrm{d}T}{\mathrm{d}t}\right|_{t_i} \cong \frac{T(t_{i+1}) - T(t_i)}{\Delta t} \tag{5.41}$$

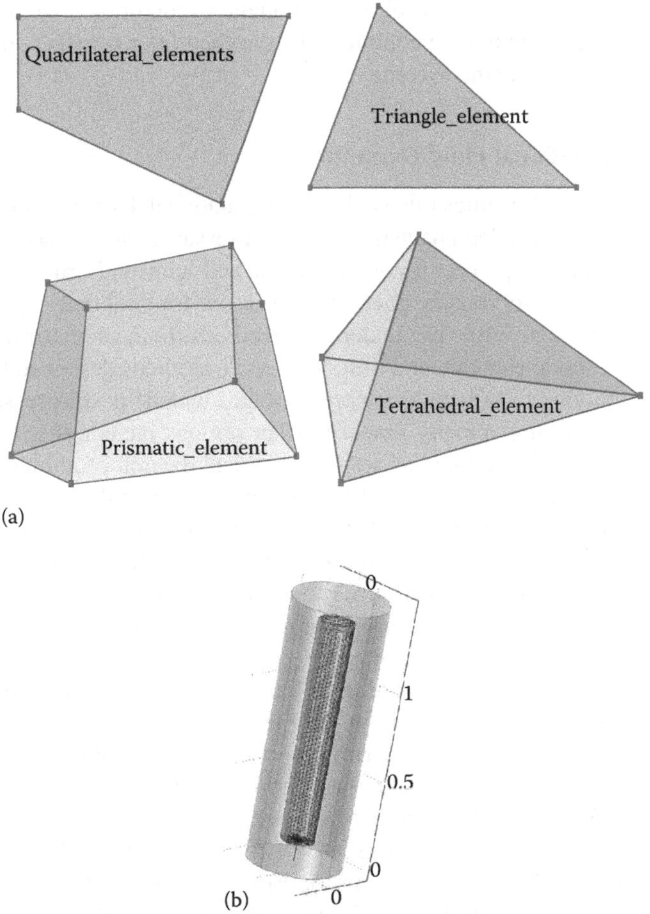


**FIGURE 5.7** Representation of finite elements: (a) typical 2D and 3D finite element shapes and (b) 3D representation of a high-pressure vessel divided into tetrahedral elements.

with $\Delta t = t_{i+1} - t_i$. This expression converges to the exact value of the derivative when $\Delta t$ decreases. A finite difference approximation can be obtained from a first-order Taylor series approximation of $T$ at time $t_i$ applied to the differential equation describing conservation of energy. Likewise, finite difference formulas can be established for second-order derivatives using the Taylor series.

For the purpose of calculating the spatial derivatives, a computational domain is subdivided into a regularly spaced grid of lines that intersect at common nodal points. Subsequently, the space and time derivatives are replaced by finite differences.

As shown in Chapter 6, the finite difference method was used (Denys et al., 2000a,b) with self-developed Delphi 3 mathematical codes to calculate conductive heat transfer in the whole high-pressure system. Ardia et al. (2004) used the finite difference method to predict compression heating of water and other water-based materials using MathCAD (Mathsoft Engineering and Education, Inc., Needham, Massachusetts, USA). Furthermore, Hartmann et al. (2004) applied this method (also self-developed) to the vessel wall boundary conditions, to represent the variable temperature due to heat conduction from the hotter compression fluid used in their finite element scheme (Section 5.4.1.1).

#### 5.4.3.2 Computational Fluid Dynamics

Computational Fluid Dynamics (also called Computational Thermal Fluid Dynamics [CTFD]) is a computer-aided analysis of fluid conservation laws (mass, momentum, and energy) to simulate a process for devices (solid and semisolid structures) that interact with fluid. It allows solving the governing equations for fluid flow and heat transfer (Equations 5.18 through 5.20). The underlying methods for the numerical analysis are the finite volume, finite element, and finite difference methods described before.

Most commercially available software packages include postprocessing features that provide temperature/velocity maps. Further options include the representation of contour, vector, and line plots as well as the visualization of animated flow and temperature fields (Norton and Sun, 2006). Postprocessing analysis tools in software packages are sometimes not sufficient, especially when trying to validate the model against real temperature data or, in some cases, when trying to use the predicted output temperature data to calculate microbial inactivation distribution data. In this case, other software routines (e.g., programmed in MATLAB) can extract the data from the solution to perform such tasks.

Numerical simulations of pressure vessels with a vertical pressure fluid inlet near the center bottom are quite common in the literature (Hartmann, 2002; Hartmann and Delgado, 2002a,b, 2003a,b; Hartmann et al., 2004; Knoerzer et al., 2007; Otero et al., 2007). In this case, 2D cross-sections are used as the computational domain (Figure 5.8) due to rotation symmetry at the central axis.

CFDs software packages such as CFX-4.4 (ANSYS CFX, ANSYS Inc., Southpointe, Canonsburg, Pennsylvania), FLUENT (FLUENT Inc., Lebanon, New Hampshire), PHOENICS (CHAM Ltd., Wimbledon Village, London,), and COMSOL Multiphysics are most commonly used in the high-pressure processing area. These software packages provide numerical algorithms for solving governing equations of fluid dynamics as well as interfaces for the implementation of special purpose software.

For example, Hartmann and Delgado (2003a,b) enhanced the software performance by using their own software routines, covering more than 5000 statements of FORTRAN 90 code, and by linking the statements at six different interfaces to the main software code. The geometry of the fluid volume of the high-pressure cell was digitized by the application of CAD-techniques. Expressions for thermophysical properties can be inserted in the software routines representing the model, by using different equations reported in the literature or determined experimentally, as a function of pressure and temperature (Hartmann et al., 2003; Knoerzer et al., 2007). Today computer speed and memory capabilities allow for reduced computational times of less than 1 h for a simulation of the entire process using a refined number of elements in the mesh (Knoerzer et al., 2007), whereas past publications report computational times of 15 h (Hartmann et al., 2004).

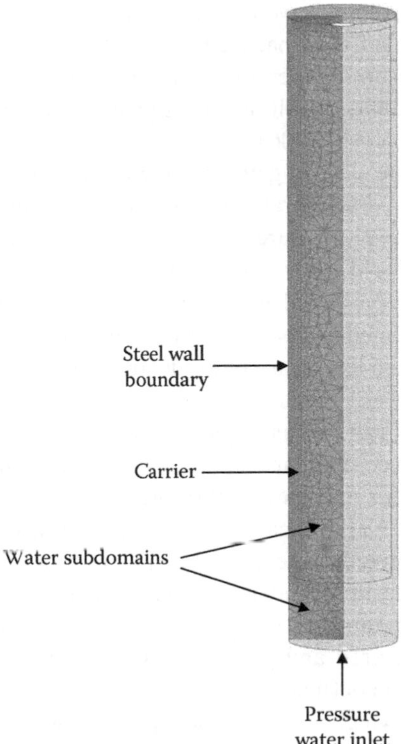


**FIGURE 5.8** Computational domain of a rotation-symmetric high-pressure vessel including a carrier for CFD modeling. (Adapted from Knoerzer, K., Juliano, P., Gladman, S., Versteeg, C., and Fryer, P., *AIChE J.*, 53, 2996, 2007.)

### 5.4.3.3 Validation of Numerical Models

Numerical methods used for predicting 2D or 3D distributions may converge suggesting solutions that might be plausible, but in fact are not accurate enough (Nicolaï et al., 2001). Therefore the numerical solution must always be validated. The validation process involves the comparison of predicted data (i.e., temperature, velocities, inactivation extent, chemical or physical change, etc.) with measured data.

Temperature validation in a high-pressure vessel has been done by using thermocouples adapted to the pressure system. For laboratory-scale systems, one or two thermocouples are sufficient to measure distribution, by changing their position at predefined points. In pilot systems, an array of thermocouples is recommended (Knoerzer et al., 2007) to reduce the number of pressure runs and the resulting inherent error. Obtaining reliable temperature measurement with thermocouples remains a challenge; hence, other options are being explored. For instance, a "thermal egg" consisting of a metal shell enclosing a temperature data logger has been developed at Food Science Australia and has successfully demonstrated accurate measurements in all high-pressure processing steps (as validated with thermocouple measurements).

It has the advantage of placing several units in different vessel locations at a single run and even inside sealed food packages without being invasive. Another device being developed to measure temperature under pressure consists of a wireless temperature probe, which emits an ultrasound signal that is read by an external data logger (Buckow and Agueeva, 2006).

Once the temperature data is gathered, there are different ways of comparing simulations with measured temperatures. The most common method of verification and validation is to compare measured and simulated temperature curves (Hartmann et al., 2004; Ghani and Farid, 2007; Knoerzer et al., 2007; Otero et al., 2007). In this case, temperature curves predicted and measured in several specific locations in 1D, 2D, or 3D can be represented in a parity plot (Knoerzer et al., 2007), where measured temperature and simulated temperatures at identical locations and selected times are represented in a graph.

Temperature curves can be easily plotted using conventional software such as Microsoft Excel. However, comparison of 2D or 3D distributions in form of parity plots requires working in matrices containing temperature data in all locations. Scripts developed in MATLAB can help with this task (Knoerzer et al., 2007). First, temperature vs. time data needs to be gathered for all locations of the modeled system. Then, measured and predicted temperature profiles are stored as vectors (generally time–temperature profiles) in 2D or 3D arrays (outlining the system geometry). Measured and predicted profiles are matched in a parity plot from which a correlation coefficient is determined.

A further approach, which is also the only possible way to investigate flow fields inside a high-pressure vessel to date, was identified by Pehl and Delgado (1999, 2002). They developed high-pressure digital particle image thermography (HP-DPIT) and high-pressure digital particle image velocimetry (HP-DPIV). Both applications involve the use of encapsulated thermochromic liquid crystals, which change their color with varying temperature. Transient color fields in the pressure/temperature domain document (photographically) the temperature and flow fields at high pressure. As of today, fluid dynamic effects can only be studied in laboratory-scale vessels. Given the complexity involved in the addition of a crystal window within the vessel structure, it is not yet possible to determine flow effects at pilot scale.

Distributions of inactivation or chemical or physical changes cannot be validated through measurements at selected points. A certain volume in packages containing an initial amount of substance at least needs to be considered. In this case, overall averages for the whole vessel or vessel areas where packages are located are calculated from the predictions in the model. While some authors have validated modeled enzyme inactivation (Hartmann and Delgado, 2003b), none have validated a model for spore inactivation at HPHT conditions. Chapter 6 provides examples of modeling distributions of $\alpha$ amylase enzyme and *Escherichia coli* solutions, and *C. botulinum* spores in fluid and solid "water-like" media.

### 5.4.4 Macroscopic Modeling

A macroscopic model is a generalized representation of the system. It is a flow-based visual programming technique that defines applications as networks of "black box" processes that exchange data across predefined connections. Black boxes are graphic representations of objects and systems of interest that can be reconnected endlessly to form different applications without having to be changed internally. The objects can be inputs, outputs, operators (e.g., a multiplier), or specific functions (e.g., an adiabatic heating equation). Different platforms allow for multidomain simulation and model-based designs of dynamic systems. Software packages such as Simulink (The Mathworks Inc.) and LabVIEW (National Instruments Corporation, Austin, Texas) provide an interactive graphical environment and a customizable set of block model libraries to design and model time-varying systems.

In the case of high-pressure systems, macroscopic modeling encompasses a thermal exchange between different parts of the equipment connected in blocks, forming a circuit. Thermal interactions between, for example, the sample/fluid and vessel, the heating coil and steel mass, the pressure intensifier and vessel inlet and other portions of the high-pressure system are some of the building blocks for the model, including variables not usually considered in heat balances. In fact, these types of models are useful in evaluating the impact of each and every component involved in ensuring the sample reaches the target temperature. The model still requires a numerical conductive/convective heat transfer model to reproduce the thermal gradients established within the sample and the compression fluid. Hence, thermophysical properties of the system are needed. Chapter 6 provides an example of a macroscopic model.

### 5.4.5 Artificial Neural Network Modeling for Process Control

An ANN is a mathematical algorithm with the capability of relating the input and output variables (Table 5.3) and learning from such examples through iteration, without requiring prior knowledge of the relationships between process variables (Torrecilla et al., 2004, 2005); the ANN learns its internal representation from the input/output data of its environment and response. Data representative of the process is gathered and put into the training algorithms for automatic learning of the structure of the data (i.e., neural networks learn by example). The network is defined by connections in parallel and sequences between neurons. Previous publications on heat transfer and thermal process predictions have employed neural networks to predict parameters characterizing thermal inactivation, such as process time, process lethality, associated quality factors, and surface heat transfer coefficients (Sablani et al., 1995, 1997; Sreekanth et al., 1999; Afaghi et al., 2001; Torrecilla et al., 2005; Chen, 2006; Chen and Ramaswamy, 2006).

In high-pressure processing, neural networks can be also employed to characterize the temperature and pressure history at specified points. For example, they can predict the maximum temperature reached in the sample after pressurization. In this case, the advantage is that thermophysical properties are not needed to perform the prediction (see examples in Chapter 6).

Neural networks can be developed either by using a specific computer language accounting for their principles, or on use of commercial neural networks software packages. Development of neural network codes involves turning the theory of a particular network model into a computer simulation implementation (Chen and Ramaswamy, 2006). Commercially available neural network software packages have been developed, for example, by NeuroDimension Inc. (Gainsville, Florida) and The Mathworks Inc.

#### 5.4.5.1 Neural Network Architecture

Neural networks consist of a set of neurons called processing units, which are arranged in several parallel layers. The most commonly used neural network architecture (as explained later) is the multilayered feed-forward network layer using back-propagation of error in the learning mechanism.

The structure of this model is based on the use of three or more layers. There are two layers for input and output data and a number of hidden layers for processing and iterating. The *input layer* receives information from an external source and passes this information on to the network for processing. The hidden layer (or layers) receives and processes information from the input layer. The number of hidden layers could be one, up to three, depending on the problem in place. The *output layer* receives process information from the network and sends the result to an external receptor. When the input layer receives the information from an external source, it is activated, emitting signals to all neurons in the first hidden layer, which will in turn transfer the signal to the next layer. Depending on the strength (weight) of the interconnections, the signals can excite or inhibit the nodes to ultimately reach the output layer and provide the prediction.

A neural network can be viewed as a "black box" into which a specific input to each node (or neuron) in the input layer is sent. The network is defined by connections in parallel and sequence between hidden nodes through which information is processed. Finally, the network gives an output from the nodes to the output layer. For example, in a high-pressure process, the input layer may include the applied pressure, pressure rate, high-pressure vessel temperature, and fluctuations in ambient temperature, whereas the output layer provides predicted variables such as the maximum temperature reached after pressurization as an average temperature or specific vessel point temperature, depending on the type of data used during the training process.

#### 5.4.5.2 Artificial Neural Network Development

The ANN can be developed according to the following steps (Torrecilla et al., 2004, 2005; Chen and Ramaswamy, 2006):

##### *5.4.5.2.1 Selection of the Number of Layers*

It has been shown that selecting one hidden layer is sufficient to approximate any continuous nonlinear function for network training purposes (Torrecilla et al., 2004). However, more hidden layers could be used in special applications.

##### *5.4.5.2.2 Selection of the Transfer Function between Neurons*

In general, neurons can be connected to each other by weighted links, $w_{ij}$, over which signals can pass. Each neuron receives multiple inputs proportional to their connection weights, generating a single output that may be propagated to several other neurons (Sreekanth et al., 1999; Torrecilla et al., 2005; Chen and Ramaswamy, 2006).

An interactive function between neurons is shown in the scheme presented in Figure 5.9. The inputs ($y_i$) into each incoming node $i$ are multiplied by their corresponding connection weights ($w_{ij}$) and added together to yield $x_i$:

$$x_i = \sum_{i=1} w_{ij} \cdot y_i \tag{5.42}$$

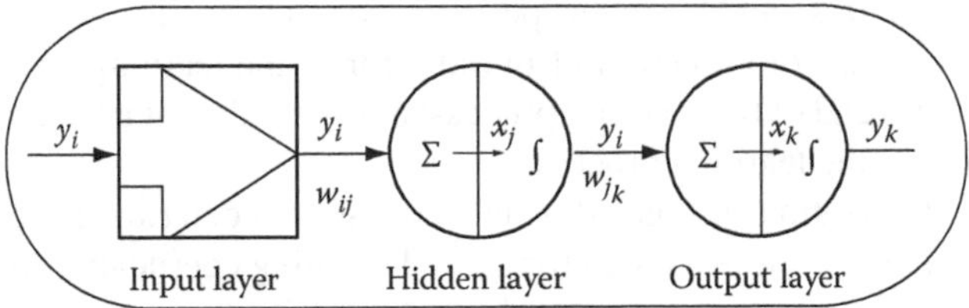


**FIGURE 5.9** Structure of neural network model: (a) a typical multilayer neural network with one hidden layer, and input and output variables representing schematic of data transfer between neurons. The inputs ($y$) into a neuron are multiplied by their corresponding connection weights ($w$) and summed together. This sum is then transformed through a selected function to produce a single output to be passed between neurons in other layers. (Adapted from Torrecilla, J.S., Otera, L., Sanz, P.D., *J. Food Eng.*, 69, 299, 2005.)

This sum is then transformed by means of an input activation function, producing a single output $y_j$, where $j$ represents a neuron in the hidden layer, which may be passed on to other neurons. The input activation function can be any continuous function and is typically a monotonic nondecreasing nonlinear function (e.g., hyperbolic, linear threshold, Gaussian function, or sigmoid function).

#### 5.4.5.2.3 ANN Training and Learning Step

For learning purposes, the training data set consists of pairs of input and desired output data. The input data is fed into the network and the estimated output is compared with the real output by calculating the input–output difference as the error signal. The training of the network is based on adjusting the connection parameters ($w$) so that the difference is minimized between the estimated output $y_k$ of each neuron $k$ at the output layer and the real output data ($r_k$). The prediction error $E_k$ can be used as a comparative parameter between ANN response and real output:

$$E_k = \frac{1}{2}\sum_k (r_k - y_k)^2 \tag{5.43}$$

The *learning rule* is a method to adjust the weight factors based on trial and error. Chapter 6 provides two examples (Torrecilla et al., 2004, 2005) that use the "error-correction learning" as a learning rule, and the back propagation algorithm to automatically adjust the weights ($w$) to minimize the estimation error $E_k$ after back distribution across the network. In these examples, $E_k$ is back distributed to the previous layers across the network until minimum error is obtained.

There are two parameters that can be used to optimize the ANN at the training step by minimizing the prediction error: the number of neurons in the hidden layer and the learning coefficient μ (see example in Chapter 6). The number of neurons in the hidden layer is related to the converging performance of the output error function during the training process of the network. It defines the "topology" of the system. For example, a topology "5, 3, 2" defines a system with five nodes in the input layer, three neurons in the hidden layer, and two neurons in the output layer.

Several parameters can be used as performance indices (Torrecilla et al., 2004): (a) initial slope or rate of reduction of initial error in the learning process; (b) final error or error at the end of the learning process; or (c) number of iterations or learning runs needed to end the learning process.

The first optimization instance during training is to determine the topology (or number of nodes in the hidden layer) for a fixed learning coefficient, which provides minimal estimated error with a minimal number of iterations through the system. Once the optimal topology is found, the learning coefficient $\mu$ can also be optimized through a trial-and-error method. In this case, the learning error and iteration number must also be minimized. Once optimization is performed, the values and distribution of output data from the model can be compared with real values using averages, standard deviation, and variance.

##### *5.4.5.2.4 Recall, Generalization, and Validation*

After the training step, the network will be subjected to a wide array of input patterns used in training and adjustments introduced to make the system more reliable and robust (recall step). In the generalization step, a set of data from an independent test can be run through the network using the selected topology and the optimized learning coefficient with the previously adjusted weights. The validation step evaluates the competence of the trained network. Statistical comparisons as well as correlation coefficients can be determined to evaluate the model performance or to validate it with known data sets.

If a high-pressure system is designed such that it is proven uniform in terms of temperature distribution inside the vessel, then ANN can be applied to retrofeed the system controls. This means that the entire volume of temperature data logged during production runs can be used to train the system in parallel. Once the neural network is well trained (i.e., producing outcomes within an acceptable error range), it can be directly applied to autocorrect the high-pressure system if any deviation from the target temperature occurs (e.g., by stopping an under processed run). Furthermore, the ANN will still be capable of learning continuously and thus improve the estimation of the autocorrect function.

## 5.5 COMPARISON BETWEEN MODELING APPROACHES

Models can be evaluated in terms of different capabilities. For example, some may require significant simplifications, leading to erroneous predictions, while others can allow for more detailed representations by also including peripheral devices. Furthermore, some models are limited to providing a single temperature value while others are capable of expressing temperature, flow, and inactivation extent/concentration distributions in three dimensions as a function of time. Other requirements can include extraction or implementation of thermophysical properties as functions of temperature and pressure, package shrinkage, heat source definition, and several food/equipment materials. In addition, the initial time invested for model development, the computational demand, and the solving speed can differ significantly among modeling approaches. Table 5.5 developed by the authors compares the features of the modeling approaches applied to high-pressure systems. Comparisons are outlined in the following paragraphs.

**TABLE 5.5**
**Comparison of Different Modeling Approaches and Their Capabilities**

| Modeling Approach → Characteristic ↓ | Analytical | Numerical: Finite Differences | Numerical: Finite Elements | Numerical: Finite Volumes | Macroscopic | ANN |
|---|---|---|---|---|---|---|
| Vessel system simplifications required | High | Medium | Low | Low | High | High |
| Temperature distribution | 1D (axis sym) | 1D/2D/3D distribution | 1D/2D/3D distribution | 1D/2D/3D distribution | No (1 value) | No (1 value) |
| Time–temperature profiles | Yes | Yes | Yes | Yes | Yes | No |
| Flow distribution | No flow (no convection) | 1D/2D/3D distribution[a] | 1D/2D/3D distribution 3D[a] | 1D/2D/3D distribution | No flow (no convection) | No flow (no convection) |
| Determination of thermophysical properties | Yes | Not directly[a] | Not directly[a] | Not directly | Not directly[a] | No |
| Insertion of properties $f(P, T)$ | Yes[a] | Yes | Yes | Yes | Yes[a] | Not required |
| Coupling with inactivation/ reaction models | External[a] | Internal/ external | Internal/ external | Internal/ external | Internal/ external[a] | Direct training[a] |
| Inactivation/ reaction distribution | 1D (axis sym)[a] | 1D/2D/3D distribution | 1D/2D/3D distribution | 1D/2D/3D distribution | 1 average value[a] | 1 value[a] |
| Package shrinkage | No | Yes[a] | Yes[a] | Yes[a] | No | No |
| Heat source | Equation 5.17[a] | Equation 5.17 | Equation 5.17 | Equation 5.17[a] | Equation 5.17 | Not required |
| Multiple materials in a model | No | Yes | Yes | Yes | No | Yes[a] |
| Multiple unit operations (peripheral devices) | No | Yes[a] | Yes[a] | Yes[a] | Yes | Yes[a] |
| Computational demand | Low | Medium | High | High | Medium | Low |
| Solver speed | Fast | Medium | Slow | Slow | Medium | Fast |
| Time investment | Depends | Medium | Medium | Medium | Medium | High |
| Adaptability to system modifications | Limited | High | High | High | Limited | Limited |

[a] Work not published.

### 5.5.1 Vessel Simplifications Required to Establish the Model

Among the modeling approaches described in previous sections, the analytical, macroscopic, and ANN models are those that require a number of assumptions and simplifications for their determination, which may reduce the accuracy of the prediction. On the other hand, numerical models can easily represent the full extent of the vessel geometry and provide the required accuracy. Subdividing the vessel structure (computational domain), by using meshes with different elements and element sizes, facilitates the easy application of partial differential equations from governing laws of thermofluid dynamics and thereby the inclusion of complex structures.

### 5.5.2 Predicting Temperature and Flow Distribution

Numerical models have the capability of providing temperature distributions throughout the food processing system. Other model types like the analytical, macroscopic, and ANN models are, according to today's capabilities, unable to provide a distribution of both temperature and flow. While analytical models are capable of representing a 1D axis symmetric temperature variation (i.e., temperature variation in radial direction), macroscopic and ANN models can only provide a single value; for macroscopic models—a temperature profile for a single point in the vessel or an average temperature value; for ANN—a single parameter (e.g., maximum or minimum temperature reached at pressure holding or temperature reached after decompression).

### 5.5.3 Predicting Time–Temperature Profiles

In HPHT processing, time–temperature profiles are essential to understanding the temperature evolution during pressure application. Numerical models not only provide temperature distribution, but also have the capability of representing distributions as a function of time. Analytical and macroscopic models can also provide variations in time while ANN is restricted to provide a single value.

### 5.5.4 Determination of Thermophysical Properties by Fitting Temperature Data

In contrast to multidimensional numerical models and macroscopic models, analytical solutions can be easier to apply in curve fitting problems for the extraction of thermophysical properties. In many cases the use of an entirely conductive heat transfer model, neglecting convective effects, may allow for back calculation of thermophysical properties from a measured temperature curve. These properties cannot directly be extracted by multidimensional numerical models; instead, they are extracted by iterative repetitions of the simulation procedure, at varying thermophysical properties themselves, until the output fits to a measured time–temperature curve (Kowalczyk et al., 2004). The use of ANN models is not applicable for this purpose, mainly because this approach (during training or prediction) does not involve thermophysical properties.

### 5.5.5 Incorporation of Thermophysical Properties as Functions of Pressure and Temperature

To increase the accuracy of a model, the variation in properties with pressure and temperature must be accounted for. Analytical, numerical, and macroscopic models can allow for the substitution of initially assumed "constant" thermophysical property values, with equations representing their dependency on temperature and pressure. Determination of these equations can be performed by surface fitting of preexisting tabulated data for water (e.g., from NIST database, Harvey et al., 1996) or from experimental data measured for other materials. As mentioned before, thermophysical properties are not required for ANN. This is an advantage over other modeling approaches because of the difficulty at present in determining these properties at high-pressure conditions in different food components (Otero and Sanz, 2003).

### 5.5.6 Coupling Thermal and Fluid Dynamic Models with Inactivation and Reaction Models

The final aim of predicting temperature distribution is to represent transformations occurring (i.e., microbial inactivation or physicochemical reactions) due to the combined application of pressure and temperature. This representation can be achieved by external or internal coupling of kinetic models with the heat transfer models, which results in a distribution of inactivation or reactant/product concentration. External coupling refers to the transformation of the model output once solved, by means of another software, into a distribution pattern solution, while internal coupling is the simultaneous combination of differential equations employed (including inactivation kinetic equations expressed in the form of a differential equation) within the same software package.

Analytical models only allow for external coupling, while the mathematical design of numerical models allows for the inclusion of kinetic equations. However, numerical models can be transformed by both external and internal coupling. In the same way, macroscopic models allow for both external and internal coupling because the blocks can be arranged at the end (external coupling) or in parallel (internal coupling) in data flow programming software. Given the nature of ANN, coupling is not possible externally or internally. It is not possible internally because there are no time-dependent equations involved in the information transfer through the network layers. For external coupling, time–temperature profiles are needed and ANN provides only single value solutions. However, ANN could be directly trained to predict microbial inactivation or reaction outputs.

### 5.5.7 Inactivation Extent or Reaction Output Distribution throughout the Vessel or Package

Once kinetic models are coupled externally or internally, a distribution representing the extent of inactivation or reaction output can be obtained in a similar way to temperature and flow distributions. While numerical models can provide distributions in three dimensions, analytical models are limited to representing variations (in microbial/enzyme reduction or concentration) in the radial direction. Distributions are not given by macroscopic or ANN models since they provide a single value for the overall kill or reaction yield.

### 5.5.8 Models Accounting for Package Shrinkage due to Hydrostatic Compression

Package shrinkage is a phenomenon inherent in the high-pressure process and, if accounted for, can provide more accurate information about the physical mechanisms of heat and momentum transfer. To date, no publication has dealt with this challenging modeling task. However, the latest developments in numerical modeling allow for representation of dynamic and transient changes due to hydrostatic compression (and recovery) of a package-containing food by means of coupling partial differential equations describing structural mechanics (application of moving meshes). So far, numerical modeling is the only approach capable of handling this problem.

### 5.5.9 Heat Source: the Compression Heating Model

As shown in Section 5.3.1, a model for compression heating (Equation 5.17) has been theoretically derived from thermodynamic equations and validated with water and other food materials (Otero et al., 2000; Ardia et al., 2004). This equation can be incorporated in all modeling approaches except for the ANN model, where it is not required due to the nature of the method. Some authors (Carroll et al., 2003; Hartmann et al., 2004) have used a constant value for compression heating in analytical and finite volume modeling, while others (Denys et al., 2000a,b; Otero et al., 2002a, 2007; Knoerzer et al., 2007) have applied Equation 5.17 to their finite difference, finite element, and macroscopic models by including thermophysical properties as functions of pressure and, in some cases, temperature. In particular, macroscopic models require a block including a numerical heat transfer model to represent temperature changes, which in turn will include Equation 5.17 as a heat source.

### 5.5.10 Materials and Their Properties in a Multicomponent Model

A model can become more realistic and better assess the design or inactivation effectiveness information when it includes all composite materials (steel, food, packaging material, water, insulating carrier polymers, etc.) playing a major or minor role in the heat, momentum, and mass transfer throughout the system. For example, models should allow for the inclusion of materials (and their properties) forming foods of varied composition, packages, and other devices like the carrier, or even the vessel walls and closures. Among all modeling approaches considered, analytical modeling is too simplistic to carry out this complex task. The macroscopic models cannot incorporate other materials either, since a very sophisticated numerical model should be included in a block to account for temperature changes occurring in such materials.

Nonetheless, due to the capability of current numerical software packages in drawing arbitrary shapes or even to importing CAD drawings and assigning thermophysical properties corresponding to the materials in place, this approach is the best option to simulate the most realistic scenario.

ANN could be trained to receive data from complex systems including several materials, for example, a carrier or packages. However, neural networks will not provide temperature distributions and profiles in the different areas. Furthermore, in order to provide more variables in the output layer (e.g., temperature of package or temperature inside the carrier wall) the training effort should be very high.

### 5.5.11 Multiple Unit Operations in a Single Model

A high-pressure system consists of not only the vessel and its contents, but also the number of peripheral units, which could play a role in the retention of uniformity in the processing system. Macroscopic models have been shown to be useful in evaluating the impact of each and every peripheral component involved to meet target-processing conditions. Data flow programming languages (e.g., LabVIEW and Simulink) allow for a block sequence that includes each peripheral device.

Moreover, numerical models have the capability of including the influence of peripheral devices through the boundary conditions (e.g., by including a 1D equation accounting for the heat conduction through the steel walls). It is also possible to include the devices themselves in the model by directly connecting several computational domains corresponding to each peripheral unit. The drawback in this case is the complexity of the design, which could lead to pronounced difficulties in model convergence as well as a high-computational demand.

In ANN, the input layer can be modified by increasing the number of nodes, to add variables corresponding to the temperature (or possibly pressure) determined by the peripheral devices. As in the case for inclusion of different materials, a significant effort would be required in training the ANN.

### 5.5.12 Computational Demand and Solver Speed Required

Regardless of the area of application, whether in process development or optimization, in academia or industry, computational demand and solver speeds are becoming more influential in the efficacy of providing an accurate and fast assessment. Depending on the modeling approach, systems of high complexity can be developed, thus requiring computers with larger memory and higher CPU speed. Thus, the time needed to converge to a model solution may depend on the solver speed, which is directly related to the computational demand.

Due to the complexity involved in the geometry, combined with the partial differential equations being solved, numerical models have a high-computational demand, which in turn decreases the solver speed. Models solved by means of finite differences require less computational power than finite elements and finite volumes, due to the difference in approximating the partial differential equations. Analytical and ANN models have low-computational requirements and can provide a solution in less time. Even though macroscopic models have a medium computational demand, due to the numerical step involved, they can still provide faster results than numerical models because of the simpler equations required.

### 5.5.13 Invested Time in Model Development

The time invested developing a model differs significantly among the modeling approaches. Assuming the person designing the model is experienced in that particular approach (use of software, knowledge of theory, etc.), the numerical and macroscopic models seem to require lower development time compared to ANN. The ANN model consumes a significant amount of time, mainly in the training phase. Development time for analytical models lies in the complexity of the mathematical solution method chosen. For example, Bessel functions applied by Carroll et al. (2003) are more complicated and thus the fitting process is more time consuming than the simple integration of the compression heating equation, as shown before, assuming the conditions are adiabatic.

### 5.5.14 Adaptability to System and Process Modifications

It is desirable that models are easily adaptable to system modifications (e.g., inclusion of a carrier, samples of different compositions, shapes, and sizes, etc.) to allow for redesigning and optimization. However, modeling approaches such as those for analytical, macroscopic, and ANN models are very limited in making readjustments when system modifications are made. For instance, analytical models can be applied at different scales as long as the same assumptions are kept; however, it does not apply when other materials are inserted. Macroscopic models allow for change in process conditions, yet scalability or the insertion of other materials cannot be directly handled by these models. System modification is not directly feasible in an ANN model because it is a system-specific technique. On the other hand, processing conditions can be easily modified after network training. Numerical models show an excellent adaptability to system modifications and allow for optimization of conditions and system structure.

## 5.6 CONCLUDING REMARKS

This contribution presented an overview of the components of an HPHT system, the underlying heat transfer and fluid-dynamic phenomena at each processing step, and the variables involved in developing a heat transfer model. Advantages and limitations of existing HPP modeling approaches have been identified.

### 5.6.1 Thermophysical Property Determination

Thermophysical properties can have a different influence at each step of the process, and this should be taken into account in the development of future models. During come-up time, it has been shown that compression heating is mainly dominated by the expansion coefficient at high-temperature conditions, which is part of the heat source term in the energy balance (Equation 5.16). At this step, even though the conduction term (given by thermal diffusivity) has some effect on compression heating, the net energy exchange is mostly influenced by the expansion coefficient. However, the source term is zero once target pressure is reached (Equation 5.15) and, in a nonadiabatic system, thermal diffusivity is mainly influential during holding time.

In decompression, given the speed in pressure drop, it is very clear that the expansivity plays a major role in the heat sink term during rapid cool down. Even if the variation effect of other thermophysical properties like specific heat and density could play a minor role, the best accuracy will be provided when all thermophysical properties are expressed as a function of pressure and temperature.

### 5.6.2 Analytical Modeling

Analytical solutions derived from a number of simplifying assumptions and adjusted to measured temperature data, by avoiding numerical methods, can directly provide model parameters that predict the process. This model may be useful for HPHT process characterization when the homogenous temperature distribution is assured inside the vessel, since this type of modeling is 1D in nature. The accuracy of such predictive parameters relies on the use of thermophysical properties as functions of pressure and temperature. These models can be combined with kinetic models from inactivation or physicochemical reaction predictions to determine inactivation/reaction parameters and to predict 1D axis symmetric inactivation or reaction profiles.

### 5.6.3 Numerical Modeling

Numerical models are the most versatile among all modeling approaches described in this chapter because they can provide complete information regarding temperature distributions and distributions of the inactivation extent throughout the high-pressure process. It has been shown that partial differential equations for flow and heat transfer can be solved numerically by means of discretization methods to determine approximate and exact solutions. In fact, different authors have employed current CFD software models that can be solved using different techniques such as finite differences, finite elements, and finite volume methods to model high-pressure systems. CFD packages today provide user-friendly applications for visualization of predicted temperature distributions and velocity maps with time. They further allow coupling of thermal and fluid dynamic equations with microbial, enzymatic, or reaction kinetic models, also with a type of differential equation, and predict the extent of inactivation/reaction distribution due to high pressure and heat in two or three dimensions. Chapter 6 describes the state of the art in numerical modeling of conductive and convective high-pressure systems used to demonstrate temperature and inactivation distribution at low temperature and sterilization conditions.

### 5.6.4 Macroscopic Models

Macroscopic models can be used to determine the influence of other components in the modeling system, and their effect on temperature predictions. Models of this type may help evaluate and quantify the advantage of a number of system accessories added to ensure temperature retention in modern high-pressure sterilization equipment, including process optimization in regards to energy requirements. However, no spatial (2D or 3D) temperature distributions can be calculated and thus no temperature gradients throughout the vessel space (e.g., between bottom and top or between pressurizing fluid and food packs) can be represented. On the other hand, given the homogeneity provided by small-scale vessels, a macroscopic model could still be applied; misrepresentation would take place in larger scale vessels.

### 5.6.5 ANN

ANNs provide a sound approach to predicting a specific temperature (not a profile) in preestablished locations of the system. However, this modeling alternative might not be of current interest to industry because, for instance, knowing the maximum temperature reached in the system at HPHT conditions would not be enough for regulatory authorities. A model able to predict temperature distributions in industrial scale high-pressure plant systems is indispensable for demonstrating temperature and inactivation uniformity.

Indeed, models like macroscopic ones and ANN would be mainly of interest for scientific, academic, or educational purposes. For example, these modeling approaches show potential in studying the influence of several process variables (e.g., pressure level, pressurization rate, ambient and target temperatures, or initial high-pressure vessel temperature) on the output average (or maximum) temperature. Optimal conditions can also be determined by using this tool and thus reduce the need for experimental trials (Torrecilla et al., 2005). Process characterization parameters predicted by ANN include the final product temperature during pressure holding (or after pressure release), without requiring the use of thermophysical properties as functions of pressure and temperature for its prediction.

### 5.6.6 Model Validation with Temperature and Microbial Measurements

Temperature and microbial inactivation validation of the above-mentioned models are critical for their adoption to assess the reliability of a high-pressure sterilization process. Accurate temperature measurement at HPHT conditions remains a challenge due to the frequent failure encountered at the closures of current thermocouple systems. Other possibilities such as wireless temperature probes or in-vessel temperature data loggers are under development to avoid this problem.

Uniformity of sterilization processes depend on intrinsic variables, such as process conditions like pressure, temperature, holding time, product properties, and pressure vessel dimensions, and to a lesser extent, extrinsic (peripheral) equipment. The pressure temperature inactivation kinetics of target barothermo resistant spore-forming microorganisms is another process variable to take into consideration. Expressions for combined pressure and temperature degradation kinetics of *C. botulinum*, or a surrogate microorganism of higher pressure/temperature resistance, should be coupled with thermofluid dynamic equations to simulate inactivation distribution. However, the kinetics should be known at dynamic conditions. Thus, for accurate representation of microbial inactivation distribution, spore inactivation kinetic models should be in the form of a differential equation dependent on pressure and temperature. Validation of these kinetic inactivation models by means of in-pack studies with several *C. botulinum* strains (in several food media samples) will be crucial to assessing the product safety of low-acid foods in high-pressure sterilization processes. Chapter 6 provides examples of the application for selected spore inactivation kinetic models in predicting *C. botulinum* inactivation distribution.

## NOMENCLATURE

| | |
|---|---|
| $A$ | relative activity or actual activity related to initial activity (%) |
| $B$ | Weibull's parameter; function of process temperature and/or pressure history |
| $C_P$ | isobaric heat capacity (J/kg/K) |
| $C_{p,\text{ mixture}}$ | specific heat capacity of mixture (J/kg/K) |
| $C_\mu$ | constant ($m^4/kg^4$) |
| $D_\upsilon$ | decimal reduction time (min) |
| $d, D$ | chamber diameter (m) |
| $D_P$ | diameter of round package (m) |
| $d_P, D_{PP}$ | thickness of packaging film (m) |
| $E$ | Euler number (=2.71828) |
| $E_a$ | activation energy (J) |
| $E_g$ | energy produced by chemical reactions (J) |
| $E_{in}$ | energy entering the system (e.g., work of compression, in J) |
| $E_k$ | prediction error of ANN |
| $E_{out}$ | energy leaving the system (e.g., decompression, cooling through walls, in J) |
| $E_{st}$ | energy accumulated/stored in the system (J) |
| $F$ | thermal death time (min) |
| $G$ | gravity constant (9.8 $m/s^2$) |
| $H$ | height of vessel (m) |
| $h_{P\text{–}F}$ | heat transfer coefficient between packaging and fluid ($W/m^2/K$) |
| $h_{PP}$ | heat transfer coefficient for conduction through packaging material ($W/m^2/K$) |
| $h_{S\text{–}F}$ | convective heat transfer coefficient at semisolid food surface ($W/m^2/K$) |
| $h_{S\text{–}Feq}$ | equivalent heat transfer coefficient between food package and compression fluid ($W/m^{2/}K$) |
| $h_{S\text{–}P}$ | heat transfer coefficient between food and packaging ($W/m^2/K$) |
| $h_{V\text{–}F}$ | heat transfer coefficient at vessel wall surface ($W/m^2/K$) |
| $i, j$ | positions on grid |
| $J$ | neuron number |
| $J_0, J_1$ | zero-order Bessel's functions |
| $K$ | turbulent kinetic energy ($kg^2/K^2$) |
| $K$ | first-order kinetic constant (pa s) |
| $K$ | inactivation rate constant (pa s) |
| $\overline{K}$ | inverse of inactivation rate constant (s) |
| $K$ | output layer of ANN |
| $k, k_1$ | thermal conductivity (W/m/K) |
| $k_P, k_{PP}$ | thermal conductivity of packaging material (W/m/K) |
| $k_{ref}$ | reference kinetic constant (/s) |

| | |
|---|---|
| $k_S$ | thermal conductivity of solid (W/m/K) |
| $k_{S/L}$ | thermal conductivity of semisolid foods (W/m/K) |
| $k_T$ | turbulent thermal conductivity (W/m/K) |
| $k_V$ | thermal conductivity of steel (W/m/K) |
| $L$ | extrachamber dimensionless heat transfer coefficient |
| $N$ | Weibull's parameter; function of process temperature and/or pressure history |
| $N$ | normal to surface |
| $N$ | order of inactivation kinetic |
| $N$ | time step number |
| $N, I$ | node number |
| $N_0, N$ | initial and final number of microbial spores |
| $P$ | pressure (Pa) |
| $P$ | probability value (statistics) |
| $P_0$ | atmospheric pressure (Pa) |
| $P_1, P_{target}$ | target pressure (Pa) |
| $P_2$ | pressure at end of holding time (Pa) |
| $P_f$ | pressure including a fluctuating term (Pa) |
| $p_{rate}$ | pressure rate (MPa/s) |
| $P_{ref}$ | reference pressure (Pa) |
| $Q$ | volumetric compression heating rate (J/m$^3$/s) |
| $R$ | universal gas constant (8.314 J/mol/K) |
| $r, R$ | radius (m) |
| $R^2$ | coefficient of determination |
| $r_k$ | real output data (for use in ANN error prediction) |
| $S$ | entropy (J/K) |
| $T$ | temperature (K) |
| $T$ | time (s) |
| $T_0$ | initial temperature of sample (K) |
| $T_c$ | temperature after cooling (K) |
| $T_f$ | temperature after decompression (K) |
| $t_f$ | time after decompression (s) |
| $T_h$ | preheating temperature (K) |
| $t_{hyd\_en}$ | dimensionless hydrodynamic compensation timescale for convection (enzyme solution) |
| $t_{hyd\_me}$ | dimensionless hydrodynamic compensation timescale for convection (pressure medium) |
| $t_{in}$ | dimensionless inactivation timescale |
| $T_L$ | temperature of fluid (K) |
| $t_p$ | process time (s) |
| $T_{p1}$ | temperature after compression heating (K) |
| $t_{p1}$ | time after compression (s) |
| $T_{p2}$ | temperature at the end of holding time (K) |
| $t_{p2}$ | time after holding stage (s) |
| $T_{ref}$ | reference temperature (K) |
| $t_s$ | start time (s) |

$T_s$, $T_i$ initial temperature (K)
$T_{S-P}$ temperature of food package (K)
$t_{th\_en}$ dimensionless timescale for thermal compensation of enzyme solution
$t_{th\_me}$ dimensionless timescale for thermal compensation of pressure medium
$t_{th\_PP}$ dimensionless timescale for thermal compensation of packaging material
$T_V$, $T_{wall}$ wall temperature (K)
$U$, v, and $w$ components of fluid velocity vector (in $x$, $y$, and $z$ direction, respec tively, m/s)
$V$ specific volume ($m^3$/kg)
$V$ volume ($m^3$)
$V_0$ specific volume of liquid water at atmospheric pressure and 0°C ($m^3$)
$V_a$ activation volume ($m^3$)
$\vec{v}_a$ average velocity vector (m/s)
$\vec{v}$ velocity vector (m/s)
$v_{in}$ inlet velocity (m/s)
$v_z$ velocity in $z$-direction (m/s)
$w_{ij}$, $w_{jk}$ connection weights between neurons in ANN
$x$, $y$, and $z$ coordinate positions (m)
$x_i$ yield of ANN
$y_i$ input of ANN
$y_j$ output of a neuron in each layer of an ANN
$y_k$ output of ANN
$Z_P$ pressure sensitivity (MPa)
$Z_T$ thermal sensitivity (°C)
[S] relative amount of solid (%)
[W] relative amount of water (%)

*Greek letters*

$\alpha$, $\alpha_p$ thermal expansion coefficient (/K)
$B$ water compressibility (0.17, dimensionless)
$\Gamma$ thermal diffusivity ($m^2$/s)
$E$ dissipation rate of turbulence energy ($m^2/K^3$)
$M$ learning coefficient of ANN
$H$ dynamic viscosity (Pa s)
$\eta_T$ turbulent viscosity (Pa s)
$\Pi_D$ dimensionless ratio of vessel
$P$ density (kg/$m^3$)
$\rho_{mixture}$ density of mixture (kg/$m^3$)
$\rho_{PP}$ density of packaging material (kg/$m^3$)
$T$ pressure come-up time

*Abbreviations*

1D one-dimensional
2D two-dimensional
3D three-dimensional
ANN artificial neural networks

| | |
|---|---|
| ASME | American Society of Mechanical Engineers |
| CAD | computer-aided design |
| CFU | colony-forming units |
| CFD | computational fluid dynamics |
| CPU | central processing unit |
| CTFD | computational thermal fluid dynamics |
| FDA | United States Food and Drug Administration |
| HP-DPIT | high-pressure digital particle image thermography |
| HP-DPIV | high-pressure digital particle image velocimetry |
| HPHT | high-pressure high-temperature |
| HPP | high-pressure process |
| IAPWS | International Association for the Properties of Water and Steam |
| NIST | National Institute of Standards and Technology |
| PATP | pressure-assisted thermal processing |
| PP | Polypropylene |
| PEEK | Polyetheretherketone |
| POM | Polyoxymethylene |
| PTFE | Polytetraflouroethylene |
| RANS | Reynolds averaged Navier–Stokes |
| UHMWPE | ultrahigh-molecular-weight polyethylene |
| USDA | United States Department of Agriculture |
| UHT | ultrahigh temperature |

*Operators*

| | |
|---|---|
| D | differential |
| $\Sigma$ | sum |
| $\partial$ | partial differential |
| $\Delta$ | difference |
| $\nabla$ | gradient (nabla-operator) |
| $\nabla$ | divergence |

## REFERENCES

Afaghi, M., Ramaswamy, H.S., and Prasher, S.O. 2001. Thermal process calculations using artificial neural network models. *Food Research International* 34:55–65.

Ahn, J., Balasubramaniam, V.M., and Yousef, A.E. 2007. Inactivation kinetics of selected aerobic and anaerobic bacterial spores by pressure-assisted thermal processing. *International Journal of Food Microbiology* 113:321–329.

Ardia, A., Knorr, D., and Heinz, V. 2004. Adiabatic heat modelling for pressure build-up during high-pressure treatment in liquid-food processing. *Food and Bioproducts Processing* 82:89–95.

Balasubramanian, S. and Balasubramaniam, V.M. 2003. Compression heating influence of pressure transmitting fluids on bacteria inactivation during high pressure processing. *Food Research International* 36:661–668.

Balasubramaniam, V., Ting, E., Stewart, C., and Robbins, J. 2004. Recommended laboratory practices for conducting high-pressure microbial inactivation experiments. *Innovative Food Science and Emerging Technologies* 5:299–306.

Barbosa-Cánovas, G.V. and Juliano, P. 2008. Food sterilization by combining high pressure and heat. In *Food Engineering: Integrated Approaches* G.F. Gutierrez-López, G. Barbosa-Canovas, J. Welti-Chanes, and E. Paradas-Arias, eds., pp. 9–46. New York: Springer.

Barbosa-Cánovas, G.V. and Rodríguez, J.J. 2005. Thermodynamic aspects of high hydrostatic pressure food processing. In *Novel Processing Technologies* (Eds.) G.V. Barbosa-Cánovas, M.S. Tapia, and M.P. Cano, 183–206. New York: CRC Press.

Campanella, O.H. and Peleg, M. 2001. Theoretical comparison of a new and the traditional method to calculate *Clostridium botulinum* survival during thermal inactivation. *Journal of the Science of Food and Agriculture* 81:1069–1076.

Caner, C., Hernandez, R.J., Pascall, M., Balasubramaniam, V.M., and Harte, B.R. 2004. The effect of high-pressure food processing on the sorption behaviour of selected packaging materials. *Packaging Technology and Science* 17:139–153.

Carroll, T., Chen, P., and Fletcher, A. 2003. A method to characterise heat transfer during high-pressure processing. *Journal of Food Engineering* 60:131–135.

Chen, X.D. 2006. Modeling thermal processing using computational fluid dynamics (CFD). In *Thermal Food Processing* D.W. Sun, ed., pp. 133–51. Boca Raton, FL: Taylor & Francis.

Chen, C.R. and Ramaswamy, H.S. 2006. Modeling food thermal processes using artificial neural networks. In *Thermal Food Processing* (Ed.) D.W. Sun, 107–32. Boca Raton, FL: Taylor & Francis.

Datta, A.K. 2001. Fundamentals of heat and moisture transport for microwaveable food product and process development. In *Handbook of Microwave Technology for Food Applications* A.K. Datta and C.A. Ramaswamy, eds., pp. 115–72. New York: Marcel Dekker.

Davies, L.J., Kemp, M.R., and Fryer, P.J. 1999. The geometry of shadows: effects of inhomogeneities in electrical field processing. *Journal of Food Engineering* 40:245–258.

de Heij, W.B.C., Van Schepdael, L.J.M.M., van den Berg, R.W., and Bartels, P.V. 2002. Increasing preservation efficiency and product quality through control of temperature distributions in high pressure applications. *High Pressure Research* 22:653–657.

de Heij, W.B.C., Van Schepdael, L.J.M.M., Moezelaar, R., Hoogland, H., Matser, A., and van den Berg, R.W. 2003. High-pressure sterilization: maximizing the benefits of adiabatic heating. *Food Technology* 57:37–42.

Denys, S. and Hendrickx, M.E. 1999. Measurement of the thermal conductivity of foods at high pressure. *Journal of Food Science* 64:709–713.

Denys, S., Ludikhuyze, L.R., Van Loey, A.M., and Hendrickx, M.E. 2000a. Modeling conductive heat transfer and process uniformity during batch high-pressure processing of foods. *Biotechnology Progress* 16:92–101.

Denys, S., Van Loey, A.M., and Hendrickx, M.E. 2000b. A modeling approach for evaluating process uniformity during batch high hydrostatic pressure processing: combination of a numerical heat transfer model and enzyme inactivation kinetics. *Innovative Food Science and Emerging Technologies* 1:5–19.

Farkas, D. and Hoover, D.G. 2000. High pressure processing. *Journal of Food Science Supplement* 47–64.

Forst, P., Werner, F., and Delgado, A. 2000. The viscosity of water at high pressures—especially at subzero degrees centigrade. *Rheologica Acta* 39:566–573.

Ghani, A.G.A. and Farid, M.M. 2007. Numerical simulation of solid–liquid food mixture in a high pressure processing unit using computational fluid dynamics. *Journal of Food Engineering* 80:1031–1042.

Gola, S., Foman, C., Carpi, G., Maggi, A., Cassara, A., and Rovere, P. 1996. Inactivation of bacterial spores in phosphate buffer and in vegetable cream treated with high pressures. *High Pressure Bioscience and Biotechnology*, Vol. 13 (Progress in Biotechnology), (Eds.) R. Hayashi and C. Balny, pp. 253–259. Amsterdam: Elsevier.

Hartmann, C. 2002. Numerical simulation of thermodynamic and fluid dynamic processes during the high pressure treatment of fluid food systems. *Innovative Food Science and Emerging Technologies* 3:11–18.

Hartmann, C. and Delgado, A. 2002a. Numerical simulation of convective and diffusive transport effects on a high-pressure-induced inactivation process. *Biotechnology and Bioengineering* 79:94–104.

Hartmann, C. and Delgado, A. 2002b. Numerical simulation of thermofluid dynamics and enzyme inactivation in a fluid food system under high hydrostatic pressure. *Trends in High Pressure Bioscience and Biotechnology*, Vol. 19 (Progress in Biotechnology), (Ed.) R. Hayashi, pp. 533–540. Amsterdam: Elsevier.

Hartmann, C. and Delgado, A. 2003a. Numerical simulation of thermal and fluid dynamical transport effects on a high pressure induced inactivation. *High Pressure Research* 23:67–70.

Hartmann, C. and Delgado, A. 2003b. The influence of transport phenomena during high-pressure processing of packed food on the uniformity of enzyme inactivation. *Biotechnology and Bioengineering* 82:725–735.

Hartmann, C., Delgado, A., and Szymczyk, J. 2003. Convective and diffusive transport effects in a high pressure induced inactivation process of packed food. *Journal of Food Engineering* 59:33–44.

Hartmann, C., Schuhholz, J., Kitsubun, P., Chapleau, N., Bail, A., and Delgado, A. 2004. Experimental and numerical analysis of the thermofluid dynamics in a high-pressure autoclave. *Innovative Food Science and Emerging Technologies* 5:399–411.

Harvey, A.H., Peskin, A.P., and A.K. Sanford. 1996. *NIST/ASTME-IAPSW Standard Reference Database 10, version 2.2.*

Hjelmqwist, J. 2005. Commercial high-pressure equipment. In *Novel Food Processing Technologies* G.V. Barbosa-Cánovas, M.S. Tapia, and M.P. Cano, eds., pp. 361–73. Boca Raton, FL: CRC Press.

Holdsworth, S.D. 1997. *Thermal Processing of Packaged Foods*. New York: Blackie Academic & Professional.

Hoogland, H., de Heij, W.B.C., and Van Schepdael, L.J.M.M. 2001. High pressure sterilization: novel technology, new products, new opportunities. *New Food* 4:21–26.

Juliano, P. 2006. High pressure thermal sterilization of egg products, PhD dissertation. Washington State University, Pullman, WA.

Juliano, P., Li, B.S., Clark, S., Mathews, J.W., Dunne, P.C., and Barbosa-Canovas, G.V. 2006a. Descriptive analysis of precooked egg products after high-pressure processing combined with low and high temperatures. *Journal of Food Quality* 29:505–530.

Juliano, P., Toldra, M., Koutchma, T., Balasubramaniam, V.M., Clark, S., Mathews, J.W., Dunne, C.P., Sadler, G., and Barbosa-Cánovas, G.V. 2006b. Texture and water retention improvement in high-pressure thermally treated scrambled egg patties. *Journal of Food Science* 71:E52–E61.

Juliano, P., Clark, S., Koutchma, T., Ouattara, M., Mathews, J.W., Dunne, C.P., and Barbosa-Canovas, G.V. 2007. Consumer and trained panel evaluation of high pressure thermally treated scrambled egg patties. *Journal of Food Quality* 30:57–80.

Juliano, P., Knoerzer, K., Fryer, P., and Versteeg, C. 2008. *C. botulinum* inactivation kinetics implemented in a computational model of a high pressure sterilization process. *Biotechnology Progress*. In Press.

Knoerzer, K., Regier, M., Erle, U., Pardey, K.K., and Schubert, H. 2004. Development of a model food for microwave processing and the prediction of its physical properties. *Journal of Microwave Power and Electromagnetic Energy* 39:67–177.

Knoerzer, K., Juliano, P., Gladman, S., Versteeg, C., and Fryer, P. 2007. A computational model for temperature and sterility distributions in a pilot-scale high-pressure high-temperature process. *AICHE Journal* 53:2996–3010.

Koutchma, T., Guo, B., Patazca, E., and Parisi, B. 2005. High pressure-high temperature sterilization: from kinetic analysis to process verification. *Journal of Food Process Engineering* 28:610–629.

Kowalczyk, W., Hartmann, C., and Delgado, A. 2004. Modelling and numerical simulation of convection driven high pressure induced phase changes. *International Journal of Heat and Mass Transfer* 47:1079–1089.

Kowalczyk, W., Hartmann, C., Luscher, C., Pohl, M., Delgado, A., and Knorr, D. 2005. Determination of thermophysical properties of foods under high hydrostatic pressure in combined experimental and theoretical approach. *Innovative Food Science and Emerging Technologies* 6:318–326.

Krebbers, B., Matser, A.M., Koets, M., and van den Berg, R.W. 2002. Quality and storage-stability of high-pressure preserved green beans. *Journal of Food Engineering* 54:27–33.

Krebbers, B., Matser, A., Hoogerwerf, S., Moezelaar, R., Tomassen, M., and Berg, R. 2003. Combined high-pressure and thermal treatments for processing of tomato puree: evaluation of microbial inactivation and quality parameters. *Innovative Food Science and Emerging Technologies* 4:377–385.

Larousse, J. and Brown, B.E. 1997. *Food Canning Technology*. New York: Wiley-VCH.

Leadley, C. 2005. High pressure sterilisation: a review. *Campden and Chorleywood Food Research Association* 47:1–42.

Margosch, D. 2005. *Behavior of bacterial endospores and toxins as safety determinants in low acid pressurized food*, PhD dissertation. Technical University Berlin, Germany.

Margosch, D., Ehrmann, M.A., Ganzle, M.G., and Vogel, R.F. 2004. Comparison of pressure and heat resistance of *Clostridium botulinum* and other endospores in mashed carrots. *Journal of Food Protection* 67:2530–2537.

Margosch, D., Ehrmann, M.A., Buckow, R., Heinz, V., Vogel, R.F., and Ganzle, M.G. 2006. High-pressure-mediated survival of *Clostridium botulinum* and *Bacillus amyloliquefaciens* endospores at high temperature. *Applied and Environmental Microbiology* 72:3476–3481.

Matser, A.A., Krebbers, B., van den Berg, R.W., and Bartels, P.V. 2004. Advantages of high pressure sterilisation on quality of food products. *Trends in Food Science and Technology* 15:79–85.

Meyer, R.S., Cooper, K.L., Knorr, D., and Lelieveld, H.L.M. 2000. High-pressure sterilization of foods. *Food Technology* 54:67–72.

Nicolaï, B.M., Scheerlinck, N., Verboven, P., and Baerdemaeker, J.D. 2001. Stochastic finite-element analysis of thermal food processes. In *Food Processing Operations Modeling. Design and Analysis* J. Irudayarai, ed., pp. 265–304. New York: Marcel Dekker.

Norton, T. and Sun, D.W. 2006. Computational fluid dynamics (CFD)—an effective and efficient design and analysis tool for the food industry: a review. *Trends in Food Science and Technology* 17:600–620.

Otero, L. and Sanz, P. 2003. Modelling heat transfer in high pressure food processing: a review. *Innovative Food Science and Emerging Technologies* 4:121–134.

Otero, L., Molina-Garcia, A., and Sanz, P. 2000. Thermal effect in foods during quasi-adiabatic pressure treatments. *Innovative Food Science and Emerging Technologies* 1:119–126.

Otero, L., Molina-Garcia, A.D., Ramos, A.M., and Sanz, P.D. 2002a. A model for real thermal control in high-pressure treatment of foods. *Biotechnology Progress* 18:904–908.

Otero, L., Molina-Garcia, A.D., and Sanz, P.D. 2002b. Some interrelated thermophysical properties of liquid water and ice. I. A user-friendly modeling review for food high-pressure processing. *Critical Reviews in Food Science and Nutrition* 42:339–352.

Otero, L., Ramos, A.M., de Elvira, C., and Sanz, P.D. 2007. A model to design high-pressure processes towards an uniform temperature distribution. *Journal of Food Engineering* 78:1463–1470.

Pehl, M. and Delgado, A. 1999. An in situ technique to visualize temperature and velocity fields in liquid biotechnical substances at high pressure. In *Advances in High Pressure Bioscience and Biotechnology* (Ed.) H. Ludwig, pp. 519–522.

Pehl, M. and Delgado, A. 2002. Experimental investigation on thermofluiddynamical processes in pressurized substances. In *Trends in High Pressure Bioscience and Biotechnology* R. Hayashi, ed., pp. 429–435. Amsterdam: Elsevier.

Peleg, M. 2006. *Advanced Quantitative Microbiology for Foods and Biosystems. Models for Predicting Growth and Inactivation.* Boca Raton, FL: Taylor & Francis.

Peleg, M., Normand, M.D., and Campanella, O.H. 2003. Estimating microbial inactivation parameters from survival curves obtained under varying conditions—the linear case. *Bulletin of Mathematical Biology* 65:219–234.

Peleg, M., Normand, M.D., and Corradini, M.G. 2005. Generating microbial survival curves during thermal processing in real time. *Journal of Applied Microbiology* 98:406–417.

Perry, R.H. 1997. *Perry's Chemical Engineers' Handbook*. New York: McGraw-Hill.

Pflug, I.J. 1987. Using the straight-line semilogarithmic microbial destruction model as an engineering design-model for determining the *F*-value for heat processes. *Journal of Food Protection* 50:342–346.

Rajan, S., Ahn, J., Balasubramaniam, V.M., and Yousef, A.E. 2006a. Combined pressure-thermal inactivation kinetics of *Bacillus amyloliquefaciens* spores in egg patty mince. *Journal of Food Protection* 69:853–860.

Rajan, S., Pandrangi, S., Balasubramaniam, V.M., and Yousef, A.E. 2006b. Inactivation of *Bacillus stearothermophilus* spores in egg patties by pressure-assisted thermal processing. *Lwt-Food Science and Technology* 39:844–851.

Rasanayagam, V., Balasubramaniam, V.M., Ting, E., Sizer, C.E., Bush, C., and Anderson, C. 2003. Compression heating of selected fatty food materials during high-pressure processing. *Journal of Food Science* 68:254–259.

Ramaswamy, H.S., Balasubramaniam, V.M., and Sastry, S.K. 2007. Thermal conductivity of selected liquid foods at elevated pressures up to 700 MPa. *Journal of Food Engineering* 83:444–451.

Raso, J., Barbosa-Canovas, G., and Swanson, B.G. 1998. Sporulation temperature affects initiation of germination and inactivation by high hydrostatic pressure of *Bacillus cereus*. *Journal of Applied Microbiology* 85:17–24.

Rodríguez, J.J. 2003. Thermodynamic aspects of high hydrostatic pressure processing. Processing food by high hydrostatic pressure, PhD dissertation. Washington State University, Pullman, WA.

Rovere, P., Gola, S., Maggi, A., Scaramuzza, N., and Miglioli, L. 1998. Studies on bacterial spores by combined pressure-heat treatments: possibility to sterilize low acid foods. In *High Pressure Food Science, Bioscience and Chemistry*, N.S. Isaacs, ed., pp. 354–363. Cambridge: The Royal Society of Chemistry.

Sablani, S.S., Ramaswamy, H.S., and Prasher, S.O. 1995. A neural-network approach for thermal-processing applications. *Journal of Food Processing and Preservation* 19:283–301.

Sablani, S.S., Ramaswamy, H.S., Sreekanth, S., and Prasher, S.O. 1997. Neural network modeling of heat transfer to liquid particle mixtures in cans subjected to end-over-end processing. *Food Research International* 30:105–116.

Sato, H., Uematsu, M., and Watanabe, K. 1988. New international skeleton tables for thermodynamic properties of ordinary water substance. *Journal of Physical and Chemical Reference Data* 17:1439–1540.

Saul, A. and Wagner, W. 1989. A fundamental equation for water covering the range from the melting line to 1273-K at pressures up to 25000 Mpa. *Journal of Physical and Chemical Reference Data* 18:1537–1564.

Schauwecker, A., Balasubramaniam, V.M., Sadler, G., Pascall, M.A., and Adhikari, C. 2002. Influence of high-pressure processing on selected polymeric materials and on the migration of a pressure-transmitting fluid. *Packaging Technology and Science* 15:255–262.

Sizer, C.E., Balasubramaniam, V.M., and Ting, E. 2002. Validating high-pressure processes for low-acid foods. *Food Technology* 56:36–57.

Sreekanth, S., Ramaswamy, H.S., Sablani, S.S., and Prasher, S.O. 1999. A neural network approach for evaluation of surface heat transfer coefficient. *Journal of Food Processing and Preservation* 23:329–348.

Ter Minassian, L., Pruzan, P., and Soulard, A. 1981. Thermodynamic properties of water under pressure up to 5 kbar and between 28 and 120°C. Estimations in the supercooled region down to −40°C. *The Journal of Chemical Physics* 75:3064–3072.

Ting, E., Balasubramaniam, V.M., and Raghubeer, E. 2002. Determining thermal effects in high-pressure processing. *Food Technology* 56:31–35.

Torrecilla, J.S., Otero, L., and Sanz, P.D. 2004. A neural network approach for thermal/pressure food processing. *Journal of Food Engineering* 62:89–95.

Torrecilla, J.S., Otero, L., and Sanz, P.D. 2005. Artificial neural networks: a promising tool to design and optimize high-pressure food processes. *Journal of Food Engineering* 69:299–306.

Versteeg, H.K. and Malalasekera, W. 1995. *An Introduction to Computational Fluid Dynamics: the Finite Volume Method*. Harlow: Longman Scientific & Technical.

Watson, J.T.R., Basu, R.S., and Sengers, J.V. 1980. An improved representative equation for the dynamic viscosity of water substance. *Journal of Physical and Chemical Reference Data* 9:1255–1290.

Zhu, S., Ramaswamy, H.S., Marcotte, M., Chen, C., Shao, Y., and Le Bail, A. 2007. Evaluation of thermal properties of food materials at high pressures using a dual-needle line-heat-source method. *Journal of Food Science* 72:E49–E56.

Zienkiewicz, O.C. 1977. *The Finite Element Method*. New York: McGraw-Hill.

# 6 High-Pressure Processes: Thermal and Fluid Dynamic Modeling Applications

*Pablo Juliano, Kai Knoerzer, and Gustavo V. Barbosa-Cánovas*

## CONTENTS

## 6.1 INTRODUCTION

Chapter 5 discussed the principles behind four modeling approaches that can be utilized to predict temperatures in a high-pressure system: analytical, numerical, macroscopic, and artificial neural networks (ANN). This chapter will highlight some applications of each modeling approach to high-pressure low-temperature (HPLT) systems and high-pressure high-temperature (HPHT) conditions reported in literature. Several authors have done extensive research in developing such models to predict transient temperature distributions, uniformity, and the loss of compression heating through high pressure vessel walls during all high pressure processing steps. In particular, some authors utilized discrete numerical modeling and computational fluid dynamics (CFD) to predict temperature and flow distribution inside the high-pressure vessel (Denys et al., 2000a,b; Hartmann and Delgado, 2002a, 2003a; Hartmann et al., 2003, 2004). Some recent models include solid materials (Knoerzer et al., 2007; Otero et al., 2007; Juliano et al., 2008), whereas some predict temperature distribution in three dimensions (Ghani and Farid, 2007) and at high pressure sterilization conditions (Knoerzer et al., 2007; Juliano et al., 2008). Special attention has been given to the distribution of enzyme and microbial inactivation throughout the chamber and packages, particularly *Clostridium botulinum* in the case of HPHT sterilization.

On the other hand, macroscopic models have been developed to integrate the effects of different portions of the high-pressure system contributing to heat transfer or heat retention to predict the final temperature inside the vessel (Otero et al., 2002a). Furthermore, a few publications have shown the application of ANN to predict temperatures in a high-pressure vessel (Torrecilla et al., 2004, 2005).

## 6.2 EXAMPLE OF AN ANALYTICAL MODEL FOR SINGLE-POINT TEMPERATURE PREDICTION

As defined in Chapter 5, analytical modeling in high-pressure processing involves the fitting of experimental temperature data from a heating or cooling curve obtained during pressurization and depressurization. This chapter presents an example of a complex analytical model (Carroll et al., 2003) developed to characterize heat transfer in a high-pressure process (HPP).

### 6.2.1 Temperature Profile Prediction

Carroll et al. (2003) deducted and solved a conductive heat transfer model for predicting temperature-drop during the come-up time and holding times in a Stansted Fluid Power Micro Food-lab plunger press (diameter 17 mm). Pure water was pressurized to 600 MPa from an initial temperature of 51.3°C at 5.81 MPa s$^{-1}$ and held for 300 s.

The heat transfer balance shown in Chapter 5 has been solved using the following initial conditions:

$$T = T_s = T_{\text{wall}}; \quad P = P_1 = \text{constant at } t = 0 \tag{6.1}$$

where the initial water temperature at all points $T_s$ in the pressure vessel coincides with the temperature in the vessel wall $T_{\text{wall}}$ at time zero. Carroll et al. (2003) assumed constant thermal conductivity and expressed the balance in the following form:

$$\rho C_p \frac{\partial T}{\partial t} = k\nabla^2 T + T\alpha_p \frac{\partial P}{\partial t} \tag{6.2}$$

The fluid cooling at holding time was obtained from Equation 6.2 by assuming a pressure constant during holding time, i.e., d$P$/d$t$ = 0, and thus gives the following expression:

$$\frac{\partial T}{\partial t} = \gamma \nabla^2 T \tag{6.3}$$

An infinite cylindrical chamber was assumed and the following general solution to Equation 6.3 was found by adopting the solution from Carslaw and Jaeger (1959) for the radial temperature profile during holding (assuming axi symmetry; vessel aspect ratio, radius:height less than 1:3):

$$T(r,t) - T_j = 2\sum_{N=1}^{\infty} e^{\frac{-4\gamma\beta_N^2}{d^2}t} \frac{\beta_N^2}{\beta_N^2 + L^2} \frac{J_0(\beta_N r)}{J_0^2(\beta_N)} \times \int_0^1 rf(r) J_0(\beta_N r)\,\mathrm{d}r \tag{6.4}$$

where

$\beta_N$ represents the roots of $\beta_N J_1(\beta_N) = LJ_0(\beta_N)$
$J_0$ and $J_1$ $z$ are zero-order Bessel functions
$\gamma$ is the thermal diffusivity
$d$ is the chamber diameter
$r$ is the dimensionless thermocouple position (from $r = 0$ at center to $r = 1$ at wall)
$L$ is the extra-chamber dimensionless heat transfer coefficient, being $L = hd/2k_v$

The solution includes the function *f*(*r*), representing the radial initial temperature distribution $T_{p1}$ at the end of pressure come-up. This general solution described the cooling curve of a fluid in the infinitely long cylindrical container cooled by a combination of conduction and convective interfacial heat transfer. Thus, no profiles were obtained in the vertical (*z*) direction.

Furthermore, Carroll et al. (2003) found a general solution (adapted from Carslaw and Jaeger, 1959) for temperature profile during pressurization (or decompression) by assuming the compression (or decompression) rate is constant, i.e., $dP/dt$ = constant. If physical properties are assumed constant during pressure come-up, the temperature profile can be given by

$$T(r,t)-T_j=\frac{2T\alpha}{k_v}\frac{dP}{dt}\frac{d^2}{4}\left[\left\{\frac{1-r^2}{8}+\frac{1}{4L}\right\}-\sum_{N=1}^{\infty}e^{\frac{-4\gamma\beta_N^2}{d^2}t}\frac{1}{\beta_N^3\left(1+\frac{\beta_N^2}{L^2}\right)}\frac{J_0(\beta_N r)}{J_1(\beta_N)}\right] \quad (6.5)$$

The authors showed from Equation 6.5, in a particular case (Carslaw and Jaeger, 1959), a way to find an expression for $f(r)$ that corresponded to the end of the pressurization and the start of holding time:

$$f(r)=T(r,\tau)-T_j=\frac{2T\alpha}{k_v}\frac{dP}{dt}\frac{d^2}{4}$$

$$\left[\left\{\frac{1-r^2}{8}+\frac{1}{4L}\right\}-\sum_{N=1}^{\infty}e^{\frac{-4\gamma\beta_N^2}{d^2}\tau}\frac{1}{\beta_N^2\left(L+\frac{\beta_N^2}{L}\right)}\frac{J_0(\beta_N r)}{J_0(\beta_N)}\right] \quad (6.6)$$

where $\tau$ is the come-up time at the pressure set point. Equation 6.6 can be solved for a general cooling curve (during holding time) by substituting the radial integrals of the mentioned Bessel functions so that the solution of Equation 6.4 yields:

$$T(r,t)-T_j=\left[\frac{2T\alpha}{k_v}\frac{dP}{dt}\frac{d^2}{4}\left(1-e^{\frac{-4\gamma\beta_1^2}{d^2}\tau}\right)\times\frac{1}{\beta_1^3\left(1+\frac{\beta_1^2}{L^2}\right)}\frac{J_0(\beta_1 r)}{J_1(\beta_1)}\right]e^{\frac{-4\gamma\beta_1^2}{d^2}t} \quad (6.7)$$

where $t \geq \tau$ corresponds to the holding time at the pressure set point, $N=1$, i.e., the first truncation of the series given in Equation 6.6 is taken considering long holding times. A particular solution to Equation 6.6 applies for $t<\tau$ at constant compression rate.

The authors validated Equation 6.7 by adjusting temperature data obtained during holding at 600 MPa/51°C using the Stansted Fluid Power Micro Food-lab plunger press with a diameter of 17 mm. Inputs to a curve-fitting algorithm were the chamber diameter $d$, the pressurization come-up time, the target pressure, and the experimental temperature data (during pressure holding).

### 6.2.2 Thermophysical Properties Prediction

The authors (Carroll et al., 2003) also used Equation 6.7 to obtain system's heat transfer properties (thermal diffusivity $\gamma$, dimensionless extra-chamber heat transfer coefficient $L$, and thermal expansivity $\alpha$) from this analytical solution corresponding to the cooling curve. Determination of heat transfer parameters from curve fitting into the analytical solution allows for predicting the temperature at all points within the chamber at all times during an HPP. Mathematical software can perform this extraction and prediction from the analytical solution as Carroll et al. (2003) showed in this example.

Modern high-pressure systems can minimize heat loss during pressure come-up, at least for relatively short come-up times, by setting the vessel wall's temperature (or internal heater) at 5°C–15°C higher than the target start temperature. The temperature of the insulating carrier and of the water surrounding the carrier would also be higher than the initial temperature of the compression fluid inside the carrier. In this case the system could be assumed adiabatic and homogeneous during pressurization, i.e., no heat is transferred to the chamber walls while fluid is pumped into the vessel. Then, Equation 6.2 (or, similarly, the thermal balance described in Chapter 5) can be rewritten in the following form by neglecting the conductive heat transfer component:

$$\frac{\partial T}{\partial t} = \frac{T\alpha_p}{\rho C_p} \frac{\partial P}{\partial t} \tag{6.8}$$

The three-dimensional (3D) representation of the thermophysical properties of water provided in Chapter 5 (Figure 5.3c) shows that $\rho$ is mainly influenced by pressure changes, while the $\alpha_p$ and $C_p$ are mostly affected by temperature change. Thus, Equation 6.8 can be rearranged in the following way:

$$\int_{T_s}^{T_{p1}} \frac{C_p}{T\alpha_p} \mathrm{d}T = \int_{P_0}^{P_1} \frac{\mathrm{d}P}{\rho} \tag{6.9}$$

In this case, functions of $\alpha_p(T)$, $C_p(T)$, and $\rho(P)$ can be determined and substituted into Equation 6.9 to obtain an integral value. $C_p(T)$ and $\rho(P)$ may not be available for certain food mixtures. Thus, further assumptions need to be made.

For example, from what has been discussed in Section 5.3.2.4.3, the ratio $\alpha_p/\rho C_p$ decreases by 28% in water when reaching 690 MPa and 105°C and is mainly attributed to the change in $\alpha_p$, which changes similarly. On the other hand, the multiplier $\rho C_p$ is maintained with a noticeable 5% change in water. In this case, $\rho C_p$ could be assumed constant and removed from the integral term. If an empirical expression is determined for $\alpha_p(T)$, Equation 6.8 could be solved by separating variables as follows:

$$\int_{T_s}^{T_{p1}} \frac{\alpha_p}{T} \mathrm{d}T = \int_{P_0}^{P_1} \frac{1}{\rho C_p} \mathrm{d}P = \rho C_p \Delta P \tag{6.10}$$

The resulting solution could predict compression heating during pressurization and provide parameters to assess heating effects during pressurization. Furthermore, performing iterative fits of analytical or numerical expressions to experimental data also allows determination of thermophysical properties for other food components as functions of temperature and pressure (Carroll et al., 2003; Kowalczyk et al., 2005).

## 6.3 CONDUCTIVE MODELS SOLVED BY APPLYING THE FINITE DIFFERENCE METHOD

Among the numerical methods shown to solve heat transfer balances, Chapter 5 explained the use of the finite difference method. This section will show examples of conductive models to predict temperature distribution and inactivation distribution.

### 6.3.1 Temperature Distribution Prediction

Denys et al. (2000a) developed a numerical model for conductive heat transfer in HPPs in agar gel (3% agar in distilled water). Agar gel was selected as a model food (or food phantom) due to its similar composition to water, which allowed using the thermophysical properties of water in the model. They utilized a 0.59 L vessel (internal diameter 50 mm) containing a sample holder (internal diameter 42 mm) wherein the agar solid was placed. The vessel used a special additive solution as compression fluid. Nine conventional high pressure batch processes were completed, each one characterized by a combination of initial temperatures, 8.9°C, 23.4°C, or 35.1°C, and applied pressures, 186, 348, or 503 MPa, and holding time of approximately 30 min.

The sample holder (a thin hollow aluminum cylinder) was used to minimize the impact of convective and conductive heat transfer between the sample holder and the inner wall. The bottom of the sample holder consisted of a movable piston, allowing for separating the samples from the incoming pressurization fluid at the bottom and thus minimizing heat exchange with the incoming fluid. Thermocouples were fixed in three known positions. An overall heat transfer coefficient at the surface of the cylinder accounted for heat transfer through the aluminum holder wall and the high-pressure medium remaining outside.

The cylindrical sample holder was the system chosen to simulate, accounting for heat transfer in the pressure medium (water) and the housing. The model was based on a numerical solution of the previously shown Fourier balance (Equation 6.2) for a conductive heating finite cylinder including the compression heating term. The model was evaluated for all high-pressure processing steps (come-up, holding, and decompression times). Thermal properties were collected from the National Institute of Standards and Technology (NIST)/American Society of Mechanical Engineers (ASME) Steam Properties Database (Harvey et al., 1996) and used in the model as a function of pressure.

Denys et al. (2000a,b) used the explicit finite difference method to derive an energy balance. These energy balances allowed calculating the temperature distribution at time $t+\Delta t\left(T_{ij}^{t+\Delta t}\right)$ as a function of temperature at time $t\left(T_{ij}^{t+\Delta t}\right)$. Details on these energy balances are shown in Denys et al. (2000a). Furthermore, the explicit finite difference numerical solution for a two-dimensional (2D) conductive heat transfer with a finite surface heat transfer coefficient (determined by iterative comparison of experimental and simulated temperature) is also shown by the same authors. Denys et al. (2000a,b) applied the computer program Delphi 3 to numerically solve the heat transfer differential equations of an enzyme solution within the 2D axi symmetric geometry of the high pressure vessel (Figure 6.1).

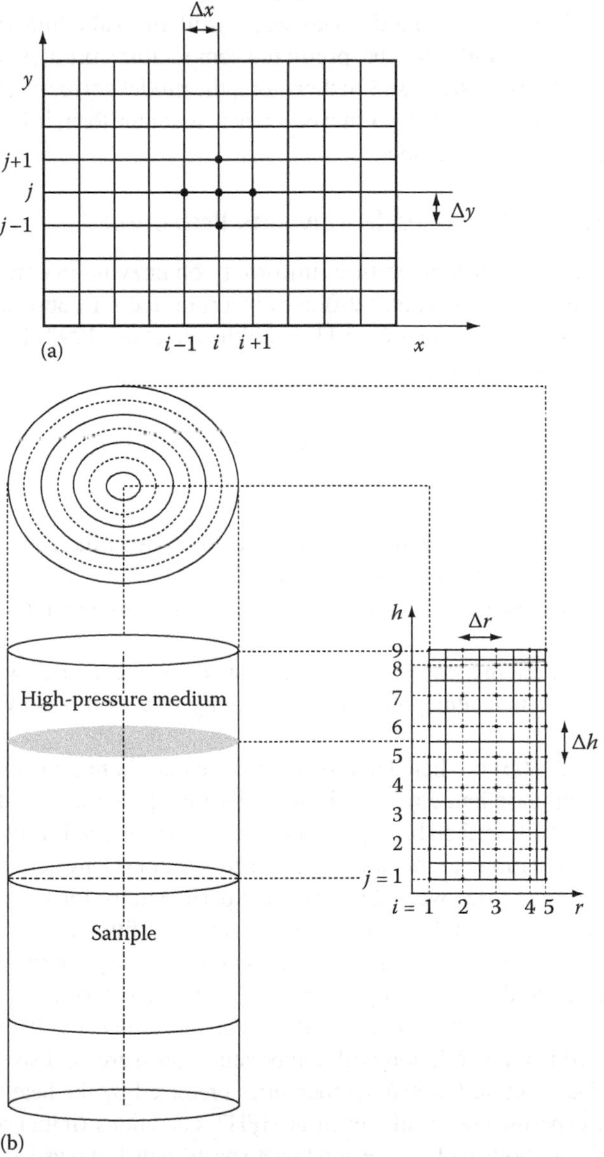


**FIGURE 6.1** Computational domain for finite difference method. (a) Finite difference grid of a 2D rectangular region. The nodes involved in computation of temperature at position $(i, j)$ are indicated by dots. (b) Representation of the mesh of the cylindrical vessel content (high pressure medium and sample) divided into volume elements appearing as layers of concentric rings having rectangular cross sections. This shows a simplified version of the mesh with 44 nodes (the real model consisted of 284 nodes). The dark shaded area is associated with the product in the sample holder and the white area with the pressure medium. (From Denys, S., Ludikhuyze, L.R., Van Loey, A.M., and Hendrickx, M.E., *Biotechnol. Prog.*, 16, 92, 2000a. With permission.)

The accuracy of the simulation was expressed by the mean associated with all data points of the relative difference (%) of measured and predicted temperatures with respect to the measured temperatures and their standard deviations. They found average deviations of 0.36%, 0.59%, and 0.14% for 186, 348, and 503 MPa, respectively. Hence, this numerical model was successful in evaluating the uniformity of the processes and validating the predicted temperature profiles with the fixed thermocouple readings in the agar. Furthermore, the model correctly predicted compression heating temperatures, even in cases where a nonuniform initial temperature distribution existed in the chamber.

### 6.3.2 Enzyme and Microbial Inactivation Prediction

To determine the effect of temperature uniformity on enzyme inactivation distribution inside the vessel, Denys et al. (2000a,b) incorporated (or coupled) the *Bacillus subtilis* α-amylase model (Equation 6.11; Ludikhuyze et al., 1997) in the numerical conductive heat transfer model.

$$\frac{\partial A}{\partial t} = -K(P,T)A \tag{6.11}$$

where

- $A$ is the relative activity or actual activity related to the initial activity, varying between 100% and values close to zero
- $K(P,T)$ is the inactivation rate constant as functions of pressure and temperature

The pressure–temperature–time profiles calculated by the model were integrated through the numerical scheme, and thus the activity retention was evaluated at any point in time.

Denys et al. (2000b) predicted and measured the distribution of activity retention of *B. subtilis* α-amylase and soybean lipoxygenase in apple sauce and tomato paste inside the previously described high-pressure vessel. They applied similar temperature and pressure combinations to the above-described system (Denys et al., 2000a). The relevant physical properties were determined experimentally for the pressure transfer medium, apple sauce, and tomato paste: density was fitted into a second-degree polynomial as a function of pressure and was assumed to vary linearly with temperature; specific heat was determined by differential scanning calorimetry, expressed as a function of temperature, and corrected with the calculated density; thermal expansivity was calculated from its relation with temperature, pressure, and specific heat.

Similar to Denys et al. (2000a), temperatures predicted by the heat transfer model coincided with experimental results even at HPHT conditions (initial temperature of 50°C at 600 MPa) for both apple sauce and tomato paste, which showed a thermal behavior similar to water. In contrast to soybean lipoxygenase, *B. subtilis* α-amylase has been found to be an appropriate system for assessing the uniformity of selected semifluid foods of varied viscosity and composition, as the numerical simulation of the inactivation of this enzyme showed a reasonable agreement with experimental results. The predicted maximum activity retention close to the walls of the vessel was 95%, whereas in the center of the vessel retention values around 55% were calculated. This showed that conduction-driven heat transfer inside the vessel may lead to strong process heterogeneities.

In addition to enzyme activity retention distribution, numerical simulations can also predict a number of other temperature-related effects (e.g., microbial inactivation, gelation, nutritional and sensorial quality degradation, crystallization, alterations in enzyme activity or protein stability, fat migration, changes in oxidation state, and energy expense); however, these predictions can be achieved to a certain degree of accuracy if the combined pressure and temperature degradation kinetics of the attribute under consideration are known (Denys et al., 2000a; Otero et al., 2002b; Otero and Sanz, 2003).

For spore inactivation prediction, de Heij et al. (2002) built a mathematical model by integrating thermodynamics and inactivation kinetics of *Bacillus stearothermophilus*, and showed the effect of the temperature distribution inside the high-pressure vessel. The model was based on a sterilization treatment consisting of two pulses at 700 MPa for 100 s, and initial temperature of 90°C. In this case, a one-dimensional (1D) finite-element model based on heat conduction was developed to predict temperature at the vessel wall and the center. Furthermore, the first-order kinetic constant $k$ was found from spore inactivation data and fit into a modified Eyring–Arrhenius equation (Equation 6.12), and expressed as a function of pressure and temperature.

$$k = k_{\mathrm{ref}} \cdot \exp\left(\frac{E_{\mathrm{a}}}{R}\left(\frac{1}{T_{\mathrm{ref}}} - \frac{1}{T}\right) - V_{\mathrm{a}}\left(P - P_{\mathrm{ref}}\right)\right) \tag{6.12}$$

where

the activation volume $V_{\mathrm{a}}$ is the characteristic parameter for the pressure dependence of the rate constant

$E_{\mathrm{a}}$ is the activation energy

$R$ is the universal gas constant (8.314 J $\mathrm{mol}^{-1}$ $\mathrm{K}^{-1}$)

Even though a good agreement was stated, the validation of the model was not included in this publication.

Ardia et al. (2004) modeled compression heating using a finite difference code, which was also based on heat conduction in radial coordinates. The model included a time-dependent expression (including d$P$/d$T$) as heat source or sink during the compression or decompression phase, respectively. Values for $\beta$, $\rho$, and $C_p$ for the actual temperature and pressure were calculated at each spatial node and for every time step of the numerical routine, taking into account the different parts constituting the system. The numerical routine was written in MathCAD by implementing the NIST formulations for regressive calculation of the expansion coefficient $\alpha_p$, the density $\rho$, and the specific heat $C_p$. Selected conditions were temperatures ranging from 5°C to 90°C in combination with 600 MPa (no pressure holding). A microbial inactivation model has been implemented into the finite difference scheme that yields the degree of inactivation for any radial position; the time-dependent inactivation was mathematically expressed assuming $n$th order inactivation kinetics:

$$\frac{\mathrm{d}N}{\mathrm{d}t} = -k \cdot N^{n} \tag{6.13}$$

In this case, the inactivation rate constant was expressed as a function of pressure and temperature by using the Eyring–Arrhenius equation (Equation 6.12). This modeling approach allowed predicting the temperature rise for water, while matching temperature and pressure values provided by NIST, but also allowed predicting temperature evolution in sucrose solutions at different concentrations. Furthermore, the model provided information about the inactivation of *Alicyclobacillus acidoterrestris* spores at 800 MPa and initial temperature of 50°C, finding a difference of 6 log cycles between predictions at the center of the pressure chamber and closer to the chamber walls.

## 6.4 CONVECTIVE MODELS SOLVED BY APPLYING COMPUTATIONAL FLUID DYNAMICS

As shown in Chapter 5, CFD (or CTFD, computational thermal fluid dynamics) can be applied to determine how a number of variables in the HPP affect heat transfer and, therefore, temperature evolution throughout the vessel. The CFD software packages can assist not only in calculating temperature evolution, but also allow coupling governing equations of fluid dynamics with kinetic inactivation equations to predict enzyme and microbial inactivation distribution from temperature and flow distribution. Coupling governing equations of fluid dynamics with kinetic inactivation models can be done either internally (solved simultaneously within the software package) or externally (by converting the temperature profile predicted by the CFD model into a parameter representing the extent of inactivation).

Among the numerical methods previously presented to solve heat transfer balances, Chapter 5 described the finite element and finite volume methods. Examples described in this section use these methods to develop convective CFD models. This section will show how CFD models were developed and applied to evaluate the effects of varying inlet velocities, vessel size, fluid viscosity, presence of packages, presence of solid foods, and sample carriers on temperature, flow, and/or inactivation distribution. Special attention will be given to the timescale analysis and the coupling of known kinetic models to predict the inactivation distribution of *C. botulinum* in two separate sections.

### 6.4.1 Prediction of Temperature Uniformity and Velocity Distribution

#### 6.4.1.1 Influence of Inflow Velocity on Temperature Distribution

In direct high-pressure systems a certain amount of fluid is pumped into the already existing fluid filling the vessel to reach the target pressure level. The velocity at the inlet where the pressure fluid is pumped into the vessel determines the come-up time. It has been observed that the velocities used in the microvessel modeled by Hartmann (2002) provided laminar flow at the inlet, whereas a 35 L pilot-scale vessel modeled by Knoerzer et al. (2007) gave turbulent conditions at the inlet region, as will be shown in the following subsections.

##### *6.4.1.1.1 Laminar Model*

Hartmann (2002) analyzed the thermodynamic and fluid dynamic effects of the pressurizing fluid (water) in a high-pressure vessel by numerical simulation. Temperature and fluid velocity profiles were modeled in a 4 mL chamber, pressurizing up to 505 MPa (initial temperature 15°C and pressure holding time 200 s) at three inflow velocities (2, 4, and 8 mm $s^{-1}$). The process simulations were based on a numerical solution of fluid dynamics, using the finite volume method explained in Chapter 5, in a CFX 4.4 commercial package from AEA technologies. The velocity profile was maintained constant at the inlet cross section for each case, giving a parabolic velocity profile further downstream in the inlet tube. In the inlet tube, velocity was zero at the surrounding wall, reaching a maximum value at its central axis.

The flow field obtained inside the vessel was governed by forced convection. Within a close region around the inlet, entering fluid underwent a strong deceleration, resulting in a nonuniform temperature distribution. Temperature gradients caused a nonhomogeneous density distribution that generated a buoyancy-induced fluid motion (natural convection due to gravitational field). Conditions of very low inlet velocities have been shown to be close to the isothermal case, i.e., no temperature rise due to compression occurring in the system. However, a faster compression rate gave a temperature increase closer to the maximum adiabatic conditions.

Modeling has also shown that both temperature and velocity fields are transient during pressure come-up and pressure holding. Free convection subsisted during the pressure holding at 500 MPa. Validation has been performed with a temperature probe, which was placed in the central plane near the wall. The numerical values were fitted within the error range of 0.7 K of the experimental values throughout the whole process, which was based on the uncertainty of the manual positioning of the temperature (±1 mm) and temporal resolution of the thermocouple (approx. 2 s), showing good agreement with experimental data.

##### *6.4.1.1.2 Turbulent Model to Predict Temperature Distribution and Flow*

Knoerzer et al. (2007) used the finite element method (COMSOL Multiphysics, COMSOL AB, Stockholm, Sweden) to model a much higher inlet velocity of 5.7 m$s^{-1}$ velocity for an empty 35 L pilot-scale vessel, which corresponded to a Reynolds number of 60,000 creating turbulent flow in the vessel bottom region. In this case, the pronounced turbulent region with arising eddies provided significant cooling during pressure come-up and holding steps. Thus, opposite from what was observed by Hartmann (2002) at laminar conditions, higher inlet velocities (at turbulent conditions) hinder the system from achieving maximum compression heating.

To account for turbulence, the equations for energy and momentum conservation were extended by further terms, taking into account the contributions of arising eddies (increased thermal conductivity due to mixing and increased dynamic viscosity due to increased shear). To solve this flow problem, an averaged representation was necessary, which was done by employing Reynolds-averaged Navier–Stokes (RANS) equations (Nicolaï et al., 2001; COMSOL Multiphysics, 2006), comprising terms that include average velocity and pressure values and a fluctuating term represented by the Reynolds stress tensor.

In particular, Knoerzer et al. (2007) applied the $k$–ε model, a commonly used stochastic turbulence model for industrial applications, by including an additional "turbulent viscosity" term to the equations expressing conservation of momentum and continuity. The turbulent viscosity $\eta_T$ is given by

$$\eta_T = \rho C_\mu \frac{k^2}{\varepsilon} \tag{6.14}$$

where

$C_\mu$ is a constant
$k$ is the turbulent kinetic energy
$\varepsilon$ is the dissipation rate of turbulence

The momentum equation, extended according to COMSOL Multiphysics (2006), is given as follows:

$$\rho\left[\frac{\partial \vec{v}_a}{\partial t} + \left(\vec{v}_a \cdot \nabla\right)\vec{v}_a\right] = -\nabla P_f + \nabla \cdot \left((\eta + \eta_T) \cdot \nabla \vec{v}_a\right) + \rho g \tag{6.15}$$

where

$\vec{v}_a$ denotes the average velocity
$P_f$ includes a fluctuating term

In addition to the continuity equation, the $k$–$\varepsilon$ closure includes two extra transport equations solved for both $k$ and $\varepsilon$, using model constants from experimental data (COMSOL Multiphysics, 2006). COMSOL Multiphysics was used to couple the $k$–$\varepsilon$ closure equations with the energy conservation equation for heat transfer through convection and conduction, assuming nonisothermal flow. The energy conservation equation (Equation 6.16) was modified by including the turbulent thermal conductivity, $k_T = \eta_T \cdot C_p$, as follows:

$$\frac{\partial\left(\rho C_p T\right)}{\partial t} + \nabla \cdot \left(\rho \vec{v} C_p T\right) = Q + \nabla \cdot ((k_1 + k_T)\nabla T) \tag{6.16}$$

where

$k_1$ is the thermal conductivity
$C_p$ is the specific heat capacity

In this way, turbulent flow and heat transfer were coupled.

#### 6.4.1.2 Influence of Fluid Viscosity on Temperature Uniformity and Flow

Hartmann and Delgado (2002b) used a similar numerical analysis, as indicated before, to study how the fluid chamber viscosity affects the spatial and temporal evolution of temperature and fluid velocity fields, under the same conditions used by Hartmann (2002). They showed that the uniformity can be disturbed by convective and conductive heat and mass transport conditions, i.e., the fluid velocity distribution strongly influenced the temperature distribution and vice versa. As expected, a less viscous (Newtonian) fluid led to a more uniform process behavior. A matrix fluid with viscosity a 100 times larger than that of water led to a strong nonuniformity. In this case, a higher viscosity provides lower motion in the fluid system, which leads to a decreased convective interaction between fluid sections and as a result lowered heat transfer. No validation information was provided in this publication.

#### 6.4.1.3 Flow Fields Predicted and Measured Inside a High-Pressure Vessel

Hartmann and Delgado (2003a,b) considered an energy balance that accounts for the fluid movement inside the chamber during pressurization and holding time. They described the moving fluid inside the chamber as "a time-dependent temperature field induced by heat exchange" between the walls of the pressure chamber, the pressure medium, the packaging material, and the packed food. Pehl et al. (2000) and Pehl and Delgado (1999, 2002) have shown that a fluid velocity field develops due to forced convection during pressurization (a result of pressurizing fluid entering the vessel) and natural convection during pressure holding time (a result of temperature differences within the fluid and vessel). They were the first to examine the dynamic effects of fluid convection and heat transfer in HPPs. This was accomplished by developing high pressure-digital particle image thermography (HP-DPIT) and high pressure-digital particle image velocimetry (HP-DPIV) using a 2 mL pressure cell with four sapphire windows. As of today, fluid-dynamic effects can only be studied in laboratory-scale vessels. Given the complexity involved in adding a large pressure-resistant crystal window within the vessel structure, it is not yet possible to determine flow effects at pilot scale.

#### 6.4.1.4 Influence of the Vessel Boundaries on Temperature Uniformity

A number of publications from Hartmann et al. on thermal fluid dynamic modeling in HPPs were based on the assumption that the inner wall temperature of the vessel remains constant when using a tempering device that surrounds the steel structure. However, later works (Hartmann and Delgado, 2003b; Hartmann et al., 2004) showed that better agreement with experimental data is obtained if temperature at the vessel inner walls are assumed as time-dependent (rather than constant), due to heat flux into the steel housing. A 1D heat conduction equation for the steel vessel wall was used to determine the transient temperature distribution of the vessel surface. Then, the model was applied as a boundary condition to represent temperature distribution inside the vessel, using a thermal fluid dynamic model. The model agreement was verified with data obtained from water pressurized at 500 MPa and initial temperature 22°C. The good agreement found with experimental data for a 3.5 L vessel at three positions led to the conclusion that a temperature decrease, from the inner vessel wall layer (in touch with the fluid) to the outer wall layers, occurs at a nonlinear rate.

In order to predict temperature profiles at the vessel boundary, Hartmann et al. (2004) developed a specific algorithm to account for heat transfer between the fluid and the steel vessel wall. It was assumed that heat conduction within the solid vessel structure occurred only in the direction normal to the wall (one-dimensionally), whereas in the pressure-transmitting fluid domain, convective and diffusive transport were accounted for both radial and axial directions. Heat conduction in the vessel was expressed in the following heat transfer equations derived for the radial direction through the vessel walls (Equation 6.17) and the top and bottom surfaces (Equation 6.18):

$$\frac{\partial T}{\partial t} = \frac{k}{\rho C_p}\left[\frac{\partial^2 T}{\partial r^2} + \frac{1}{r}\frac{\partial T}{\partial r}\right] \tag{6.17}$$

$$\frac{\partial T}{\partial t} = \frac{k}{\rho C_p}\left[\frac{\partial^2 T}{\partial z^2}\right] \tag{6.18}$$

Solutions to Equations 6.17 and 6.18 were determined by converting the expressions into a second-order explicit finite difference scheme, as expressed in Equations 6.19 and 6.20 (Hartmann et al., 2004):

$$T_i^{n+1} = T_i^n + \frac{k}{\rho C_p}\frac{\Delta t}{(\Delta r)^2}\left[T_{i+1}^n - 2T_i^n + T_{i-1}^n + \frac{\Delta r}{r_i}\left(T_{i+1}^n - T_i^n\right)\right] \tag{6.19}$$

$$T_i^{n+1} = T_i^n + \frac{k}{\rho C_p}\frac{\Delta t}{(\Delta z)^2}\left[T_{i+1}^n - 2T_i^n + T_{i-1}^n\right] \tag{6.20}$$

where

$i$ is the position on the 1D grid
$n$ is the time step number
$\Delta z$ and $\Delta r$ represent the distances of neighboring grid points

A boundary condition algorithm was created with a software package using a time step size $\Delta t$ prescribed via superior routines, incorporating the numerical solutions shown above.

Figure 6.2 shows the finite difference grid used for calculating temperature profiles at the inner vessel wall. The fluid cell "a" touching the wall is taken as the inner boundary of the finite difference scheme, which is applied to $N+1$ cells. A constant temperature is assumed at the outer boundary corresponding to the vessel wall. Integration to the next time step yields a temperature profile in the steel wall at cells ranging from 1 to $N$. Node $N$ (temperature at solid cell "b") is also a boundary

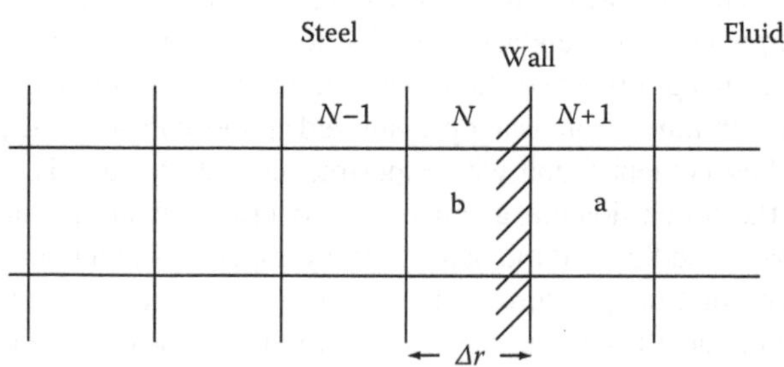


**FIGURE 6.2** Finite difference grid used for the determination of a numerical model representing temperature variation at inner vessel wall (compression fluid) boundary. (From Otero, L., Molina-Garcia, A.D., Ramos, A.M., and Sanz, P.D., *Biotechnol. Prog.*, 18, 904, 2002a. With permission.)

condition in the fluid domain. Hence, simultaneous calculations of temperature distributions in both vessel wall and fluid domain were done.

Otero et al. (2007) and Juliano et al. (2008) did not require this assumption since they integrated the steel vessel walls into the finite element modeling scenario using COMSOL Multiphysics software (as mentioned). As a result, direct heat up from the hotter compression fluid (due to compression heating) could be represented in the vessel wall regions.

#### 6.4.1.5 Vessel Boundary for Turbulent Conditions: The Logarithmic Wall Function

When using a turbulent model, a logarithmic wall function condition can be assumed (Nicolaï et al., 2001, 2007) at both vessel and package carrier walls (Knoerzer et al., 2007; Juliano et al., 2008). The $k$–ε turbulent model allows accounting for turbulent transport in states close to isotropic. However, close to the solid walls, turbulence transport is no longer isotropic, as fluctuations resulting from turbulence vary greatly in magnitude and direction. At these regions, deviations from isotropy are no longer negligible and must be accounted for in a proper model (COMSOL Multiphysics, 2006).

In this case, an empirical relation between the value of velocity parallel to the wall and wall friction replaces the thin boundary layer near the wall. This empirical relation is the logarithmic wall function. It expresses the wall velocity as a function of friction velocity and length scale, which depends on wall shear stress, density of the fluid, and viscosity. It also includes the constants characteristic of a wall surface. Part of the analysis of this condition includes identifying relevant scales for the turbulent kinetic energy $k$ and the turbulence energy dissipation rate ε used in the $k$–ε turbulent model, and analyzing the equilibrium boundary layer using a logarithmic profile. The logarithmic wall function model is accurate for high Reynolds numbers and situations where pressure variations along the wall are not very large. However, the method can often be used outside its frame of validity with reasonable success (COMSOL Multiphysics, 2006). A model representing a pilot-scale 35 L vessel (Knoerzer et al., 2007; Juliano et al., 2008) used the logarithmic wall function condition to account for turbulence at the vessel walls.

#### 6.4.1.6 Temperature and Flow in Vessels with Packages at Various Scales

To evaluate the presence of packages inside a laboratory-scale vessel, two 0.8 L chambers were modeled (Hartmann and Delgado, 2003a), one with compression fluid and the other containing five packages, with a net volume of packed fluid of 0.25 L. Process conditions were 500 MPa, holding time 20 min, and initial temperature 40°C. The chamber without packages showed a homogenous temperature distribution due to an intensive fluid motion that yielded efficient mixing and convective heat transfer to the vessel walls. Temperature was comparatively low at the inlet region of the vessel without packages due to the forced convection of inflowing cold pressure transmission liquid; this is a common problem in HPPs.* In the other system, the packaging material represented a barrier to flow and heat transfer, which increased thermal heterogeneities inside the vessel due to reduced convective fluid motion.

Hartmann et al. (2003) also studied the influence of five packages containing milk on heat transfer in the same high pressure system (Hartmann and Delgado, 2003a), but focused on the holding period at 400 MPa and 20°C (fluid temperature inside vessel) or 30°C (vessel wall). Apart from solving conservation equations for mass, momentum, and energy, they solved for temperature conduction through the packaging material (polypropylene, PP) by coupling the following Fourier equation (see Chapter 5) in selected directions:

$$\frac{\partial T}{\partial t} = \frac{k_{pp}}{\rho C_{pp}} \nabla^2 T \tag{6.21}$$

where

$k_{pp}$ represents the thermal conductivity
$\rho$ is the density
$C_{pp}$ is the specific heat capacity of the packaging material

A correction was incorporated for the density of milk with respect to water. A scale-up analysis going from a 0.8 to a 6.3 L vessel was performed with standard packaging material parameters and with increased package material conductivity $k_{pp}$ by a factor of 10. This is equivalent to using a package material of a thickness reduced by a factor of $1/\sqrt{10}$ as derived from the second-order derivative in Equation 6.21.

Heat retention occurred to a different extent in the two vessels, with the smaller 0.8 L vessel yielding higher heat loss inside each package. As expected the model showed that when using a thinner packaging material, with a tenth of the thermal conductivity, the temperature loss increased per package inside the larger 6.3 L vessel. Hartmann et al. (2003) validated their simulations with inactivation data obtained in the smaller 0.8 L vessel, as explained later in Section 6.4.2.

Hartmann et al. (2003) also analyzed temperature distribution and flow per package, and compared average temperatures inside each package located at selected heights inside the vessel, finding lower temperatures closer to the pressure fluid inlet. A vortex pattern was observed in the upper part of the package (Figure 6.3), while a uniform flow was observed in the lower part, therefore, creating regions of temperature differences due to convective transport mechanisms. It was concluded that if the fluid motion is of very low intensity (e.g., if the food is highly viscous or semisolid), thermal heterogeneities may be preserved and give rise to process nonuniformities, including differences between survivors per package.

Hartmann and Delgado (2003b) further studied the geometrical scale and heat transfer characteristics of the packaging material, accounting for both the compression

* The heat generated in the fluid leaving a high-pressure pump due to compression heating is commonly lost through the pipelines before the liquid reaches the bottom of the pressure vessel. This is common in many HPHT systems since their design does not contain proper insulation or a heat source to maintain the inlet fluid at the intended temperature under pressure.

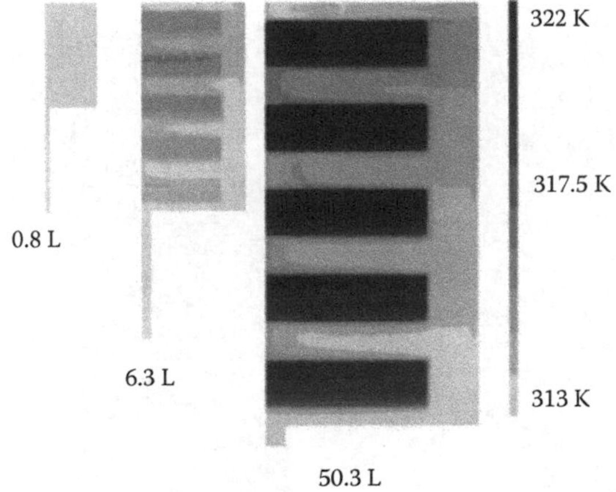


**FIGURE 6.3** Temperature distribution in vessels of different scales (0.8, 6.3, and 50.3 L) at the end of the process (1200 s). (From Hartmann, C. and Delgado, A., *Biotechnol. Bioeng.*, 82, 725, 2003b. With permission.)

and holding times. They modeled using five packages with enzyme solution treated in vessels with volumes in microscale (0.8 L), pilot scale (6.3 L), and semi-industrial scale (50.3 L).

The authors verified findings (as mentioned) by determining that more heat retention existed in packages in the 50.3 L vessel, showing an average temperature difference per package of around 7 K with the 0.8 L vessel and 4 K with the 6.3 L vcssel (Figure 6.3). Figure 6.4 shows the temperature and velocity distribution after 27 and 450 s (7 min) at 550 MPa ($T_s$ = 40°C or 313 K), representing the thermofluid dynamic effects during the complete holding phase in the same configuration. As explained before (Hartmann and Delgado, 2002a; Hartmann et al., 2003), temperature and fluid velocity gradients are mainly seen due to the incoming "cold" pressure medium. Figure 6.4a shows how the pressure medium between the package and the chamber wall exhibits a vortical motion. Hartmann et al. (2003) explained that this flow is driven by "instable density stratification" (the authors would simply describe this phenomenon as natural convection). They found a difference of 7 K between the lower and the upper part of the vessel after compression up to 400 MPa. One possible explanation for stratification could be that the colder region at the bottom of the vessel causes layers of water with higher density to develop. However, the authors (Hartmann et al., 2003) did not explain why during holding time the symmetry axis shows regions with much higher velocity (Figure 6.4b). One possibility would be higher buoyancy forces created due to hotter regions and nonuniform heat distribution.

Located at the top of the vessel, package 5 (Figure 6.4) is close to the cool chamber walls but exposed to warm pressure medium, which then rises to the top of the chamber due to natural convection. However, the authors did not clarify if the effect of the colder lid at the top of the vessel, which is made of steel and does not comprise any form of heat insulation or heat source, was accounted for. Package 4 showed higher heat retention than package 5, and the lowest heat retention was found in package 1 as a result of the inflowing cold pressure medium (Hartmann and Delgado, 2003b).

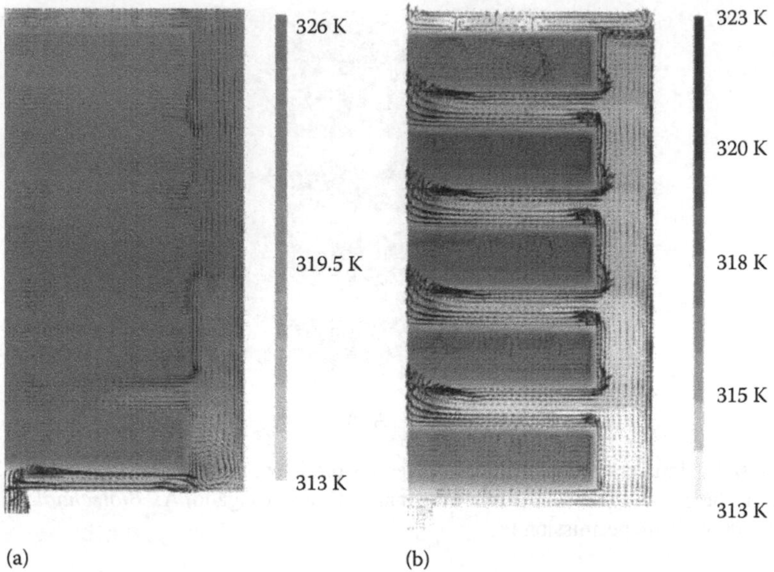


**FIGURE 6.4** Temperature and velocity distribution in a 6.3 L vessel after (a) 27 s and (b) 450 s. (Note: arrows proportional to the maximum velocity at specific time.). (From Hartmann, C., Delgado, A., and Szymczyk, J., *J. Food Eng.*, 59, 33, 2003. With permission.)

#### 6.4.1.7 Temperature Uniformity in a Pressure Vessel Containing Solid Food Materials

Given the difference in composition and structure, solid food materials provide compression heating values different from water or pressure medium. Therefore, presence of solid materials inside the vessel can influence the outcome of temperature and flow distribution throughout the vessel and within packages. Otero et al. (2007) developed a mathematical model to describe the phenomena of heat transfer taking place during the high-pressure treatment of solid foods, with particular focus on the influence of free convection arising in the pressure medium on the thermal evolution of processed samples. They fitted samples of agar gels containing 99% water as solid model foods with similar physical properties to water. Two sample amounts, 1.2 and 0.2 L (corresponding to 71% and 12% filling ratio), were each inserted in a container with a rubber cap, and then placed into a 2.4 L vessel to investigate the relative influence of free convection currents from the pressure fluid. Trials were done at room temperature (20°C) at a pressure of 350 MPa. The model prediction was validated with two thermocouples, one placed at the surface of the agar, the other located inside. The pressure was held until reaching a total cooling down of initial temperature (160 min for bigger sample; approx. 90 min for smaller sample). The computational domain included the sample and rubber cap, the pressure medium, and the steel domain of the pressure vessel.

They investigated two scenarios: one considering only conduction and the other considering equations for free convection and conduction. In the latter scenario, the model was able to predict temperature rise and profiles during pressurization in both samples (error was 0.1°C–0.2°C). However, when only conduction was considered, the model gave an error of 3°C in the small sample at both the surface and center, while a smaller error was found in the large sample. With more pressure media (small sample, low sample filling ratio) present in the vessel, free convection becomes important in the cooling down of the solid. Nevertheless, no effect due to free convection was seen when a small amount of pressure medium remains inside the vessel (big sample). They also investigated a carbohydrate-based sample (tylose) to illustrate that samples with different material properties can yield a significantly different heating behavior. The properties for these materials were obtained from a previous work on the determination of thermophysical properties of tylose at high pressure conditions (Lemmon et al., 2005; Otero et al., 2006).

Ghani and Farid (2007) simulated the temperature distribution and velocity profiles during high-pressure processing of water and a mixture of beef fat and water in three dimensions. The modeled process was performed at temperatures 20°C–25°C and pressure of 500 MPa for up to 1000 s. Two beef fat samples of approximately 10 mL were fitted in a 0.3 L high-pressure vessel (filling ratio 6%). They predicted the compression heating in two solid fat volumes and water in the vessel with insulated walls. The model showed that the heat transfer from the solid to the water by free and forced convection was due to the higher temperature achieved by the beef fat. It also showed the cooling down caused by the pumped water through the top inlet of the vessel in the fluid area and the boundaries of the solid. Due to the short come-up time (30 s) of this process no heat loss was observed throughout the bulk of the solid. This paper did not validate temperatures in the solid:liquid system by direct measurements.

#### 6.4.1.8 Effect of Adding a Carrier on Temperature Uniformity and Flow

Modern pilot-scale, high-pressure sterilization systems (described in Chapter 5) require a means of preheating and transporting preheated products into the system. Cylindrical carriers provide not only a means of preheating the food packages but also act as insulating barriers, which helps to at least partially retain heat at the package regions. Carriers also act as barriers to flow of incoming colder fluid into the package area. Furthermore, the carrier helps in insulating the cooling down at the vessel's lid area, given that the metallic lid acts as a heat sink.

Knoerzer et al. (2007) investigated the effect of placing two types of food carriers, i.e., a metal composite carrier and a Teflon polytetrafluoroethylene (PTFE) carrier, inside a pilot-scale sterilization vessel on temperature uniformity. They simulated all processing steps in a 35 L vessel filled with water at a pressure of 600 MPa, 415 s pressure holding, and initial temperature of 90°C. Three vessels were used in the simulation: (1) one without carrier, (2) one with metal composite carrier, and (3) one with Teflon PTFE carrier. The turbulent conditions at the compression fluid inflow region were described by coupling the $k$–$\varepsilon$ turbulence model (Section 6.4.1.1.2) with the governing equations for mass, energy, and momentum conservation. To validate the simulated temperatures, a 3×3 thermocouple array placed in an axi symmetric slice was set up in randomized form at several runs. An example of the modeling domain and temperature distribution is shown in Figure 6.5a.

Compression heating was predicted using Equation 5.12 where expressions of density, specific heat, and thermal expansivity were substituted as functions of temperature and pressure. Simulations in the empty vessel showed that conduction and convection causes excessive cooling. However, when a carrier was inserted, cooling down was observed mainly at the water inlet region below, as seen in papers by Hartmann et al., which was due to the colder turbulent incoming fluid. The metal carrier provided a barrier against cooling down but the lower region was still colder and, therefore, an uneven temperature distribution was observed and validated. An excellent correlation ($R^2=0.97$) was obtained by comparing predicted and measured values in a parity plot, which included all locations of the thermocouples at all time steps throughout the process.

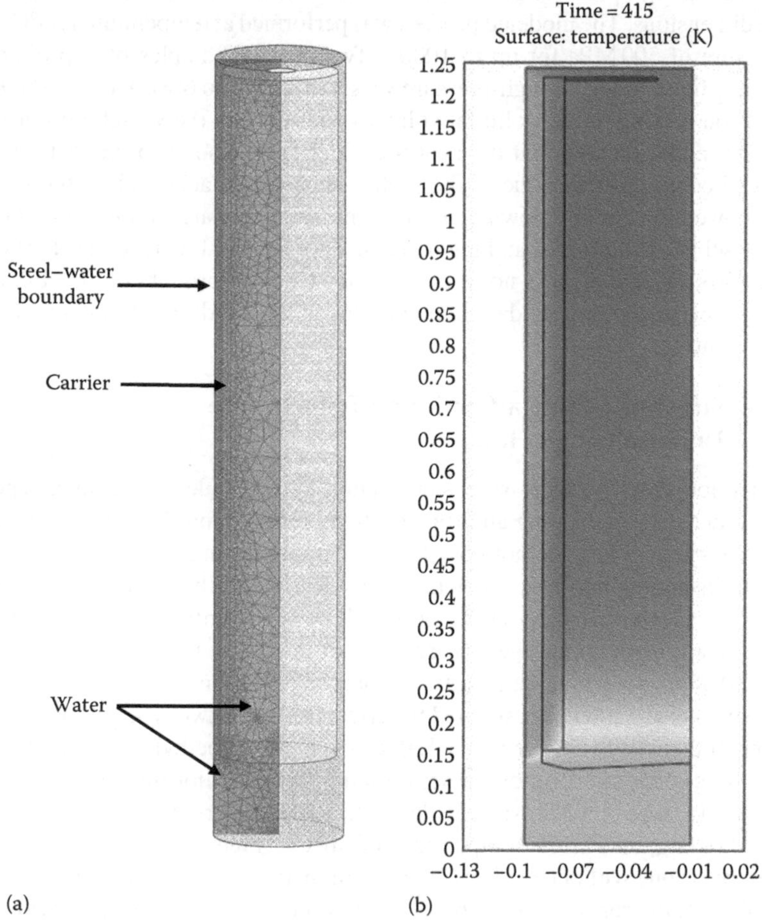


**FIGURE 6.5** CFD model of a 35 L vessel including a carrier: (a) computational domains of the high pressure vessel structure, (b) thermal profile in the vessel including a metal carrier at end of holding time (415 s) at 600 MPa and (c) thermal profile in the vessel including a PTFE carrier at end of holding time. (From Knoerzer, K., Juliano, P., Gladman, S., Versteeg, C., and Fryer, P., *AIChE J.*, 53, 2996, 2007. With permission.)

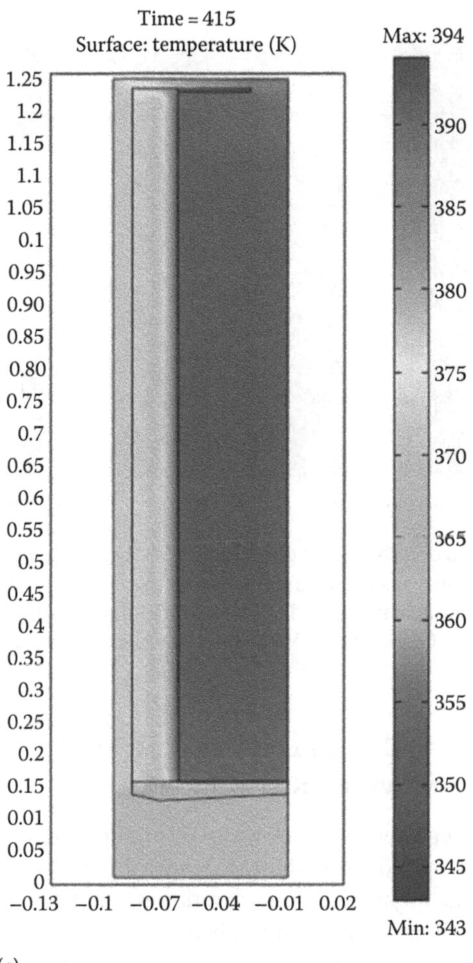


**FIGURE 6.5 (continued).**

On the other hand, the carrier made of PTFE was able to retain most of the compression heat generated even during holding time throughout its entire volume, providing a much more uniform temperature distribution. Figure 6.5b and c depicts the insulating effect of the PTFE carrier with respect to the metal, one at the end of holding time (415 s at 600 MPa). This model was further developed (Juliano et al., 2008) by including the steel vessel walls, the vessel lid, and cylindrical packages containing a "water-like" solid, and also by still assuming rotation symmetry (Figure 6.6a). The initial temperature of the vessel lid was considered lower since it does not include a heating source for the actual process. As expected, this addition lowered the temperature at the upper region of the vessel but did not affect the end temperature inside the packages (Figure 6.6b). This model was designed as a platform to evaluate the performance of *C. botulinum* kinetic models, which will be covered in Section 6.4.2.2.

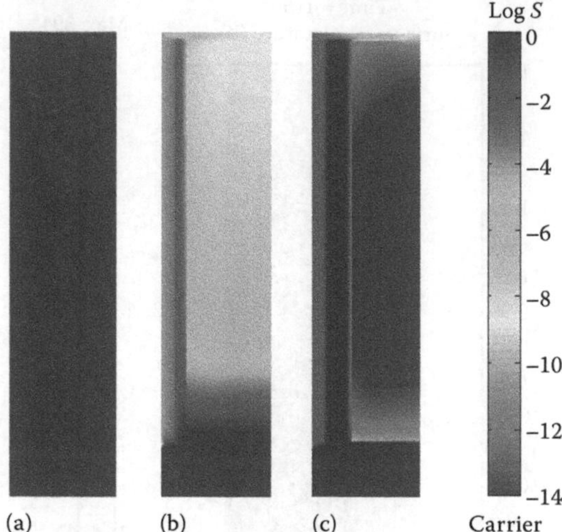


**FIGURE 6.6** Predicted distribution of the extent of *C. botulinum* log reduction extent inside a pilot-scale high pressure vessel in three scenarios: (a) vessel without carrier, (b) vessel including a metal composite carrier, and (c) vessel including a Teflon (PTFE) carrier. (From Knoerzer, K., Juliano, P., Gladman, S., Versteeg, C., and Fryer, P., *AIChE J.*, 53, 2996, 2007. With permission.)

### 6.4.2 Coupling of CFD Models with Enzyme and Microbial Inactivation Kinetic Models

Several publications reported the uniformity of high pressure processing vessels in terms of temperature and flow distribution. Some of those also combined equations for mass, energy, and momentum conservation with inactivation kinetic models to predict a pattern of enzyme or microbial inactivation distribution with time. As explained in Chapter 5, this "coupling" of equations can be done either internally, within the model itself, or externally, by using another software package to transform the temperature output provided by the model into inactivation values.

An equation describing the temporal and spatial enzyme activity or microbial inactivation distribution (Equation 6.22) can be coupled to determine the inactivation distribution or relative retention throughout the vessel volume at different times (Hartmann et al., 2003).

$$\frac{\partial A}{\partial t} + u\frac{\partial A}{\partial x} + v\frac{\partial A}{\partial y} + w\frac{\partial A}{\partial z} = K(P,T)A \quad (6.22)$$

where

- $A$ is the relative enzyme activity or microbial load (actual value related to initial value)
- $K(P,T)$ is the inactivation rate constant
- $u$, $w$, and $v$ are the components of the fluid velocity vector in the $x$-, $y$-, and, $z$-directions, respectively

The left-hand side contains the coupling between *A* and the flow field, i.e., the velocity of the solution. The right-hand side represents the coupling of *A* and the temperature distribution.

The following subsections describe the prediction of α-amylase, *Escherichia coli*, and *C. botulinum* inactivation distribution, in vessels of various sizes, accounting for some of the effects discussed in the sections above.

#### 6.4.2.1 Prediction and Quantification of Residual α-Amylase and *E. Coli* Inactivation

To understand the influence of the presence of five packages (0.25 L altogether) contained in a 0.8 L vessel on the retention of *B. subtilis* α-amylase (Hartmann and Delgado, 2003a), two cases were considered: (1) a chamber filled with only enzyme solution and (2) a chamber filled with enzyme solution and packages. The model (also described in Section 6.4.1.5) included Equation 6.22 coupled with mass, energy, and momentum conservation equations. The enzyme activity retention in the vessel with no packages was variable (28%–48%); the lowest activity retention was seen in the vessel's top region and the highest was seen in the core and inlet region due to incoming cold fluid. As mentioned before, the presence of packages influenced the temperature distribution and, therefore, the activity retention, which ranged between 39% and 50% depending on the location of the packages. No enzyme inactivation validation was reported by Hartmann and Delgado (2003a).

Hartmann et al. (2003) also studied the influence of packaging material on heat transfer and inactivation of *E. coli* suspended in ultrahigh temperature (UHT) treated milk. In this case the inactivation equation (Equation 6.22) included an inactivation constant for *E. coli*. The study mainly discussed the inactivation effects during the holding period at 400 MPa and 20°C (fluid temperature inside vessel) or 30°C (at vessel wall).

Hartmann et al. (2003) validated their simulations by comparing numerical simulation results with experimental data obtained for *E. coli* in UHT milk at an initial concentration of $10^8$ cfu $mL^{-1}$. In this case, validation was done with inactivation data obtained in a small 0.2 L vessel. They also compared numerical and experimental inactivation results with predictions from a two-parameter model (pressure and temperature), describing inactivation with an ordinary differential equation (Hinrichs, 2000); the model was valid between 5°C and 40°C for a pressure range 300–500 MPa. The numerical simulation, including dynamic effects, found good agreement with the experimental data and the two-parameter model. Differences between the numerical simulation and experimental results were attributed to the assumption of adiabatic pressurization set for the numerical model, resulting in a temperature overshoot of about 10°C.

In their scale-up study, going from a 0.2 L vessel to 0.8 and 6.3 L vessels, there was much higher inactivation of *E. coli* (8 log reductions in 6.3 L; 6 log in 0.8 L) in the larger vessel due to higher heat retention, as explained in Section 6.4.1.5. In both vessels, there was a maximum inactivation in package 4 (Figure 6.4) since it was further away from the pressurizing fluid inlet region and top vessel walls. When the thermal conductivity of the packaging material was increased ten times, packages located in both vessels had less heat retention and, thereby, achieved much lower inactivation per package (2 log reductions less inactivation for each vessel case).

Hartmann and Delgado (2003b) completed the study in geometrical scale, but in this case, modeled the use of five packages containing enzyme solution in the 0.8, 6.3, and 50.3L vessels (Section 6.4.1.5). Equation 6.22 was used for predicting the inactivation of *B. subtilis* α-amylase. The study confirmed that the higher inactivation occurring in the larger scale vessels was mainly due to increased heat retention in the packages. More temperature gradients were also observed in larger packages containing enzyme solution. Process uniformity Λ was defined as the ratio of the minimum ($A_{ave\_min}$) and maximum ($A_{ave\_max}$) average activity retention per package among all five packages:

$$\Lambda = \frac{A_{ave\_min}}{A_{ave\_max}} \tag{6.23}$$

Maximum activity retention was found in package 1 and minimum activity retention was found in package 4 (Figure 6.4). The heat transfer coefficient for conduction through the packaging material $h_{pp}$, where $h_{pp} = k_{pp}/D_{pp}$, was varied from the standard value for PP to simulate changes in the material properties as well as modifications in the packaging material thickness.

Table 6.1 describes the uniformity values obtained for each vessel volume while varying $h_{pp}$. Process uniformity for the 0.8L vessel did not depend on $h_{pp}$, whereas the other larger vessels were greatly affected. The lowest heat transfer coefficients provide more inactivation uniformity between packages because heat is retained better inside the packages for a major part of the process. For instance, a uniformity of 0.97 for the 50.3L vessel indicates that heat is retained inside the package throughout most of the process, leading to a high degree of inactivation and uniformity. In order to explain the heat mechanisms determining the degree of uniformity a timescale analysis can be done, as will be explained further in Section 6.4.2.3.

#### 6.4.2.2 Distribution of *C. botulinum* Inactivation in a Pilot-Scale Vessel

So far little has been published on CFD modeling of HPPs at high temperature sterilization conditions. There are two publications where some authors in this chapter (Knoerzer et al., 2007; Juliano et al., 2008) characterize the CFD models described in Section 6.4.1.7 by externally coupling the temperature output with *C. botulinum* inactivation models in the form of a differential equation.

**TABLE 6.1**
**Process Uniformity for Simulated Vessel Volumes and Heat Transfer Coefficients**

| | | $h_{pp}$ (W m$^{-2}$ K$^{-1}$) | | |
|---|---|---|---|---|
| Λ (–) | | $1 \times 10^{-4}$ | $1 \times 10^{-3}$ | $1 \times 10^{-2}$ |
| Vessel volume (L) | 0.8 | 0.86 | 0.84 | 0.84 |
| | 6.3 | 0.90 | 0.78 | 0.74 |
| | 50.3 | 0.97 | 0.81 | 0.69 |

*Source:* Hartman, C., Delgado, A., and Szymczyk, J., *J. Food Eng.*, 59, 33, 2003. With permission.

###### 6.4.2.2.1 *Adding Carriers with Different Compositions and Their Effect*

Knoerzer et al. (2007) evaluated the *C. botulinum* spore inactivation distribution inside a vessel, as shown in Figure 6.5 for the three scenarios described before (vessel 1 without carrier, filled with only water; vessel 2 with metal carrier; vessel 3 with PTFE carrier). The temperature component was used to calculate the *F*-value, which was transformed into inactivation values by using the expression $F = D \log(N/N_0)$, where $F$ is the thermal death time (min) and $D$ the decimal reduction time (min). In the vessel without carrier, inactivation distribution yielded less than 1 log reduction, while an inhomogeneous distribution was seen in the metal carrier (between 2 and 9 log reductions). For the insulated PTFE carrier, over 94.6% of the carrier's length showed more than 12 (up to 14) log reductions in *C. botulinum* spores (Figure 6.6). Based on the increased extent of inactivation observed when using an insulated carrier, this type of model can be of aid in finding an optimum carrier design that produces thermal uniformity and uniform spore inactivation throughout the carrier as well as optimum processing times.

###### 6.4.2.2.2 *Prediction of Inactivation Distribution Using Selected* C. Botulinum *Inactivation Kinetic Models*

Juliano et al. (2008) developed a more detailed CFD model, as described in Figure 6.7. This model was applied as a platform to compare the performance of known *C. botulinum* inactivation kinetic models in predicting inactivation distribution. Apart from the previously considered linear kinetic model including the *F*-value (Knoerzer et al., 2007), the Weibullian model (described in Chapter 5), *n*th order model (Equation 5.13), and a combination of the *n*th order model and log-linear model were selected.

The kinetic models were expressed in the form of an ordinary differential equation (Equation 6.24) to externally couple them with the CFD model temperature output through the MATLAB® routine also used by Knoerzer et al. (2007), which allowed solving for a nonisothermal scenario.

$$\frac{\mathrm{d}\left[\log S(t)\right]}{\mathrm{d}t} = f\left[\log S(t)\right] \tag{6.24}$$

where

$f$ is a generic function of $\log S(t)$
$S$ being the survival ratio $N/N_0$
$N$ and $N_0$ the final and initial number of spores

The routine fitted the temperature data to give a function $T(t)$, and was subsequently substituted in the differential equation, which was then solved for the total time of the process. The output of this routine was capable of providing: (a) $\log S(t)$ vs. $t$ plots at different locations, (b) $\log S(x,y)$ distributions at specific times, and (c) $\log S(x,y,t)$ vs. $t$ animations.

Table 6.2 summarizes the microbial inactivation models used (A, traditional first-order log-linear kinetics model; B, Weibullian model; C, *n*th order kinetics model; D, combined log-linear/*n*th order model) expressed in Equation 6.24 and includes the respective parameter values for *C. botulinum*. The log-linear first order and Weibullian models originated from *C. botulinum* inactivation data at near atmospheric conditions and, therefore, only accounted for the temperature variation. On the other hand, the *n*th order model was obtained from *C. botulinum* inactivation data corresponding to several combinations of temperature and pressure (Margosch et al., 2006) and thus

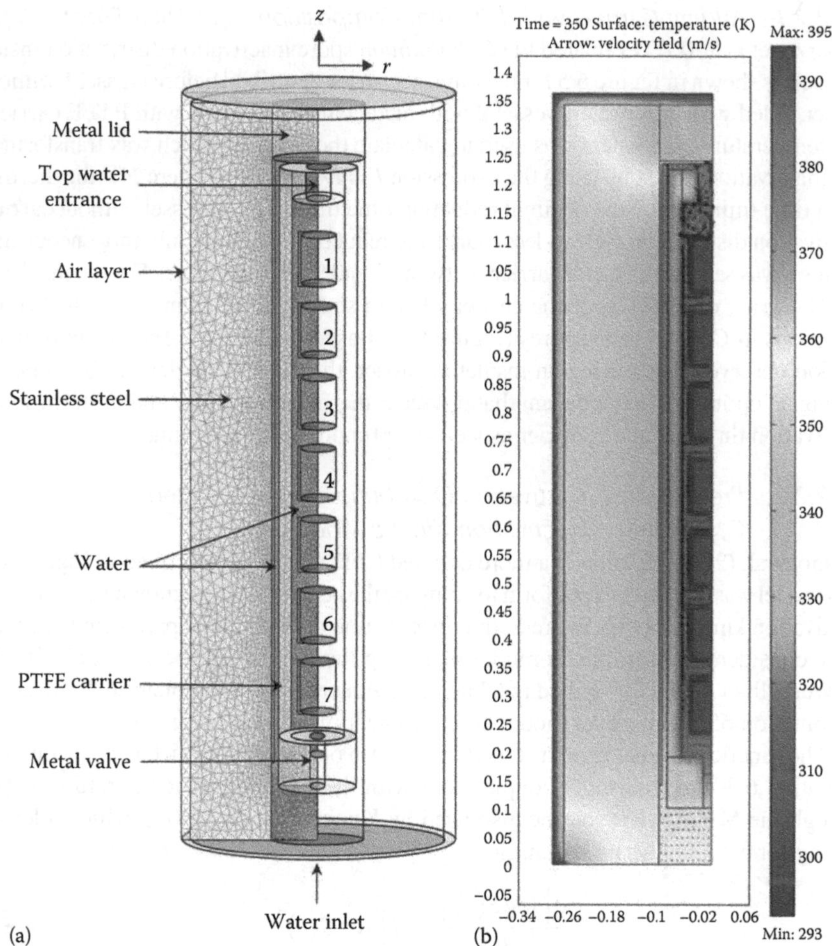


**FIGURE 6.7** CFD model of a 35 L vessel including carrier, packages, steel walls, and metal lid: (a) computational domains of the model structure and (b) thermal and flow profile in the vessel at the end of the holding time (315 s) at 600 MPa. (Note: arrows proportional to the maximum velocity at specific time.). (From Juliano, P., Knoerzer, K., Fryer, P., and Versteeg, C., *Biotechnol. Prog.,* 2008. With permission.)

depended on both temperature and pressure. Considering that the *n*th order model was valid above 100°C in the pressure range of 0.1–600 MPa, a combined discrete model was tested, which solved for log-linear kinetics at temperatures equal to or less than 100°C and for *n*th order kinetics at temperatures higher than 100°C (Table 6.2).

The distribution of *C. botulinum* inactivation log reduction predicted by each model from the CFD platform for the 35 L vessel is shown in Figure 6.8. The final inactivation calculated inside each package when using model A (log-linear kinetics) was 16.5 log reductions; model C (*n*th order kinetics) and model D (*n*th order and combined log linear/*n*th order) achieved approximately 12.0 log reductions; model B (Weibull) achieved only around 9.4 log reductions at the end of the process. Hence, the conventional thermal processing kinetics (not accounting for pressure) required shorter holding times to achieve a 12D reduction of *C. botulinum* spores compared to

**TABLE 6.2**
**Description of Selected *C. botulinum* Inactivation Models and Their Parameter Values for Comparison in a CFD Model Platform of a 35 L High-Pressure System**

| Model | $\frac{d[\log S(t)]}{dt}$ | Parameters for *C. botulinum* |
|---|---|---|
| A—linear kinetics | $-\frac{10^{\frac{T(t)-T_{ref}}{z_T}}}{D_{@T=T_{ref}}}$ | $T_{ref}=121.1°C$<br>$z_T=10°C$<br>$D_{@T=T_{ref}}=12.6\,s$ |
| B—Weibull distribution | $-b[T(t)]\cdot t^{n'[T(t)]}\cdot\left\{-\frac{\log S(t)}{b[T(t)]}\right\}^{\frac{n'[T(t)]-1}{n'[T(t)]}}$ | $b[T(t)]=\ln\left\{1+e^{0.3\cdot[T(t)-102.3]}\right\}$<br>$n'[T(t)]=0.325+\frac{0.425}{1+e^{0.0994\cdot[T(t)-101]}}$ |
| C—*n*th order kinetics | $-10^{(n-1)\cdot\log s(t)}\cdot\frac{k_i'[T(t),P(t)]}{\ln(10)}$ | $k_i'[T(t),P(t)]=e^{A_0+A_1\cdot P+A_2\cdot T+A_3\cdot P^2+A_4\cdot T^2+A_5\cdot P\cdot T+A_6\cdot P\cdot T^2}$<br>where $n=1.35$<br>$A_0=2.465$<br>$A_1=-0.023$<br>$A_2=-0.149$<br>$A_3=2.259\times10^{-5}$<br>$A_4=1.462\times10^{-3}$<br>$A_5=1.798\times10^{-4}$<br>$A_6=1.806\times10^{-7}$ |
| D—combined linear *n*th order | $\text{if}\left[T\le T_C,-\frac{10^{\frac{T(t)-T_{ref}}{z_T}}}{D_{@T=T_{ref}}},-10^{(n-1)\cdot\log s(t)}\cdot\frac{k_i'[T(t),P(t)]}{\ln(10)}\right]$ | Refer to models A and C<br>$T_c=100°C$ is the critical temperature when model A switches to model C |

*Source:* Juliano, P., Knoerzer, K., Fryer, P., and Versteeg, C., *Biotechnol. Prog.*, 2008. With permission.

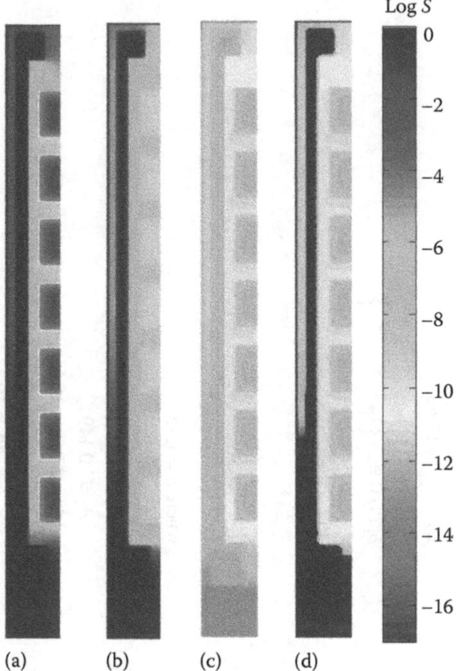


**FIGURE 6.8** Predicted distribution of *C. botulinum* log reduction extent according to four selected kinetic inactivation models using a CFD model platform for a high pressure 35 L sterilization system: (a) traditional log-linear kinetic model, (b) Weibull distribution model, (c) *n*th-order kinetic model, and (d) combined discrete log-linear and *n*th order kinetic model. (From Juliano, P., Knoerzer, K., Fryer, P., and Versteeg, C., *Biotechnol. Prog.*, 2008. With permission.)

the other models. Moreover, the temperature distribution inside the vessel resulted in a more uniform inactivation distribution when using the Weibull model or *n*th order kinetics model, compared to the linear kinetics model (Figure 6.8).

The lower inactivation and more uniform distribution provided by models B, C, and D (Table 6.2) were attributed to (1) the tailing of curves given by the model parameters and (2) the fact that data from different *C. botulinum* strains were used in each model. Furthermore, the Weibull model (B) only accounted for the temperature variation, not pressure, which also might have influenced the outcome. This CFD platform became quite useful in evaluating and comparing the inactivation extent and uniformity provided by different *C. botulinum* inactivation models in the same system. Inactivation models used in different food media, or in accounting for the effect of package size during the preheating step, to be developed in the future, could be evaluated through such CFD platforms and used as an aid for regulatory filing of the technology as well as in process and equipment design.

#### 6.4.2.3 Timescale Analysis: Influence of Pressure Vessel Size on Temperature Distribution

Hartmann and Delgado (2002a) introduced a timescale analysis to explain the effect of scale on process nonuniformities as applied to two vessel sizes (0.8 and 6.3 L).

This kind of analysis can be used both to explain the simulated results and to predict the uniformity of inactivation distribution. In this sense, Hartmann and Delgado (2003b) used timescales to explain nonuniformities in high-pressure processing with respect to convective and diffusive transport effects and the presence of packages in the system.

As explained earlier, the temporal and spatial distribution of residual *B. subtilis* α-amylase activity (Hartmann and Delgado, 2003b) and *E. coli* concentration (Hartmann et al., 2003) was used to account for the extent of combined temperature and pressure inside the pressure chamber. This distribution was described by a first-order inactivation kinetics equation with a constant $K$ in terms of continuum mechanics in Equation 6.22.

The dimensionless inactivation timescale $t_{\text{in}}$, generally represented by the inverse of the inactivation constant $1/K$, is conveniently expressed as 1 (Equation 6.25) for comparative purposes (Hartmann and Delgado, 2003b), since the ratio between different timescales is of stronger physical significance than the absolute values.

$$t_{\text{in}} = \frac{\overline{K}}{\overline{K}} = 1 \tag{6.25}$$

Field distributions of activity and temperature relate to the ratio between the thermal or hydrodynamic timescale with respect to the inactivation timescale, as they indicate the prevalence of heat transfer mechanisms (conduction and convection, respectively).

#### 6.4.2.3.1 Thermal Compensation (Conductive) Timescales

The dimensionless timescales for thermal compensation of the packaging material $t_{\text{th\_pp}}$, enzyme solution $t_{\text{th\_en}}$, and pressure medium $t_{\text{th\_me}}$ can be defined as follows (Hartmann and Delgado, 2003b):

$$t_{\text{th\_pp}} = \frac{\rho_{\text{pp}} C_{\text{pp}} D_{\text{pp}}^2 \overline{K}}{k_{\text{pp}}} \tag{6.26}$$

$$t_{\text{th\_en}} = \frac{\rho C_p D_{\text{p}}^2 \overline{K}}{k} \tag{6.27}$$

$$t_{\text{th\_me}} = \frac{\rho C_p D^2 \overline{K}}{k} \tag{6.28}$$

where

$\overline{K}$ is the average inactivation rate constant
$C_{\text{pp}}$ is the specific heat capacity
$\rho_{\text{pp}}$ is the density
$k_{\text{pp}}$ the thermal conductivity of the packaging material
$D_{\text{pp}}$ is the length scale of the package wall (thickness)
$\rho$ and $k$ are the fluid's density and thermal conductivity
$C_p$ is the specific heat capacity of the enzyme solution
$D_{\text{p}}$ is the length scale of the round package (diameter)
$D$ is the length scale of the chamber (diameter)

Large thermal timescales (greater than 1) represent a slow decay of temperature differences by conductive heat transfer (i.e., temperature perturbation vanishes slowly), whereas small thermal timescales (lower than 1) imply fast conduction of temperature and decay of temperature gradients. Thus, when the timescale is small, temperature disturbances are rapidly compensated (Hartmann and Delgado, 2002a).

#### 6.4.2.3.2 *Hydrodynamic Compensation (Convective) Timescales*

Furthermore, the dimensionless hydrodynamic compensation timescale for convection (hydrodynamic process), i.e., for the enzyme solution $t_{\text{hyd\_en}}$ and pressure medium $t_{\text{hyd\_me}}$ were defined as

$$t_{\text{hyd\_en}} = \frac{\rho D_{\text{p}}^2 \bar{K}}{\eta} \tag{6.29}$$

$$t_{\text{hyd\_me}} = \frac{\rho D^2 \bar{K}}{\eta} \tag{6.30}$$

where $\eta$ represents the dynamic viscosity of the enzyme solution and the pressure medium. This definition resulted from a dimensional analysis of all parameters to determine the fluid motion (Hartmann and Delgado, 2002a). Large hydrodynamic timescales imply intensive fluid motion and strong convective transport of heat and suspended substances. Small hydrodynamic timescales imply a rapid decay of a previously generated fluid motion and thus less intensive convective transport. In this case, the previously generated fluid motion decays rapidly due to the damping effect of the viscosity. Hartmann and Delgado (2002b, 2003a) found that a higher viscosity led to larger decay timescales for both heat (thermal) and momentum transfer (hydrodynamic), both timescales being the same order as the typical inactivation timescale.

Kowalczyk et al. (2004) established the dimensionless ratio of the vessel $\Pi_{\text{D}}$ (Equation 6.31) and the Fourier number *Fo* (Equation 6.32) as an alternative to characterizing the vessel size and timescale of conductive heat transfer, respectively.

$$\Pi_{\text{D}} = \frac{D}{H} \tag{6.31}$$

where

- $D$ is the vessel diameter
- $H$ is the vessel height

$$Fo = \frac{k}{\rho C_p D^2} t_{\text{p}} \tag{6.32}$$

where

- $k$ is the thermal conductivity of the food
- $\rho$ is the density
- $C_p$ is the specific heat of the food
- $t_{\text{p}}$ is the process time

The Fourier number *Fo* was found to increase with increasing pressure.

Hartmann and Delgado (2002a) presented an analysis of the timescales of fluid motion, heat transfer, and inactivation to explain results on the simulation of temperature and enzyme (*B. subtilis* $\alpha$-amylase) activity distributions. They determined that the ratios of the timescales for the involved physical processes are of stronger

**TABLE 6.3**
**Analysis of Dimensionless Timescales. Comparison of Two Vessel Geometries (No Packaging Inside) and the Effect of Highly Viscous Fluid on Temperature Uniformity**

| Volume (L) | Viscosity | Relation to Inactivation Timescale $t_{in} = 1$ | Damping of Fluid Motion (Velocity Gradients) | Damping of Thermal Gradients (to Isothermal Conditions) | Dominating Mechanism | Uniformity in Vessel Space (Activity Distribution) |
|---|---|---|---|---|---|---|
| 0.8 | Normal | $t_{hyd\_me} < t_{in}$<br>$t_{th\_me} < t_{in}$ | Rapid | Rapid | Convection and conduction | Achieved |
| 0.8 | High | $t_{hyd\_me} < t_{in}$<br>$t_{th\_me} \gg t_{in}$ | Rapid | Slow | Conduction | Not achieved |
| 6.3 | Normal | $t_{hyd\_me} \gg t_{in}$<br>$t \gg t_{in}$ | Very slow | Very slow | Convection | Achieved |
| 6.3 | High | $t_{hyd\_me} < t_{in}$<br>$t_{th\_me} \gg t_{in}$ | Rapid | Very slow | Conduction | Not achieved |

*Source:* Hartmann, C., *Innov. Food Sci. Emerg. Technol.*, 3, 11, 2002. With permission.

physical significance than their absolute values, explaining the occurrence of nonuniform inactivation results. As mentioned before, an enzyme solution was the single fluid component (both compression fluid and product) used during the whole process. In this case, timescales indicated in Equations 6.27 through 6.30 can be considered the same for the thermal and hydrodynamic compensation timescales, respectively.

Table 6.3 summarizes the characteristics of the mechanisms and vessel uniformity according to timescale values. If the hydrodynamic timescale is significantly smaller and the thermal compensation timescale is significantly larger (due to a highly viscous fluid or a semisolid food inside the vessel), the process uniformity can be strongly disturbed. Thus, large thermal compensation timescales can represent significant nonuniformities.

If the hydrodynamic timescale were larger than the inactivation timescale, the fluid motion and convective heat transfer would be very intensive, allowing spatial temperature gradients to be damped out quickly. In consequence, each portion of the vessel would have similar temperature values throughout the process and the distribution of residual enzyme or microbes would be uniform at the end of the process, as seen in 0.8 and 6.3 L vessels containing a fluid with normal viscosity. Table 6.3 also shows that timescales for hydrodynamic (convective) and conductive heat transport vary considerably with the volume of the vessel. As long as both timescales are significantly smaller than the timescale for inactivation, the thermal nonuniformity due to compression heating and heat transfer has no influence on the results of the inactivation (activity retention), as seen in a 0.8 mL microscale chamber.

Hartmann and Delgado (2003b) also used timescales (see Equations 6.28 through 6.30) to explain the mechanisms of generation for thermal/hydrodynamic nonuniformities at different vessel volumes (0.8, 6.3, and 50.3 L). They used these dimensionless numbers to identify which heat transfer mechanisms dominated inside the package and within the fluid medium for different vessel sizes and heat transfer coefficients ($h_{pp}$). The problem increased in complexity as timescales for the package and enzyme solution inside the package were considered. Table 6.4 summarizes the conclusions drawn in comparing each timescale with the inactivation timescale $t_{in}$.

The microscale machine (0.8 L) showed that conduction was dominant throughout the packaging material, as thermal compensation in the material was quicker than

**TABLE 6.4**
**Analysis of Dimensionless Timescales. Comparison of Three Vessel Geometries Including Five Equally Distributed Packages**

| Volume (L) | System | Relation to Inactivation Timescale $t_{in} = 1$ | Damping of Fluid Motion (Velocity Gradients) | Damping of Thermal Gradients (to Isothermal Conditions) | Dominating Mechanism | Uniformity (Activity Distribution) |
|---|---|---|---|---|---|---|
| 0.8 | Package/ enzyme solution | $t_{th\_pp} < t_{in}$ | Rapid | Very rapid (due to package) | Conduction (through package) | Not achieved |
| | | $t_{th\_en} < t_{in}$ | | | | |
| | | $t_{hyd\_en} < t_{in}$ | | | | |
| | | $<< t_{in}$ | | | | |
| | Medium | $t_{th\_me} > t_{in}$ | Slow | Slow | Convection | Achieved |
| | | $t_{hyd\_me} > t_{in}$ | | | | |
| 6.3 | Package/ enzyme solution | $t_{th\_pp} \approx t_{in}{}^*$ | Slow | Slow | Conduction (through package) | Achieved (temperature maintained in package) |
| | | $t_{th\_en} > t_{in}{}^*$ | | | | |
| | | $t_{hyd\_en} \leq t_{in}$ | | | | |
| | Medium | $t_{th\_me} > t_{in}$ | Slow | slow | convection | Not achieved |
| | | $t_{hyd\_me} > t_{in}$ | | | | |
| 50.3 | Package/ enzyme solution | $t_{th\_pp} \approx t_{in}{}^*$ | Slow | Slow | Conduction (through package) convection (in-package) | Depends |
| | | $t_{th\_en} > t_{in}{}^*$ | | | | |
| | | $t_{hyd\_en} \approx t_{in}$ | | | | |
| | Medium | $t_{th\_me} < t_{in}$ | Slow | Slow | Conduction | Not achieved |
| | | $t_{hyd\_me} > t_{in}$ | | | | |

*Source:* Hartman, C. and Delgado, A., *Biotechnol. Bioeng.*, 82, 725, 2003b. With permission.

in the enzyme solution or pressure medium. Furthermore, intensive convection in the pressure medium led to a rapid thermal compensation, as previously shown. The pilot scale vessel (6.3 L), as shown in Fig. 6.3, better maintained compression heating inside the packages. This can be explained by the similar magnitude observed between the heat conduction through the packages and the inactivation timescales.

For the industrial size machine (50.3 L), heat removal through the package dominates the overall heat transfer throughout the process, as seen in the equivalence between $t_{th\_pp}$ and $t_{in}$ (i.e., $t_{th\_pp}$ is closer to 1). Consequently, the compression heating obtained will be maintained during holding time in each of the five packages. Since the compensation timescales of the enzyme solution $t_{hyd\_en}$ inside the package is also closer to $t_{in}$, convection becomes the governing heat transfer mechanism inside the package. By considering convection as the driving mechanism inside the package, Hartmann and Delgado (2003b) concluded that the inactivation per package is approximately homogeneous, but differences were found between packages due to the cooling down provided by the pressure medium in different vessel positions.

The disadvantage of these timescales is that they require knowledge of thermal properties, such as specific heat $C_p$, density ρ, thermal conductivity $k$, and viscosity η at HPHT sterilization conditions. In the case of liquids, a correcting factor can be used (Hartmann et al., 2003) to adapt values known for water with values known for the specific food fluid at ambient pressure. This is not as direct as in the case of semi-solid materials, which represent the bulk of low-acid foods. Thus, further research is needed for the determination of these properties at high pressure conditions.

## 6.5 MACROSCOPIC MODEL TO REPRESENT PROCESSING CONDITIONS IN AN ENTIRE HIGH-PRESSURE SYSTEM

Chapter 5 provided some basic principles of macroscopic modeling as applied to high-pressure processing. Otero et al. (2002a) proposed a model to control temperature in a 2.35 L high-pressure system using water as pressure media (Figure 6.9). The model allowed predicting temperature conditions for a given high pressure treatment and, by evaluating the impact of different components in the system, showed potential for process optimization.

The model (Otero et al., 2002a) took into account the following thermal exchanges (Figure 6.10):

- Sample fluid/vessel steel mass
- Vessel steel mass/heating coil surrounding the vessel/ambient air
- Heating coil surrounding the high-pressure vessel/pipes coming out of the bath
- Thermoregulation bath (fluid)/ambient air

The output variables accounted for by the model were the following:

- Temperatures inside the vessel (at center and 1/4 distance from vessel surface)
- Vessel steel temperature
- Temperature at the entrance and exit of the surrounding coil
- Temperature inside thermoregulating bath

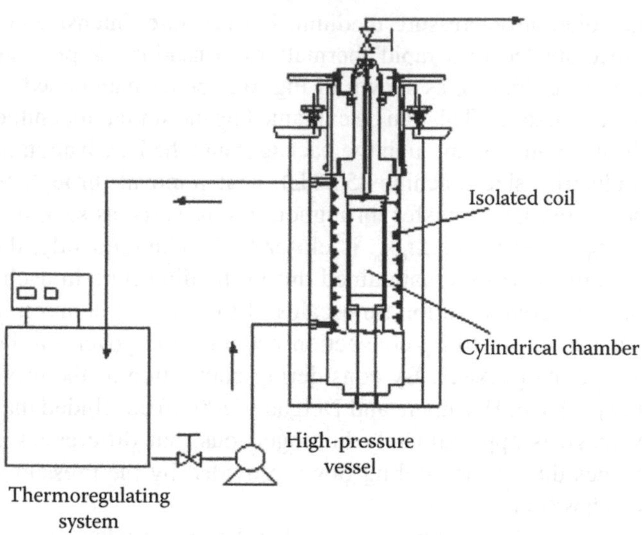


**FIGURE 6.9** Example of high pressure system used by Otero et al. (2002a) for macroscopic modeling. (From Otero, L., Molina-Garcia, A.D., Ramos, A.M., and Sanz, P.D., *Biotechnol. Prog.*, 18, 904, 2002a. With permission.)

The model allows testing not only the effects of the initial sample temperature, but also the effects of temperature variations in the thermoregulation system, heating powers, and number of baths (Otero and Sanz, 2003). The heat transfer mechanisms considered by Otero et al. (2002a) were convection (ambient air, fluid inside bath, and fluid inside vessel) and conduction throughout the different layers present. This modeling approach allowed handling of the nonadiabatic conditions occurring during compression, where a significant part of the energy in the system transfers to the water bath and surroundings.

This model was established using MATLAB 6 software, Release 12 (The MathWorks Inc., Natick, MA). The additional toolbox Simulink was used to perform simulations based on the Euler method for solving ordinary differential equations. Heat transfer equations were employed, taking into consideration the different layers of steel comprising the vessel, pipes, fluid, and sample (Figure 6.10). Heat capacities and thermal conductivities were employed from literature. The values of the thermophysical properties employed under pressure were calculated using the software implementation by Otero et al. (2002b). Also, temperature variations with pressure were calculated as described by Otero et al. (2000) and included in the model. The method was based on ordinary differential equations, which did not represent spatially distributed quantities inside the vessel, but average quantities instead. To calculate the error inside the vessel an average of two thermocouple readings (one at center of vessel; another at one-quarter from surface) were taken.

The authors simulated compression of water in a 2.35 L vessel at different temperatures (30°C, 37°C, and 40°C), pressures (0.1 to 400 MPa), and holding time necessary to recover temperature in the sample. Simulated and experimental values

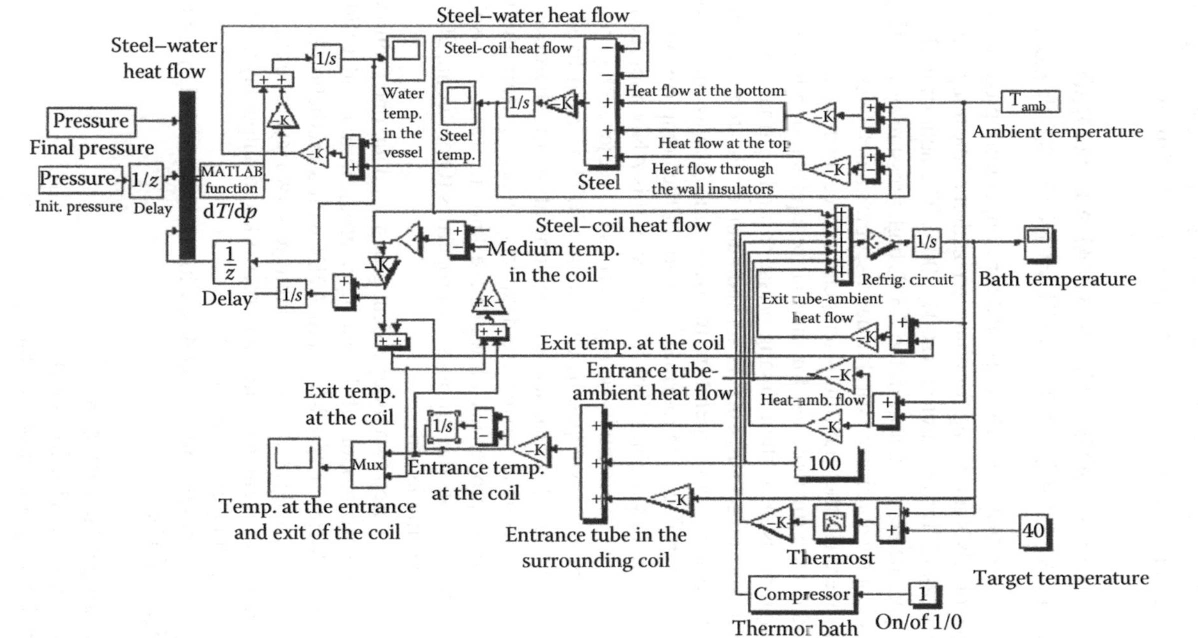


**FIGURE 6.10** Simulation of thermal exchanges in a complete high pressure equipment system. (From Otero, L., Molina-Garcia, A.D., Ramos, A.M., and Sanz, P.D., *Biotechnol. Prog.*, 18, 904, 2002a. With permission.)

of temperature in the vessel sample appeared to agree with the compression, holding time, and pressure release, with a maximum error of 2.3%. Agreement between the simulated and experimental temperature values at the entrance and exit of the surrounding coil and temperature of the thermoregulating bath showed much lower absolute errors than inside the vessel. The model revealed the massive significance of the high pressure vessel's steel mass in the thermal control of the process.

## 6.6 APPLICATION OF ARTIFICIAL NEURAL NETWORKS FOR HIGH-PRESSURE PROCESS TEMPERATURE PREDICTION

Chapter 5 describes the basics of ANN applied for temperature prediction during a HPP. One example of ANN modeling for characterization of the temperature history of an HPP is the work of Torrecilla et al. (2004, 2005). The authors evaluated ANN models to predict parameters of interest in the 2.35 L high-pressure system (Figure 6.9) and conditions previously considered for the macroscopic model (Otero et al., 2002a). These parameters were (1) the maximum temperature reached in the sample after pressurization and (2) the time needed for thermal re equilibration to ambient temperature in the high-pressure system during pressure holding time. The maximum temperature reached at target pressure is an indicator of the temperature performance in an HPP, whereas the temperature recovery time (due to compression heat loss through the vessel) reflects the thermal behavior of the complete system due to its dependence on the input variables. The advantage of this ANN approach was that it did not require the use of thermophysical properties to predict such parameters.

The macroscopic model developed by Otero et al. (2002a) was utilized to generate data to train the ANN. In this case, a computer simulation was chosen as opposed to experimental data, to generate extensive data sets needed to train the network. Given that thermophysical properties of water are known, water was selected as the pressure medium. Thus, its thermophysical properties (Otero et al., 2002b) were used in the macroscopic simulation.

Torrecilla et al. (2004) used two independent groups of data: one group was used to train the neural network model while the other group tested and validated the trained model. They selected a "feed-forward ANN with a prediction horizon and supervised learning," characterized by three main layers: two layers connected to the outside world (input layer for entering data into network and output layer for network response) and one hidden layer (Figure 6.11).

The input layer was chosen with five nodes, which corresponded to the five input variables (Y in Figure 6.11):

- Applied pressure (250–450 MPa)
- Pressure increase rate (1–2 MPa $s^{-1}$)
- Initial set point temperature (40°C–80°C)
- High-pressure vessel temperature (between set point temperature and ambient temperature, depending on thermoregulation time elapsed)
- Ambient temperature (10°C–30°C)

These variables were selected because they mainly set the temperature performance during the process. Initial temperatures of the liquid water (sample), thermoregulating

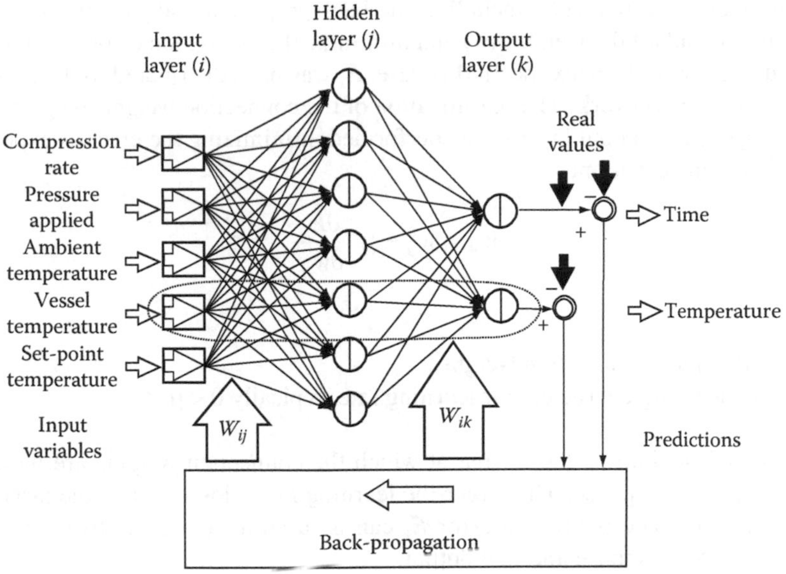

**FIGURE 6.11** Structure of neural network model developed by Torrecilla et al. (2005). A multilayer neural network with one hidden layer, as well as input and output variables, from a high pressure process. A standard back-propagation algorithm is used as a training step. (From Torrecilla, J.S., Otero, L., and Sanz, P.D., *J. Food Eng.*, 69, 299, 2005. With permission.)

bath, and entrance and exit of the surrounding coil were the same as the set point temperature.

The output layer consisted of one neuron (Torrecilla et al., 2004), for the time needed to dissipate compression heat and to recover the initial temperature, or two neurons (Torrecilla et al., 2005) ("❍" in Figure 6.11); the other neuron predicted the maximum temperature reached in liquid water once the target pressure was achieved.

The ANN used a standard back-propagation algorithm with the sigmoid transfer function using QuickBasic 4.5 software. The sigmoid function was used to determine the output from a neuron in each layer, which is bound between 0 and 1 (Torrecilla et al., 2004, 2005):

$$y_j = f(x_j) = \frac{1}{1+e^{-x_j}} \tag{6.33}$$

In this way, prediction values ($y_k$) (output layer $k$) were calculated taking as inputs the corresponding output values of the previous hidden layer neurons ($y_j$) and multiplying them by the connection weights ($w_{jk}$). The inputs ($y$) into a neuron are multiplied by their corresponding connection weights ($w$) and added together. This sum is then transformed through a selected function to produce a single output to be passed between neurons in other layers. Given the selected sigmoid function oscillates between 0 and 1, both the input and output values must be normalized between 0 and 1. In particular, Torrecilla et al. (2004, 2005) normalized input values in the 0–1 range by dividing each by its maximum value.

An error-correction rule, including the back-propagation algorithm, was set to automatically adjust the weights ($w$) and minimize the estimation error $E_k$ after back distribution across the network. In this case, $E_k$ was back distributed to the previous layers across the network. The optimization of the connection weights ($w_{jk}$) between hidden layer $j$ and output layer $k$ was performed, minimizing the error as expressed in the following equation:

$$w_{jk} = w_{jk}^0 - \mu \cdot \frac{\partial E_k}{\partial w_{jk}} \tag{6.34}$$

where

$w_{jk}^0$ is the initial connection weight
$\mu$ is the learning coefficient (or learning rate, typically $0 < \mu < 1$)

This coefficient controls the degree at which the connection weights are modified during the training phase. The larger the learning rate (close to 1) is, the larger the weight changes. The prediction error $E_k$ can be used as a comparative parameter between ANN's response and real output:

$$E_k = \frac{1}{2}\sum_k (r_k - y_k)^2 \tag{6.35}$$

Coupling and rearranging Equations 6.34 and 6.35 gives the connection weights ($w_{jk}$) between hidden $j$ and output layer $k$:

$$w_{jk} = w_{jk}^0 - \mu \cdot (r_k - y_k) \cdot y_k \cdot (1 - y_k) \cdot y_j \tag{6.36}$$

Similarly, the weights ($w_{ij}$) between the input $i$ and hidden layer $j$ can be found:

$$w_{ij} = w_{ij}^0 - \mu \cdot \sum_k (r_k - y_k) \cdot y_k \cdot (1 - y_k) \cdot w_{jk} \cdot y_i \cdot (1 - y_i) \cdot y_i \tag{6.37}$$

Torrecilla et al. (2004) tested from 1 to 8 neurons in the hidden layers, and the sum of errors of 202 predictions for each iteration obtained from the training set were calculated. As explained before, the learning process ended when the minimum number of iterations for minimum overall error was found.

After training, the network tried to find the minimal number of iterations needed to minimize prediction errors while changing learning coefficients for temperature and time predictions. Once optimization was performed, the values and distribution of output data from the model were compared with real values using averages, standard deviation, and variance.

The minimum overall error with minimum number of iterations was reached with 3 and 7 neurons in the hidden layers (topologies 5, 3, 1 and 5, 7, 1); topology 5, 3, 1 was selected due to the small number of weights required ($\mu = 0.5$). Different learning processes were performed for different coefficients from 0.1 to 1.0 and the selected topology, 5, 3, 1, finding the optimum learning coefficient of $\mu = 0.5$ (Torrecilla et al., 2004). On the other hand, Torrecilla et al. (2005) found the optimum topology to be

5, 7, 2. They calculated the minimum overall error and number of iterations for this topology and found a learning coefficient of $\mu = 0.9$. The end of the learning process was verified by showing similar numbers between the average standard deviation and variance of desired and predicted data.

For the validation/testing step (see Chapter 5), the selected topology and learning coefficient were utilized and the ANN was only used to predict process data. The ANN model accurately predicted the time needed to re equilibrate the temperature in liquid water with a relative error of only ±2.5% and a high correlation coefficient ($R^2 = 0.98$); values were evenly distributed throughout the entire range (Torrecilla et al., 2004). In the five-input model evaluated by Torrecilla et al. (2005), the authors observed excellent correlations between high values ($R^2 = 0.98$ for temperature and $R^2 = 0.99$ for time data) in the prediction curve values. Mean error values were accurate enough to predict and model (lower than ±1.6%) for both temperature and time predictions. In both models, statistical analyses also showed no significant difference ($p > 0.05$) between the distribution of the macroscopic and predicted data. In this case, no comparison with "real" measured temperature data was made.

This example of an ANN developed for liquid water illustrates the capability of the model to predict the maximum process temperature, or other HPHT processing parameters, without the need for thermophysical properties under pressure. Although no information is available on the relationships between the initial and final conditions in different high-pressure systems (i.e., data sets of inputs and outputs of the process), experimental observations can help in ANN training. However, this needs to be done for each new product and/or high pressure system tested.

## 6.7 COMPARISON OF CAPABILITIES OF EXISTING MODELS

Several advances have been made in terms of modeling temperature profiles in high-pressure processing, mainly at low temperature conditions. Most of the approaches presented in this contribution use water as a baseline to predict temperature homogeneity and processing history at high pressure conditions, primarily due to the simplicity of the modeled system, as opposed to systems containing food components of varied composition, porosity, and volume, among other thermophysical properties of unknown value at high pressure conditions.

The analytical model provided by Carroll et al. (2003) has the capability of providing temperature variations in time in the radial direction. It assumes an infinite and, therefore, cannot account for the effect of the entrance of a colder incoming pressurization fluid. Furthermore, the model assumes the thermophysical properties do not vary with pressure and temperature. The inclusion of thermophysical properties as functions of pressure and temperature would probably improve the accuracy of the model. In a uniform vessel system (i.e., with negligible temperature gradients) this model could also be used to predict the extent of microbial/enzyme inactivation or physicochemical reactions by externally coupling conduction equations with kinetic models, as shown with numerical models.

The numerical models presented in this chapter not only provide the temperature distribution throughout the vessel contents but also are capable of representing such distributions as a function of time. Although the models provided by Denys et al.

(2000a,b) were adjusted to real measurements using the finite difference method, they only represented conductive heat transfer inside the vessel. Otero et al. (2007) showed the importance of using both conductive and convective components in a model for improved accuracy.

The finite volume method has been applied in models developed by Hartmann and others to predict the (a) influence of inflow velocity (Hartmann, 2002), fluid viscosity (Hartmann and Delgado, 2002b); (b) presence of packages inside the vessel (Hartmann et al., 2003); and (c) influence of geometrical scale (Hartmann and Delgado, 2003b) on temperature distribution. These CFD models were able to show the influence of incoming colder pressure fluid in the vessel. This capability was also observed in other CFD models applying the finite volume method (Ghani and Farid, 2007) and finite element method (Knoerzer et al., 2007; Otero et al., 2007). In fact, one particular CFD model, using the finite element method, was able to predict the cooling down at the inlet region under turbulent conditions resulting from the intensifier pump in the 35 L pilot vessel (Knoerzer et al., 2007; Juliano et al., 2008).

Another aspect shown in the numerical examples was the capability of modifying and improving the thermal boundary conditions. The initial models were simpler and assumed a constant temperature value at the vessel wall. Accuracy of models could be enhanced by either including an equation representing heat conduction from the compression heated fluid into the vessel walls (Hartmann and Delgado, 2003b; Hartmann et al., 2004), using a logarithmic wall function condition with turbulent stochastical components (Knoerzer et al., 2007), or directly including an axi symmetric domain for the steel vessel wall in the CFD model (Otero et al., 2007; Juliano et al., 2008).

The CFD examples discussed before could also evaluate the effect of including materials (other than liquid water) in the vessel's modeling domain on the temperature and flow distribution of all the vessel contents. Some of these materials included (a) only packages of similar size filled with water (Hartmann et al., 2003), (b) solid samples (agar gel and rubber cap) at different fill ratios with the compression medium (Otero et al., 2007), (c) water and a beef fat sample (Ghani and Farid, 2007), (d) sample carriers of different compositions stopping inlet flow (Knoerzer et al., 2007), and (e) cylindrical packages contained inside the carrier (Juliano et al., 2008). It is worth noting that most models assumed axi symmetry, whereas the finite volume model developed by Ghani and Farid (2007) evaluated the cooling down of the fat sample in three dimensions.

As mentioned in Chapter 5, CFD models also allow for predicting temperature, flow, and inactivation distribution in three dimensions by including samples of asymmetrical shapes distributed at different corners of the vessel. There is little published on 3D models due to the computational demand required for their operation. Today's numerical software packages have user friendly tools for the development of 3D models as well as the capability to draw arbitrary shapes or even to import computer-aided design (CAD) drawings. Furthermore, they include databases with thermophysical properties corresponding to the materials in place (at atmospheric conditions). Last but not least, 3D models can provide predictive information on the uniformity of horizontal high-pressure vessels, in which case gravitational forces do not allow for an axi symmetric fluid–solid system.

CFD modeling can also assist in scale-up studies as shown here (Hartmann and Delgado, 2002a, 2003b; Hartmann et al., 2003), some of which have proven there is higher heat retention in machines of industrial size. Commercial size vessels require short come-up times and, therefore, high inflow velocities, which provide turbulent flow conditions at the inlet areas. As shown in the example of a 35 L pilot-scale vessel (Knoerzer et al., 2007), CFD packages allow including a turbulent component to better represent flow in large-scale systems.

Other CFD examples simulated the uniformity inside a sample, where transport mechanisms of heat and substrate were represented (Hartmann et al., 2003). In particular, these CFD models were applied to evaluate the effect of package thickness in the retention of heat and as a barrier to flow.

Some of the CFD models developed to predict temperature distribution have also been applied to predict the extent of the inactivation distribution of microorganisms or enzymes. Kinetic models for *B. subtilis* α-amylase, *E. coli*, or *C. botulinum* have been internally coupled with conduction equations (Denys et al., 2000a,b) and conservation equations for momentum and heat (Hartmann, 2002; Hartmann and Delgado, 2002a,b, 2003a,b; Hartmann et al., 2003, 2004). The models were also externally coupled (Knoerzer et al., 2007; Juliano et al., 2008) by means of software routines, which transformed the temperature distribution output with time into animated maps showing the inactivation extent throughout the process. This allowed evaluating the range of inactivation at certain locations of the vessel and the average inactivation per package to determine the overall process uniformity.

The macroscopic model presented by Otero et al. (2002a) included all elements involved in the thermal regulation of an HPP, with information on materials, dimensions, cooling and heating power, among other components. Particularly, this model in predicting the temperature during pressurization had the unique characteristic of accounting for temperatures in the external parts of the vessel, such as the surrounding heating coil (including the pipes in and out of the thermal regulation bath controlling the heating fluid). Unlike the previously mentioned numerical models, this modeling approach is much more limited and cannot calculate spatial (2D or 3D) temperature distributions; thus, temperature gradients throughout the vessel space cannot be represented, which diminishes the model's capabilities.

Among the predictive capabilities of the numerical models discussed, the effect of the cold inlet temperature could be modeled in a macroscopic model by accounting for the external devices controlling temperature in the high-pressure pump and pipes entering the vessel. However, due to the limitation of representing temperature profiles in more than one point, the effect of adding other materials inside the vessel cannot be addressed by these models. In this case, a very sophisticated numerical model should be included in a block to account for temperature changes occurring in added materials.

Unlike numerical models discussed before, the macroscopic model example did not predict microbial or enzyme inactivation. To predict the approximate extent of microbial/enzyme inactivation (from a single point in the vessel) with a macroscopic model, a block can be added comprising an inactivation kinetic model in the form of an ordinary differential equation, which then transforms the temperature output into inactivation extent.

Like macroscopic models, numerical models have the capability of including the influence of peripheral devices through the boundary conditions. It is also possible to include the specific devices in the numerical model by directly connecting several computational domains corresponding to each peripheral unit. The drawback in this case is the complexity of the design, which could lead to pronounced difficulties in model convergence as well as a high computational demand.

Among the types of models presented so far, ANN is the most limited, since it is restricted to providing a single value (e.g., temperature or time). The model developed by Torrecilla et al. (2004, 2005) is the only ANN model at present described in the high-pressure processing literature; it provides the maximum temperature value reached at the end of pressurization as well as the time necessary for complete dissipation of the compression heat. The unique advantage of this type of modeling technique is that it does not require thermophysical properties to be functions of temperature and pressure, as in analytical, numerical, or macroscopic models.

In comparison to macroscopic models, the ANN can also include the influence of peripheral devices by increasing the number of nodes in the input layer where the ANN receives the temperature (and possible pressure). The ANN is also capable of further training to predict a concentration value (microbial, enzymatic, or chemical), and furthermore deal with situations when several materials are included in the system or when multiple unit operations are considered. However, the evaluation of maximum temperature (or cooling time) due to the inclusion of different materials inside the vessel seems challenging, since a significant effort would be required in training the ANN.

## 6.8 CONCLUDING REMARKS

The state-of-the-art in modeling HPP has been introduced by summarizing the works of several experts, in which four modeling approaches applied mainly at HPLT conditions have been evaluated. Among these approaches there are examples of analytical, macroscopic, and ANN models in the literature, which indicates that much more research is needed to advance the development of these models in high pressure processing applications. However, numerical modeling showed the most applications, in the prediction of temperature, flow, and inactivation extent distribution. These models were able to illustrate the cold regions in the vessel and packages at several scales, and can assist in the design of devices to prevent the loss of compression heat.

Temperature and enzyme/microbial inactivation validation of the abovementioned models are critical for their adoption in assessing the reliability of the model. Validation will also assist in the development of more detailed models. A must in the accurate determination of process uniformity is finding the kinetic constants or expressions for the enzyme/microbial inactivation as a function of pressure and temperature. Expressions for thermophysical properties of materials other than water (e.g., food, packaging, and carrier materials) as functions of temperature and pressure will be essential in the development of further analytical, numerical, and macroscopic models.

Numerical models have shown excellent adaptability to system modifications and allow for optimization of processing conditions and system structure. The development of 3D models will be of further aid in modeling horizontal high pressure vessels and the effect of packages inside on temperature distribution, flow, and inactivation distribution. So far no model has been able to include the shrinkage of packages (to account for significant compression of package headspace, food content releasing of gas at pressure), and packaging material. In this sense, multiphysics modeling software packages used today in chemical and soil engineering applications include the governing equations accounted for structural mechanics and can be applied to address these issues. Another requirement in high pressure processing is that vessels (or carriers) should be used at maximum package capacity, which was not been considered in previous models as yet.

Macroscopic models have proven useful in understanding the performance of external devices in thermal control. On the other hand, as more advanced computers with higher memory and central processing unit (CPU) speeds become available, peripheral devices (e.g., a sample preheater) can easily be included in the design of a numerical model.

## NOMENCLATURE

*Latin letters*

| | |
|---|---|
| $A$ | relative or actual activity related to initial activity (%) |
| $b$ | Weibull parameter; function of process temperature and/or pressure history |
| $C_{pp}$ | isobaric heat capacity of packaging material (J $kg^{-1}$ $K^{-1}$) |
| $C_{\mu}$ | constant ($m^4$ $kg^{-4}$) |
| $D$ | decimal reduction time (min) |
| $d$, $D$ | chamber diameter (m) |
| $D_p$ | diameter of round package (m) |
| $d_p$, $D_{pp}$ | thickness of packaging film (m) |
| $e$ | Euler number (= 2.71828) |
| $E_a$ | activation energy (J) |
| $E_k$ | prediction error of ANN |
| $F$ | thermal death time (min) |
| $g$ | gravity constant (9.8 m $s^{-2}$) |
| $H$ | height of vessel (m) |
| $h_{pp}$ | heat transfer coefficient for conduction through packaging material (W $m^{-2}$ $K^{-1}$) |
| $i, j$ | positions on grid |
| $J$ | neuron number |
| $J_0$, $J_1$ | zero-order Bessel functions |
| $k$ | turbulent kinetic energy ($kg^2$ $K^{-2}$) |
| $k$ | first-order kinetic constant ($s^{-1}$) |
| $K$ | inactivation rate constant ($s^{-1}$) |
| $\bar{K}$ | inverse of inactivation rate constant (s) |
| $k$ | output layer of ANN |
| $k$, $k_1$ | thermal conductivity (W $m^{-1}$ $K^{-1}$) |
| $k_p$, $k_{pp}$ | thermal conductivity of packaging material (W $m^{-1}$ $K^{-1}$) |
| $k_{ref}$ | reference kinetic constant ($s^{-1}$) |
| $k_T$ | turbulent thermal conductivity (W $m^{-1}$ $K^{-1}$) |

| | |
|---|---|
| $k_V$ | thermal conductivity of steel (W m$^{-1}$ K$^{-1}$) |
| $L$ | extra-chamber dimensionless heat transfer coefficient |
| $n$ | Weibull parameter; function of process temperature and/or pressure history |
| $n$ | normal to surface |
| $n$ | order of inactivation kinetic |
| $n$ | time step number |
| $N$, $i$ | node number |
| $N_0$, $N$ | initial and final number of microbial spores |
| $P$ | pressure (Pa) |
| $p$ | probability value (statistics) |
| $P_0$ | atmospheric pressure (Pa) |
| $P_1$, $P_{target}$ | target pressure (Pa) |
| $P_2$ | pressure at end of holding time (Pa) |
| $P_f$ | pressure including a fluctuating term (Pa) |
| $P_{ref}$ | reference pressure (Pa) |
| $R$ | universal gas constant (8.314 J mol$^{-1}$ K$^{-1}$) |
| $r$, $R$ | radius (m) |
| $R^2$ | coefficient of determination |
| $r_k$ | real output data (for use in ANN error prediction) |
| $S$ | entropy (J K$^{-1}$) |
| $T$ | temperature (K) |
| $T$ | time (s) |
| $t_{hyd\_en}$ | dimensionless hydrodynamic compensation timescale for convection (enzyme solution) |
| $t_{hyd\_me}$ | dimensionless hydrodynamic compensation timescale for convection (pressure medium) |
| $t_{in}$ | dimensionless inactivation timescale |
| $T_L$ | temperature of fluid (K) |
| $t_p$ | process time (s) |
| $T_{p1}$ | temperature after compression heating (K) |
| $T_{ref}$ | reference temperature (K) |
| $t_s$ | start time (s) |
| $T_s$, $T_i$ | initial temperature (K) |
| $T_{S-P}$ | temperature of food package (K) |
| $t_{th\_en}$ | dimensionless timescale for thermal compensation of enzyme solution |
| $t_{th\_me}$ | dimensionless timescale for thermal compensation of pressure medium |
| $t_{th\_pp}$ | dimensionless timescale for thermal compensation of packaging material |
| $T_{wall}$ | wall temperature (K) |
| $u$, $w$, $v$ | components of fluid velocity vector (in $x$-, $y$-, $z$-directions, respectively, m s$^{-1}$) |
| $V_a$ | activation volume (m$^3$) |

| | |
|---|---|
| $\underline{v}$ | average velocity vector (m $s^{-1}$) |
| $\vec{V}$ | velocity vector (m $s^{-1}$) |
| $v_z$ | velocity in *z*-direction (m $s^{-1}$) |
| $w_{jk}$ | connection weights between neurons in ANN |
| $x, y, z$ | coordinate positions (m) |
| $y_j$ | output of a neuron in each layer of an ANN |
| $y_k$ | output of ANN |
| [S] | relative amount of solid (%) |

*Greek letters*

| | |
|---|---|
| $\alpha, \alpha_p$ | thermal expansion coefficient ($K^{-1}$) |
| $\beta$ | water compressibility (0.17, dimensionless) |
| $\gamma$ | thermal diffusivity ($m^2$ $s^{-1}$) |
| $\varepsilon$ | dissipation rate of turbulence energy ($m^2$ $K^{-3}$) |
| $\mu$ | learning coefficient of ANN |
| $\eta$ | dynamic viscosity (Pa s) |
| $\eta_T$ | turbulent viscosity (Pa s) |
| $\Pi_D$ | dimensionless ratio of vessel |
| $\rho$ | density (kg $m^{-3}$) |
| $\rho_{pp}$ | density of packaging material (kg $m^{-3}$) |
| $\tau$ | pressure come-up time |

*Abbreviations*

| | |
|---|---|
| 1D | one-dimensional |
| 2D | two-dimensional |
| 3D | three-dimensional |
| ANN | artificial neural networks |
| ASME | American Society of Mechanical Engineers |
| CAD | computer-aided design |
| cfu | colony-forming units |
| CFD | computational fluid dynamics |
| CPU | central processing unit |
| CTFD | computational thermal fluid dynamics |
| HP-DPIT | high pressure-digital particle image thermography |
| HP-DPIV | high pressure-digital particle image velocimetry |
| HPHT | high pressure high temperature |
| HPP | high pressure process |
| NIST | National Institute of Standards and Technology |
| PP | polypropylene |
| PTFE | polytetraflourethylene |
| RANS | Reynolds-averaged Navier–Stokes |

*Operators*

| | |
|---|---|
| d | differential |
| $\Sigma$ | sum |
| $\partial$ | partial differential |

| | |
|---|---|
| $\Delta$ | difference |
| $\nabla$ | gradient (nabla-operator) |
| $\nabla$ | Divergence |

## REFERENCES


Ardia, A., Knorr, D., and Heinz, V. 2004. Adiabatic heat modelling for pressure build-up during high-pressure treatment in liquid-food processing. *Food and Bioproducts Processing* 82:89–95.

Campanella, O.H. and Peleg, M. 2001. Theoretical comparison of a new and the traditional method to calculate *Clostridium botulinum* survival during thermal inactivation. *Journal of the Science of Food and Agriculture* 81:1069–1076.

Carroll, T., Chen, P., and Fletcher, A. 2003. A method to characterise heat transfer during high-pressure processing. *Journal of Food Engineering* 60:131–135.

Carslaw, H.S. and J.C. Jaeger. 1959. *Conduction of Heat in Solids*. London: Oxford University Press.

COMSOL Multiphysics. 2006. *Chemical Engineering Module*. Stockholm, Sweden: COMSOL AB.

de Heij, W.B.C., Van Schepdael, L.J.M.M., van den Berg, R.W., and Bartels, P.V. 2002. Increasing preservation efficiency and product quality through control of temperature distributions in high pressure applications. *High Pressure Research* 22:653–657.

Denys, S., Ludikhuyze, L.R., Van Loey, A.M., and Hendrickx, M.E. 2000a. Modeling conductive heat transfer and process uniformity during batch high-pressure processing of foods. *Biotechnology Progress* 16:92–101.

Denys, S., Van Loey, A.M., and Hendrickx, M.E. 2000b. A modeling approach for evaluating process uniformity during batch high hydrostatic pressure processing: Combination of a numerical heat transfer model and enzyme inactivation kinetics. *Innovative Food Science and Emerging Technologies* 1:5–19.

Ghani, A.G.A. and Farid, M.M. 2007. Numerical simulation of solid–liquid food mixture in a high pressure processing unit using computational fluid dynamics. *Journal of Food Engineering* 80:1031–1042.

Hartmann, C. 2002. Numerical simulation of thermodynamic and fluiddynamic processes during the high pressure treatment of fluid food systems. *Innovative Food Science and Emerging Technologies* 3:11–18.

Hartmann, C. and Delgado, A. 2002a. Numerical simulation of convective and diffusive transport effects on a high-pressure-induced inactivation process. *Biotechnology and Bioengineering* 79:94–104.

Hartmann, C. and Delgado, A. 2002b. Numerical simulation of thermofluiddynamics and enzyme inactivation in a fluid food system under high hydrostatic pressure. *Trends in High Pressure Bioscience and Biotechnology*, Vol. 19 (Progress in Biotechnology), R. Hayashi, (Ed.). Amsterdam: Elsevier, pp. 533–540.

Hartmann, C. and Delgado, A. 2003a. Numerical simulation of thermal and fluiddynamical transport effects on a high pressure induced inactivation. *High Pressure Research* 23:67–70.

Hartmann, C. and Delgado, A. 2003b. The influence of transport phenomena during high-pressure processing of packed food on the uniformity of enzyme inactivation. *Biotechnology and Bioengineering* 82:725–735.

Hartmann, C., Delgado, A., and Szymczyk, J. 2003. Convective and diffusive transport effects in a high pressure induced inactivation process of packed food. *Journal of Food Engineering* 59:33–44.

Hartmann, C., Schuhholz, J., Kitsubun, P., Chapleau, N., Bail, A., and Delgado, A. 2004. Experimental and numerical analysis of the thermofluiddynamics in a high-pressure autoclave. *Innovative Food Science and Emerging Technologies* 5:399–411.

Harvey, A.H., Peskin, A.P., and Sanford, A.K. 1996. *NIST/ASTME—IAPSW Standard Reference Database 10, version 2.2.*

Hinrichs, J. 2000. Ultrahochdruckbehandlung von Lebensmitteln mit Schwerpunkt Milch und Milchprodukte-Phaenomene, Kinetik und Methodik (Ultrahighpressure treatment of foods with focus on milk and dairy products—phenomena, kinetics, and methodology). *Fortschritt-Berichte VDI* 3:656.

Juliano, P., Knoerzer, K., Fryer, P., and Versteeg, C. 2008. *C. botulinum* inactivation kinetics implemented in a computational model of a high pressure sterilization process. *Biotechnology Progress*, In press.

Knoerzer, K., Juliano, P., Gladman, S., Versteeg, C., and Fryer, P. 2007. A computational model for temperature and sterility distributions in a pilot-scale high-pressure high-temperature process. *AIChE Journal* 53:2996–3010.

Kowalczyk, W., Hartmann, C., and Delgado, A. 2004. Modelling and numerical simulation of convection driven high pressure induced phase changes. *International Journal of Heat and Mass Transfer* 47:1079–1089.

Kowalczyk, W., Hartmann, C., Luscher, C., Pohl, M., Delgado, A., and Knorr, D. 2005. Determination of thermophysical properties of foods under high hydrostatic pressure in combined experimental and theoretical approach. *Innovative Food Science and Emerging Technologies* 6:318–326.

Lemmon, E.W., McLinden, M.O., and Friend D.G. 2005. *Thermophysical Properties of Fluid Systems. NIST Chemistry WebBook, NIST Standard Reference Database*. Gaithersburg MD, 20899: National Institute of Standards and Technology. Available from http://webbook.nist.gov.

Ludikhuyze, L.R., Van den Broeck, I., Weemaes, C.A., and Hendrickx, M.E. 1997. Kinetic parameters for pressure–temperature inactivation of *Bacillus subtilis* alpha-amylase under dynamic conditions. *Biotechnology Progress* 13:617–623.

Margosch, D., Ehrmann, M.A., Buckow, R., Heinz, V., Vogel, R.F., and Ganzle, M.G. 2006. High-pressure-mediated survival of *Clostridium botulinum* and *Bacillus amyloliquefaciens* endospores at high temperature. *Applied and Environmental Microbiology* 72:3476–3481.

Nicolaï, B.M., Scheerlinck, N., Verboven, P., and Baerdemaeker, J.D. 2001. Stochastic finite-element analysis of thermal food processes. In: *Food Processing Operations Modeling. Design and Analysis*, J. Irudayarai (Ed.). New York: Marcel Dekker, pp. 265–304.

Nicolaï, B.M., Verboven, P., and Scheerlinck, N. 2007. Modeling and simulation of thermal processes. In: *Thermal Technologies in Food Processing*, (Ed.) P. Richardson, Boca Raton, FL: CRC Press. 91–112.

Otero, L. and Sanz, P. 2003. Modelling heat transfer in high pressure food processing: A review. *Innovative Food Science and Emerging Technologies* 4:121–134.

Otero, L., Molina-Garcia, A., and Sanz, P. 2000. Thermal effect in foods during quasi-adiabatic pressure treatments. *Innovative Food Science and Emerging Technologies* 1:119–126.

Otero, L., Molina-Garcia, A.D., Ramos, A.M., and Sanz, P.D. 2002a. A model for real thermal control in high-pressure treatment of foods. *Biotechnology Progress* 18:904–908.

Otero, L., Molina-Garcia, A.D., and Sanz, P.D. 2002b. Some interrelated thermophysical properties of liquid water and ice. I. A user-friendly modeling review for food high-pressure processing. *Critical Reviews in Food Science and Nutrition* 42:339–352.

Otero, L., Ousegui, A., Guignon, B., Le Bail, A., and Sanz, P. 2006. Evaluation of the thermophysical properties of tylose gel under pressure in the phase change domain. *Food Hydrocolloids* 20:449–460.

Otero, L., Ramos, A.M., de Elvira, C., and Sanz, P.D. 2007. A model to design high-pressure processes towards an uniform temperature distribution. *Journal of Food Engineering* 78:1463–1470.

Pehl, M. and Delgado, A. 1999. An in situ technique to visualize temperature and velocity fields in liquid biotechnical substances at high pressure. In: *Advances in High Pressure Bioscience and Biotechnology* H. Ludwig (Ed.). Heidelberg: Springer, pp. 519–522.

Pehl, M. and Delgado, A. 2002. Experimental investigation on thermofluiddynamical processes in pressurized substances. In: *Trends in High Pressure Bioscience and Biotechnology*, R. Hayashi (Ed.). Amsterdam: Elsevier, pp. 429–35.

Pehl, M., Werner, F., and Delgado, A. 2000. First visualization of temperature fields in liquids at high pressure using thermochromic liquid crystals. *Experiments in Fluids* 29:302–304.

Pflug, I.J. 1987. Using the straight-line semilogarithmic microbial destruction model as an engineering design-model for determining the F-value for heat processes. *Journal of Food Protection* 50:342–346.

Torrecilla, J.S., Otero, L., and Sanz, P.D. 2004. A neural network approach for thermal/pressure food processing. *Journal of Food Engineering* 62:89–95.

Torrecilla, J.S., Otero, L., and Sanz, P.D. 2005. Artificial neural networks: A promising tool to design and optimize high-pressure food processes. *Journal of Food Engineering* 69:299–306.

# Part II

## Modeling and Simulation

# 7 Direct Calculation of Survival Ratio and Isothermal Time Equivalent in Heat Preservation Processes

*Maria G. Corradini, Mark D. Normand, and Micha Peleg*

## CONTENTS

## 7.1 INTRODUCTION

Thermal inactivation of microbial cells, bacterial spores, and enzymes has been and will remain one of the most effective ways to assure food safety and stability. The governing philosophy has been to identify the most heat-resistant agent of spoilage or of potential health risk and then create time–temperature conditions that will eliminate it even at the coldest point of the food. Once accomplished, this would guarantee that "all" potentially harmful or undesirable agents have been destroyed "everywhere" in the product and hence, that the food will remain biologically stable and safe for as long as its container remains hermetically sealed. Finding the coldest point requires knowledge of heat transfer theories, or a series of measurements. Identification of the pertinent target, be it a bacterial endospore, microbial cell or an enzyme, and determination of its heat resistance, requires knowledge of food microbiology and biochemistry. We will assume that knowledge of both kinds already exists or can be created by collection of pertinent experimental data. Therefore, we will limit our discussion to methods of calculating or estimating the efficacy of thermal processes in terms of the theoretical extent of the inactivation at the coldest point, regardless of whether it has been determined directly or through heat transfer models.

## 7.2 FIRST-ORDER INACTIVATION KINETICS

Traditionally, and according to almost all textbooks on general food microbiology, isothermal microbial inactivation follows first-order kinetics (Jay, 1996; Holdsworth, 1997). Expressed mathematically, the survival ratio has a log linear relationship with time, i.e.,

$$\log_{10} S(t) = -k(T)t \tag{7.1}$$

where

$S(t)$ defined as $N(t)/N_0$ is the momentary number after time $t$, $N(t)$, divided by the initial number, $N_0$

$k(T)$ is a temperature-dependent (exponential) rate constant

Equation 7.1 is also written in the form:

$$\log_{10} S(t) = -\frac{t}{D(T)} \tag{7.2}$$

where $D(T)$, the "$D$-value," is the time needed to increase or reduce the number of surviving cells or spores by a factor of 10.

The temperature dependence of the "$D$-value" has been assumed to obey a log linear relation, i.e.,

$$\log_{10} D(T) = \frac{T - T_{\text{ref}}}{z} \tag{7.3}$$

The temperature dependence of the $k(T)$ has been assumed to follow the Arrhenius equation, i.e.,

$$\log\left[\frac{k(T)}{k(T_{ref})}\right] = \frac{E_a}{R}\left(\frac{1}{T} - \frac{1}{T_{ref}}\right) \tag{7.4}$$

In Equation 7.3, $z$ is the "$z$-value," the increase or decrease of the temperature in degree centigrade or Fahrenheit that will shorten or prolong the "$D$-value" by a factor of 10. In Equation 7.4, $T_{ref}$ is a reference temperature in degree Kelvin, $E_a$, the "energy of activation" and $R$, the universal gas constant. Mathematically, Equations 7.3 and 7.4 are mutually exclusive models and hence cannot be used interchangeably. They also have a variety of consistency and other problems (Peleg, 2006). For example, either can be applicable if and only if all the isothermal survival curves in the lethal temperature regime follow first-order kinetics. And, both require that the momentary rate constant is only a function of temperature and therefore is totally unaffected by the exposure time. Thus, according to either model, spores at 115°C, say, must have exactly the same heat resistance, regardless of whether they have just reached this temperature or had "cooled" to it after being held at 125°C for 10 min! Why all the complex biophysical and biochemical processes that cause a microbial cell's mortality or a spore's destruction should be universally coordinated to produce a first-order inactivation with a single temperature-independent "energy of activation" has never been satisfactorily explained. It has also never been proven that there must always be a "limiting reaction rate" that controls the inactivation and there is no reason to assume that such a "reaction" exists. Moreover, the relevance of the gas constant, $R$, and the "mole" to microbial inactivation kinetics is unclear and so is the rationale for compressing the temperature scale by converting the temperatures into their degree Kelvin reciprocals. Stated differently, despite the wide use of the Arrhenius equation in heat process calculations, the assumed analogy between microbial thermal inactivation and a reaction between two gases, for which the model was originally developed, has never been convincingly demonstrated (Peleg, 2006).

The cumulative effect of any nonisothermal heat preservation process is currently assessed in terms of an "$F_0$-value" defined by

$$F_0(t) = \int_0^t 10^{\frac{T(t)-T_{ref}}{z}}\, dt \tag{7.5}$$

or, alternatively, by integration of the Arrhenius equation.

According to Equation 7.5, the "survival ratio," $S(t)$, is related to $F_0(t)$ by

$$\log_{10} S(t) = -\frac{F_0(t)}{D(T_{ref})} \tag{7.6}$$

Where $D(T_{ref})$ is the "$D$-value" at the reference temperature, $T_{ref}$.

This, however, "will not be the case" if either the isothermal survival curves are not all log linear and/or if the temperature dependence of the "$D$-value" is not log linear. In other words, had the Arrhenius equation been a valid model, there would be no clear relation between the survival ratio and the calculated "$F_0$-value." What really matters, theoretically and in practice, is "the accomplished survival ratio" or "the number of decades reduction" and not any lethality parameter like the "$F_0$-value," whose calculated magnitude might depend on the gap between the chosen reference temperature and the product's momentary temperature—see Datta (1993).

Many in the food industry claim, and with some justification, that an "$F_0$-value" having time units is an intuitively understood parameter and hence a useful sterility measure despite its theoretical shortcomings. For those adhering to this view, it represents an equivalent time (usually in minutes) to an isothermal process at an agreed upon reference temperature. In low-acid food heat sterilization, the reference temperature has been 121.1°C (250°F) and the target for elimination has been the heat-resistant spores of *Clostridium botulinum*. Thus, if it is known that the time to achieve "commercial sterility" (thermal death time) at the reference temperature is, for example, 2 min, a 6 min process will be three times the "theoretical requirement." Even without considering the question of what constitutes commercial sterility and if and how the "thermal death time" can be determined unambiguously, one can legitimately ask whether the theoretical compromise needed to achieve this convenience is really necessary. Stated differently, can one replace the "$F_0$-value" by an isothermal equivalent time parameter that requires neither the first-order kinetics assumption nor that the $D$-value must have log linear temperature dependence? The answer to this question is yes, and what follows will describe how this can be done. It will also be shown that the described alternative method applies to microbial inactivation patterns that follow the traditional models as well as to those that do not. The nonlinearity that characterizes the latter can be manifested either in the targeted spores' isothermal survival curves themselves or in the temperature dependence of their exponential rate constants, if they are not a function of time.

## 7.3 NONLINEAR INACTIVATION KINETICS

Most isothermal microbial survival curves when plotted on semilogarithmic coordinates are curvilinear with a noticeable upward or downward concavity. This is true for both bacterial vegetative cells and bacterial endospores. Some curves can be considered linear, at least as a first approximation, but these can be considered as a special case of "zero concavity"; see below. A very small minority of inactivation curves are sigmoid and these have two basic shapes; downward concavity changing to upward concavity or vice versa (Figure 7.1). Notice that the "shoulder" that appears in the first kind can be an artifact of the come-up time (Peleg, 2004). In this chapter, we will only deal with the first three types.

According to Peleg and Penchina (2000), upper concavity, or "tailing," indicates fast elimination of the heat-sensitive members of the microbial populations, leaving behind members of higher and higher heat resistance. Downward concavity, according to these authors, indicates that damage accumulation sensitizes the survivors and consequently that it takes progressively less time to reduce their number by the same ratio. A log linear survival curve might be a manifestation of a "balance" between the roles of damage accumulation and culling of the weak and hence must be rare.

Other special cases are survival curves having a "flat" or "activation" shoulder—see Figure 7.1—the latter almost exclusively found in *Bacilli* spores. The shoulder's implications in nonisothermal inactivation are assessed in detail elsewhere (Peleg, 2002, 2006). Being rather uncommon, they will not be further discussed. Suffice

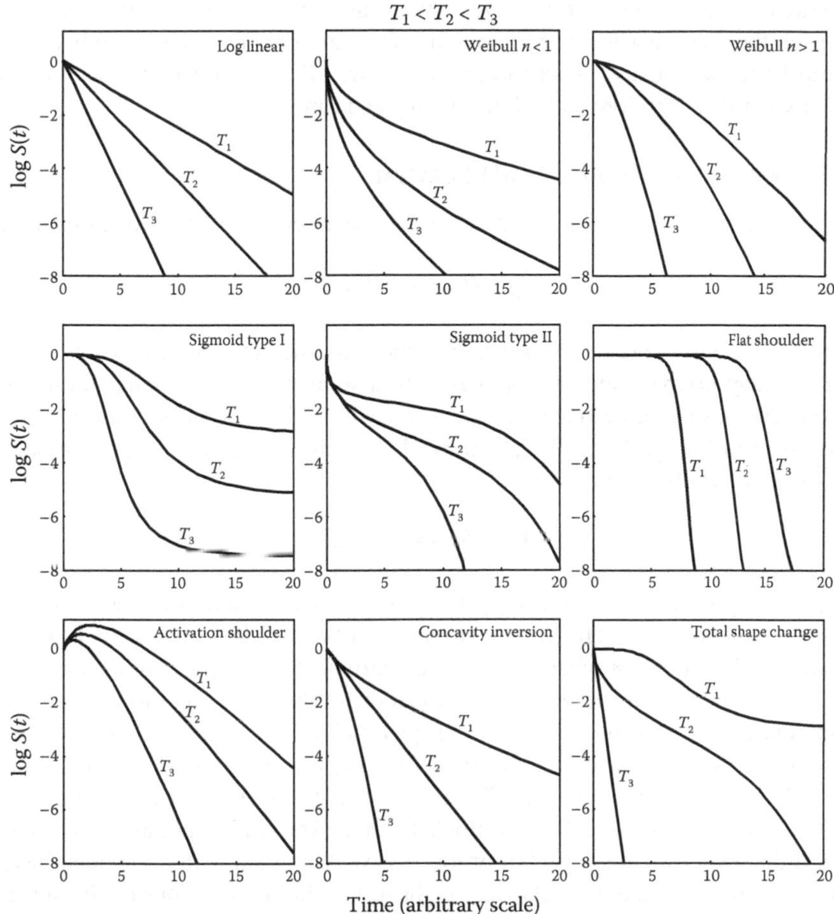


**FIGURE 7.1** Typical shapes of isothermal microbial inactivation curves.

it to say here that the methods to estimate or predict nonisothermal inactivation patterns from isothermal survival data, which were developed for the case of monotonic isothermal curves—see below—are also applicable to the two "sigmoid types" and to curves exhibiting either a flat or activation shoulder.

Another possible departure from the standard model of microbial inactivation is the effect of temperature on the "overall shape" of the isothermal survival curves. In principle, at least (see Figure 7.1) the shape need not be the same at the lower and higher ends of the lethal temperature range. Therefore, one can expect that "tailing," for example, which is noticeable at a relatively low temperature, will disappear as the temperature increases. This will result in the loss of the survival curve's upper concavity or even in a concavity direction reversal—as shown in the figure. Similarly, a "shoulder" can not only be shortened as the temperature increases, but also can disappear altogether. The same can be said about sigmoid survival curves which, theoretically at least, can become monotonic at either end of the lethal temperature range, i.e., they can be clearly upward at low temperatures and clearly concave downward at very high temperatures.

The above assortment of possible isothermal survival patterns and the manners by which they can be affected by temperature cannot be effectively described by the standard theories of heat inactivation, which are all based on first-order kinetics. Their expansion, therefore, is both necessary and timely.

### 7.3.1 Weibullian ('Power Law') Survival

Most monotonic isothermal survival curves can be described by the Weibullian model

$$\log_{10} S(t) = -b(T)t^{n(T)} \tag{7.7}$$

where $b(T)$ and $n(T)$ are temperature-dependent coefficients (Peleg and Cole, 1998).

For many microorganisms, $n(T)$ is only a weak function of temperature (van Boekel, 2002), or can be fixed as a constant with only little effect on the fit (e.g., Corradini and Peleg, 2005; Peleg, 2006). By fixing the power $n(T)$, the model is considerably simplified, i.e.,

$$\log_{10} S(t) = -b(T)t^{n} \tag{7.8}$$

Notice that upper concavity (tailing) is expressed by $n<1$ and downward concavity by $n>1$—see Figure 7.2. The log linear model, or "first-order kinetics," is just a special case of Equation 7.8 where $n = 1$. The coefficient $b(T)$, whose units are time$^{1/n}$, serves as a "rate parameter," albeit a nonlinear one. When $n=1$, $b(T)$ would be exactly the exponential inactivation rate $k(T)$ in Equation 7.1. The name 'Weibullian' comes from the observation that an organismic population whose members' heat resistances have a Weibull distribution will exhibit a survival pattern characterized by Equations 7.7 or 7.8 (Peleg and Cole, 1998; van Boekel, 2002). The reader should be alerted that when fitting experimental $\log_{10} S(t)$ or $\log_e S(t)$ vs. time data with the model, instead of those of $S(t)$ vs. time, the relative weight of the different regions of the survival curve and with it the associated errors are altered. Therefore, if one uses the equation $\log_e S(t) = -(t/\beta)^{\alpha}$ as a regression model, the calculated parameters, $\alpha$ and $\beta$,

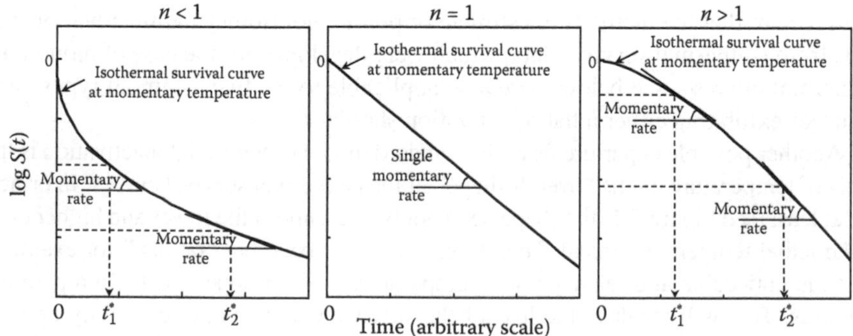


**FIGURE 7.2** Schematic view of three Weibullian semilogarithmic isothermal survival curves; linear ($n = 1$), and with upward ($n < 1$) or downward ($n > 1$) concavity. Notice that when the isothermal survival curve is not log linear, the momentary inactivation rate is a function not only of temperature, but also of the exposure time.

might be only "rough approximations" of the resistance distribution's shape and scale (or location) factors, respectively, depending on the experimental data's scatter pattern (Peleg, 2003, 2006). Moreover, most of the survival curves that can be described by the Weibullian model can also be fitted with other mathematical models (e.g., Peleg and Penchina, 2000; Peleg, 2006). Still, because of its simplicity, and the intuitive interpretation of its parameters, we will use this model almost exclusively in this chapter. Also, notice that we are using the base 10 logarithm in order to conform to the manner microbial survival data are most commonly presented in the literature on the subject and used by the food and pharmaceutical industries in sterility calculations.

### 7.3.2 Log Logistic Temperature Dependence of $b(T)$

Growing evidence supports the notion that the temperature dependence of $b(T)$ can be effectively described by the log logistic model (e.g., Corradini et al., 2005; Pelag, 2006):

$$b(T) = \log_e\{1+\exp[k(T-T_c)]\} \tag{7.9}$$

where $k$ and $T_c$ (together with $n$) are the heat resistance characteristics of the organism or spores at hand. When $n$ is a constant or only a weak function of temperature, a high heat resistance is manifested by small $k$ and high $T_c$, while heat sensitivity by a large $k$ and low $T_c$, as shown in Figure 7.3. According to this model, at $T \ll T_c$ $b(T) \approx 0$ and at $T \gg T_c$ $b(T) \approx k(T - T_c)$. The model expressed by Equation 7.9 has two main advantages over the traditional Arrhenius equation or the log linear model. First, it accounts for the fact that there is a qualitative difference between the effects of heat at low and high temperatures, i.e., well above and well below $T_c$. Second, this model's use also eliminates the unnecessary logarithmic conversion of the rate parameter,

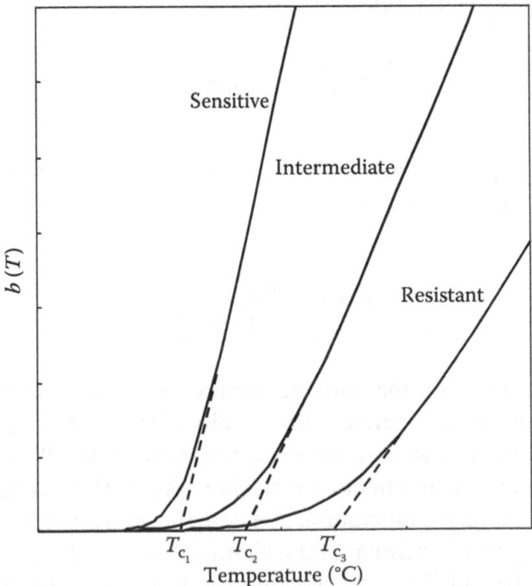


**FIGURE 7.3** Schematic view of the log logistic temperature dependence of $b(T)$.

which rarely if ever varies by orders of magnitude in the pertinent lethal temperature range. The unnecessary compression of the temperature scale that the Arrhenius model requires, i.e., through the conversion of degree Celsius into degree Kelvin reciprocals, is also avoided (Peleg, 2006).

### 7.3.3 Calculation of the Survival Ratio in Isothermal and Nonisothermal Heat Treatment

When it comes to the efficacy of heat preservation processes (and also nonthermal methods of microbial inactivation), the "bottom line" is how many decades of reduction in the targeted microbial population have been achieved. As already stated, what really counts is not the "$F_0$-value," whose calculation is model dependent, but the "actual residual survival ratio or its logarithm." If properly calculated, every combination of a primary and secondary model that fits the isothermal survival data will yield a very similar result. This is provided that the models are only used for time–temperature conditions covered by the experimental data from which they have been derived (Peleg, 2003; Peleg et al., 2005; Peleg, 2006). In other words, as long as the models are not used for extrapolation, they need not be unique. Thus, everything that will be said about the Weibullian–log logistic (WeLL) model is equally applicable to any other primary and secondary inactivation models if they correctly capture the organism or spores' survival patterns in the pertinent temperature range.

#### 7.3.3.1 Nonisothermal Conditions

The momentary isothermal inactivation rate of an organism or spore that obeys the Weibullian survival model with a fixed power $n$ (Equation 7.8) is (Peleg and Penchina, 2000; Peleg, 2003, 2006):

$$\frac{\mathrm{d}\log_{10}S(t)}{\mathrm{d}t} = -b(T)nt^{*n-1} \tag{7.10}$$

The time $t^*$ is the time that corresponds to the momentary ratio $\log_{10} S(t)$. It is defined by the inverse of Equation 7.10, i.e.,

$$t^* = \left[-\frac{\log_{10} S(t)}{b(T)}\right]^{\frac{1}{n}} \tag{7.11}$$

In a nonisothermal process, the temperature profile can be either described by an algebraic expression, $T(t)$, or entered numerically when a data logger is used. Let us examine the first case and we will return to the second later. We assume that when the temperature varies, the momentary inactivation rate, $\mathrm{d}\log_{10} S(t)/\mathrm{d}t$, is the isothermal rate at the momentary temperature, at a time that corresponds to the momentary survival ratio (Peleg and Penchina, 2000; Corradini et al., 2005; Peleg, 2003, 2006). If so, Equations 7.10 and 7.11 can be combined to produce the nonisothermal rate equation:

$$\frac{\mathrm{d}\log_{10} S(t)}{\mathrm{d}t} = -b[T(t)]n\left\{-\frac{\log_{10} S(t)}{b[T(t)]}\right\}^{\frac{n-1}{n}} \tag{7.12}$$

or, when $b(T)$ is defined by Equation 7.9, the rate model becomes:

$$\frac{\mathrm{d}\log_{10}(t)}{\mathrm{d}t} = -\log_e[1+\exp\{k[T(t)-T_c]\}]\cdot n\cdot\left[-\frac{\log_{10} S(t)}{\log_e[1+\exp\{k[T(t)-T_c]\}]}\right]^{\frac{n-1}{n}} \tag{7.13}$$

Despite its cumbersome appearance, Equation 7.13 is an ordinary differential equation (ODE), which can be easily solved numerically by a program like Mathematica® (Wolfram Research, Champaign, IL) and similar advanced mathematical software. The boundary condition is that at $t = 0$, $\log_{10} S(t) = 0$. (Sometimes, it is necessary to replace $\log_{10} S(0)$ by a small negative value, e.g., −0.0001, to make sure that the program will work.) The solution of Equation 7.13, in the form of a $\log_{10} S(t)$ vs. time relationship, enables the calculation of the "survival ratio" after any given time $t$—see Figure 7.4. Moreover, the model's rate equation can be solved for almost any conceivable temperature profile, $T(t)$, which may include heating and cooling, oscillating temperature, and even discontinuities (Corradini et al., 2005; Peleg, 2006). In canning, because of inertia, really abrupt temperature changes are expected to be rather rare. Still, the possibility to calculate the survival ratio, when the temperature profile is described by a model that contains "If" statements, suggested that a solution

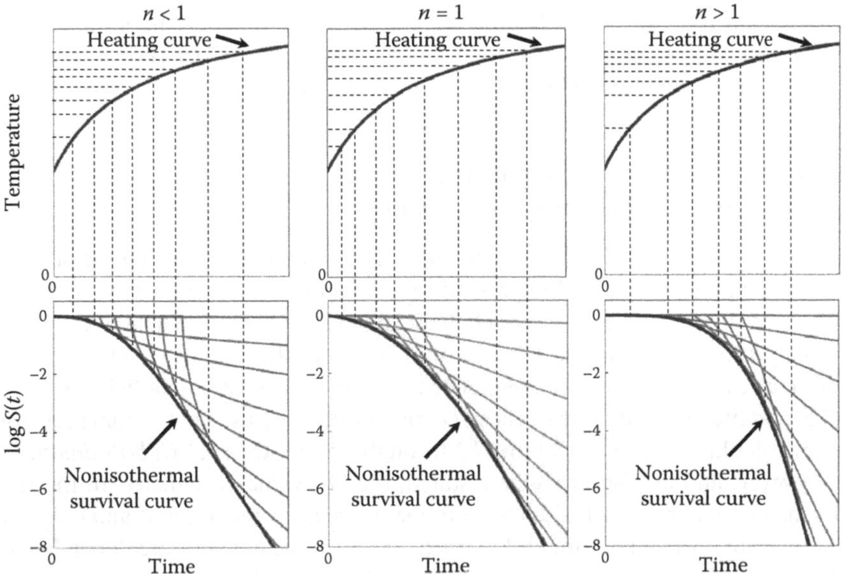


**FIGURE 7.4** Schematic view of the construction of a nonisothermal survival curve from isothermal inactivation data. Notice that during heating, the survival curve can be concave downward despite the fact that the isothermal curves from which it is constructed all have upward concavity.

will always be found for scenarios where a trend is reversed in a smoother fashion as and when heating is followed by cooling. Notice that if the isothermal survival curves are described by an alternative "non-Weibullian" mathematical model, the procedure to construct the nonisothermal rate model will be the same. If, in general, the isothermal curves are described by $\log_{10} S(t) = f(t)$, then $d \log_{10} S(t)/dt = df(t)/dt|_{T=\text{const.}}$ and $t^* = f^{-1}(t)$. Thus, when the two are combined, the rate equation will be

$$\frac{d \log S(t)}{dt} = \frac{df(t)}{dt}\bigg|_{t=t^*} \tag{7.14}$$

where the rate model equation's "coefficients" are nested functions of time, i.e., $k_1(t) = k_1[T(t)]$, $k_2(t) = k_2[T(t)]$, etc.

There is also a way to write and solve the differential rate equation even when the function used to describe the isothermal survival curves does not have an analytical inverse (Peleg, 2002, 2006). However, most if not all the commonly encountered survival curves are monotonic. Therefore, finding a model that has an analytic inverse to describe them should not be too difficult. If all the isothermal survival curves in the lethal range happen to be log linear, i.e., they all follow "first-order kinetics" (or the Weibullian model with $n = 1$) and their exponential rate constant's temperature dependence can be described by the log logistic model, then the rate model equation will be

$$\frac{d \log S(t)}{dt} = -\log_e[1 + \exp\{k[T(t) - T_c]\}] \tag{7.15}$$

This equation too can be easily solved numerically to produce the nonisothermal survival curve for almost any conceivable temperature profile $T(t)$. [For a linear heating profile, it also has an analytical solution (Peleg, 2006).]

#### 7.3.3.2 Calculation of the Equivalent Time at a Constant Reference Temperature

For better or worse, the $F_0$-value's appeal to food scientists and technologists stems from its being calculated and expressed in time units. Its value conveys a sense of how a given thermal process compares with an isothermal treatment at a lethal reference temperature known to produce "commercial sterility." For low-acidity canned foods, as already mentioned, the reference temperature has traditionally been 121.1°C (250°F) and the targeted spores (whose *D*- and *z*-values are used in thermal processes' efficacy calculation—see Equations 7.2 through 7.5) are those of *C. botulinum*.

However, and as also previously stated, the "$F_0$-value" calculated in the traditional manner can be translated into a number of decades reduction if and only if the isothermal survival curves are all log linear and so is the temperature dependence of the "*D*-value." Once either condition is not satisfied, the relationship between the calculated "$F_0$-value" and the survival ratio becomes obscure and with it the rationale of using the "$F_0$-value" as a sterility criterion.

A way out of this impasse is to monitor a heat process's progress in terms of the "equivalent isothermal time," $t_{eq}$, at a chosen reference temperature, which can be

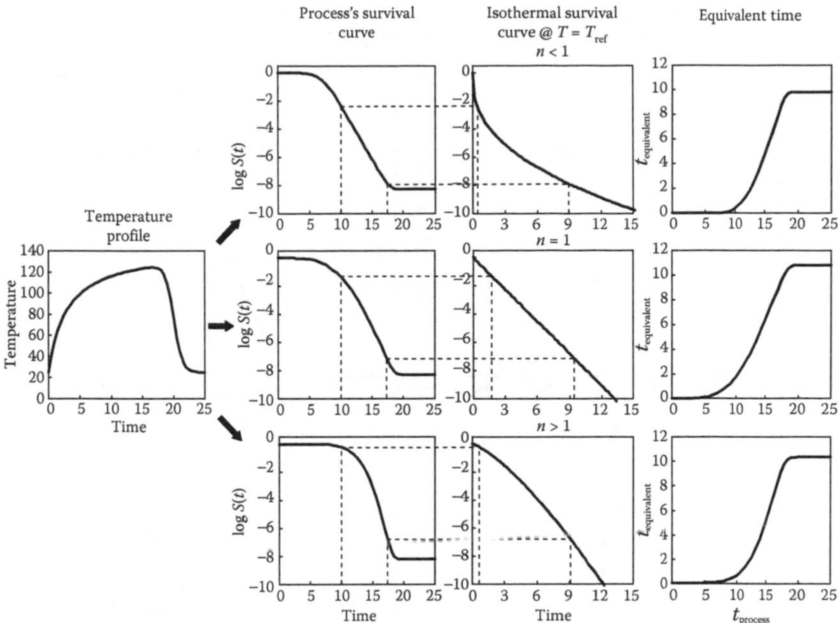


**FIGURE 7.5** Schematic view of the construction of the equivalent isothermal time ($t_{eq}$) vs. process time curve in heat treatments.

determined directly (Corradini et al., 2006). The reference temperature can remain the traditional 121.1°C (250°F). The equivalent isothermal time, $t_{eq}$, vs. the actual process time, $t_{proc}$, relationship can be constructed in the manner shown schematically in Figure 7.5. The survival ratios, determined during the process, having a temperature profile $T(t)$, are plotted side by side with the isothermal survival curve of the targeted organism or spore at the chosen reference temperature. The equivalent isothermal and nonisothermal times, as shown in the figure, are the times in the two plots that have the "same survival ratio." Notice that, in principle, the shown construction does not require any survival model and it can be based solely on the experimentally observed survival patterns under the isothermal and nonisothermal conditions. Preferably, both curves should be determined with the actual targeted organism or spores and in the actual food. This might not be always a feasible option and hence a surrogate organism and a substitute medium might be used. The implications of this issue are outside the scope of this chapter. However, some of the uncertainties introduced by the use of a surrogate organism and medium can be addressed by the introduction of a safety factor.

Suppose that an isothermal heat treatment of certain duration is known to result in a safe product, by reducing the targeted population by at least eight orders of magnitude, for example. (Eight orders of reduction seem to be a limit set by the current methods of detection, but there is no reason why their sensitivity would not be improved in the future.) Suppose that for the spores of *C. botulinum* in a given medium, 3 min at 121.1°C will result in at least eight orders of magnitude reduction. If so, see Figure 7.6, then when $t_{eq}$ of the planned or actual process exceeds 3 min,

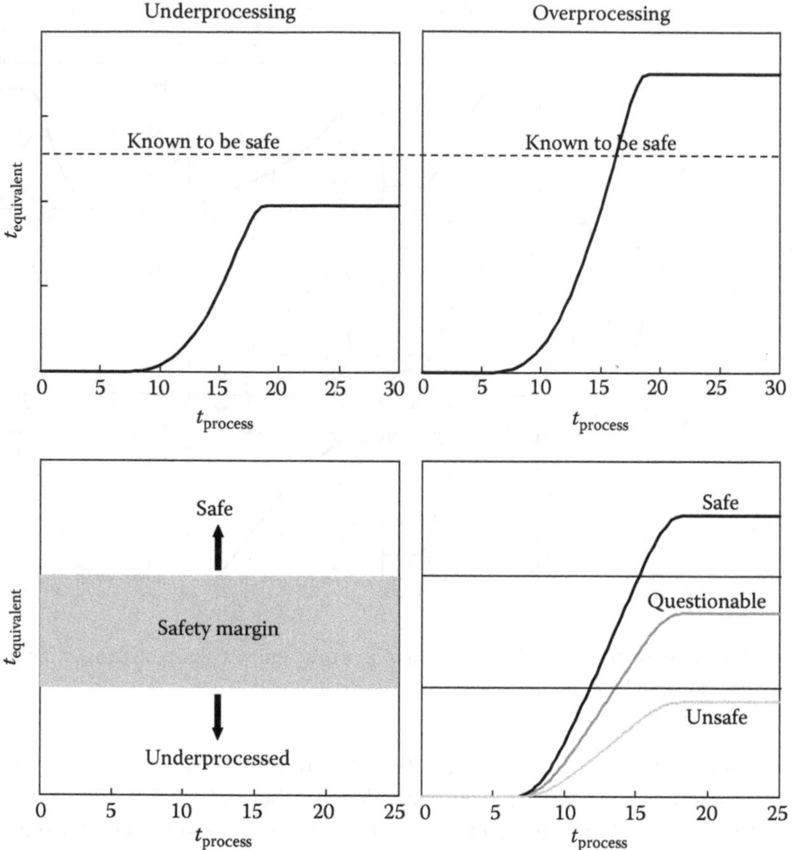


**FIGURE 7.6** Schematic view of the progress of two heat sterilization processes expressed in terms of their equivalent isothermal time at a reference temperature (top) and the effect of a safety margin addition on their efficacy (bottom).

there will be at least eight orders of magnitude reduction and the product will be considered as safe. If $t_{eq}$ is less than 3 min, the number of decades' reduction is smaller than eight and the product will be considered "underprocessed." A "safety factor" can be added by setting the required equivalent time to 4 min, say, as in the above example, or to any level deemed appropriate in light of experience, as shown schematically in Figure 7.6. Notice that assessing a thermal process's efficacy in this way does not require any assumption concerning the isothermal inactivation pattern, let alone that it must follow first order or any other particular kinetics (Corradini et al., 2006; Peleg, 2006). However, if the targeted organism or spore's isothermal inactivation is known to follow the Weibullian model, then $t_{eq}$ can be generated from the corresponding mathematical expression (Corradini et al., 2006; Peleg, 2006), i.e.,

$$t_{eq}(t) = \left[ -\frac{\log S(t)}{b(T_{ref})} \right]^{\frac{1}{n(T_{ref})}} \tag{7.16}$$

where $b(T_{ref})$ and $n(T_{ref})$ are the Weibullian parameters at the reference temperature, $T_{ref}$.

When the targeted organism or spore's survival parameters in the pertinent medium are known or can be estimated from pertinent experimental data, $t_{eq}$ can be calculated continuously or iteratively. This makes it possible to monitor the progress of an actual or simulated thermal process in order to determine its efficacy. Such calculations can be used to assess the safety of existing or contemplated industrial treatments.

### 7.3.4 Non-Weibullian Inactivation

Similar expressions to those given above can be derived for non-Weibullian's isothermal survival patterns and these too can be used to predict nonisothermal inactivation. Suppose that a certain organism or spore exhibits a strong degree of "tailing," manifested in measurable residual survival under all practical processing conditions. Heat-resistant *Bacilli* spores are a case in point. A possible model for isothermal survival curves that show a residual ratio is (Peleg and Penchina, 1998; Corradini et al., 2005; Peleg, 2006):

$$\log_{10} S(t) = -\frac{t}{k_1(T) + k_2(T)t} \tag{7.17}$$

where $k_1(T)$ and $k_2(T)$ are temperature-dependent coefficients.

This model entails that as $t \to \infty$, $\log_{10}S(t) \to 1/k_2(T)$, i.e., that there is an asymptotic residual survival ratio that decreases as $k_2(T)$ increases. (Notice that, here again, "first-order kinetics" is just a special case of a more general nonlinear model. When in Equation 7.17 $k_2(T) = 0$, $k_1(T)$ will be the familiar "*D*-value" of the traditional log linear model.)

According to Equation 7.17, the momentary nonisothermal inactivation rate during a thermal process with a temperature profile $T(t)$ is (Corradini et al., 2005; Peleg, 2006):

$$\frac{d\log_{10} S(t)}{dt} = -\frac{k_1[T(t)]}{\{k_1[T(t)] + k_2[T(t)]t^*\}^2} \tag{7.18}$$

where

$$t^* = -\frac{k_1\left[T(t)\right] \log_{10} S(t)}{1 + k_2[T(t)] \log_{10} S(t)} \tag{7.19}$$

Like Equation 7.13, Equation 7.18 too is an ODE that can be solved numerically with any advanced mathematical program or even general-purpose software (Peleg et al., 2004).

The equivalent time for an organism or spores whose inactivation follows the above model is

$$t_{eq}(t) = -\frac{k_1(T_{ref}) \log_{10} S(t)}{1 + k_2(T_{ref}) \log_{10} S(t)} \tag{7.20}$$

where $T_{ref}$ is the chosen reference temperature.

## 7.4 WEB PROGRAM TO CALCULATE STERILITY

A continuous rate equation such as Equations 7.13 or 7.18 can be solved numerically by its conversion into a "difference equation." Although the solution will be only an approximation, it can be very close to the "correct" one if the chosen time intervals are sufficiently short. Consider Equation 7.13 as an example. It can be approximated by (Peleg et al., 2004; Peleg, 2006):

$$\frac{\log_{10} S(t_i)-\log_{10} S(t_{i-1})}{t_i - t_{i-1}} = -\frac{b\left[T\left(t_{i-1}\right)\right]+b\left[T\left(t_{i-1}\right)\right]}{2}\,\frac{n\left[T\left(t_i\right)\right]+n\left[T\left(t_{i-1}\right)\right]}{2}\left\{\frac{-\left[\log_{10} S\left(t_i\right)+\log_{10} S\left(t_{i-1}\right)\right]}{b\left[T\left(t_{i-1}\right)\right]+b\left[T\left(t_{i-1}\right)\right]}\right\}^{1-\frac{2}{n\left[T\left(t_i\right)\right]+n\left[T\left(t_{i-1}\right)\right]}} \tag{7.21}$$

where $t_i$ and $t_{i-1}$ are successive times separated by a short interval.

If we know that initially, i.e., at $i = 0$, the temperature is $T_0$ and $\log S(0) = 0$, we can calculate the corresponding $b(T_0)$ and if necessary $n(T_0)$ too. For simplicity, we will assume that $n(T)$ is constant, i.e., $n(T) = n$.

For time $t_1$, Equation 7.21 then becomes:

$$\frac{\log_{10} S\left(t_1\right)}{t_1} = -\frac{b\left[T\left(t_1\right)\right]+b\left[T_0\right]}{2}\,n\left\{\frac{-\log_{10} S\left(t_1\right)}{b\left[T\left(t_1\right)\right]+b\left[T_0\right]}\right\}^{\frac{n-1}{n}} \tag{7.22}$$

But this is an ordinary algebraic equation with $\log_{10} S(t_1)$ being the only unknown. It can be solved analytically or by the GoalSeek function in MS Excel to produce the value of $\log_{10} S(t_1)$. This calculated value of $\log_{10} S(t_1)$ can now be inserted back to the equation in order to calculate $\log_{10} S(t_2)$ i.e.,

$$\frac{\log_{10} S(t_2)-\log_{10} S(t_1)}{t_2 - t_1} = -\frac{b\left[T\left(t_2\right)\right]+b\left[T\left(t_1\right)\right]}{2}\,n\left\{\frac{-\log_{10} S\left(t_2\right)-\log_{10} S\left(t_1\right)}{b\left[T\left(t_2\right)\right]+b\left[T\left(t_1\right)\right]}\right\}^{\frac{n-1}{n}} \tag{7.23}$$

Here again, the only unknown is $\log_{10} S(t_2)$ and it too can be calculated in the same manner as $\log_{10} S(t_1)$. The procedure can now be reiterated to produce $\log_{10} S(t_3)$, $\log_{10} S(t_4)$, and so forth, until the whole survival curve is generated. Similar solutions can be found for non-Weibullian's inactivation patterns except that in the case of Equation 7.18, for example, the changing parameters at each iteration would be $k_1[T(t_i)]$ and $k_2[T(t_i)]$ (Corradini et al., 2005; Peleg, 2006).

### 7.4.1 Creating Survival Curves with Model-Generated Data

When one knows the targeted organism's Weibullian-log logistic survival parameters, namely $n$, $k$, and $T_c$, then the value of any $b[T(t_i)]$ can be calculated by inserting the temperature at the particular $t_i$. If $T(t)$ is also expressed algebraically, then $T(t_i)$ can also be readily calculated. The procedure has been implemented in MS Excel® (Microsoft Corp., Seattle, WA) and the program is available as freeware on the Web (see http://www-unix.oit.umass.edu/~aew2000/GrowthAndSurvival/Sterilize/CBotSurvival.html). The user has to insert (or leave the default values of) the targeted organism or spores' $n$, $k$, and $T_c$ and the temperature profile parameters which include the initial temperature, the target temperature, a "heating rate parameter," the time of the cooling onset, a "cooling rate parameter," and the cooling water temperature. A detailed explanation of these parameters is included in the program and in each pertinent cell of the spreadsheet. Once all parameters are entered, all that remains to be done is to click on the Solve button. The program will then plot the temperature profile and below it the corresponding survival curve (Figure 7.7).

The program also allows the user to choose a reference temperature. (The default value in the sterilization version is 121.1°C.) At each step, the program also calculates the corresponding isothermal equivalent time at this reference temperature, $t_{eq}$, using the $n$, $k$, and $T_c$ entered by the user. The program starts with Equations 7.22 and 7.23 and then proceeds to calculate the other $\log_{10} S(t_i)$'s as previously explained. These $\log_{10} S(t_i)$s are used to calculate the corresponding equivalent isothermal times using Equation 7.16. The program then plots the $t_{eq}$ vs. process time relationship below the process's survival curve (see Figure 7.7).

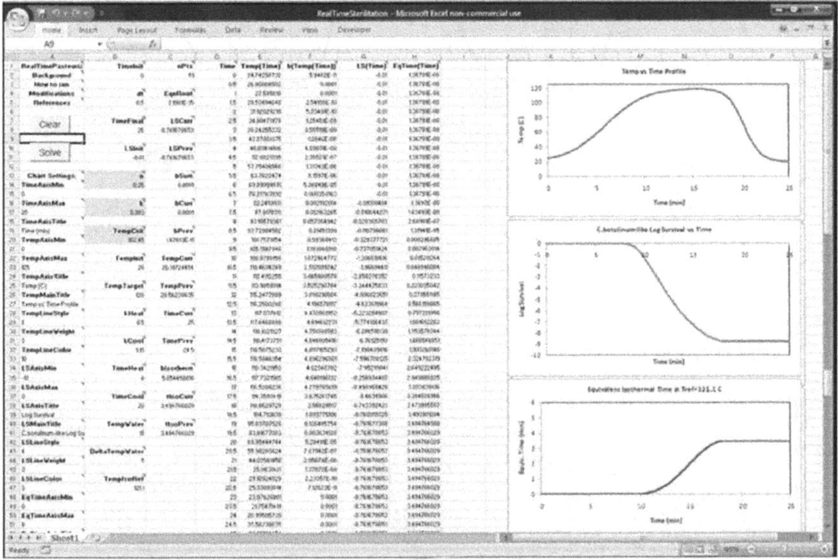

**FIGURE 7.7** Typical spreadsheet to calculate the efficacy of a nonisothermal 'dynamic' heat treatment. The plots on the right are the temperature profile and corresponding survival curve and equivalent isothermal times at the reference temperature, which are generated by the program.

#### 7.4.1.1 Examples of the Program's Performance

A typical spreadsheet of the sterilization version of the program is shown in Figure 7.7. On the right (columns I to Q) are the plots of the chosen temperature profile, the semilogarithmic survival curve and the $t_{eq}$ vs. process time relationship. They were generated with the temperature profile and survival parameters, listed in columns A and B, which can be set by the user. Plots of this kind would be a very helpful tool in screening heat treatments by showing which of the contemplated profiles has a good chance to be effective for achieving either the desired inactivation level or the equivalent isothermal time. All that is required is to set the temperature profile parameters to the appropriate values and rerun the program to produce the new survival and equivalent time curves. The program can also be used to evaluate the potential implications of a change in the targeted organism or spore's heat resistance by resetting the $n$, $k$, and $T_c$ entries to values that reflect the alteration. The number of possible modifications in the parameter settings is huge, of course, but it takes very little time to generate a new survival curve with a new set of parameters.

Another version of the program, also posted as freeware on the Web, see http://www-unix.oit.umass.edu/~aew2000/GrowthAndSurvival/Sterilize/CBotSurvival.html, allows the user to generate and examine the outcome of up to "five different scenarios simultaneously." An example of this program's output is shown in Figure 7.8. It shows (in a grossly exaggerated manner) how differences in a retort

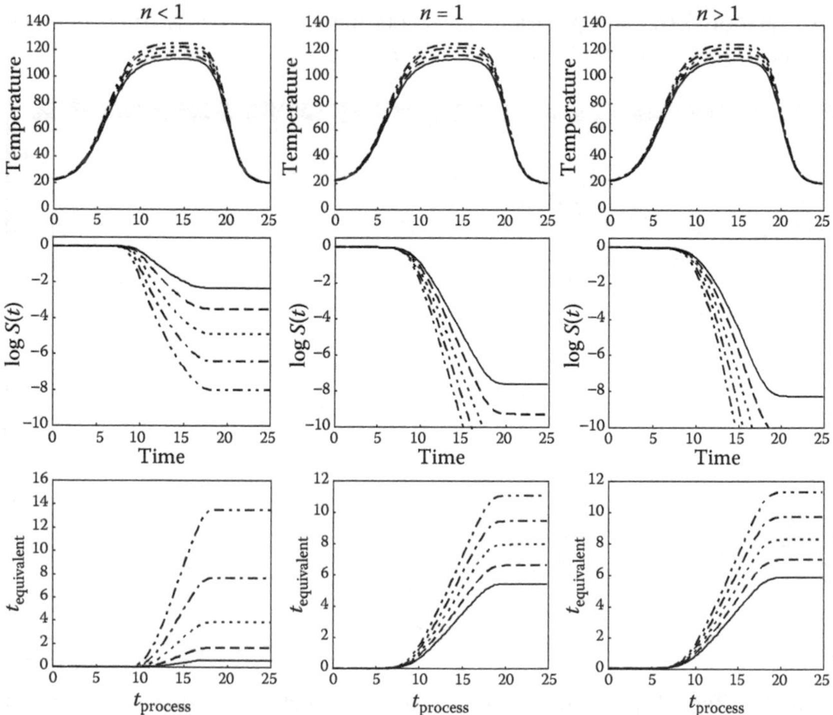


**FIGURE 7.8** Simulated (grossly exaggerated) temperature profiles, corresponding survival curves, and equivalent isothermal times calculated with a version of the program that allows the user to generate and examine up to five different scenarios simultaneously.

or retorts' temperature histories would affect the uniformity of an industrial thermal process's efficacy. The same program can also be used to produce plots that will show how spores having different heat resistances will respond to the same process, by simply entering the corresponding survival parameters, $n$, $k$, and $T_c$ in the appropriate cells of the spreadsheet. Since the kinetics of chemical nutrients and pigment degradation can be analyzed in a manner similar to that of microbial survival (Corradini and Peleg, 2005; Corradini et al., 2006; Peleg, 2006), the program can also be used to analyze both on a single spreadsheet. This will allow the user to examine the quality implications of a contemplated or existing process at the same time.

It should be emphasized, however, that as in any process simulation, the generated plots should be used as a screening device or guideline only. The actual determination of a food or pharmaceutical product's safety should always come from incubation tests and not from computer simulations alone.

#### 7.4.1.2 Running the Program with Experimental Data

Several versions of the program allow the user to paste his/her own time–temperature data rather than to generate them with a formula. This enables the user to assess the efficacy of actual pilot plant runs in the development of a thermal process. The same can be done with records of an industrial operation. Here too, all the options mentioned before are open for the user, i.e., the program can be employed to assess the roles of variations in the organism's resistance, deviation from log linearity, temperature variations within the retort, etc. One could then assess the difference between the calculated sterility by the traditional and the proposed method, which may lead to improvement in the product's safety and/or quality. Since the user has direct access to the posted time–temperature entries he or she can simulate problems or accidents, like those that are caused by an irregular steam supply or a complete shutdown (Peleg, 2004, 2006). Following the simulation, the user can test the efficacy of various corrective measures, like prolonging the holding time once the steam supply has been restored. By running the program with relevant kinetic parameters, the user can also examine the quality implications of the added processing time, which could affect the product's grading and marketing strategy. This version of the program also has numerous additional potential applications. Not the least is in training retort operators and shift managers on how to respond to emergency situations in the plant. An example of a modified temperature profile produced in this way, and its manifestation in the survival and equivalent isothermal time curves is given in Figure 7.9.

#### 7.4.1.3 Calculation of the Survival Ratios in Real Time

In principle at least, the time–temperature entries can come from data logger or a digital thermometer. Therefore, the program can be used to monitor a thermal process's progress "in real time." But unlike the conventional method, the output will be a series of theoretical survival ratios and equivalent isothermal times rather than "$F_0$-values," whose deficiencies have been explained earlier. Consequently, the program can be incorporated into industrial control systems where it will serve not only as a means to monitor thermal processes, but also to adjust them in order to guarantee a safe product even under changing circumstances.

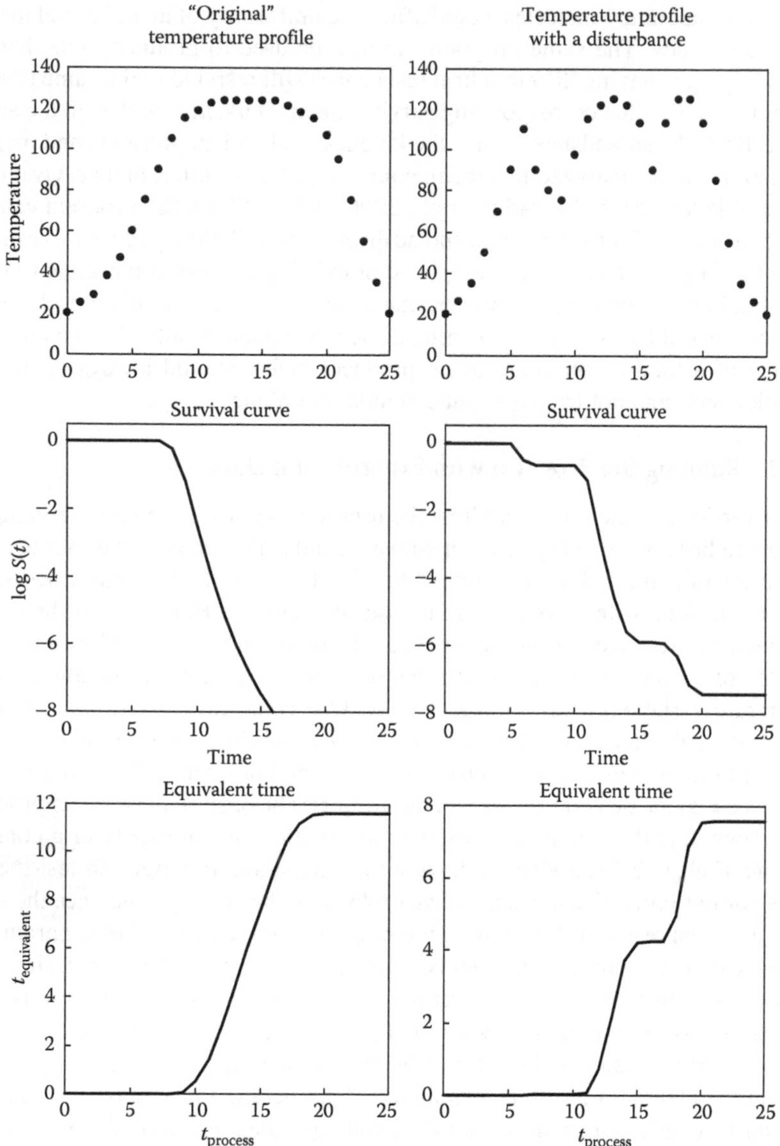


**FIGURE 7.9** Simulated experimental time–temperature records pasted in the spreadsheet by the user, their corresponding survival curves, and equivalent isothermal times. Notice that each record can be edited or modified to create and examine the effect of irregularities on the spores' inactivation and the efficacy of corrected measures.

## 7.5 CONCLUDING REMARKS

Microbial heat inactivation, many recent studies show, need not follow the first-order kinetics model as has been traditionally assumed. This chapter and the works on which it is based demonstrate that the nonlinear aspects of microbial inactivation kinetics can be taken into account in programs that are written in general-purpose software like MS Excel®. The use of these programs, which are available as freeware on the Web, would remove most of the conceptual problems to calculate sterility that the current methods have. The implementation of the new programs may result in improved product quality without compromising safety. This is because the same kind of mathematical models can be used not only to assess microbial inactivation, but also, and simultaneously, the loss of nutrients and pigments or the formation of undesirable chemical compounds (Corradini et al., 2005; Peleg et al., 2005; Peleg, 2006). The described methodology can be used to establish commercial sterility criteria that are based on the survival ratio. Unlike the "$F_0$-value" it is a measurable quantity that has physical meaning, regardless of the inactivation model. No part of the proposed method to calculate sterility is based on extrapolation—always a risky endeavor, and especially when done over several orders of magnitude. All the above also pertain to the establishment of a safety factor, which also ought to be based on measurable survival ratios and not on an "$F_0$-value" that has a physical meaning only under very restrictive conditions. The applicability of the described methodology has been demonstrated in several organisms. This was done not only by demonstrating the Weibull–log logistic model's fit to experimental isothermal survival data, but also by showing the ability of the resulting rate model (Equation 7.13) to correctly predict nonisothermal survival patterns (Peleg, 2006). We hope that future research will demonstrate the usefulness of the approach to additional microorganisms and bacterial spores. We also hope that the methodology will be validated by incubation tests, which have been and will always remain the ultimate test of a food product's safety.

As the traditional methods to assess sterility, the ones described in this chapter only address what happens at a "point," which naturally should be the coldest in the product. The rationale is clear; if the coldest point is safe, so is the rest of the food. Future research, however, may combine the new models with heat transfer theories (Teixeira et al., 1969a,b, 1999; Teixeira and Mason, 1982; Datta and Teixeira, 1987; Ghani and Farid, 2006). This will enable the researcher and technologist to estimate the implications of various strategies for achieving sterility and their potential effect on other qualities of the product "throughout the container". This refers to nutrient and pigment loss, texture degradation and the generation of potentially harmful compounds, which might be formed by excessive heating.

Although this chapter only deals with heat preservation, the described methodology can be applicable to nonthermal methods of preservation, like chemical disinfection with volatile agents such as ozone, and combined processes, like those that involve heat and ultrahigh pressure (Peleg, 2006). Here too, there is growing evidence that the inactivation does not follow first-order kinetics and that there is no reason that it should. Consequently, the equivalency of different preservation methods' safety will have to be established on the basis of theories that take into account the inactivation's nonlinear kinetics. Thus, although sterility calculation has been one of the oldest topics in food technology, it might become once more an exciting and challenging field of new research.

## ACKNOWLEDGMENT

Contributions of the Massachusetts Agricultural Experiment Station at Amherst.

## REFERENCES

Corradini, M.G., Normand, M.D., and Peleg, M. 2005. Calculating the efficacy of heat sterilization processes. *Journal of Food Engineering*, 67: 59–69.

Corradini, M.G., Normand, M.D., and Peleg, M. 2006. On expressing the equivalence of non-isothermal and isothermal heat sterilization processes. *Journal of the Science of Food and Agriculture*, 86(5): 785–792.

Datta, A.K. 1993. Error estimates for approximate kinetic parameters used in food literature. *Journal of Food Engineering*, 18: 181–199.

Datta, A.K. and Teixeira, A.A. 1987. Numerical modeling of natural convection heating in canned liquid foods. *Transactions of the American Society of Agricultural Engineers*, 30: 1542–1551.

Ghani Al-Baali, A.G.A. and Farid, M.M. (2006). *Sterilization of Food in Retort Pouches*. Springer Science + Business Media, New York, NY.

Holdsworth, S.D. 1997. *Thermal Processing of Packaged Foods*. Blackie Academic, London.

Jay, J.M. 1996. *Modern Food Microbiology*. Chapman & Hall, New York, NY.

Peleg, M. 2002. A model of survival curves having an "activation shoulder." *Journal of Food Science*, 67: 2438–2443.

Peleg, M. 2003. Microbial survival curves: Interpretation, mathematical modeling and utilization. *Comments on Theoretical Biology*, 8: 357–387.

Peleg, M. 2006. *Advanced Quantitative Microbiology for Food and Biosystems: Models for Predicting Growth and Inactivation*. CRC Press, Boca Raton, FL.

Peleg, M. and Cole, M.B. 1998. Reinterpretation of microbial survival curves. *Critical Reviews in Food Science and Nutrition*, 38: 353–380.

Peleg, M. and Penchina, C.M. 2000. Modeling microbial survival during exposure to a lethal agent with varying intensity. *Critical Reviews in Food Science and Nutrition*, 40: 159–172.

Peleg, M., Normand, M.D., and Corradini, M.G. 2005. Generating microbial survival curves during thermal processing in real time. *Journal of Applied Microbiology*, 98: 406–417.

Teixeira, A.A. and Manson, J.E. 1982. Computer control of batch retort operations with on-line correction of process deviations. *Food Technology*, 36: 85–90.

Teixeira, A.A., Dixon, J.R., Zahradnik, J.W., and Zinsmeister, G.E. 1969a. Computer determination of spore survival distributions in thermally-processed conduction-heated foods. *Food Technology*, 23: 78–80.

Teixeira, A.A., Dixon, J.R., Zahradnik, J.W., and Zinsmeister, G.E. 1969b. Computer optimization of nutrient retention in thermal processing of conduction-heated foods. *Food Technology*, 23: 137–142.

Teixeira, A.A., Balaban, M.O., Germer, S.P.M., Sadahira, M.S., Teixeira-Neto, R.O., and Vitali, A.A. 1999. Heat transfer model performance in simulation of process deviations. *Journal of Food Science*, 64: 488–493.

van Boekel, M.A.J.S. 2002. On the use of the Weibull model to describe thermal inactivation of microbial vegetative cells. *International Journal of Food Microbiology*, 74: 139–159.

# 8 New Kinetic Models for Inactivation of Bacterial Spores

*Arthur Teixeira and Alfredo C. Rodriguez*

## CONTENTS



## 8.1 INTRODUCTION

In the United States, The Code of Federal Regulations includes the law that defines the application of "Commercial Sterilization" to food products. One of the critical concepts included there is that of a "Process Authority" as the person that carries most of the responsibility for technical decisions related to sterilization processes. The need for a qualified Process Authority to perform or direct the development, validation, and troubleshooting of commercial sterilization processes is evident. Availability of a qualified person to lead this work is of top importance to reduce the risk of failure implicit in the nature of commercial sterilization processes. Currently the availability of computer software programs that perform curve fitting and other related calculations may mislead some of their potential users to depart from the legal requirement to have a Process Authority leading the corresponding tasks. This is a risky approach that should be avoided. Therefore, this chapter provides information that should prove useful to process authorities but does not intend to replace them. Although much of the chapter is devoted to thermal inactivation in response to lethal heat treatments, kinetic models for inactivation in response to high pressure as well as curing salts are also addressed later in the chapter.

The manner in which populations of microorganisms increase or decrease in response to controlled environmental stimuli is fundamental to the engineering design of bioconversion processes (fermentations) and thermal inactivation processes (pasteurization and sterilization) that are important in the food, pharmaceutical, and bioprocess industries. In order to determine optimum process conditions and controls to achieve desired results, the effect of process conditions on rates (kinetics) of population increase or decrease needs to be characterized and modeled mathematically. This chapter will focus on the development and application of mechanistic models capable of more accurately predicting thermal inactivation of bacterial spores important to sterilization and pasteurization treatments in food, pharmaceutical, and bioprocess industries.

The complex physiology and morphology of spore-forming bacteria is often responsible for the observation of growing evidence that traditional microbial inactivation models do not always fit the experimental data. In many cases the data plotted on semilogarithmic graphs showing logarithm of survivors over time at constant lethal temperature (survivor curves) reveal various non-log-linear behaviors, such as shoulders and tails. These non-log-linear survivor curves often confound attempts to accurately model thermal inactivation behavior on the basis of assuming a traditional single first-order reaction. Although it is frequently possible to adequately reproduce these non-log-linear curves mathematically with empirical polynomial curve-fitting models, such models are incapable of predicting response behavior to thermal treatment conditions outside the range in which experimental data were obtained, and have limited utility. In contrast to such empirical models, mechanistic models are derived from a basic understanding of the mechanisms responsible for the observed behavior, and how these mechanisms are affected by thermal treatment conditions of time and temperature. As such, mechanistic models are capable of accurately predicting the microbial inactivation response to thermal treatment conditions outside the range in which experimental data were obtained. This capability is highly valued by process engineers responsible for the design of ultrahigh temperature (UHT) or high-temperature short-time (HTST) sterilization and pasteurization processes that operate at temperatures far above those in which experimental data can be obtained.

This chapter begins by attempting to establish a strong case for the use of mechanistic models with a summary of background information on the history of mechanistic, vitalistic, and probabilistic model approaches to thermal inactivation of bacterial spores. This is followed by a presentation of the scientific rationale in support of first-order kinetic models as true mechanistic models for thermal inactivation of microbial populations. Following sections describe the development of mechanistic models capable of dealing with non-log-linear survivor curves exhibiting shoulders and tails in the case of spore-forming bacteria, followed by validation, applications, and comparisons of model performance in the case of selected spore-forming bacteria. Subsequent sections address the use of similar modeling approaches to development and application of kinetic models in response to high-pressure sterilization and stabilization by curing salts.

## 8.2 HISTORY OF SURVIVOR CURVE MODEL APPROACHES

The history of survivor curve models provides vital information to address current important problems related to the characteristics observed in semilogarithmic inactivation curves of microbial populations. Basic mathematical models were proposed very early in the twentieth century (Chick, 1908, 1910). Watson (1908) examined two main approaches used to study the inactivation of bacterial spores at that time: mechanistic and vitalistic. He presented Chick's model mathematically as a first-order reaction (Chick–Watson equation) describing the exponential decay commonly observed in microbial survivor curves. This model is the basis of the mechanistic approach and defines the inactivation of bacterial spores as a pseudo first-order molecular transformation. The temperature dependency of the corresponding rate constant was found to be appropriately described by the Arrhenius equation.

The vitalistic approach was based on the assumption that the observed exponential decay could be explained by differences in resistance of the individuals. Watson noted that in order for this argument to hold, most of the spores would have to be at the low extreme of resistance, instead of following the normal distribution that would be expected from natural biological variability. He concluded that the vitalistic approach could not be correct. Kellerer (1987) stated that the vitalistic approach is naïve and ignores the rigorous stochastic basis for the inactivation transformation (Maxwell–Boltzman's distribution of speed of molecules or random radiation "hits" on DNA) by using the simplistic assumption that the biological variability of the resistance can explain the observed behavior correctly.

A third option is the approach developed by Aiba and Toda (1965), where the life span probability of spores was mathematically defined and used to develop expressions describing inactivation of single and clumped spores. This analysis postulated that for a population of spores ($N_i$ in number and $t_i$ in life span), the probability associated with the distribution of life span is

$$P_r = \frac{N_0!}{\prod_i N_i!} \frac{1}{\kappa} \tag{8.1}$$

where κ is a normalizing factor.

Based on this probability function, $N_i$ could be described as a function of the life span ($t_i$) provided that the initial number of spores ($N_0$) was large:

$$N_i = \alpha' \exp(-\beta' t_i) \tag{8.2}$$

or, modifying the discrete equation into its continuous form (for large $N_0$):

$$-\frac{dN}{dt} = \alpha' \exp(-\beta' t) \tag{8.3}$$

Thus, the probabilistic approach led to the same response equation as that of the mechanistic approach. This revelation leads to several important concepts:

1. Arbitrary use of frequency distribution equations, such as the Weibull distribution, is an extensive "jump" in logic that cannot be rigorously justified.
2. Clumping induces a delay that depends on the number of spores per clump, and can be described using the probabilistic approach. However, clumping will not lead to increments in the number of survivors such as those induced by spore activation.
3. The probabilistic approach established a direct relationship between the average life span and the rate constant used in the mechanistic approach. Both approaches lead to the same basic mathematical model.

## 8.3 SCIENTIFIC BASIS FOR FIRST-ORDER KINETIC MODELS

### 8.3.1 Normal Log-Linear Survivor Curves

The nature of the inactivation transformation can be addressed and explained in terms of Eyring's transition-state theory and the Maxwell–Boltzman distribution of the speed of molecules from molecular thermodynamics. For a given configuration (i.e., mass and degrees of freedom), the fraction of molecules that have enough kinetic energy to overcome the energetic barrier (i.e., the activation energy) for a transformation such as inactivation of spores depends mostly on temperature. Therefore, at a given temperature, the fraction of molecules that will reach the level of energy required for the transformation (in this case inactivation) to happen is constant. The fraction of molecules that have enough energy to react will increase with temperature. For instance, if at a given temperature 10% of the molecules have enough energy for inactivation to occur, the percentage inactivated during a time interval will remain constant. This explains why the instantaneous rate of inactivation is proportional to the number of surviving spores present at that moment. This can be envisioned by random swatting of flies by a blind person trapped in a phone booth. The number of flies successfully swatted per unit of time will depend upon (and be proportional to) the number still in flight at that moment, and will decrease exponentially with time, just as with any concentration-dependent (first-order) reaction. Variability from biological or environmental factors is reflected by the effect they have on values of the rate constants. For example, an increase in temperature (speed of molecules) could be represented by an increase in rate of swatting (speed of fly swatter).

### 8.3.2 Non-Log-Linear Survivor Curves

Activation of dormant spores and the presence of subpopulations with different resistance give rise to the "shoulders and tails" appearing as deviations from log-linearity in some survivor curves exhibited by spore-forming bacteria. Models that include these concepts based on systems analysis of population dynamics were developed in the late 1980s and early 1990s (Rodriguez et al., 1988, 1992; Sapru et al., 1992, 1993), and verified experimentally under extreme-case conditions. The cases of "shoulders" caused by activation of dormant spores as well as that of "biphasic tailing" (early rapid inactivation of a subpopulation of relatively low heat resistance followed by a slower inactivation of a subpopulation of relatively high heat resistance) are shown in Figure 8.1.

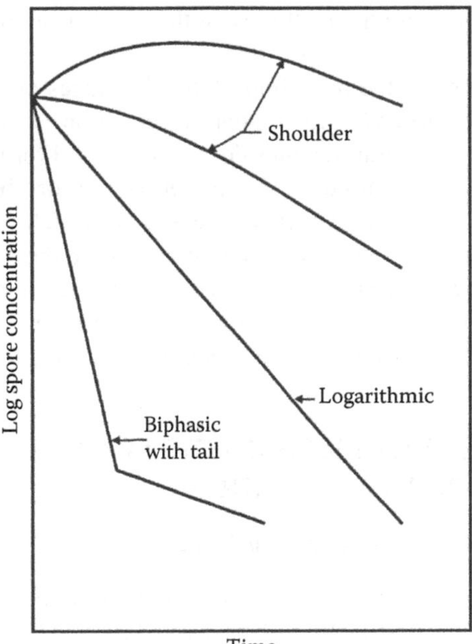


**FIGURE 8.1** Schematic nonlinear survivor curves for bacterial spores illustrating "shoulders" and "biphasic" population responses to exposure at constant lethal temperatures. (Reprinted from Teixeira, A.A., *Modeling Microorganisms in Food.* Woodhead Publishing Limited. Cambridge, UK, 2007. With permission.)

These models may be described by a summation of first-order terms as in the following expression:

$$N(t) = \sum_i a_i \exp(-b_i t) \tag{8.4}$$

where

$N(t)$ is the density of population $j$ (i.e., CFU/mL) as a function of time ($t$)

$a_i$ is the parameter related to initial population density of the different subpopulations in the system. Subindex $i$ may refer to subpopulations of active spores, dormant spores, or other spores with different resistance.

$b_i$ is the parameter related to rate constants for the different transformations in the system. Subindex $i$ may refer to subpopulations of active spores, dormant spores, or other spores with different resistance.

The parameters needed for each term ($a_i$, $b_i$) can be estimated using nonlinear regression or the successive residuals method, similar to analysis of stress relaxation curves. A step-by-step description of the use of nonlinear regression for this purpose can be found in Rodriguez (1987). The simulation of curves where two or more subpopulations with different resistances to the lethal agent must be taken into consideration is explained by letting each of the subpopulations be described by one of the exponential terms, and the total effect by their summation. Moreover, when only tails are present, parameter estimation by the method of successive residuals will automatically reveal the number of terms (subpopulations) required in the model.

Finally, perhaps one of the most important advantages of the models presented here is that they have succeeded to simulate inactivation of bacterial spores under transient conditions at temperatures outside the range used for parameter estimation where extrapolation must be used. HTST pasteurization or UHT sterilization processes operate in temperature ranges far above those in which survivor curves can be generated. Models derived from survivor curves for use in the design/specification of such processes should be mechanistic models that are based upon an understanding of the mechanisms responsible for the behavior patterns observed in the survivor curves. Empirical (curve fitting) models generally should not be used for this purpose.

## 8.4 DEVELOPMENT OF MECHANISTIC MODELS FOR MICROBIAL INACTIVATION

### 8.4.1 Normal Log-Linear Survivor Curves

The case has now been made for newfound confidence in the long-standing assumption that thermal inactivation of bacterial spores can be modeled by a single first-order reaction, as exemplified by Stumbo (1973). As explained earlier, this can be described as a straight-line survivor curve when the logarithm of the number of surviving spores is plotted against time of exposure to a lethal temperature, as shown in Figure 8.2 for survivor curves obtained at three different lethal temperatures. The decline in population of viable spores can be described by the following exponential or logarithmic equations:

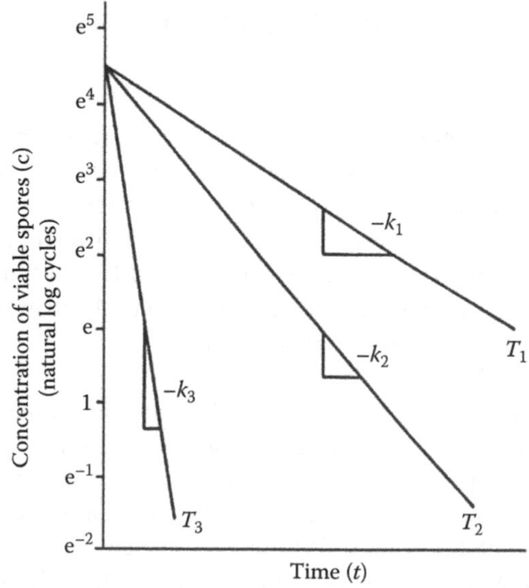


**FIGURE 8.2** Family of spore survivor curves on semilog plot showing viable spore concentration vs. time at different lethal temperatures. (Reprinted from Teixeira, A.A., *Modeling Microorganisms in Food.* Woodhead Publishing Limited. Cambridge, UK, 2007. With permission.)

$$C = C_0(e^{-kt}) \tag{8.5}$$

$$\ln(C/C_0) = -kt \tag{8.6}$$

where

$C$ is the concentration of viable spores at any time ($t$)
$C_0$ is the initial concentration of viable spores
$k$ is the first-order rate constant and is temperature dependent

This temperature dependency is illustrated in Figure 8.2, in which $T_1$, $T_2$, and $T_3$ represent increasing lethal temperatures with corresponding rate constants, $-k_1$, $-k_2$, and $-k_3$.

The temperature dependency of the rate constant is also an exponential function that can be described by a straight line on a semilog plot when the natural log of the rate constant is plotted against the reciprocal of absolute temperature. The equation describing this straight line is known as the Arrhenius equation:

$$\ln[k/k_0] = -(E_a/R)\,[(T_0 - T)/T_0T] \tag{8.7}$$

where

$k$ is the rate constant at any temperature $T$
$k_0$ is the reference rate constant at a reference temperature $T_0$

The slope of the line produces the term $E_a/R$, in which $E_a$ is the activation energy, and $R$ is the universal gas constant. Thus, once the activation energy is obtained in this way, Equation 8.7 can be used to predict the rate constant at any temperature. Once the rate constant is known for a specified temperature, Equations 8.5 or 8.6 can be used to determine the time required for exposure at that temperature to reduce the initial concentration of viable bacterial spores by any number of log cycles, or conversely, predict the degree of log cycle reduction in population at any point in time during lethal exposure. The objective in most commercial sterilization processes is to reduce the initial spore population by a sufficient number of log cycles so that the final number of surviving spores is one millionth of a spore, interpreted as probability for survival of one in a million (in the case of spoilage-causing bacteria), and one in a trillion in the case of pathogenic bacteria such as *Clostridium botulinum*.

### 8.4.2 Non-Log-Linear Survivor Curves

Actual survivor curves plotted from laboratory data often deviate from a straight line, particularly, during early periods of exposure. These deviations can be explained by the presence of competing reactions taking place simultaneously, such as heat activation causing "shoulders" and early rapid inactivation of less heat-resistant spore fractions causing biphasic curves with tails, as shown previously in Figure 8.1. Activation is a transformation of viable, dormant spores enabling them to germinate and grow in a substrate medium. This transformation is part of the life cycle of spore-forming bacteria shown in Figure 8.3. Under hostile environments, vegetative cells undergo sporulation to enter a protective state of dormancy as highly heat-resistant spores.

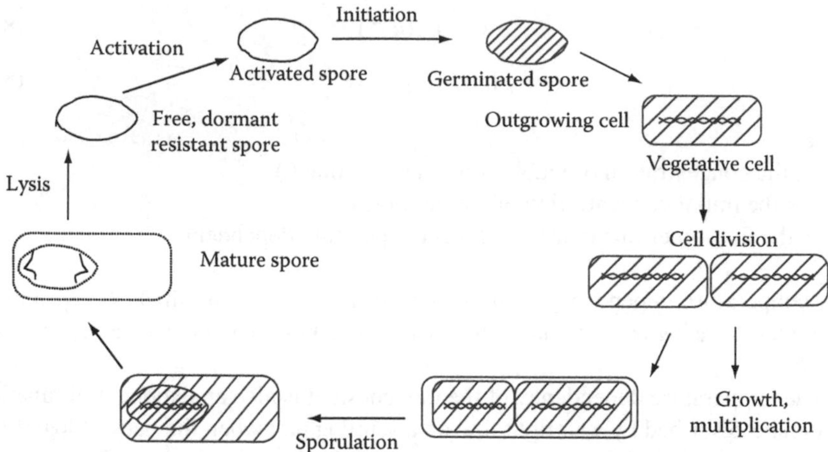


**FIGURE 8.3** Cycle of spore formation, activation, germination, and outgrowth. (Reprinted from Teixeira, A.A., *Modeling Microorganisms in Food*. Woodhead Publishing Limited. Cambridge, UK, 2007. With permission.)

When subjected to heat in the presence of moisture and nutrients, they must undergo the activation transformation in order to germinate once again into vegetative cells and produce colonies on substrate media to make their presence known (Keynan and Halvorson, 1965; Lewis et al., 1965; Gould, 1984).

In recent years, several workers have approached the mechanistic modeling of shoulders in survivor curves of bacterial spores subjected to lethal heat (Shull et al., 1963; Rodriguez et al., 1988, 1992; Sapru et al., 1992, 1993). The most recent model of Sapru et al. encompasses the key features of the earlier models, and is based on the system diagram presented in Figure 8.4. The total population at any time is distributed among storehouses (stores) containing the fraction of population in a given

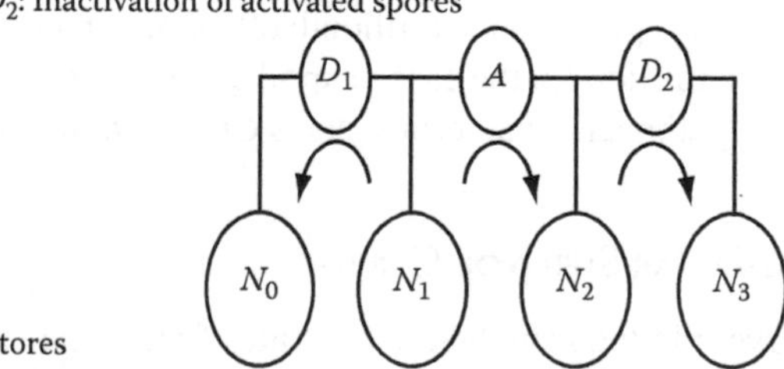


**FIGURE 8.4** Process diagram of the Sapru model of bacterial spore populations during heat treatment. (Reprinted from Teixeira, A.A., *Modeling Microorganisms in Food*. Woodhead Publishing Limited. Cambridge, UK, 2007. With permission.)

stage of the life cycle. Dormant, viable spores, potentially capable of producing colonies in a growth medium after activation, comprise population $N_1$. Activated spores, capable of forming colonies in a growth medium, comprise population $N_2$. Inactivated spores, incapable of germination and growth, comprise populations $N_3$ and $N_0$. Activation, $A$, transforms many members of $N_1$ to $N_2$ while inactivation, $D_1$, transforms other members of $N_1$ to $N_0$. Inactivation, $D_2$, transforms activated spores from $N_2$ to $N_3$. It is assumed that $A$, $D_1$, and $D_2$ are independent, concomitant, and first order with respective rate constants $K_a$, $K_{d1}$, and $K_{d2}$. The Rodriguez model (Rodriguez et al., 1988, 1992) is obtained from the Sapru model by assuming that inactivation transformations $D_1$ and $D_2$ have identical rate constants, i.e., $K_{d1}=K_{d2}=K_d$. The Shull model (Shull et al., 1963) is obtained from the Sapru model by deleting $D_1$ ($K_{d1}=0$) and $N_0$, and setting $K_{d2}=K_d$. Finally, the conventional model for a normal straight-line survivor curve is obtained by deleting $N_1$, $N_0$, $D_1$, and $A$ ($K_{d1}=K_a$), so that only $N=N_2$, $D=D_2$, and $N_3$ remain, and setting $K_d=K_{d2}$ (see Table 8.1).

**TABLE 8.1**
**Comparison of Mathematical Models of Sapru, Rodriguez, Shull, and Conventional Models**

| Model | Mathematical Model |
|---|---|
| Sapru | $\frac{dN_1}{dt}=-(K_{d1}+K_a)N_1$ |
| | $\frac{dN_2}{dt}=K_aN_1-K_{d2}N_2$ |
| Rodriguez | $\frac{dN_1}{dt}=(K_d+K_a)N_1$ |
| | $\frac{dN_2}{dt}=K_aN_1-K_dN_2$ |
| Shull | $\frac{dN_1}{dt}=-K_aN_1$ |
| | $\frac{dN_2}{dt}=K_aN_1-K_dN_2$ |
| Conventional | $\frac{dN}{dt}=-K_dN$ |

*Source:* Reprinted from Teixeira, A.A., *Modeling Microorganisms in Food*. Woodhead Publishing Limited. Cambridge, UK, 2007. With permission.

*Note:* Dormant population is $N_1$, activated population is $N_2$, and survivors $N = N_2$.

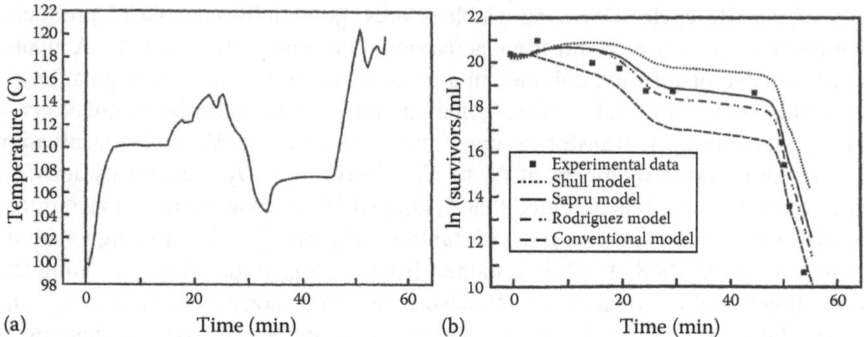


**FIGURE 8.5** (a) Temperature history experienced by spore population during experimental heat treatment. (b) Experimental data points and model-predicted survivor curves in response to temperature history above for the Sapru, Rodriguez, Shull, and conventional models. (Reprinted from Teixeira, A.A., *Modeling Microorganisms in Food.* Woodhead Publishing Limited. Cambridge, UK, 2007. With permission.)

## 8.5 MODEL VALIDATION AND COMPARISON WITH OTHERS

The mechanistic models described in Section 8.4 were tested by comparing model predicted with experimental survivor curves in response to constant and dynamic lethal temperature histories. For dynamic temperatures, the curves were predicted by simulation, using Arrhenius equations to vary rate constants with temperature. In separate experiments with dynamic temperature, sealed capillary tubes containing bacterial spores (*Bacillus stearothermophilus*) suspended in phosphate buffer solution were placed in an oil bath, and the temperature was varied manually. Tubes were removed at times spaced across the test interval and analyzed for survivor counts. Temperature histories of the oil bath were recorded (Sapru et al., 1993). The dynamic, lethal temperature for one of these experiments is shown at the top of Figure 8.5. Corresponding survivor data and survivor curves predicted by the various models in response to this dynamic temperature are given at the bottom of Figure 8.5. Note that activation of the significant subpopulation of dormant spores was predicted well by the early increase in population response shown by the Rodriguez and Sapru models. Both models performed much better overall than the other two. The conventional model, being limited to a single first-order death reaction, was incapable of predicting any increase in population.

## 8.6 APPLICATIONS OF MECHANISTIC MODELS TO THERMAL STERILIZATION

Advantages of the Rodriguez/Sapru model make it preferable to the classic model for many applications; this section provides an overview of its application. First, suitability of the model for representing a specific situation is assessed by ascertaining that it involves lethal heating of a homogeneous, single species/strain population of spores in potentially mixed dormant/activated states. If a different possibly more complex, situation exists, a new model should be developed consistent with the concepts and methods embodied in the Rodriguez/Sapru model as was exemplified herein and the model of a complex of normal and injured spores by Rodriguez et al. (1988). In the sequel, the

Rodriguez/Sapru model is assumed to be suitable to the situation addressed. Quantitative application of the Rodriguez/Sapru model requires species/strain specific, numerical values of rate constants $K_a$, $K_d$, and $K_{dl}$ and initial subpopulations $N_{10}$ and $N_{20}$, respectively, of dormant and activated spores. Initial subpopulations in a sample of untreated suspension or product are determined under the common assumption that all spores in a direct microscopic count (DMC) of the sample are viable (Gombas, 1987). Incubation of the sample and enumeration of CFU yield the initial number of activated spores, $N_{20}$, and the initial number of dormant spores, $N_{10}$, is calculated from

$$N_{10} = \text{DMC} - N_{20} \tag{8.8}$$

Rate constants for a specific species/strain of spores at specified temperature are obtained by analysis of an experimental, isothermal survivor curve for a suspension of those spores and known values of $N_{10}$ and $N_{20}$. Specifically, $K_a$, $K_c$, and $K_{d2}$ are estimated by nonlinear regression, using the procedure in SAS (SAS, 1985; Sapru et al., 1992) or by Levenburg–Marquardt (Press et al., 1986; Rodriguez et al., 1988, 1992), fitting survivor curve of the model to data defining the experimental curve. Initial estimates of $K_a$, $K_d$, and $K_{dl}$ required by the nonlinear procedures are appropriately calculated from the data by the method of successive residuals (Rodriguez et al., 1992).

Rate constants estimated at a single temperature apply only to that temperature; applications of the model for other constant and, especially, dynamic temperature regimes require estimates of the rate constants and, preferably, continuous functions describing them as functions of temperature over a range of temperature. It follows that isothermal experiments and estimations of rate constants must be performed at several temperatures over the prescribed range, and expressions relating rate constants to temperature must be found by regressing graphs of rate constants vs. temperature. This was done for *Bacillus subtilis* spores over 87°C–99°C by Rodriguez et al. (1992) and for *B. stearothermophilus* spores over 105°C–120°C by Sapru et al. (1992). In both cases, dependencies of activation and inactivation rate constants on temperature were described well by the empirical Arrhenius equation (Williams and Williams, 1973),

$$K = Ae^{-(E_a/RT)} \tag{8.9}$$

where

$K$ denotes a rate constant
$A$ is the frequency constant (time$^{-1}$)
$E_a$ is the activation energy (J/mol)
$T$ is the absolute temperature (K)
$R$ is the universal gas constant 8.314 (J/mol K)

Regression to estimate $A$ and $E_a$ is better done with

$$\ln(K) = \ln(A) - E_a/RT \tag{8.10}$$

when shown on an Arrhenius plot of rate constant data (semilogarithmic plot of $K$ against the reciprocal of absolute temperature).

Use of the Rodriguez/Sapru model at UHT conditions is contingent upon the ability to estimate valid rate constants in that range, and generation of isothermal, UHT survivor curves for parameter estimation is difficult. An approach to the matter is extrapolation to UHT of results obtained at lower, lethal temperature, using the Arrhenius equations established for that range. In a series of UHT experiments with *B. stearothermophilus* spores over 123°C–146°C by Sapru et al. (1992), rate constants in that range estimated with Arrhenius equations established at 105°C–120°C gave very good agreement between model predicted (by simulation) and experimentally enumerated numbers of survivors at the conclusion of UHT heating.

With rate constants and their dependencies on temperature for a specific species or strain of spores known, and with $N_{10}$ and $N_{20}$ known for dormant and activated subpopulations of those spores in a specific suspension or product, the dynamics of those subpopulations and the survivor curve caused by specific, lethal heating of the suspension or product may be estimated by computer simulation or analytical solution of the Rodriguez/Sapru model given in Table 8.1. Both simulation and analytical solution are readily accomplished by computer, and graphs of the temperature regime and response variables, e.g., $N_1$ and $N = N_2$, over the exposure interval are the most useful forms of output. During a simulation or analytical solution, rate constants are varied as temperature varies by means of the Arrhenius equations; those variations of rate constants are the only way temperature enters the model and affects population dynamics. Temperature may be constant or dynamic and in the low or UHT lethal range. Such analyses of the behavior of the model enable one to predict, understand, and interpret the dynamics and effectiveness of existing and proposed sterilization processes, and the Rodriguez/Sapru model should be a tool in the design and validation of new, thermal sterilization processes.

## 8.7 APPLICATION TO HIGH HYDROSTATIC PRESSURE STERILIZATION

It is well known that dormant bacterial spores are very resistant to high pressure. As stated by Gould (1983), "It is evident that ungerminated spores are resistant to hydrostatic pressures of well over 8000 atm, and yet much lower pressures than these, of the order of a few hundred atm, can initiate spore germination." Bacterial spores lose their resistance early in the germination process once they have been activated (i.e., heat shock), and become sensitive to high pressure when they exist as active vegetative cells.

The resistance of bacterial spores is lost half-way through the interval when birefringence is lost. Just as with thermal inactivation, mathematical models of high-pressure sterilization processes need to account for the fact that activated spores have a significantly different resistance than dormant spores. Therefore, a model of the inactivation and activation of bacterial spores undergoing high-pressure sterilization would be the same as the conceptual model for thermal inactivation shown previously in Figure 8.4. Equations for the inactivation rate and the dependency of the rate constant with respect to pressure and temperature can be derived and presented as follows (Rodriguez, 2005):

$$\frac{dN}{dt} = -kN \tag{8.11}$$

$$d\ln k = \left(\frac{\partial \ln k}{\partial T}\right)_T dT + \left(\frac{\partial \ln k}{\partial P}\right)_P dP \tag{8.12}$$

$$\left(\frac{\partial \ln k}{\partial T}\right)_P = \frac{E_a}{RT^2} \left(\frac{\partial \ln k}{\partial P}\right)_T = -\frac{\Delta V^*}{RT} \tag{8.13}$$

$$k = k_0 \exp\left\{-\left[\frac{\Delta V^*}{RT_0}(P - P_0) + \frac{E_a}{R}\left(\frac{1}{T} - \frac{1}{T_0}\right)\right]\right\} \tag{8.14}$$

The corresponding mathematical model for high pressure is

$$\mathrm{d}N_1/\mathrm{d}t = -(k\mathrm{d}D + kA)\, N_1 \ldots 1 \tag{8.15}$$

$$\mathrm{d}N_2/\mathrm{d}t = kA\, N_1 - k\mathrm{d}A\, N_2 \ldots 2 \tag{8.16}$$

The isothermal/isobaric response equations are

$$N_1[t] = \mathrm{e}^{-(kA + k\mathrm{d}D)^t} N_{10} \tag{8.17}$$

$$N_2 = \frac{\mathrm{e}^{-k\mathrm{d}At}\left(kA\left(\left(-1 + \mathrm{e}^{-(kA - k\mathrm{d}A + k\mathrm{d}D)t}\right)N_{10} - N_{20}\right) + (k\mathrm{d}A - k\mathrm{d}D)N_{20}\right)}{-kA + k\mathrm{d}A - k\mathrm{d}D} \tag{8.18}$$

These equations were used to predict the number of surviving spores of *C. botulinum* over time when subjected to a constant high hydrostatic pressure of 680 MPa at a temperature of 110°C. The smooth survivor curve shown in Figure 8.6 was predicted by the model (Equation 8.18), and can be compared with the actual experimental

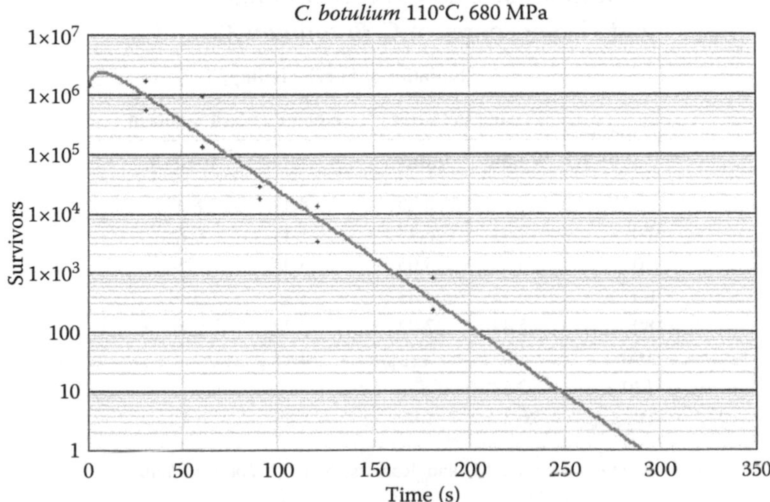


**FIGURE 8.6** Correspondence between experimental data points and calculated data from mathematical model (smooth curve) for spores of *C. Botulinum* exposed to combined high temperature (680 MPa) and temperature (110°C). Notice the activation "hump" predicted by the mathematical model.

data points. It is interesting to note how the model accounts nicely for the activation "hump" that should be expected from the initial "heat shock" experienced at the very onset of exposure to the combined lethal conditions.

## 8.8 APPLICATION TO STABILIZATION BY CURING SALTS (SPORE INJURY)

Spore injury is another type of transformation that can occur with bacterial spores that does not allow them to germinate and grow under normal conditions, but will allow them to "recover" when exposed to different (abnormal) conditions. Preservation of meats by salt curing is a process that takes advantage of this "injury" transformation. Review of the literature shows that injured spores cannot grow when curing salts are present in the growth medium. Spores that have not been injured (not exposed to sublethal moist-heat treatment) are not inhibited by the salts. Adding salts to the heating medium/substrate does not change the rate of injury/inactivation significantly.

Table 8.2 (Ingram and Roberts, 1971) shows that injured spores of *C. botulinum* will not grow in the presence of curing salts; that uninjured spores will grow regardless of the presence of curing salts; and that curing salts have no significant effect during heat treatment. Data show the need to injure the spores as a requisite to achieve shelf-stability. Therefore, the kinetics of the injury transformation determines the safety of cured-meat shelf-stable products.

**TABLE 8.2**
**Effect of Curing Salts (NaCl and $NaNO_2$) on Heat Resistance and Recovery of Injured[a] and Viable[a] Spores of *C. Botulinum* 33A**

| | Number of Survivors ($\log_{10}$) | | | |
|---|---|---|---|---|
| | Heated in Water | | Heated in Salt Solution | |
| Time (min at 95°C) | Viable | Injured | Viable | Injured |
| 0 | 8.36 | 8.36 | 8.42 | 8.38 |
| 20 | 7.41 | 6.73 | 7.45 | 6.89 |
| 40 | 6.39 | 5.05 | 6.98 | 5.54 |
| 66 | 5.75 | 3.33 | 5.95 | 3.61 |
| 80 | 5.42 | 2.72 | 5.20 | 2.66 |
| 100 | 4.05 | — | 4.19 | — |
| 120 | 3.52 | — | — | — |
| $D_{95°C\ (min)}$ | 26.3 | 14.9 | 24.3 | 14.1 |

*Source:* Ingram, M. and Roberts, T.A., *J. Food Technol.*, 6, 21, 1971.

[a] Plating medium for viable spores was RCA/Bicarbonate, and for injured spores was RAC/Bicarbonate plus 2% NaCl and 50 ppm $NaNO_2$.

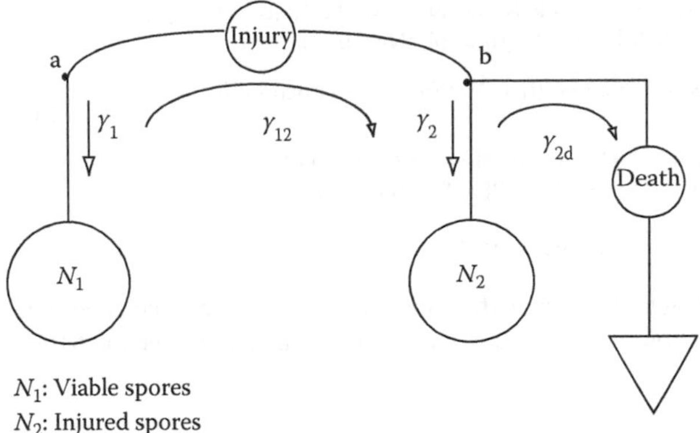


**FIGURE 8.7** Conceptual model describing the dynamics of a population of spores undergoing the transformations of injury and inactivation in a sequential form. This model describes the behavior presented in Table 8.2.

A conceptual model describing the dynamics of a population of spores undergoing the transformations of injury and inactivation in a sequential form is illustrated in Figure 8.7, and describes the behavior presented in Table 8.2. The mathematical model derived from the conceptual model in Figure 8.7 consisted of a system of differential equations as follows (Rodriguez, 2007):

$$\begin{aligned} \frac{dN_1}{dt} &= -K_1 N_1 \\ \frac{dN_2}{dt} &= K_D N_2 - K_1 N_1 \end{aligned} \tag{8.19}$$

With the following initial conditions:

$N_1(t = 0) = N_0$ (initial population density of the viable spores)

$N_2(t = 0) = 0$ (initial population density of the injured spores is zero)

The response equation corresponding to an isothermal process was developed by solving the above system of differential equations based upon initial conditions that the initial number of injured spores was zero, and the initial number of viable spores was found by plate count (enumeration) from a sample taken at zero time just prior to exposure ($N_0$).

The resulting solution (response equations) was obtained using Mathematica version 6.0.1.0 (Wolfram Research Inc., 2007), and is given in Figure 8.8 for isothermal conditions. Note that the number of viable spores is a function of the rate of injury, and the rate of injury determines the kinetics of the process as long as the curing salts are present at a constant effective concentration. Figure 8.9 shows the calculated values predicted by the model for injured spores (+) and viable spores in the presence of curing salt.

Injury is a transformation of bacterial spores that renders them incapable to germinate and grow under normal conditions. Injured spores may however germinate

```
In[1]:= DSolve [{Nv' [t] == -ki Nv [t], Ni' [t] == kd Ni [t] - ki Nv [t],
        Nv [0]==N0, Ni [0] ==0}, {Nv, Ni}, t]
```

$$\text{Out[1]= \{\{ Nv} \rightarrow \text{Function [\{t\}, } e^{-kit} \text{ N0] , Ni} \rightarrow \text{Function [\{t\}, } - \frac{e^{-kit} (-1 + e^{kd\,t + kit})\, ki\, N0}{kd + ki} \text{]\}\}}$$

```
In[2]:= {Nv' [t] == -ki Nv [t], Ni' [t] == kd Ni [t] - ki Nv [t],
      Nv [0] == N0, Ni [0] ==0} / . % //Simplify

Out[2]:= { {True, True, True, True} }
```

**FIGURE 8.8** Solution of the mathematical model for isothermal conditions, and the verification of the solution (response equations) using Mathematica version 6.0.1.0.

and grow under different conditions that compensate for some sublethal damage that prevents germination or growth or both from happening properly. In the case of *C. botulinum* in cured meats, injured spores will not grow in the presence of the appropriate concentration of curing salts. Injured spores will grow if the salts are not present as described, and uninjured spores will grow in the presence of the curing salts. This phenomenon has enabled industry to develop high pH (low acid) shelf-stable products that undergo minimal heat treatments with an accumulated $F_0$ less than 1 min.

## 8.9 SOURCES OF OTHER INFORMATION

Much of the material presented in this chapter serves as a summary of much more detailed work carried out by others (Rodriguez et al., 1988, 1992; Sapru et al.,

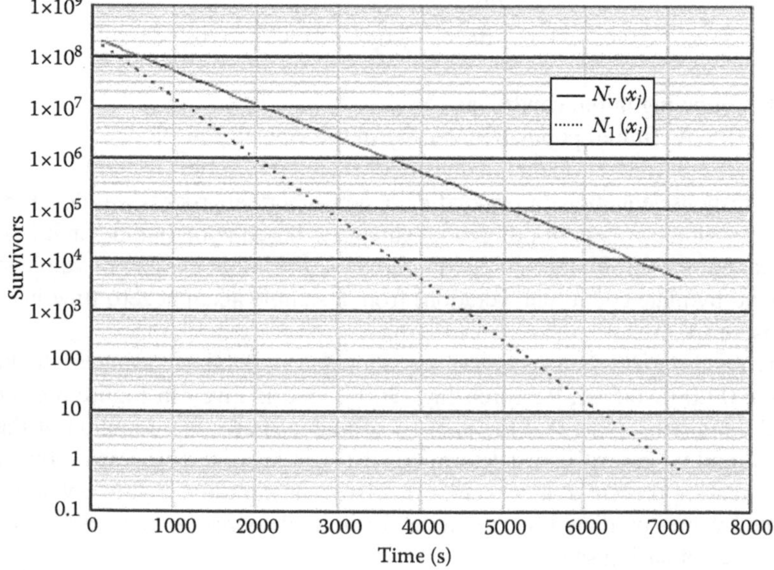


**FIGURE 8.9** Model-predicted survivor curves for injured (+) and viable spores of *C. botulinum A33* in the presence of curing salt solution (sodium chloride and sodium nitrite).

1992, 1993), and can be found in the cited references that are listed at the end of the chapter. Of particular interest should be a recent Institute of Food Technologists Summit Conference report, edited by Heldman (2003), on *Kinetics for Inactivation of Microbial Populations Emphasizing Models for Non-Log-Linear Microbial Survivor Curves*. This was a well-attended conference with considerable informal discussion and dialog concerning the different modeling approaches being promoted by different conference participants. The report highlights the various arguments presented, and concludes with recommendations on how kinetic data should be reported and presented in published scientific literature.

## REFERENCES


Aiba, S. and Toda, K. 1965. An analysis of bacterial spores' thermal death rate. *J. Ferment. Technol.* (Japan), 43(6): 528–533.

Chick, H. 1908. An investigation into the laws of disinfection. *J. Hygiene*, 8: 92–158.

Chick, H. 1910. The process of disinfection by chemical agencies and hot water. *J. Hygiene*, 10: 237–286.

Gombas, D.E. 1987. Bacterial sporulation and germination. In: Montville, T.J. (Ed.), *Topics in Food Microbiology*, Vol. I (Concepts in Physiology and Metabolism). CRC Press, Boca Raton, FL, pp. 131–132.

Gould, G.W. 1983. Mechanisms of resistance and dormancy. In: Hurst, A. and Gould, G.W. (Eds.), *The Bacterial Spore*, Vol. 2. Academic Press, London, p. 181.

Gould, G.W. 1984. Injury and repair mechanisms in bacterial spores. In: Andrew, M.H.E. and Russel, A.D. (Eds.), *The Revival of Injured Microbes*. Academic Press, New York, p. 199.

Heldman, D.R. 2003. Kinetics for inactivation of microbial populations emphasizing models for non-log-linear microbial survivor curves, IFT Summit Conference Report, Institute of Food Technologists, Orlando, FL, January 12–14, 2003.

Ingram, M. and Roberts, T.A. 1971. Application of the '*D*-concept' to heat treatments involving curing salts. *J. Food Technol.*, 6: 21–28.

Kellerer, A.M. 1987. Models of cellular radiation action. In: Freeman, R.G. (Ed.), *Kinetics of Nonhomogeneous Processes*. John Wiley & Sons, New York, pp. 305–375.

Keynan, A. and Halvorson, H. 1965. Transformation of a dormant spore into a vegetative cell. In: Campbell, L.L. and Halvorson, H.O. (Eds.), *Spores III*. American Society of Microbiology, Washington, DC, p. 174.

Lewis, J.C., Snell, N.S., and Alderton, G. 1965. Dormancy and activation of bacterial spores. In: Campbell, L.L. and Halvorson, H.O. (Eds.), *Spores III*. American Society for Microbiology, Washington, DC, p. 47.

Press, W.H., Flannery, B.P., Teukolsky, S.A., and Vetterling, W.T. 1986. *Numerical Recipes: The Art of Scientific Computing*. Cambridge University Press, Cambridge.

Rodriguez, A.C. 1987. Biological validation of thermal sterilization processes, PhD dissertation, Agricultural and Biological Engineering Department, University of Florida, Gainesville, FL.

Rodriguez, A.C. 2005. High pressure sterilization of pharmaceutical products. Presented in Session C3 of the 2005, Annual Meeting/Conference of the Parenteral Drug Association, Chicago, IL, April 4–6.

Rodriguez, A.C. 2007. Introduction to the Food Safety Objective (FSO) approach. Presented at the 26th Annual Conference of the Institute for Thermal Processing Specialists, San Antonio, TX, February 2007.

Rodriguez, A.C., Smerage, G.H., Teixeira, A.A., and Busta, F.F. 1988. Kinetic effects of lethal temperatures on population dynamics of bacterial spores. *Trans. ASAE*, 31(5): 1594.

Rodriguez, A.C., Smerage, G.H., Teixxeira, A.A., Lindsay, J.A., and Busta, F.F. 1992. Population model of bacterial spores for validation of dynamic thermal processes. *J. Food Proc. Eng.*, 15: 1–30.

Sapru, V., Teixeira, A.A., Smerage, G.H., and Lindsay, J.A. 1992. Predicting thermophilic spore population dynamics for UHT sterilization processes. *J. Food Sci.*, 57(5): 1248–1252, 1257.

Sapru, V., Smerage, G.H., Teixeira, A.A., and Lindsay, J.A. 1993. Comparison of predictive models for bacteria spore population response to sterilization temperatures. *J. Food Sci.*, 58(1): 223–228.

SAS Institute, Inc. 1985. SAS User's Guide: Statistics, 5th edn., SAS Inst., Inc., Cary, NC.

Shull, J.J., Cargo, G.T., and Ernst, R.R. 1963. Kinetics of heat activation and thermal death of bacterial spores. *Appl. Microbiol.*, 11: 485.

Stumbo, C.R. 1973. *Thermobacteriology in Food Processing*, 2nd edn. Academic Press, New York.

Teixeira, A.A. 2007. Mechanistic Models of Microbial Inactivation Behavior in Foods, chapter 10. In: Bruhl, S., Van Gerwen, S., and M. Zwietering (Eds.), *Modeling Microorganisms in Food.* Woodhead Publishing Ltd, Cambridge, UK, pp. 198–213.

Watson, H.E. 1908. A note on the variation of the rate of disinfection with change in the concentration of the disinfectant. *J. Hygiene*, 8: 536.

Williams, V.R. and Williams H.B. 1973. *Basic Physical Chemistry for the Life Sciences*, 2nd edn. W.H. Freeman and Co, San Francisco.

Wolfram Research, Inc. 2007. Champaign, IL. http://www.wolfram.com/services/.

# 9 Modeling Heat Transfer in Thermal Processing: Retort Pouches

*Alejandro Amézquita and Sergio Almonacid*

## CONTENTS



## 9.1 INTRODUCTION

The retort pouch is a flexible laminate that can be thermally processed in a similar way that traditional cans are processed. Consequently, food products packaged in retort pouches are shelf stable although in some cases, the product and process are designed for stability at chilled temperatures. However, pouches provide partial oxygen/water insulation, and the seaming integrity is weaker than that of cans. The laminate materials have barrier properties to reduce oxygen and moisture transfer between the packaged food and the storing environment and shelf life mainly depends on pouch integrity and permeability; in general, it goes from 6 months to 3 years, depending on the kind of food, pouch materials, and construction.

Development of the pouch began in the United States in early 1950s, mainly by the research done in the U.S. Army Natick Development Center, for military use in packaged combat rations. Reynolds Metal Co. and Continental Can Co. were also developing the retort pouch at the same time. During the 1960s, test runs were made with encouraging results. In the late 1960s, the first food products in retort pouches were produced in Japan and Europe, they used the technology developed in the United States under a licensing arrangement. In 1977, Food and Drug Administration (FDA) gave approval to Continental Group Inc. (former Continental Can Co.) and Reynolds Metal Co. for their respective developed retort pouch. During the 1980s, low-acid foods thermally sterilized in retort pouches were successfully marketed in Japan and Europe where the government regulations were less stringent than in the United States (López, 1981). However, in the last decade, the retort pouch sector in the United States, which is not as advanced as in Europe and Japan, has been growing at double and even at triple digits. These numbers are far above those of a decade ago when the military's Meal, Ready to Eat (MRE) rations and a few pet foods were at the periphery of a "risky" technology. In Japan, the production is estimated to be 1.5–3 billion pouches per year. In 2006, global retort pouch consumption exceeded 10 billion pouches per year, and it is expected to reach nearly 19 billion pouches per year in 2011 (FLEX-NEWS, 2007). With boosts from increased military and humanitarian needs, offshore processing/packaging facilities and consumer acceptance, retort pouch/tray technologies appear to be entering the mainstream of food preservation and packaging. Moreover, hurdles for retort pouches to enter the market, such as line speed, cost, and pouch size availability, are being eliminated and are replacing cans at an ever higher rate; in Japan, sterilized foods in retort pouch are already half of the market, sharing it with traditional cans. Retort pouches are already in the market with tuna, pet foods, seafood products, rice and pasta products, liquid soups and sauces, meats, and vegetables. It is evident that value is being added through packaging (Huston, 2005).

## 9.2 MATERIAL DEVELOPMENTS, POUCH STRUCTURE, AND CRITICAL ASPECTS

Retort pouches are normally multilayer laminated packs having three or four layers. There are approximately 16 basic laminating materials with 100 different possible combinations (CFIA, 2002). The laminate constructions typically have an inner food contact heat seal layer (normally polypropylene [PP]), a barrier layer (aluminum foil, silicon oxide [SiOx], aluminum oxide [AlOx], or ethylene vinyl alcohol [EVOH]), an optional nylon layer for increased mechanical strength, and an outer polyester [PET] layer, which is often reverse printed. The layers are combined by an adhesive, usually a polyurethane-based adhesive. The selection and specification of the film for a particular application depends on many factors, including but not limited to (1) pouch type (e.g., institutional pillow pack, gusseted stand-up pouch [sometimes called doypack], etc.), (2) type of product to be packed, (3) barrier properties required, (4) intended heat treatment (e.g., sterilization, pasteurization, hot-fill, etc.), (5) type of filling and sealing equipment, (6) intended consumer usage (e.g., microwaveability, ease of opening, etc.), (7) supply chain requirements (e.g., ambient, refrigerated, etc.), and (8) target shelf life.

Commercially available films for longest product shelf life at ambient conditions typically consist of four-ply constructions, where aluminum is used as the barrier layer. The four layers in these laminates are usually as follows (from inside to outside) (1) PP (40–100 μm), (2) oriented polyamide (OPA, 12–25 μm), (3) aluminum (ALU, 8–12 μm), and (4) PET (12 μm). PP on the inside is used as a retort-stable heat sealing layer. The thickness of the PP layer mainly influences the mechanical strength, seal strength, and stiffness of the film and it must allow tight seals. The OPA layer provides mechanical strength (e.g., puncture resistance, drop resistance, and seal strength), and protects the bond to aluminum against potential aggressive ingredients in the food product. ALU is the barrier material mainly against oxygen, light, water vapor, and aroma. Finally, PET on the outside acts as a protecting layer, provides stiffness and tensile strength, and serves as a sealing-resistant substrate for printing. Generally, the thickness of these layers needs to be adapted according to specific applications (i.e., pouch size, filling machine, and mechanical strength and stiffness requirements). A good example of this construction is included in the performance specification MIL-PRF-44073F of the U.S. Army for preformed ration pouches, which have four layers consisting of, from inside to outside, PP (76–102 μm thick), OPA type 6 (15 μm thick), ALU foil (9–18 μm thick), and PET (12 μm thick) (USDOD, 2001). This performance specification stipulates that the ALU foil layer and the OPA layer may be in either order. For some applications, three-ply ALU-based laminates without the OPA layer can be used. However, these constructions have lower mechanical strength and therefore it is essential that the entire process from packaging to final usage is carefully monitored and controlled. For stand-up pouches, it is advisable to use at least a four-ply construction for the gusset.

In some applications, where the intended use of the product is to be heated in a microwave oven, or where the product needs to be visible, there are transparent films that do not include the ALU barrier. In this case, the barrier is provided by a coated PET layer. The coating usually consists of a very thin inorganic SiOx- or AlOx-based layer, or an organic coating (e.g., modified polyacrylic acid). In general, these films do not allow longer shelf lives than aluminum-based constructions. The structure of these transparent laminates is typically as follows (from inside to outside): (1) PP (40–100 μm), (2) OPA (12–25 μm), and (3) coated PET (12 μm). It is also common to find structures where the OPA layer is coated with SiOx as barrier instead of the coated PET layer. As with the ALU-based laminates, the thickness of each layer is determined by the application.

Additional information about potential pouch and laminate defects, examination and evaluation procedures, and pack integrity factors is beyond the scope of this chapter. However, the authors recommend some sources of additional information for interested readers. A comprehensive document covering in detail the critical aspects mentioned above is available free of charge via the Internet from the Canadian Food Inspection Agency Web site (CFIA, 2002). Additionally, there are other guidelines on good manufacturing practices for heat processing of foods in retort pouches available for purchase from research institutions and industry associations such as Campden & Chorleywood Food Research Association (CCFRA, 2006) and Food Products Association (FPA, 1985).

## 9.3 MAIN ADVANTAGES AND DISADVANTAGES OF RETORT POUCHES COMPARED WITH METAL CANS OR GLASS JARS

Retort pouches offer benefits for food manufacturers, consumers, and retailers alike. When compared to tin cans or glass jars, some of the main advantages of retort pouches can be listed as follows:

- A thin profile (flat geometry) that provides a small cross-sectional dimension that enables rapid heat transfer during processing (i.e., sterilization or pasteurization).
- Reduction of retort cycle times, which results in energy savings and reduced overcooking of the product (with improvement of organoleptic attributes and nutritional quality).
- Low weight and reduced storage space of empty pouches before processing, which results in lower transportation and storage costs for manufacturers.
- Reduced weight and volume of finished food product, which facilitates transportation and storage by consumers with limited space available (e.g., astronauts during space missions, army soldiers in the field, general public practicing outdoor activities, etc.).
- Consumer convenience (e.g., safe to open, microwaveability [transparent pouches], minimum requirement for opening tools, etc.).
- Comparable shelf lives to foods packed in metal cans.
- Savings in shelf space during retail display.

However, retort pouches also have some disadvantages when they are compared against metal cans or glass jars; some of the main disadvantages are

- Filling of pouches and loading into crates is usually slower and more complex than cans or glass containers.
- Establishing the scheduled process and controlling critical factors is more complex.
- Pouches are relatively sensitive to snagging and cutting impacts, as well as to a softening effect during heating; therefore, the pack integrity can be more easily compromised than in rigid containers.
- High capital investment, especially in equipment necessary for appropriate pouch handling during loading and unloading of the retort (e.g., stackable trays, racks) and during postthermal process handling, to ensure pack integrity.

## 9.4 HEAT TRANSFER MODELING IN RETORT POUCHES

Mathematical models are tools that can be used to describe and understand physical processes; in combination with experiments can save time and money. When adequately validated, they can provide process optimization, predictive capability, improved process automation, and control possibilities. Scientists and engineers now have access to modern modeling technologies and to a wide range of techniques and solving tools. Modern computers allow the use of mathematical models as integral part of most studies in food science and engineering.

Processing of heat-preserved foods in flexible pouches has gained considerable commercial relevance in recent years worldwide. The market of ambient-stable soups, sauces, microwave-ready rice, and other popular food products has been progressively "reinvented" to respond to the growing consumer demand for convenient, fresher, and healthier products, with improved quality and nutritional content. This has led to a rapid growth of the market of foods in retort pouches, both for retail and foodservice operations.

From a heat transfer viewpoint, retort pouches offer advantages over conventional cylindrical containers, thereby affording improved product quality. This is mainly due to simple geometrical considerations. That is, pouches loaded into retorts have a characteristic thin and "flat" profile when compared to traditional pack formats (e.g., cans or glass jars), and therefore the surface area-to-volume ratio is much larger for pouches than for those traditional (i.e., cylindrical) geometries. This leads to a faster rate of heat transfer to the slowest heating zone in the pack, which allows the achievement of equivalent lethality targets in shorter retort cycle times under identical retort temperature conditions. This generally also results in lower product cook values, higher retention of heat-sensitive nutrients and pigments, and overall better quality.

During in-pack heat processing of conduction-heating foods, the knowledge about the location of the slowest heating zone in the container facilitates the collection of temperature history data by direct measurement. This has led to various methods of thermal process evaluation such as the General Method or the Ball formula method(s), whereby time–temperature data collected during heat penetration tests are used directly to assess whether or not the process delivers the minimum lethality requirements.

However, in some instances, heat transfer modeling has proven to be a useful tool to gain a more thorough understanding of temperature distribution and flow behavior inside packs where heating within the pack is governed by natural convection (i.e., buoyancy-driven heat transfer), rather than conduction. In this case, the location of the slowest heating zone is shifted from the geometric center, and model predictions can save a considerable amount of experimentation time. Heat transfer modeling is also useful for irregularly-shaped containers where the location of the slowest heating zone is not obvious. These applications of heat transfer modeling are particularly useful in the case of liquid foods that are heat processed in retort pouches, such as soups, bouillons, sauces, fruit juices, among other products. Sections 9.5 and 9.6 address the principles and main considerations for modeling heat transfer in retort pouches.

## 9.5 MATHEMATICAL MODEL

When modeling heat transfer of biomaterials, all the relevant dependent variables of interest involved in the fundamental heat, mass, and momentum conservation equations seem to obey a generalized conservation principle. If the dependent variable is denoted by the Greek letter $\phi$, the general differential equation can be written (in vector notation) as

$$\frac{\partial}{\partial t}(\rho\phi)+\nabla\cdot(\rho\mathbf{v}\phi)=\nabla\cdot(\Gamma\nabla\phi)+S \tag{9.1}$$

where
$t$ is the time
$\rho$ is the density of the material
$\mathbf{v}$ is the velocity vector
$\Gamma$ is the diffusion coefficient
$S$ is the source term

The quantities $\Gamma$ and $S$ are specific to a particular meaning of $\phi$. On the left-hand side of Equation 9.1, the first term is the unsteady or transient term, and the second is the convection term. On the right-hand side, the first term is the diffusion term and the second is the source term. The dependent variable $\phi$ can stand for a variety of different quantities, such as the temperature or the enthalpy of the material, a velocity component, or the mass fraction of a chemical species. Accordingly, for each of these variables, an appropriate meaning will have to be assigned to the diffusion coefficient ($\Gamma$) and the source term ($S$). The use of the diffusion term ($\Gamma\nabla\phi$) does not limit the general equation to gradient-driven diffusion processes, though. In such cases, an appropriate diffusion quantity can be expressed as part of the source term ($S$). The equations presented below for different heat transfer mechanisms occurring during the processing of foods in retort pouches are specific forms of Equation 9.1.

## 9.6 HEAT TRANSFER MECHANISMS

When modeling heat transfer in foods packed in retort pouches, there are two main mechanisms that can govern this phenomenon: natural convection (i.e., buoyancy-driven heat transfer) and conduction. Natural convection occurs in liquid foods, in which the product will initially heat up by conduction during the early stages of the retort cycle, but as the temperature increases, the density of the liquid decreases at the pouch walls, and buoyancy forces are created that lead to liquid movement. Throughout heating, these buoyancy forces are opposed by the viscosity of the liquid. The velocity of the convective current depends on the strength of the buoyancy forces and the magnitude of resistance to flow by the liquid's viscosity. The movement of liquid continues as long as the buoyancy forces are higher than the viscous forces. Conversely, heat transfer by conduction occurs in solid foods or in liquid products with a viscosity sufficiently high to prevent buoyancy-driven flow.

### 9.6.1 Governing Equations and Assumptions

To model natural convection phenomena in a pouch, the temperature and flow fields need to be coupled; hence, the principles of conservation of mass, momentum, and energy in the liquid need to be considered together. The mass continuity equation and Navier–Stokes equations are used to describe flow field. The energy equation is used to solve for the temperature field. Liquid foods are relatively incompressible, and therefore, to solve for natural convection in retort pouches, the assumption of incompressible flow is used to describe the heat transfer model. Under the incompressible flow assumption, the density ($\rho$) is constant, and the following basic equations are used (in vector notation):

Mass conservation equation (continuity equation)

$$\nabla \cdot \mathbf{v} = 0 \tag{9.2}$$

Momentum (Navier–Stokes equations, three equations in mutually perpendicular directions)

$$\rho\left(\frac{\partial \mathbf{v}}{\partial t} + \mathbf{v} \cdot \nabla \mathbf{v}\right) = -\nabla p + \mu \nabla^2 \mathbf{v} + \mathbf{F} \tag{9.3}$$

Energy equation

$$\rho c\left(\frac{\partial T}{\partial t} + \mathbf{v} \cdot \nabla T\right) = \nabla \cdot (k \nabla T) \tag{9.4}$$

where

- $p$ is the pressure
- $\mu$ is the apparent dynamic viscosity
- $\mathbf{F}$ is the vector of body forces
- $c$ is the specific heat of the food material
- $k$ is the thermal conductivity of the food material
- $T$ is the temperature

The other terms in Equations 9.2 through 9.4 have been defined earlier.

The left-hand side of Equation 9.3 represents the transient and inertial terms. On the right-hand side of this equation, the first term is due to pressure gradients, the second to viscous forces, and the third term to body forces (i.e., gravity force only in the case of heat transfer in retort pouches). In Equation 9.4, the left-hand side represents the transient and convective terms. The term on the right-hand side of this equation is due to heat conduction (diffusion) through the fluid.

As mentioned before, under the incompressible flow assumption, the density ($\rho$) is assumed to be constant. However, in buoyancy problems, such as natural convection heat transfer, it is common to use the Boussinesq approximation, by which the buoyancy force caused by density differentials is only applied to the momentum equations. In this case, density variations are considered sufficiently small to be neglected in the inertial terms, and are only applied in the body force term in Equation 9.3. In the Boussinesq approximation, the density can be expressed as

$$\rho = \rho_{\text{ref}}\left[1 - \beta\left(T - T_{\text{ref}}\right)\right] \tag{9.5}$$

where

- $\beta$ is the thermal expansion coefficient of the liquid (units of 1/absolute temperature)
- $\rho_{\text{ref}}$ is the density at reference condition
- $T_{\text{ref}}$ is the temperature at reference condition (absolute units)

In Equation 9.5, it is a common approach to use the initial conditions as reference values for $T_{ref}$ and $\rho_{ref}$. The density variation described by Equation 9.5 is applied to the vertical motion equation (i.e., typically the $y$-coordinate equation in Cartesian coordinates) in the body force term, where ρ multiplied by $g$, the acceleration due to gravity. The Boussinesq approximation is accurate as long as the changes in actual density are small.

Equations 9.2 through 9.4 are valid under the following assumptions: (1) there is no heat generation due to viscous dissipation, (2) no-slip condition at the pouch inside wall, and (3) the Boussinesq approximation is valid.

In the case of foods where conduction governs the heat transfer phenomenon, Equations 9.2 through 9.4 reduce to a single equation, as follows:

$$\rho c \frac{\partial T}{\partial t} = \nabla \cdot (k \nabla T) \tag{9.6}$$

Boundary conditions for Equations 9.2 through 9.4 and Equation 9.6 will be discussed separately later on this chapter. Initial conditions for the same equations can be written as $T(x, y, z, t=0) = T_{ref}$, $v_x = 0$, $v_y = 0$, and $v_z = 0$.

### 9.6.2 Thermophysical Properties

In addition to density (described above), the other thermophysical properties involved in Equations 9.2 through 9.6 can be taken as constants or as functions of temperature. When modeling heat transfer in retort pouches, it is a common approach to keep constant the value of the specific heat ($c$), the thermal conductivity ($k$), and the thermal expansion coefficient (β). This assumption can be considered valid because variations in the values of these properties in food materials over the range of temperatures of interest (i.e., 20°C–130°C) are relatively small to affect considerably the accuracy of predictions. However, many of the commercially available partial differential equation (PDE) solvers and computational fluid dynamics (CFD) software packages allow for temperature dependency of $c$ and $k$ by means of user-defined polynomial equations or by using interpolation (e.g., piecewise linear) between input values at temperatures specified by the user. In this case, usually four or five values are enough to account for the temperature dependency. These values at specific temperature points can be obtained either by experimental measurements or estimated by using empirical correlations based on mass or volume fractions of the main food components (see, for instance, Choi and Okos, 1986).

When natural convection is the governing heat transfer mechanism, the temperature dependency of the apparent viscosity (μ) of the liquid food must be considered, because the variation of μ with temperature is high when compared with the other thermophysical properties mentioned above. In some instances, μ is described as a function of temperature only (Newtonian approximation) even if the food is a non-Newtonian fluid. This assumption is valid for foods with a pseudoplastic (i.e., shear-thinning) rheological behavior. The validity of this assumption is based on the fact that in buoyancy-driven flow (i.e., natural convection) the shear rates are relatively small, and under these conditions, the apparent viscosity is constant with changing shear rates (Steffe, 1996). In this case, the temperature dependency of μ is incorporated in the model either by allowing the PDE or CFD software to interpolate between known input values at specified temperatures, or by fitting experimental data to a polynomial equation, where temperature is the only independent variable (see, for instance, Abdul Ghani et al., 2001, 2003). When non-Newtonian behavior is considered in the heat transfer model, then it is necessary to use a rheological model that takes account of the effects of temperature and shear rate on the apparent viscosity of the food. One such model described by Steffe (1996), which has commonly been used to model natural convection heat transfer in canned liquid foods (see, for instance, Kumar et al., 1990; Kumar and Bhattacharya, 1991), can be written as follows:

$$\mu = f(T,\dot{\gamma}) = K_T \exp\left(\frac{E_a}{RT}\right)(\dot{\gamma})^{\bar{n}-1} \tag{9.7}$$

where

- $\bar{n}$ is the average flow behavior index (based on all temperatures used for model development—no units)
- $E_a$ is the energy of activation for viscosity (determined from experimental data—units of kJ/kg mol in SI system)
- $R$ is the universal gas constant (units of kJ/kg mol K in SI system)
- $K_T$ is the consistency coefficient (units of Pa s$^n$ in SI system)
- $\dot{\gamma}$ is the shear rate (1/s)

### 9.6.3 Boundary Conditions

When modeling heat transfer in retort pouches, the boundary conditions for Equations 9.2 through 9.4 and Equation 9.6 are usually of the first kind (i.e., Dirichlet boundary conditions) or the second kind (i.e., Neumann boundary conditions). As mentioned before, a no-slip condition is assumed at the inside walls, and therefore the velocity boundary conditions are $v_x = 0$, $v_y = 0$, and $v_z = 0$. Regarding the temperature boundary conditions, they can be easily specified. The most common approach is to prescribe either a temperature (i.e., the wall temperature value) or a convection flux (via a heat transfer coefficient and the temperature of the surrounding medium). As the retort temperature is variable during the come-up and cooling periods, a time-varying temperature boundary condition can be defined. This is a sensible approach as it is relatively easy to obtain retort temperature data from direct measurements during temperature distribution and/or heat penetration studies. An example of this approach is discussed in Section 9.7. In general, the thermal boundary conditions are applied directly to the liquid boundaries (i.e., this assumption is valid based on the very small thickness of film laminates used in retort pouches).

## 9.7 APPLICATIONS OF HEAT TRANSFER MODELS IN PROCESSING OF FOODS PACKED IN RETORT POUCHES

Efforts to model natural convection heating during sterilization of liquid foods in cans have been reported in the literature since the mid-1980s (see, for instance, Datta and Teixeira, 1988). Many of the early publications were limited to the relatively simple problems due to limitations in computational capabilities. Over the years, along with the rapidly increasing computing power, other more sophisticated CFD approaches have been applied to study heat transfer in more complex systems, dealing, for instance, with non-Newtonian foods, agitated systems, and irregular pack formats or geometries. Tattiyakul et al. (2001) modeled non-Newtonian flow and heat transfer in an axially rotating can. The authors concluded that for rotating cans, the slowest heating zone in the container was different than for stationary cans, with a strong influence of the rotation speed. Abdul Ghani and Farid (2006) modeled the effect of natural convection on the heating of a solid–liquid mixture (pineapple slices in juice) in a cylindrical can, showing that the slowest heating zone was located below the geometrical center of the can (about 30%–35% of the can height from the bottom), but at a higher distance from the bottom than if the can were filled with liquid only. End-over-end rotation in retort processes is a common application for heat processing of foods in metal cans or glass jars (either sterilization or pasteurization processes) and successful attempts to model coupled flow and heat transfer have been reported (James et al., 2006).

Regarding specific models for foods in retort pouches, there are a series of publications from the University of Auckland, New Zealand, which report modeling of natural convection heat transfer in retort pouches for a beef–vegetable soup (Abdul Ghani et al., 2002) and a carrot–orange soup (Abdul Ghani et al., 2001, 2003), using a three-dimensional transient model solved in a commercial CFD package (PHOENICS, Cham Ltd., London). Predictions from these models showed a shifting of the slowest heating zone from the geometric center toward the bottom of the pouch, as previously shown by numerical simulations of natural convection in cans (see, for instance, Datta and Teixeira, 1988; Kumar et al., 1990; Abdul Ghani et al., 1999). Simpson et al. (2004) developed and validated a heat conduction model for fish sterilized in retort pouches using an axisymmetric transient model (cylindrical coordinates) solved by an explicit finite difference (FD) scheme. This validated model was subsequently used for process optimization by searching for a retort temperature profile that resulted in the shortest process time and the minimum quality variation inside the pack.

Results of simulations carried out in our laboratory are shown in Figure 9.1 as temperature contours for a conduction-heating (Figure 9.1a) and a natural convection-heating (Figure 9.1b) foods in a stand-up pouch (or doypack). The objective of this simulation was to compare the effect of the governing heat transfer mechanism on intrapouch lethality distributions resulting from temperature gradients in the product during a typical sterilization cycle in a water spray retort (operated in still mode). For both cases, the come-up time was 30 min, and the retort temperature was 122°C during the sterilization period. The sterilization times at 122°C required to achieve the minimum lethality design target (i.e., $F_0$ = 5 min at the end of the sterilization period or "steam-off") were 20 and 7 min for the conduction-heating and convection-heating foods, respectively. Equations 9.2 through 9.6 were solved in a commercial CFD package (Fluent 6.1, Fluent Inc., Lebanon, New Hampshire), using a three-dimensional meshed geometry consisting of 106,000 cells. The thermophysical properties (i.e., $\rho$, $c$, and $k$) were considered constants and estimated from the composition of the foods by the thermal properties predictor COSTHERM (Miles et al., 1983). For the natural convection case (i.e., chicken bouillon), viscosity values were measured experimentally at four temperatures between 30°C and 90°C

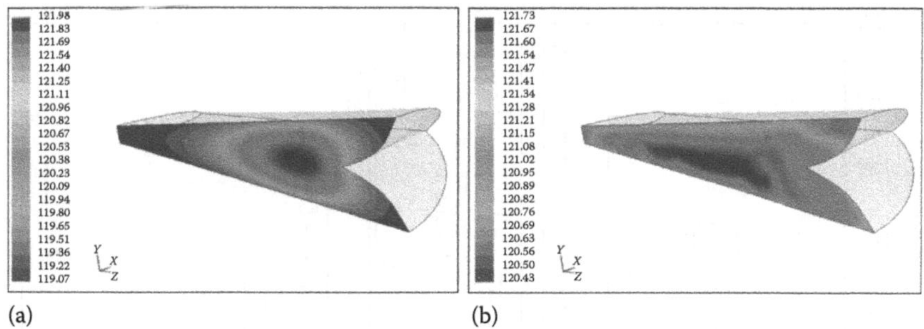


**FIGURE 9.1** Temperature (°C) contour plots at steam-off in a stand-up pouch sterilized a water spray retort operated in still mode (come-up time = 30 min, retort temperature during sterilization = 122°C) containing (a) a conduction-heating food (i.e., creamy white sauce) and (b) a natural-convection heating food (i.e., chicken bouillon). Contour plots are shown on a vertical plane located at the center of the pouch. Identical boundary conditions were used in both cases to simulate the same location inside the retort.

and input into the CFD software for piecewise linear interpolation. The Boussinesq approximation was used to model the buoyancy-driven flow, using a constant thermal expansion coefficient ($\beta = 1.8 \times 10^{-4}$ K$^{-1}$). The temperature boundary condition was defined as a prescribed convection flux, using a time-varying profile for the external temperature (from retort temperature data collected experimentally), and a heat transfer coefficient of 400 and 100 W/m$^2$·K for the heating and cooling periods, respectively. Boundary conditions were identical in both cases to simulate the same pouch location inside the retort.

The results of these simulations offer valuable insights about the location of the slowest heating zone in the pouch. It is clearly noticeable that in the conduction-heating food (i.e., a creamy white sauce), as expected, the slowest heating zone is located in the horizontal center plane (the *XZ* plane) of the pouch and at a distance of approximately 20%–25% from the gusset area (along the pouch length or *z*-direction). This information can be useful as a guide for accurate location of temperature probes during heat penetration tests for assurance of food safety. Furthermore, in the natural convection case, there is a depression of the slowest heating zone to a region located at approximately 20%–25% of the pouch height from the bottom face (which is in contact with the tray holding the pouch). The models were validated comparing the predictions at the slowest heating zone against heat penetration data, finding good agreement between predicted and measured values.

Additionally, the simulation results offer very valuable information about the effect of the governing heat transfer mechanism on the distribution of integrated lethality values. This effect can be visualized in Figure 9.2, which depicts histograms of integrated lethality values (expressed as $F_0$ values) at steam-off for the two scenarios shown in Figure 9.1. The $F_0$ values were calculated using the nodal temperature outputs (°C) from 2000 randomly selected cells at each time step ($\Delta t$ = 1 s) as inputs to the following equation:

$$F_0 = \int 10^{\left(\frac{T-121.1}{10}\right)} \cdot \mathrm{d}t \tag{9.8}$$

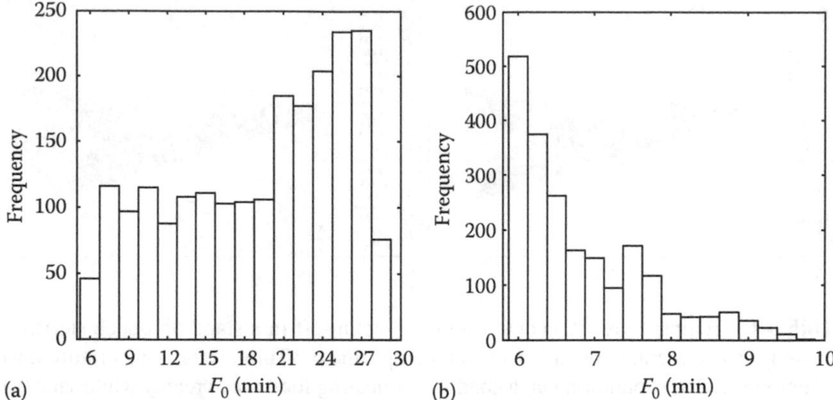


**FIGURE 9.2** Histograms of $F_0$ values at steam-off within a stand-up pouch sterilized in a water spray retort operated in still mode (come-up time = 30 min, retort temperature during sterilization = 122°C) containing (a) a conduction-heating food (i.e., creamy white sauce) and (b) a natural convection-heating food (i.e., chicken bouillon). Identical boundary conditions were used in both cases to simulate the same location inside the retort.

The $F_0$ values inside the conduction-heating pouch (Figure 9.2a) ranged from 5.4 to 28.2 min, whereas the $F_0$ values for the natural convection-heating case (Figure 9.2b) ranged from 5.6 to 9.6 min, with more than 90% of values below 8.0 min. Clearly these differences will also be reflected on the quality gradients for these two foods.

Finally, heat transfer modeling in retort pouches can also be useful in assessing the effect of temperature variations on lethality distributions throughout a retort vessel. Figure 9.3 illustrates an example of temperature data collected during a temperature distribution study in a water spray retort operated in still mode, and the effect of pouch location inside the retort on $F_0$ value distributions. As the hot water used

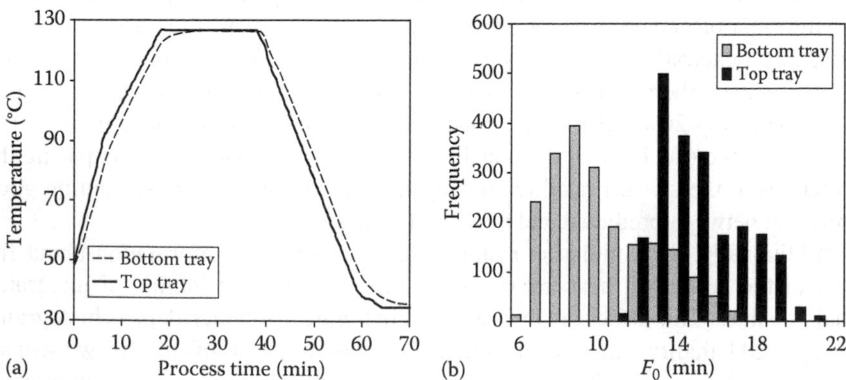


**FIGURE 9.3** Effect of pouch location within a retort basket on intrapack distribution of $F_0$ values at steam-off during sterilization in a water spray retort operated in still mode (come-up time = 18 min, retort temperature during sterilization = 126°C). Panel (a): retort temperature profiles collected during a temperature distribution study and used to define the boundary conditions of the heat transfer model. Panel (b): histograms of $F_0$ values calculated inside each stand-up pouch.

as heating medium comes from the top of the retort, the fastest heating tray in each basket is usually the top tray. Conversely, the "cold spot" in each basket is usually located around the bottom trays, and the exact location could vary slightly depending on the specific conditions at each operation (e.g., loading pattern, orientation of pouches, tray design, etc.). Following a similar approach to that one presented in Figures 9.1 and 9.2, the two retort temperature profiles shown in Figure 9.3a were used to define the thermal boundary conditions of a heat transfer model for pouches containing a 2% modified corn starch solution. As described before, the output nodal temperatures were used to calculate intrapouch distributions of $F_0$ values (Figure 9.3b). It can be observed from this figure that, at steam-off (i.e., end of sterilization period), when the slowest heating pouch in the basket (i.e., a pouch located in the bottom tray) reaches the minimum lethality value ($F_0$=5.9 min), a pouch located on the top tray will have an $F_0$ value of 10.4 min at its slowest heating point, yielding a ratio between these two $F_0$ values of 1.7. This ratio is likely to be reduced at the end of the process due to the lethality accumulated during the cooling period (greater for the pouch located on bottom tray as it cools slower). However, cooling lethality is not often used to establish a scheduled heat process design target due to difficulties in controlling the temperature variations throughout the retort. The ratio between the maximum and the minimum $F_0$ value throughout a retort should, ideally, be as close to 1.0 as reasonably possible to ensure product quality uniformity.

## 9.8 MODEL SOLUTION AND IMPLEMENTATION

The equations presented in the previous sections are typically solved using numerical methods. For conduction-dominated heat transfer, techniques such as the finite element (FE) and the FD methods are commonly used. These techniques can be easily implemented for relatively simple problems (e.g., one- or two-dimensional problems involving regular geometries), using commercially available spreadsheet software packages, basic PDE solvers, or writing an "in-house" code using a suitable mathematical programming language. For more complex problems (e.g., problems considering nonisotropic thermophysical properties, handling of irregularly-shaped geometries, time-dependent boundary conditions, nonuniform initial temperature conditions, etc.), there are a wide range of commercially available FE or FD solvers, with user-friendly interfaces, and good meshing and postprocessing capabilities.

However, in recent years, the finite control volume method (Patankar, 1980) has gained more interest over FE and FD methods in the solution of coupled flow and energy problems and has become the main computational scheme in commercially available CFD software packages. Modeling heat transfer in retort pouches is a classical case of time-dependent natural convection inside closed domains. As mentioned before, the Boussinesq approximation is the preferred technique to deal with the buoyancy-driven nature of this problem. Using this approximation usually leads to faster convergence of the numerical solution than if the problem is set up with fluid density as a function of temperature. Most commercial CFD packages include the Boussinesq model as the standard method for solving natural convection and buoyancy-driven flow problems. Detailed information about CFD techniques and applications in thermal processing of foods is presented in a separate chapter in this book. Some commercial CFD solvers commonly used by the food industry include (all Web sites accessed in February 2008)

- Fluent (http://www.fluent.com/software/fluent/index.htm)
- Star-CD (http://www.cd-adapco.com)
- PHOENICS (http://www.cham.co.uk)
- ANSYS CFX (http://www.ansys.com/products/cfx.asp)
- Flow-3D (http://www.flow3d.com)

Each of these software packages offers a wide range of different models and capabilities for different transport scenarios. Successful usage of any of these CFD packages requires a well-trained and dedicated user, with sufficient knowledge about fundamental modeling principles and very good understanding of the actual physical phenomenon to be modeled. An informed decision for choosing any of these packages depends mainly on the balance of license cost and functionality for the desired application.

## 9.9 COMBINATION OF HEAT TRANSFER MODELS WITH MICROBIAL INACTIVATION MODELS

The primary purpose when designing a heat process for the preservation of foods is to destroy all microorganisms capable of growth during subsequent normal storage of the food (commercial sterility). For ambient-stable low-acid canned foods, the minimum process criterion required for assurance of food safety is a heat treatment of 121.1°C for 3 min ($F_0$ = 3 min), which is based on a 12$D$ process targeting spores of proteolytic *Clostridium botulinum*. In practice, a higher heat treatment will often be necessary to assure product stability in addition to product safety. Other performance criteria are used for different heat-preserved foods, and these are either set by local or international food safety authorities or recommended by relevant food industry organizations. In order to determine what process criteria (e.g., time and temperature of heating) are needed to meet a given performance criterion, robust and reliable microbiological thermal inactivation data are required. These data are derived from research that strives to understand and describe microbial death kinetics. In the food industry this is typically achieved by count reduction procedures, which are designed to follow the inactivation of specific microbial populations exposed to specified lethal temperatures (either constant or predictably changing) under well-controlled conditions, by recovering and enumerating surviving organisms. This is one of the key elements involved in designing thermal processes and these data are commonly described by mathematical models.

Thermal inactivation was probably the first microbial response to be modeled and predictive models of one form or another have been applied for more than 80 years, principally in the sterilization of low-acid canned foods. The kinetics of inactivation are still the subject of much debate but overall, application of log-linear kinetic models that employ $D$- and $z$-values are accepted and applied successfully throughout food processing. The "classic" microbial heat inactivation models describe the survival curve as a first-order decay of the microbial population, given by the following initial value problem (Chick, 1908):

$$\frac{\mathrm{d}C}{\mathrm{d}t} = -k_{\max} C \tag{9.9}$$

where

$C$ is the concentration of bacteria (cells or spores/mL or g)

$k_{max}$ is the specific inactivation rate (units of 1/time)

With initial condition $C = C_0$ (initial microbial population).

The temperature dependence of $k_{max}$ is then modeled by either an Arrhenius-type equation, or, more commonly, by the classic model of Bigelow (1921), which can be written as

$$k_{max} = \frac{\ln 10}{D_{ref}} \exp\left( \ln 10 \frac{T - T_{ref}}{z} \right) \tag{9.10}$$

where

$D_{ref}$ is the decimal reduction time ($D$-value) at the reference temperature

$T_{ref}$ is the reference temperature (e.g., 121.1°C for commercial sterilization)

$z$ is the temperature change required to increase or decrease the $D$-value by a factor of 10 (e.g., 10°C for spores of proteolytic *C. botulinum*)

However, unlike growth, where the shape of the microbial response (growth curve) is generally the same (i.e., sigmoid), the kinetics of thermal inactivation are not easily predicted. Oftentimes, survival curves show a steady (log-linear) decline in numbers as described by Equation 9.9; however, there are many cases where survival curves may show an initial rapid decrease with a subsequent tail, or an initial shoulder followed by a phase of rapid decrease, or may have a biphasic shape. Many models to describe these deviations from log-linearity are available, and these are covered separately in chapter 8 of this book.

Combining CFD approaches with microbial heat inactivation models offers a robust assessment of the lethality delivered by a commercial sterilization process. Abdul Ghani et al. (2002) used the general conservation equation (Equation 9.1) to model the concentration distribution and destruction of spores of *Geobacillus stearothermophilus* within a retort pouch filled with a beef–vegetable soup (Equation 9.11).

$$\frac{\partial C_r}{\partial t} + \nabla \cdot (\mathbf{v} C_r) = D_s \nabla^2 C_r + S \tag{9.11}$$

where

$C_r$ is the relative concentration of spores per unit volume ($C/C_0$)

$D_s$ is the diffusion coefficient of spores (units of m$^2$/s in SI system)

$S$ is the source term ($S = -k_{max} C_r$)

This equation was coupled with temperature and velocity profile predictions obtained from a natural convection CFD three-dimensional model. Bacterial inactivation kinetics were considered in the CFD model as the source term (see above) where the temperature dependence of the specific rate of inactivation of spores was described by an Arrhenius-type equation.

Probabilistic modeling techniques can also be applied to microbial heat inactivation. These techniques, when combined with heat transfer models, afford real advantages in the assessment of the efficacy of a heat treatment. This approach allows for consideration of the uncertainty and variability of the microbial response to the heat treatment, resulting in more realistic estimates of surviving populations. A good example is reported by Membré et al. (2006), who used the nodal time–temperature history and spatial distribution from a three-dimensional transient heat conduction model in a retort pouch as one of the inputs into a probabilistic model for prediction of the prevalence and concentration of *Bacillus cereus* spores after a high-temperature pasteurization process of a refrigerated food of extended shelf life.

## 9.10 HEADSPACE PRESSURE MODELING AND PREDICTION OF OVERRIDING PRESSURE PROFILE

Retort pouches impact on the world market has a straightforward explanation based on a number of advantages, mentioned already in Section 9.3. However, an important processing disadvantage, in comparison with traditional cans, is the need of closer and more accurate control during the thermal processing. During thermal processing of retort pouches, internal pressure may be greater than the saturation pressure of the steam used to heat the product. The internal pressure is in part, a result of internal water vapor plus expansion of air present in the container with increasing temperature. This high internal pressure may cause serious deformation of containers if not properly counterbalanced with external pressure (Holdsworth and Simpson, 2007). The total internal pressure which develops at a given product temperature must be continuously counterbalanced by providing external pressure. The external overpressure, which is provided by compressed air, must be carefully controlled to counterbalance the internal pressure. The package may expand as a result of too little overpressure or it may crush because of too great overpressure. During processing, the retort temperature changes rapidly during come-up time and cooling, as well as through an eventual process deviation. During those periods the internal package pressure changes dramatically and proper overpressure control is most important. Each package and product system will need a unique set of overpressure requirements during various phases of a thermal process (Gavin and Weddig, 1995).

Since the walls of retort pouches are flexible, the resistance to expansion due to increasing internal pressure is negligible, and this could lead to swelling and possible bursting of pouches during the retort cycle. Expansion of retort pouches beyond allowable limits could lead to serious problems that may affect the microbiological safety and stability of the food product during shelf life. Some of these problems include, but are not limited to (1) weakening of seals, (2) excessive stretching and rupture of the barrier layer in the laminated film, especially when the barrier lacks flexibility (e.g., SiOx), which may not be evident during visual inspection, and (3) uncertainty about the location of the slowest heating zone inside the pack (i.e., pouch expansion changes the characteristic dimension or minimum thermal path of the geometry, which should have been established as a critical factor during heat penetration tests). Campbell and Ramaswamy (1992) demonstrated experimentally that, in addition to reduced heating rates and lethality values, the slowest heating zone can be shifted from the geometric center in a pouch (21 × 14 cm side area) containing a conduction-heating product with >20 mL of entrapped air when insufficient superimposed air overpressure was used in the retort. Similar observations have been reported by Ramaswamy and Grabowski (1996) in semirigid plastic containers processed in a still steam/air retort.

Therefore, expansion of pouches needs to be prevented and controlled during the retort cycle. In general, this is achieved by balancing the internal pressure in the pack with the pressure outside the pack, at any given time. The use of overriding pressure is a technically successful and practical solution to prevent ballooning and potential bursting of pouches during the retort cycle. Hence, establishing a correct overriding pressure profile in the retort becomes one of the most important critical factors that must be controlled during heat processing of retort pouches. Another, more uncommon, way of preventing and controlling pouch expansion during the retort cycle is by containing the pouches (e.g., using preformed trays that clamp the pouches to a fixed-volume cavity), thereby physically restraining any possible volume changes. This option, though effective, implies higher capital investment (mainly the cost of trays) and limits versatility of pouch sizes and shapes. However, it is the only processing alternative for some specific applications (e.g., microwave-ready rice products in retort pouches, which must be retorted under end-over-end rotation to prevent clumping).

As mentioned previously, during heat processing, the internal pressure inside the pouch increases as a result of (1) the increase in the vapor pressure of water in the food with increasing temperature, (2) the increase of the pressure of gases in the headspace with increasing temperature, (3) the release of additional occluded gas from the food (if the product has not been deaerated before filling) due to a decrease in gas solubility with increasing temperature, and (4) the thermal expansion of the product (Davis et al., 1960).

During the initial stages of the processing cycle (i.e., come-up period), the pressure differential ($\Delta P$) between the internal pressure in the pouch at a given time and the retort pressure at the same time is relatively small. As the retort temperature gradually rises, so does the product temperature. Because the pouch walls are flexible, the increase in pressure inside the pouch corresponds to the steam pressure surrounding it. The boiling point of the food product is governed by this increase, and in the case of pure water (i.e., $a_w = 1$), it is equal to the steam temperature. When the product reaches the retort temperature, the internal pressure reaches its maximum value, which is equal to the retort pressure plus the pressure due to expansion of product and entrapped gas. Assuming the pouch contains water, then at this point the vapor pressure of the water inside the pouch is equal to the retort pressure, and $\Delta P$ is due entirely to expansion of the product (as a liquid) and of the entrapped gas. At the beginning of the cooling stage, $\Delta P$ reaches its maximum value because the steam pressure is cut off, so the retort pressure starts falling more rapidly than the internal pressure in the pouch. Therefore, it is necessary to maintain the overriding pressure during cooling until the temperature inside the pouch is substantially reduced to maintain seal integrity.

Estimation of internal pressure in flexible packages is critical for product integrity as well as for process development and processing control. In the U.S. case, due to the importance of seam integrity, most of the tests required from FDA are related to this characteristic. Processors must operate in strict compliance with the U.S.

Food and Drug Administration's Low-Acid Canned Food (FDA/LACF) regulation (i.e., Code of Federal Regulations, Title 21, Part 113).

### 9.10.1 Mathematical and Physicochemical Considerations

Equations to predict the required overriding pressure during heat processing of pouches in retorts are usually based on the ideal gas law. Assuming that a pouch is filled with a measured volume of water (i.e., $a_w = 1$), then by knowing the total internal volume and the volume of entrapped gas inside the pouch, the overriding pressure can be calculated as follows (Whitaker, 1971):

$$\frac{P_1 V_1}{T_1} = \frac{P_2 V_2}{T_2} \tag{9.12}$$

where
- $P_1$ is the initial pressure in the pouch (absolute)
- $V_1$ is the initial volume of gas including occluded gas
- $T_1$ is the initial product temperature (absolute units)
- $P_2$ is the retort pressure without overriding air (absolute)
- $V_2$ is the volume of pouch at pressure $P_2$ and temperature $T_2$
- $T_2$ is the processing temperature, i.e., target retort temperature during sterilization period (absolute units)

To calculate the overriding pressure due to vapor ($P_{OV}$), the combined volume of enclosed gas and vapor must also be considered, and this involves calculating the expansion of liquid product during heat processing. The expansion of liquid inside the pouch will be equal to the specific volume of water at $T_2$ minus the specific volume of water at $T_1$ (under the assumption that the pouch is filled with water). Equation 9.12 can be therefore expanded to calculate $P_{OV}$, as follows:

$$P_{OV} = \frac{P_1 V_1 T_2}{T_1 V_3} \tag{9.13}$$

where $V_3$ = volume of gas and vapor combined at the process temperature. This is equal to the allowable expansion ($E_{allowable}$) minus the volume expansion of liquid product. Additionally, following Dalton's law of partial pressures, the volume of gas in the headspace at the process temperature must be also added to $V_3$ (see Equation 9.14).

$$V_3 = E_{allowable} - \left[\left(v_{T_2} - v_{T_1}\right) \cdot m\right] + V_{gas\ at\ T_2} \tag{9.14}$$

where
- $E_{allowable}$ is the allowable expansion (units of volume)
- $v_{T_2}$ is the specific volume of product at $T_2$ (units of volume/mass)
- $v_{T_1}$ is the specific volume of product at $T_1$ (units of volume/mass)
- $m$ is the weight of product inside the pouch (units of mass)
- $V_{gas\ at\ T_2}$ is the volume of gas in the headspace at the process temperature (units of volume, see example below)

The volume of residual gas in the pouch decreases at the process temperature due to the pressure of the steam within the retort. For example, in a pouch filled under atmospheric conditions (i.e., $P_1$ taken in this example as 101.3 kPa), at an initial temperature of 30°C, with 30 cm$^3$ of residual gas represented as the headspace, and processed in a retort at a temperature $T_2$ = 122°C, the volume of gas at $T_2$ will decrease to

$$V_{\text{gas at } T_2} = \frac{V_1 P_1 T_2}{P_2 T_1} = \frac{30\ \text{cm}^3 \times 101.3\ \text{kPa} \times (122 + 273.15)\ \text{K}}{211.4\ \text{kPa} \times (30 + 273.15)\ \text{K}} = 18.7\ \text{cm}^3$$

Finally, the overriding pressure must also consider any potential temperature variations during the sterilization period. Using the same conditions in the example above, a retort temperature drop of 0.5°C (from 122°C to 121.5°C) will cause a steam pressure drop of 3.3 kPa (from 211.4 to 208.1 kPa) and a reduction in the enthalpy of saturated liquid from 512.3 to 510.2 kJ/kg. This energy will be available to convert water (from the food) to steam, and will cause an additional volume expansion. Hence, the required overriding pressure due to controlling temperature variations ($P_{\text{OT}}$) during the sterilization period can be expressed as

$$P_{\text{OT}} = P_2 - P_{2\text{D}} \tag{9.15}$$

where $P_{2\text{D}}$ is the retort pressure at the lowest process temperature due to control variation (Absolute). The other terms in Equation 9.15 have been defined earlier.

The final combined equation to calculate overriding pressure can be written as (Whitaker, 1971)

$$P_{\text{O}} = P_{\text{OV}} + P_{\text{OT}} = \frac{P_1 V_1 T_2}{T_1 V_3} + (P_2 - P_{2\text{D}}) \tag{9.16}$$

The operating retort pressure can be then calculated (in gauge units) as follows:

$$P_{\text{R}} = P_2 + P_{\text{O}} = (2P_2 - (P_{\text{ATM}} + P_{2\text{D}})) + \frac{P_1 V_1 T_2}{T_1 V_3} \tag{9.17}$$

where $P_{\text{ATM}}$ is the atmospheric pressure (Absolute). The other terms in Equations 9.16 and 9.17 have been defined earlier.

Although Equation 9.17 appears as a simple "plug and play" method to calculate the operating retort pressure during heat processing of foods in retort pouches, it must be used with caution, understanding the assumptions in which it was based, namely: (1) the product temperature reaches the maximum retort temperature, (2) pressure drops due to variation in temperature control occur when the product temperature is maximum, (3) the pouch is filled with water (i.e., $a_w$ = 1). Furthermore, it does not consider the come-up and cooling periods, where the conditions are of transient nature, especially at the beginning of cooling, where the retort pressure can drop rapidly when the steam collapses. However, the assumptions listed above will lead to conservative estimates of pressure requirements and therefore these will tend to be fail-safe. Each processor, however, must evaluate the suitability of this method according to the specific conditions of their process.

Regarding actual food matrices, where the water activity is <1 (i.e., $a_w < 1$), it is worth mentioning a study reported by Patel et al. (1991). In this study, the authors proposed a correlation to estimate the internal pressure in semirigid containers based on the ideal gas law, but using dimensionless numbers to simplify the development and validation of the model (Equation 9.18). The experiments developed to estimate the parameters α, *a*, *b* in Equation 9.18 were carried out with water-filled containers. However, the validation of model performance was carried out with containers filled with (1) water and (2) a 3% starch solution (Thermflo, a waxy-maize modified starch, National Starch and Chemical Company, Bridgewater, New Jersey). The fitted values for these three parameters ($\alpha = 1.679$, $a = 0.515$, and $b = 0.206$) resulted in suitable predictions of internal pressure during sterilization of semirigid containers filled with both water and the 3% starch solution.

$$\left(\frac{P}{P_v}\right) = \alpha \left(\frac{(P_1/T_1)}{(P_v/T)}\right)^a \left(\frac{H}{V}\right)^b \tag{9.18}$$

where

$P$ is the internal pressure in the container (absolute)
$P_v$ is the vapor pressure at product temperature (absolute)
$T$ is the product temperature (absolute units)
$H$ is the headspace volume
$V$ is the total inside volume of container
$a$, $b$ is the parameters fitted to experimental data
The other terms in Equation 9.18 have been defined earlier

In Equation 9.18, the vapor pressure ($P_v$) was determined with the following empirical correlation (with $T$ in units of K):

$$P_v = \exp\left(16 - \frac{4967}{T}\right) \tag{9.19}$$

Although Equation 9.18 predicted the internal pressure for a 3% starch solution in a semirigid container, where the value of $a_w$ is less than (but close to) one, there are at present no studies reporting equations to predict the internal pressure during heat processing of foods in retort pouches filled with real food matrices, where the $a_w$ could be as low as 0.85 for low-acid foods. This is a research void that needs to be filled. A similar approximation was used by Awuah (2003) who used a dimensionless group correlation to relate initial pressure in the container to product headspace, initial temperature and pressure, during thermal processing of thin-walled aluminum containers.

One attempt to incorporate $a_w$ as an independent variable to predict the overriding pressure during heat processing of liquids in semirigid containers was reported by Zhang et al. (1992), and it is presented in Equation 9.20. The nomenclature in Equation 9.20 has been adapted from the original paper to match the nomenclature used in this chapter.

$$P_O = T_2 \cdot \left( \frac{P_{ATM}}{T_1} - 1.0944 \right) - 399.94 \cdot (1 - a_w) + 1.0944 \cdot T_2 \cdot a_w - \delta \quad (9.20)$$

Where $\delta$ is the normal tensile seal stress (units of pressure). The other terms in Equation 9.20 have been defined earlier.

Although Equation 9.20 includes $a_w$ of the product as an independent variable, this study validated the model with packages filled with water only, and it failed to use liquids with different $a_w$ levels to assess the performance of the predictions. In Equation 9.20, if the value of $P_O$ is positive, air overpressure is needed in addition to the steam pressure, to obtain sufficient total pressure in the retort to assure seal integrity.

## 9.11 FUTURE TRENDS

As the food canning industry continues to remain competitive in an ever-expanding global market, the need for technological advances that lead to increasing productivity, better product quality with enhanced safety assurance, and all at lower and lower cost advances in automation and intelligent online control will inevitably continue at a rapid pace (Simpson et al., 2006). New developments that are likely to occur soonest will be the application of computer-based retort control systems. Commercial retorts for pouches and trays are offered by number of companies and the crucial variable is maintenance of pressure uniformity to minimize pouch or tray expansion (Brody, 2006).

Now that retort pouches of low-acid foods have obvious commercial acceptance, recognition of superior quality and more convenient packaging, the expectation is that more heat-sterilized foods will appear in pouches, creating a new segment within the canned foods category (Brody, 2003). Most of the new developments are focused on transparent barrier films for retort packaging. It is driven by marketing convenience characteristics, such as microwaveability, ability to use metal detection equipment, visual inspection of seal area, product visibility, and environmental concerns and cost (Carespodi, 2005).

As mathematical models continue to evolve as a tool for research and development, a better understanding of problems related to the retort pouch section will be elucidated, encompassing problems such as slowing heating point (or volume) mobility, required overpressure, especially models that take into account the effect of food water activity on the pressure generated on the headspace. Similarly, as transparent pouches continue to grow, research requirements related to shelf-life modeling, such as effect of thermal processing on oxygen and water permeability and its impact on shelf life are going to be needed.

## REFERENCES

Abdul Ghani, A.G. and M.M. Farid. 2006. Using computational fluid dynamics to analyze the thermal sterilization of solid–liquid food mixture in cans. *Innovative Food Science & Emerging Technologies* 7: 55–61.

Abdul Ghani, A.G., M.M. Farid, X.D. Chen, and P. Richards. 1999. Numerical simulation of natural convection heating of canned food by computational fluid dynamics. *Journal of Food Engineering* 41(1): 55–64.

Abdul Ghani, A.G., M.M. Farid, X.D. Chen, and P. Richards. 2001. Thermal sterilization of canned food in a 3-D pouch using computational fluid dynamics. *Journal of Food Engineering* 48(2): 147–156.

Abdul Ghani, A.G., M.M. Farid, and X.D. Chen. 2002. Theoretical and experimental investigation of the thermal inactivation of *Bacillus stearothermophilus* in food pouches. *Journal of Food Engineering* 51(3): 221–228.

Abdul Ghani, A.G., M.M. Farid, and X.D. Chen. 2003. A computational and experimental study of heating and cooling cycles during thermal sterilization of liquid foods in pouches using CFD. *Proceedings of the Institution of Mechanical Engineers, Part E: Journal of Process Mechanical Engineering* 217(1): 1–9.

Awuah, G.B. 2003. Dimensionless correlation for estimating internal pressure in thin-walled aluminum containers during thermal processing. *Journal of Food Process Engineering* 26: 223–236.

Bigelow, W.D. 1921. The logarithmic nature of thermal death time curves. *Journal of Infectious Diseases* 29: 528–636.

Brody, A. 2003. Food canning in the 21st Century. *Food Technology* 56: 75–79.

Brody, A. 2006. Availableat: www.ift.org/divisions/food_pack/featurearticles/0406packaging.pdf (accessed May 2006).

Campbell, S. and H.S. Ramaswamy. 1992. Heating rate, lethality and cold spot location in air-entrapped retort pouches during over-pressure processing. *Journal of Food Science* 57(2): 485–489.

Carespodi, D. 2005. *Transparent Barrier Films for Retort Pouches* Principal, Keymark Associates. The Packaging Group Inc. Retort Pouch/Tray-2005. Princeton, NJ, April 13–14.

CCFRA. 2006. Guidelines on Good Manufacturing Practice for Heat Processed Flexible Packaging. Guideline No. 50, 75 p. Gloucestershire, UK.

CFIA. 2002. Flexible Retort Pouch Defects—Identification and Classification Manual. Available at: http://www.inspection.gc.ca/english/anima/fispoi/manman/pousac/toctdme.shtml (accessed February 10, 2008).

Chick, H. 1908. An investigation of the laws of disinfection. *Journal of Hygiene Cambridge* 8: 92–158.

Choi, Y. and M.R. Okos. 1986. Effects of temperature and composition on the thermal properties of foods. In: *Food Engineering and Process Applications*, M. Le Maguer and P. Jelen (Eds.), Elsevier Science, New York, pp. 93–101.

Datta, A.K. and A.A. Teixeira. 1988. Numerically predicted transient temperature and velocity profiles during natural convection heating of canned liquid food. *Journal of Food Science* 53: 191–195.

Davis, E.G., M. Karel, and B.E. Proctor. 1960. The pressure–volume relation in film packages during heat-processing. *Food Technology* 14(3): 165–169.

FLEXNEWS. 2007. Retort Pouch—A Fast-Growing Packaging Technology in Today's Consumer World. Available at: http://www.flex-news-food.com/console/PageViewer.aspx?page = 12653&str = (accessed, January 2008).

FPA. 1985. *Guidelines: Thermal Process Development-Flexible Containers*, 24 p. Washington, DC.

Gavin, A. and L. Weddig. 1995. *Canned Foods Principles of Thermal Process Control, Acidification and Container Closure Evaluation*, 6th edn. The Food Processors Institute. Washington, DC.

Holdsworth, S.D. and R. Simpson. 2007. *Thermal Processing of Packaged Food*, 2nd edn. Springer, New York.

Huston, K. 2005. *Market Opportunities for Retort Pouches Will Worldwide Pouch Growth Continue to Spread to the United States?* Principal, Keymark Associates. The Packaging Group Inc. Retort Pouch/Tray-2005. Princeton, NJ, April 13–14.

James, P.W., J.P. Hughes, T.E.R. Jones, and G.S. Tucker. 2006. Numerical simulation of non-isothermal flow in off-axis rotation of a can containing a headspace bubble. *Transactions of the Institution of Chemical Engineers Part A* 84: 311–318.

Kumar, A. and M. Bhattacharya. 1991. Transient temperature and velocity profiles in canned non-Newtonian liquid food during sterilization in a still-cook retort. *International Journal of Heat and Mass Transfer* 34(4/5): 1083–1096.

Kumar, A., M. Bhattacharya, and J. Blaylock. 1990. Numerical simulation of natural convection heating of canned thick viscous liquid food products. *Journal of Food Science* 55(5): 1403–1411, 1420.

López, A. 1981. *A Complete Course of Canning*, 11th edn. Canning Trade Inc., Baltimore, MD.

Membré, J.-M., A. Amézquita, J. Bassett, P. Giavedoni, C. de W. Blackburn, and L.G.M. Gorris. 2006. A probabilistic modeling approach in thermal inactivation: Estimation of postprocess *Bacillus cereus* spore prevalence and concentration. *Journal of Food Protection* 69(1): 118–129.

Miles, C.A., G. van Beek, and C.H. Veerkamp. 1983. Calculation of thermophysical properties of foods. In: *Physical Properties of Foods*, R. Jowitt, F. Escher, B. Hallström, H.F.T. Meffert, E.L. Spiess, and G. Vos (Eds.), Applied Science Publishers, Barking, Essex, United Kingdom, pp. 269–312.

Patankar, S.V. 1980. *Numerical Heat Transfer and Fluid Flow*. Hemisphere Publishing Corporation, Washington, DC.

Patel, P.N., D.I. Chandarana, and A. Gavin III. 1991. Internal pressure profile in semi-rigid food packages during thermal processing in steam/air. *Journal of Food Science* 56(3): 831–834.

Ramaswamy, H.S. and S. Grabowski. 1996. Influence of entrapped air on the heating behavior of a model food packaged in semi-rigid plastic containers during thermal processing. *Lebensmittel Wissenschaft und Technologie* 29(1–2): 82–93.

Simpson, R., S. Almonacid, and M. Mitchell. 2004. Mathematical model development, experimental validation and process optimization: Retortable pouches packed with seafood in cone frustum shape. *Journal of Food Engineering* 63(2): 153–162.

Simpson, R., A. Teixeira, and S. Almonacid. 2007. Adwances with intelligent on-line retort control and automation in thermal processing of canned foods. *Food control* 18:821–833.

Steffe, J.F. 1996. *Rheological Methods in Food Process Engineering*. 2nd edn. Freeman Press, East Lansing, MI.

Tattiyakul, J., M.A. Rao, and A.K. Datta. 2001. Simulation of heat transfer to a canned corn starch dispersion subjected to axial rotation. *Chemical Engineering and Processing* 40: 391–399.

USDOD. 2001. Packaging of food in flexible pouches. *Performance Specification*. MIL-PRF-44073. Available at: http://assist.daps.dla.mil/quicksearch/ (accessed February 5, 2008).

Whitaker, W.C. 1971. Processing flexible pouches. *Modern Packaging* 44(2): 83–88.

Zhang, J., C.H. Mannheim, S.G. Gilbert, and C. Zheng. 1992. Prediction of overpressure required for seal integrity during sterilization of semi-rigid food and drug packages. *Packaging Technology and Science* 5: 271–280.

# 10 Heat Transfer in Rotary Processing of Canned Liquid/Particle Mixtures

*Mritunjay Dwivedi and*
*Hosahalli S. Ramaswamy*

## CONTENTS

## 10.1 INTRODUCTION

Thermal processing, or canning, has proven to be one of the most effective methods of preserving foods while ensuring the product is safe and remains safe from harmful bacteria. While heat is used to ensure a safe food supply, it can also have effects on the sensory characteristics of the product, such as color, texture, and nutritional value. For products heated by convection, such as soups, sauces, vegetables in brine, meat in gravy, and some pet foods, high-temperature short-time (HTST) processing has proven to be useful in acquiring a balance between a safe product and a product of high quality. This method is successful because, compared with microorganisms, the nutrients in foods have a higher resistance to thermal destruction and a lower sensitivity to temperature changes, making it possible to apply the HTST technique to sterilize the product while retaining high quality. A typical HTST process is the aseptic processing and packaging of processed foods. In this system, food products are heated to a high temperature and held for a short time, cooled and then packed into presterilized containers inside a sterile chamber. Since these products are heated outside of the packaging materials, product sterilization is not limited by container configurations and it would thus be possible to optimize/enhance the heat transfer process within the product. Some high-efficiency heat exchangers such as plate, tubular, and scraped surface heat exchangers have good and rapid temperature equalization inside a narrow flow section. A very short heat treatment is required and the quality can be optimized; however, uncertain residence time distribution of particles and uncertain fluid to particle heat transfer coefficient ($h_{fp}$) have limited its use to liquids such as milk and juice and liquids that contain only small particles like that of soups. The technology is yet to be fully realized for canned liquid foods that contain large particles (Ramaswamy et al., 1997). In an effort to solve these problems, several technologies such as microwave and ohmic heating have been used (Willhoft, 1993). HTST processes are not beneficial for conduction-heating food products that heat slowly by comparison, exhibiting large temperature differences between the surface and the center of the container, and so some alternatives have been used to enhance the heat transfer rates in solid and semisolid products. Agitating the container to enhance mixing (in particulate fluids that normally heat by conduction) and the use of thin profile packages (retort pouches) are some other approaches used to promote better quality. In these products, the overall rate of heat transfer to the packaged food is enhanced either by product mixing or by keeping the heat transfer distance short. Rotary retorts can increase the convection in containers containing liquid–particle mixtures such as high-quality peas, corn, asparagus, mushrooms, and a variety of semisolid or viscous foods such as sauces or soups containing meat chunks or vegetables. Rotary processing leads to the rapid heating and uniform temperature distribution inside the product, therefore requiring less energy and shorter process times and providing higher quality retention. Rotary retort processing is more suitable to semisolid products (liquid with particulates) because of the faster heat transfer to the liquid and particles by enhanced convection. Since particulate liquid canned foods are not ideal candidates for aseptic processing, rotary retort processing is a potential alternative to aseptic processing for such products and is not limited by the problems associated with thin profile processing such as slow filling and sealing speed, high manual labor, etc., although it needs a special retort in order to agitate the containers.

Presently, designing a thermal process for canned foods in rotary retort processing requires experimentally gathered heat penetration data. Mathematical models based on the heat transfer studies can predict the transient heat penetration of canned foods, hence reducing the number and the cost of the experiments required to achieve product safety and quality (Teixeira et al., 1969). Many thermally processed foods heat by convection to some degree, which can be used to the processor's advantage in reducing process times, increasing production efficiency, and in some instances minimizing the ruinous effects of heat. This is achieved by agitating the container of food during the process by rotation, and in doing so, inducing forced convection currents that mix and heat the food more effectively. A key factor in mixing the container contents is the headspace bubble that sits above the food until the container is rotated.

For particulate liquid canned foods in rotary retort processing, both the overall heat transfer coefficient from the retort heating medium to the canned liquid, $U$, and the fluid to particle heat transfer coefficient, $h_{fp}$, are needed to predict heat transfer rates to the particle at the coldest point inside the can. Because of the practical difficulty in monitoring the transient temperature history of the particle moving inside an agitated liquid, the associated $h_{fp}$ is one of the important gaps in our knowledge of heat transfer (Maesmans et al., 1992).

The majority of studies on rotational processing (Clifcorn et al., 1950; Conley et al., 1951; Houtzer and Hill, 1977; Berry et al., 1979; Berry and Bradshaw, 1980; Berry and Kohnhorst, 1985) deal with the effect of agitation on the specific heat penetration parameters of the product. Rao and Anantheswaran (1988) provide an overview on $U$ values for canned liquids in rotary retorts. The more recent studies have focused on the fluid to particle heat transfer coefficients, $h_{fp}$, and overall heat transfer coefficients, $U$, both of which are important parameters influencing the heating rate of the liquid particulate mixture. A number of studies have been published to evaluate physical parameters that influence the heat transfer coefficients, and it was found that rotational speed, retort temperature, headspace volume, system geometry, liquid viscosity, rotation radius, particle size, and particle density were all key factors in end-over-end (EOE) mode (Ruyter and Brunet, 1973; Naveh and Kopelman, 1980; Anantheswaran and Rao, 1985a,b; Lekwauwa and Hayakawa, 1986; Rao and Anantheswaran, 1988; Britt, 1993; Sablani and Ramaswamy, 1995, 1996, 1997, 1998; Meng and Ramaswamy, 2007a,b). Other studies evaluated the effect of EOE agitation on nutrient, texture, and color retention of the food products (Abbatemarco and Ramaswamy, 1994, 1995).

Very few studies have been conducted on the determination of $U$ and $h_{fp}$ in free axially rotating cans (Lenz and Lund, 1978; Hassan, 1984; Deniston et al., 1987;

Fernandez et al., 1988; Stoforos and Reid, 1992). Furthermore, there were several limitations associated with these methods applied to predict lethality during heating for the real food particle system with finite internal and surface resistance to heat transfer ($0.1 < B_i < 40$). Therefore, there is still considerable opportunity for further research in this area. Dwivedi (2008) carried out a detailed study on the heat transfer of canned Newtonian liquid particulate system in the axial mode rotation and developed an empirical methodology for evaluating heat transfer coefficients $U$ and $h_{fp}$ during free axial agitation. Evaluation of the $h_{fp}$ associated with canned liquid–particle mixtures, while they are under free axial motion is challenging because of the difficulties involved with attaching temperature measuring devices to particles without affecting their normal motion. The methodology involved first correlating $U$ and $h_{fp}$ as a function of input variables for cans in fixed axial mode rotation in which both particle and fluid temperatures were measured experimentally using conventional flexible ultrathin wire thermocouples. Subsequently, liquid temperatures were measured using wireless sensors in free axial mode, and $h_{fp}$ values were computed from the developed correlation and the measured $U$ values. With the new approach developed, it was possible to study the response of the model in free axial mode to modifications of individual inputs by the so-called response analysis.

Some studies (Tattiyakul et al., 2002a,b; Hughes et al., 2003; James et al., 2006; Varma and Kannan, 2006) used computational fluid dynamics as well as experimental work to examine the underlying mechanism of heat transfer in liquid food during rotary processing. The published research related to the determination of the heat transfer coefficients in retort processing technique is reviewed in this chapter.

## 10.2 TYPES OF RETORT

### 10.2.1 Still Retorts

A still retort is a batch type, vertical or horizontal, nonagitating pressure vessel, used for processing food packaged in hermetically sealed containers. Generally, containers are stacked into racks, baskets, or trays for loading and unloading the retort. The high temperature required for sterilization is obtained from steam or superheated water under pressure.

### 10.2.2 Agitating Retorts

Mechanical agitation is commonly used to improve the rate of heat transfer in the product being processed. Rotational retorts have tremendous advantages over still retorts for processing of viscous foods in large containers. However, there are still several roadblocks with reference to the repeatability and reproducibility of heat transfer results. Prediction of temperature history in a particle undergoing conduction or convection heat transfer during an agitated cook is a complex phenomenon (Ramaswamy and Marcotte, 2006). Predictable heat transfer rates are necessary in order to produce a high-quality product with minimal overcooking and without compromising public health safety. New products, new packages, and new processes demand thorough testing and predictable performance during retorting operations to assure minimal thermal treatments. Headspace, fill of the container, solid–liquid ratio, consistency of the product, and the speed of agitation are crucial factors to be standardized in agitated processing, as well as the type of agitation imparted to a can. Some common types of agitation used are EOE, free axial, fixed axial, and Shaka systems.

#### 10.2.2.1 End-Over-End Mode

Clifcorn et al. (1950) suggested the use of EOE rotation to increase the heat transfer in canned food products. In EOE rotation, the sealed can is rotated around a circle in a vertical plane. As the can rotates, the headspace bubble moves along the length of the can and brings about agitation of the content of the can. Cans in EOE mode are placed as shown in Figure 10.1a.

#### 10.2.2.2 Free Axial Mode

Continuous container handling types of retorts are constructed with at least two cylindrical shells, in which processing and cooling takes place (two shells are

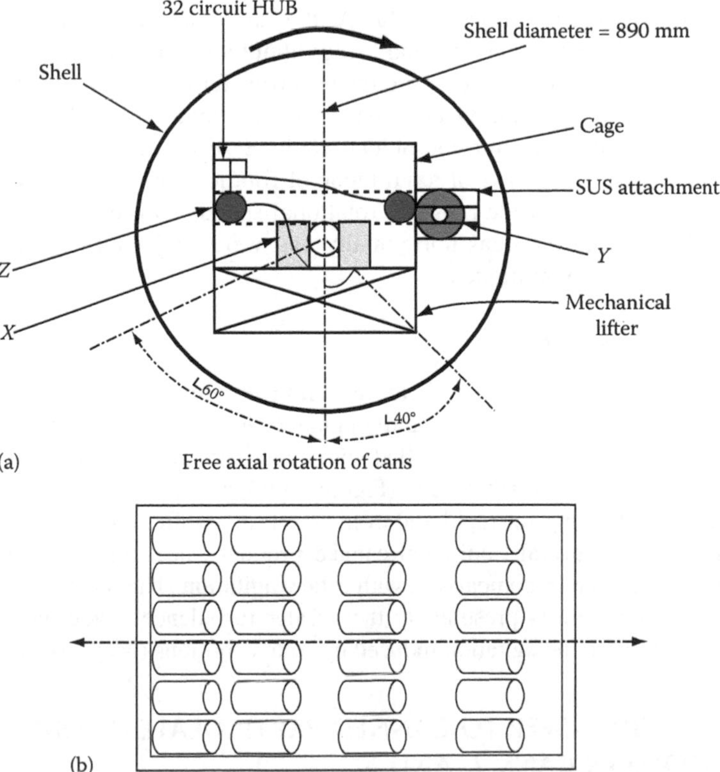


**FIGURE 10.1** (a) Placements of can in different modes inside Stock retort: *X*, EOE mode; *Y*, free axial mode; and *Z*, fixed axial mode and (b) Shaka system with back-and-forth motion.

sometimes used for cooling) and cans are subjected to axial agitation. The shell can be used for pressure processing in steam or cooling with or without pressure. The Sterilmatic (FMC Corp., San Jose, California) is an extensively used continuous agitating retort in which an entering can is carried along by a revolving reel. The Steritort is a pilot scale simulator of the thermal process in the Sterilmatic series. The motion of a can in a Steritort takes place in three phases consisting of fixed reel, transitional, and free reel motions across the bottom of the retort (Figure 10.1a). The fixed reel motion takes place over 220° of a cycle, the free rotation over the bottom 100°, and the transitional motion takes place 20° on either side of the free rotation. Some advantages of continuous retorts over batch retorts are increased production rate, reduced floor space (as fewer auxiliary types of equipment are required), reduced consumption of steam and water (caused by regeneration), and reduced labor requirements. Continuous retorts are a system in which the cans entering the retort are indexed into a revolving reel and are moved through the machine in a spiral pattern. Agitation in continuous retort is provided by allowing the cans to roll freely across the bottom of the retort.

#### 10.2.2.3 Fixed Axial Mode

In fixed axial rotation, the sealed can is rotated around a circle in a horizontal plane in a single direction (Figure 10.1a). As the can rotates, the headspace bubble moves along the length of the can and brings about agitation of the can contents. Naveh and Kopleman (1980) measured the heat transfer rates for a variety of rotational configurations, including headspace and rotational speed. Their findings revealed that the heat transfer coefficient ($U$) in EOE rotation was two or three times greater than in the case of axial mode of rotation for high-viscosity fluids, which were used over a wide range of rotational speeds. However, their findings were based on fixing the cans horizontally and were subjected to a fixed axial rotation (not in free axial mode).

#### 10.2.2.4 Shaka System

Shaka technology is a method of batch retorting in which packaged foods, pharmaceuticals, or nutraceuticals are rapidly agitated during the retort process. When using Shaka technology, a retort is filled with containers and the retort then shakes vigorously in a back-and-forth motion (Figure 10.1b), mixing the container contents thoroughly and allowing for rapid and even heat transfer throughout. Results seem to indicate that for several traditional canned products, heat transfer rates can be greatly improved even in comparison with rotary agitation. The mechanism for this extra efficient agitation is presumably the greater turbulence inside the packaged fluids compared with the agitation induced by rotary motions (May, 2001).

## 10.3 HEAT TRANSFER TO CANNED PARTICULATE LIQUID FOODS IN CANS: $U$ AND $h_{fp}$

During the agitation process, the heat transfer process is governed by an unsteady state due to the transient changes in the temperature of the product inside the can.

The heating process can be treated as two connected stages, acting on the assumption that the heat the particles receive is from the liquid surrounding them and not from contact with the can wall. The first stage involves the heating of the liquid inside the can from the heating or cooling medium inside the retort. When the heating medium is a liquid, the heat transfer mode in this stage is convection + conduction + convection, and when the heating medium is steam, the heat is transferred by condensation on the outer surface of the can, which offers no resistance to heat transfer (Holdsworth, 1997). With other heating media, the main mode of heat transfer is convection on the outer surface of the wall, and heat resistance must be taken into consideration. After this, the heat is transferred through the wall by conduction; for a metallic container of normal thickness and high thermal conductivity, there is little if any resistance to heat transfer, while glass bottles and plastic containers may offer more resistance. The final part of this stage is when the heat transfers from the interior wall of the can to the liquid by means of convection. This whole process is represented by $U$, the overall heat transfer coefficient, and is used to describe the temperature exchange between the heating or cooling medium and the liquid inside the can. The second stage is heat transfer to the particle from the liquid. In this stage, heat transfer mode is convection + conduction. Heat is transferred from the liquid inside the can to the particle surface by convection, and then transferred to the particle center by conduction. It is normally expressed by $h_{fp}$, which is the fluid to particle heat transfer coefficient. In the case of particulate liquid canned foods, heat transfer coefficients $U$ and $h_{fp}$ have been considered crucial for heat transfer models (Deniston et al., 1987). Most heat transfer studies are focussed on the quantification of $U$ and $h_{fp}$ (Lenz and Lund, 1978; Lekwauwa and Hayakawa, 1986; Densiston et al., 1987; Fernandez et al., 1988; Stoforos and Merson, 1990, 1991, 1992a and b; Sablani and Ramaswamy, 1995, 1996, 1997, 1998, 1999).

## 10.4 DETERMINATION OF $U$ AND $h_{fp}$

The proper estimation of $U$ and $h_{fp}$ under simulated processing conditions is vital for the accurate prediction of the temperature at the particle center. The temperature responses of the particle and liquid under initial and boundary conditions are traditionally used to determine $U$ and $h_{fp}$ (Maesmans et al., 1992), and the temperatures are measured by using thermocouples. Should the particle be attached too firmly inside the can, it will restrict the movement inside the can that would normally take away from real-life processing conditions. Since factors like centrifugal, gravitational, drag, and buoyancy forces, each and all, can have an effect on the particle motion and therefore on $U$ and $h_{fp}$, to not represent these in a simulation would cause deviations in the measured heat transfer coefficients. Recently, attempts have been made to measure the temperature of a particle moving freely inside a can without inhibiting its natural motion during processing. Asides from the difficulties encountered while trying to measure particle temperature, the governing equation of the energy balance inside the can with both liquid and particles is complex due to time variant temperatures of the canned liquid. The procedures used for the determination of $U$ and $h_{fp}$ are classified into two groups, based on the motion of experimental particle whose temperatures are monitored: (1) fixed particle and (2) moving particle during the agitation.

### 10.4.1 Theory for $U$ and $h_{fp}$

The overall thermal energy balance to a particulate liquid food system is used to calculate associated convective heat transfer coefficients. The governing equation for heat transfer in such systems can be written as (all symbols are detailed in nomenclature)

$$UA_c(T_R - T_l) = m_l C_{pl} \frac{dT_l}{dt} + m_p C_{pp} \frac{dT_p}{dt} \tag{10.1}$$

The following assumptions are made in the solution of Equation 10.1: uniform initial and transient temperatures for the liquid, constant heat transfer coefficients, constant physical and thermal properties for both liquid and particles, and no energy accumulation in the can wall.

The second term of the right-hand side of the equation is equal to the heat transferred to particles from liquid through the particle surface:

$$m_p C_{pp} \frac{dT_p}{dt} = h_{fp} A_p (T_l - T_{ps}) \tag{10.2}$$

Heat penetration by conduction is based on Fourier's equation, established by the French physicist Jean Baptiste Joseph Fourier (1768–1830) and written as

$$\frac{\partial T}{\partial t} = \alpha_p \nabla^2 T, \quad \alpha_p = \frac{k}{\rho C_p} \tag{10.3}$$

where

- $\rho$ is the density (kg/m$^3$)
- $c$ is the specific heat or heat capacity (J/kg °C)
- $k$ is the thermal conductivity (W/m °C)
- $\alpha$ is the thermal diffusivity (m$^2$/s)
- $\nabla^2$ is the Laplace operator, given by

$$\nabla^2 = \frac{\partial^2}{\partial x^2} + \frac{\partial^2}{\partial y^2} + \frac{\partial^2}{\partial z^2} \tag{10.4}$$

It is also assumed that the particle received heat only from the liquid and not from the can wall when it impacts, that is, heat is transferred first from can wall to liquid and then to particle. For example, heat flow in a spherical particle immersed in liquid can be described by the following partial differential equation:

$$\frac{\partial T}{\partial t} = \alpha_p \left( \frac{\partial^2 T}{\partial r^2} + \frac{2}{r} \frac{\partial T}{\partial r} \right) \tag{10.5}$$

The initial and boundary conditions are

$$T(r,0) = T_{\mathrm{i}} \text{ at } t=0 \tag{10.6}$$

$$k_{\mathrm{p}}\frac{\partial T}{\partial r} = h_{\mathrm{fp}}(T_{\mathrm{l}} - T_{\mathrm{ps}}) \text{ at } r = a \tag{10.7}$$

If the particle and liquid transient temperatures are available, the $h_{\mathrm{fp}}$ can be back calculated by solving Equation 10.5 and using $h_{\mathrm{fp}}$, $U$ can be obtained by solving Equations 10.1 and 10.2. The analytical solution for Equation 10.5, with a convective boundary condition, is complex due to time-varying liquid temperatures. Numerical solutions based on finite differences were used to solve this partial differential equation (Teixeira et al., 1969; Weng et al., 1992; Sablani, 1996, Meng, 2006).

### 10.4.2 Experimental Procedures with Restricted Particle Motion

A numerical solution was developed by Lenz and Lund (1978) using the fourth-order Runge–Kutta method and Duhamel's theorem to determine $U$ and $h_{\mathrm{fp}}$ for low Biot number (<0.1) situations. To do this, they used a lead particle fixed at the center of a can with liquid moving around it and measured its transient temperature, and verified that at all points the particle quickly reached one temperature. They proposed the following solution of Equation 10.5, for the temperature at any given point in the particle, assuming that the retort temperature reached its operating conditions instantly:

$$\left(\frac{T_{\mathrm{R}} - T}{T_{\mathrm{R}} - T_{\mathrm{i}}}\right) = \frac{2B_{\mathrm{i}}}{(r/a)}\sum_{n=1}^{\infty}\left[\frac{\beta_n^2 + (B_{\mathrm{i}} - 1)^2}{\beta_n^2 + B_{\mathrm{i}}(B_{\mathrm{i}} - 1)}\right]\frac{\sin\beta_n}{\beta_n^2}\sin\beta_n\left(\frac{r}{a}\right)$$
$$X\left\{\exp(-\tau_{\mathrm{p}}t) + \left(\frac{\tau_{\mathrm{p}}}{\tau_{\mathrm{p}} - \tau_1}\right)\left[\exp(-\tau_1\tau) - \exp(-\tau_{\mathrm{p}}\tau)\right]\right\} \tag{10.8}$$

By minimizing the sum of the squared deviations between measured and predicted particle temperature profiles, they estimated $h_{\mathrm{fp}}$. From Equation 10.8, they were also able to find the particle's average temperature, and then use it in Equation 10.1 to calculate the overall heat transfer coefficient.

By integrating Equations 10.1 and 10.2 while respectively allowing heating time to approach infinity ($<T_{\mathrm{p}}(\infty)> = T_{\mathrm{l}}(\infty) = T_{\mathrm{R}}$), Hassan (1984) was able to derive the Equations 10.9 and 10.10 for $U$ and $h_{\mathrm{fp}}$:

$$U = \frac{m_{\mathrm{l}}C_{p_{\mathrm{l}}}}{A_{\mathrm{c}}}\frac{T_{\mathrm{R}} - T_{\mathrm{li}}}{\int_0^{\infty}(T_{\mathrm{R}} - T_{\mathrm{l}})\,\mathrm{d}t} + \frac{m_{\mathrm{p}}C_{pp}}{A_{\mathrm{p}}}\frac{(T_{\mathrm{R}} - T_{\mathrm{pi}})}{\int_0^{\infty}(T_{\mathrm{R}} - T_{\mathrm{l}})\,\mathrm{d}t} \tag{10.9}$$

$$h_{\mathrm{fp}} = \frac{m_{\mathrm{p}}C_{pp}\left(<T_{\mathrm{p}}>_{\mathrm{final}} - <T_{\mathrm{p}}>_{\mathrm{initial}}\right)}{A_{\mathrm{p}}\int_0^{\infty}(T_{\mathrm{l}} - T_{\mathrm{ps}})\mathrm{d}t} \tag{10.10}$$

Using an overall heat balance equation and combining it with an equation for the transient heat conduction in a particle, Lekwauwa and Hayakawa (1986) were able to develop a model for cans subjected to EOE rotation, and considered the probability function representing the statistical particle volume distribution. Using Duhamel's formula, as well as empirical formulae describing the heat transfer to spherical, cylindrical, or oblate spheroidal-shaped particles in a liquid of constant temperature, they were able to obtain the temperature distribution for the individual particles. In their numerical solution, liquid temperatures within each time step were assumed to be a linear function of time, the coefficients of these functions being determined iteratively such that the resulting particle and liquid temperatures satisfied the overall heat balance equation.

Using Equations 10.9 and 10.10, Deniston et al. (1987) was able to determine $U$ and $h_{fp}$ in axially rotating cans. To satisfy the infinite time limits in the Equations 10.9 and 10.10, the heating time of their experiments was long enough to allow the average temperature of both liquid and particles to reach the medium heating temperature, and recognized the errors and difficulties associated with measuring surface temperature using rigid-type thermocouples.

For bean-shaped particles in cans processed in an agitated retort, Fernandez et al. (1988) were able to determine the convective heat transfer coefficients. In order to give a low Biot number ($B_i < 0.01$) condition, they preferred to use aluminum material with a high conductivity, and for $U$ and $h_{fp}$ evaluations, they used the lumped capacity method. They used rigid thermocouples to measure the time–temperature data for both the particles and the liquid, and solved the heat balance equations by using the scheme developed by Lenz and Lund (1978). In a more recent study, in an attempt to solve the differential equations governing heat transfer to axially rotating cans containing both liquid and particles, Stoforos and Merson (1995) proposed a new solution. They used an analytical solution Duhamel's theorem for the particle temperature and a numerical solution based on the fourth-order Runge–Kutta scheme for the temperature of the liquid. When they compared predicted values against the experimental data from Hassan (1984), they showed that although it deviated for the particle surface temperature, it showed a good agreement for the liquid temperature, and their solution avoided the need for empirical formulae or a constant heating medium with short time intervals.

### 10.4.3 Experimental Procedure Allowing Particle Motion

Although the particle temperature is difficult to measure due to the problems associated with attaching measuring devices, it is imperative to establish a thermal process. Many researchers, in an attempt to properly monitor particle temperature, have neglected to understand the importance of free particle motion inside the can. Since factors like centrifugal, gravitational, drag, and buoyancy forces all can have an effect on the particle motion and therefore $U$ and $h_{fp}$, to not represent these in a simulation would cause deviations in the measured heat transfer coefficients. Measurements of $h_{fp}$ in a free motion situation, therefore, have been divided into three categories.

#### 10.4.3.1 From Liquid Temperature Only

To estimate $h_{fp}$ for solid particles heated in rotating cans, Stoforos and Merson (1990) proposed a mathematical procedure using the liquid temperature as the only input parameter. They were able to estimate both $U$ and $h_{fp}$, since the fluid temperature depends on both these coefficients. They had only to systematically vary these two coefficients and minimize the difference between the experimental and predicted liquid temperature to find their two unknowns. The authors reported that the calculated $h_{fp}$ values did not always coincide with those determined from particle surface measurements.

#### 10.4.3.2 Indirect Particle Temperature Measurement

In order to monitor the surface temperature of the particles, Stoforos and Merson (1991) coated their particles with a liquid crystal coating that changed color depending on temperature, but their experiments were only carried out in the range of 20°C–50°C. To determine the convective heat transfer coefficients, a combination of time temperature integrators (microorganisms, enzymes, or chemicals) and a mathematical model has been proposed. To gather the temperature history of particles in the pasteurization process, Weng et al. (1992) used immobilized peroxidase as a time temperature integrator. Using α-amylase at reduced water content as a time temperature integrator and residual denaturation enthalpy as response, Haentjens et al. (1998) and Guiavarc'h et al. (2002) developed enzyme systems for the purpose of sterilization.

#### 10.4.3.3 Direct Particle Temperature Measurement

In order to measure the transient particle temperature during EOE rotation, Sablani and Ramaswamy (1995, 1996, 1997, 1998, 1999) used a flexible thin wire thermocouple as a measuring device that allowed adequate particle movement inside the can. They were able to study the effects of system parameters on $U$ and $h_{fp}$ based on this technique, and they developed dimensionless correlations and neural network models to predict both $U$ and $h_{fp}$.

The methodology designed by Sablani and Ramaswamy (1995) was used to evaluate the lethality of the particle with the overall heat transfer coefficient $U$ and the fluid to particle heat transfer coefficient $h_{fp}$, in EOE agitation processing of particulate Newtonian fluids. This technique was also used by Meng and Ramaswamy (2006) to extend the work to other viscous fluids. They adapted it to suit liquids of high viscosity, since previously it was only used in low viscosity situations. When using highly viscous fluids, $h_{fp}$ calculation becomes impractical. In order to predict the temperature lethality of the particle during the process, an apparent heat transfer coefficient, $h_{ap}$, was proposed and evaluated, and was feasible because of the uniformity in lethality distribution of the liquid even when highly viscous.

#### 10.4.3.4 Using $U$ and $h_{fp}$ Correlation

Dwivedi (2008) developed a new methodology to measure fluid to particle heat transfer coefficient ($h_{fp}$) and overall heat transfer coefficient ($U$) for a liquid particulate system in free axially rotating cans for thermal processing. The models developed estimated $U$ and $h_{fp}$ as a function of input process and system variables for fixed axial mode, and were coupled with experimentally measured fluid temperatures of a simulated particulate system using wireless sensors in free axial mode. This new approach of calculating $U$ and $h_{fp}$ in free axial mode is effective enough to predict the variability of the input parameters on the response variables.

## 10.5 FACTORS AFFECTING HEAT TRANSFER COEFFICIENTS

With enhanced heat transfer rates to both liquid and particles, and the potential to improve quality retention and to reduce processing time, rotary retorts possess several advantages over their still counterparts. The majority of the investigations of convective heat transfer in the presence of the particulate matter focused their attention on the liquid of the canned food (Conley et al., 1951; Hiddink, 1975; Berry et al., 1979, 1980, 1981, 1982). From previous studies, it was discovered that the most relevant factors effecting heat transfer rates to liquid particulate canned foods processed under agitation are the mode of agitation, the rotational speed, headspace, fluid viscosity, particle size, particle properties, and particle concentration (Lenz and Lund, 1978; Hassan, 1984; Lekwauwa and Hayakawa, 1986; Deniston et al., 1987; Fernandez et al., 1988; Stoforos and Merson, 1991, Sablani and Ramaswamy, 1995, 1996, 1997, 1998, 1999; Meng and Ramaswamy, 2006). Due to the large variations that characterize the properties of biological systems (foods), many researchers did their studies using model systems.

This section discusses the effects of these various parameters for both liquid and particles during agitation.

### 10.5.1 Rotational Speed

Early literature already documented the impact of rotational speed on heat transfer rates and the resulting processing times (Conley et al., 1951). In industries, the prediction of time–temperature profiles of particles processed in axial or EOE rotation is still restricted due to the lack of information of the film coefficient $h_{fp}$. The heat transfer and the lethality of canned liquid foods containing particles processed in Steritort (reel type retort–axial rotation) have been studied by Lenz and Lund (1978), and it was found that changing the reel speed from 3.5 to 8 rpm resulted in an average increase in $h_{fp}$ by 150 W/m$^2$ °C. Using Teflon, aluminum, and potato spheres in cans rotating axially in a simulator, Hassan (1984) measured the convective heat transfer coefficients of all three types of particles, and found that varying the can speed from 9.3 to 101 rpm had more of an effect on $U$ than on $h_{fp}$ but was unable to explain his results. For the potato spheres of 34.9 mm in diameter at a concentration of 30% and for the Teflon spheres at 25.4 mm in diameter at a concentration of 20%, both of which were submersed in water, $h_{fp}$ was reported to be highest at the lowest rotational speed (9.3 rpm). Consequently, it was found that at the highest rotational speed (101 rpm), $h_{fp}$ was intermediate and at it was lowest at an intermediate speed of 55.5 rpm (Maesmans et al., 1992). The fluid to particle heat transfer coefficient for potato particles in water during EOE rotation was determined by Lekwauwa and Hayakawa (1986), and it was found that the values of $h_{fp}$ ranged between 60 and 2613 W/m$^2$ °C. The heat transfer process of nylon particles in distilled water and bath oil at a high temperature (100 cst at 38°C) in EOE mode was studied by Sablani (1996). He observed that when the rotational speed increased from 10 to 20 rpm, on average, $h_{fp}$ value increased by 56% for oil and 53% for water, and the $U$ value increased by 24% for oil and 13% for water.

Stoforos (1988) explained the effect of rotational speed on $h_{fp}$ when he reported that at high speeds (100 rpm) his canned contents behaved as a solid mass and therefore provided little agitation inside the can, and further explained that revolutions per minute would affect $h_{fp}$ so long as the increase affected the relative particle to fluid velocity. Stoforos and Merson (1992a and b) showed that by increasing the rotational speed, the $U$ and $h_{fp}$ values could be increased as well. This effect was most notable at lower rotational speeds when using deionized water and silicone fluids together with 1 in. diameter spheres of aluminum or Teflon as liquid/particulate food systems; increasing rotational speed from 9.3 to 101 rpm, $U$ increased about 1.2–2.0 times. The $h_{fp}$ determined, however, was found to be insensitive to these rotational speeds. Because of the restricted movement of the particles attached to rigid thermocouples and the minimal settling effect of gravity for the same reason, and since the centrifugal forces were minimal as the particle was attached in the center of the can, the author attributed the unchanging $h_{fp}$ to the small relative particle to liquid velocity. The effect of various parameters on convective heat transfer in axially rotating cans that contained liquid and particulates was studied by Fernandez et al. (1988), and the results were represented in the form of dimensionless correlations. Increasing the Reynolds number, which is based on rotational speed, improved the Nusselt number, which is based on $h_{fp}$.

Meng and Ramaswamy (2006) reported that while using canned particles/Newtonian fluids in high-viscosity glycerine aqueous solutions (from 75% to 100%), the effect of rotational speed was significant, and the associated $h_{ap}$ values increased as well. Dwivedi (2008) reported in his study of canned Newtonian (glycerine: 80%–100%) particulate food (30% [v/v], nylon φ 19 mm) that $U$ and $h_{fp}$ in free axial mode increased from 448 to 907 and 477 to 1075 W/m$^2$ °C, respectively, with the increase in revolutions per minute from 4 to 24.

### 10.5.2 Fluid Viscosity

Hassan (1984) found that increasing the fluid viscosity decreased the overall heat transfer coefficient, and that the same was true for the particle–liquid film heat transfer coefficients, except in the case of large aluminum spheres (3.17 cm diameter), processed with particle volume fraction ($\varepsilon = 0.21$). For these, he reported increasing values for $h_{fp}$ with increasing viscosity (1.5, 50, and 350 cst), and found by studying the motion of the particle that the particle to fluid relative velocity increased in high viscous fluids. When he compared silicon fluids and deionized water, both containing Teflon particles, he found that the heating rate for the silicon fluids was slower, and he attributed this to the fluid lower thermal diffusivity. Lenz and Lund (1978) found that water had higher values for both $U$ and $h_{fp}$ than a 60% aqueous sucrose solution, and Sablani (1996) studied $U$ and $h_{fp}$ values with nylon particles in water and oil (100 cst at 38°C), and found that larger $h_{fp}$ and $U$ values were obtained from water. Sablani (1996) reported in his study with single particle in the Newtonian liquid that the $U$ and $h_{fp}$ values improved with the decrease in liquid viscosity. Using conventional methods for calculating $h_{fp}$, Meng and Ramaswamy (2006) computed the heat transfer coefficient and found that the associated $h_{fp}$ values ranged from 215 to 376 W/m$^2$K and $U$ values ranged from 112 to 293 W/m$^2$K. The $U$ values decreased with an increase of the liquid viscosity, which could be explained by the thickness of the associated boundary layer around the particle and inside the surface of the can in higher viscosity fluids. Dwivedi (2008) reported that with an increase in glycerine concentration, it was observed that $U$ and $h_{fp}$ decreased by 26% and 40%, respectively. He explained that the decrease related to concentration could be attributed to the larger thickness of the associated boundary layer in the higher viscosity liquids. He went on to further explain that the more viscous the fluid becomes the less the fluid can agitate, that is to say, it will have a low particle to fluid relative velocity, and thus a lower rate of heat transfer. For $h_{fp}$ calculated in highly viscous fluids, there is less motion inside the can and therefore less mixing, resulting in lower heat transfer rates from the liquid to particle (Stoforos, 1988).

### 10.5.3 Mode of Rotation

Working with sucrose solutions of Newtonian behavior and carboxy-methyl-cellulose (CMC) solutions of non-Newtonian behavior in a pilot plant spin cooker–cooler, Quast and Siozawa (1974) reported that the heat transfer rates with axial rotation increased two to four times compared to stationary processing. During the cooing of tomato paste at 8–60 rpm, Tsurkerman et al. (1971) found that the highest heat transfer rates were obtained when there was a 45° angle for the axis of rotation. While working with 84°Bx 70 D.E. glucose syrup (a highly viscous liquid) at 20–120 rpm, Naveh and Kopelman (1980) noticed the superiority of EOE mode when they found that the heat transfer coefficient was two to three times higher than that found in axial rotation using Stork autoclave positioning the cans horizontally and axially simultaneously; however, this study was done using Stork retort, which was not capable of providing similar biaxial rotation as with FMC continuous sterilizer. Under the intermittent agitation, when the direction of rotation was reversed every 15–45 s, Hotami and Mihori (1983) reported that heating rates and uniformity were increased and that there was no significant difference between EOE rotation and axial rotation. Quast and Siozawa (1974) found that there were no significant differences between overall heat transfer coefficients calculated with and without reversing the direction for the CMC experiments. The experimental conditions of their study remain unknown, but we can contribute their conclusion to the fact that the temperature of the fluid was generally between 69°C and 96°C with a medium temperature of 98°C. This means that they used last portion of the heating cycle to calculate the overall heat transfer coefficients and to draw their conclusions. The parameters affecting heating characteristics, however, are generally the most predominant at the early stages of the heating where fluid temperature gradients are higher. Using canned Newtonian liquid (glycerine: 0%–100%), Dwivedi (2008) reported in his study that the overall heat transfer coefficient ($U$) was significantly higher in the case of free axial mode (340 W/m$^2$K) than in the EOE mode (253 W/m$^2$K), which was contrary to the study done by Naveh and Kopleman (1980). This contradiction was due to the difference in the method used to study the axial rotation. The fixed axial mode of rotation of the can was used by Naveh and Kopleman (1980), which is not comparable to the manner the cans are processed in the FMC continuous Steritort.

### 10.5.4 Particle Concentration

The flow pattern and level of mixing in rotational processing may be influenced by the presence of particulate inside the canned liquid, which also could affect the heat transfer coefficient. The motion of these particles could cause secondary agitation of the liquid, resulting in better mixing and distribution of heat inside the can. The fraction of total effective can volume occupied by the particles is defined as the ratio of the particle volume to the sum of the particle and liquid volume, and can be expressed as $\varepsilon$. Keeping the headspace constant at 6.4 mm for 303 × 406 cans closed under 50 cm Hg vacuum, Hassan (1984) and Deniston (1984) separately studied the effects of $\varepsilon$ on the heat transfer rates. Hassan used Teflon spheres, 2.54 cm diameter in water and reported higher $U_o$ and $h_{fp}$ values, while Deniston used potato spheres at 2.86 cm in diameter in water processed at 29.1 rpm over a $\varepsilon$ range between 0.107 and 0.506 and found that for $\varepsilon$ at 0.107, the average $U_o$ value for triplicate experiments was 1300 W/m$^2$K. $U_o$ exhibited a maximum value of 1640 W/m$^2$K for $\varepsilon = 0.400$ and subsequently $U_o$ declined, having value of 1360 W/m$^2$K at $\varepsilon = 0.506$. Analogous results were observed for $h_{fp}$ values. At $\varepsilon = 0.107$, $h_{fp}$ was 175 W/m$^2$K, slightly increased up to a value of 190 W/m$^2$K at $\varepsilon = 0.400$, and sharply decreased to a value of 127 W/m$^2$K at $\varepsilon = 0.506$. Deniston et al. (1987) summarized that since the spheres were tightly packed in the can at high $\varepsilon$ values, the particles were not free to move and the functional dependency of $U_o$ and $h_{fp}$ probably differed compared to loosely packed particles. When the particle concentration was increased from one particle to 29%, Sablani and Ramaswamy (1997) discovered that $U$ increased 20% for oil and 5% for water, but a further increase in concentration to 40% resulted in a decrease of $h_{fp}$ by 12% for oil and 7% for water. Meng and Ramaswamy (2006) used nylon particles at 22.225 mm in diameter to study the effect of particle concentration using full factorial design, with a rotational speed of 15 rpm, headspace of 9% (v/v), retort temperature of 120°C, and a rotation radius of 120 mm in 0.9% CMC concentration, and they found that particle concentration had significant effect ($p < 0.05$). Dwivedi (2008) reported that the particle concentration has a significant influence ($p < 0.05$) on the heat transfer coefficients $U$ and $h_{fp}$. For a particle size of 25 mm, $U$ increased with particle concentration ranging from 20% to 30%, while further increasing the concentration to 40% would decrease $U$ and $h_{fp}$. Such a decrease could be due to restricted free movement attributed to shirking space inside the can with increasing particle concentration.

### 10.5.5 Particle Size

Hassan (1984) found that, when processed in water, process times for fluid and particles increased when the diameter of potato spheres increased from 2.22 to 3.49 cm. He was also able to determine that for the larger particles, the effect of rotational speed on the particle heating times was higher than for smaller particles, and found similar results when he used Teflon and aluminum particles processed in water or silicon liquids. When the particle liquid film transfer coefficient was examined by Deniston in 1984, he found analogous results for his potato spheres in water. The overall heat transfer coefficient, however, was reported to have the highest values at all rotational speeds when particles of an intermediate size (2.86 cm) were used. When the rotational speeds were lower, between 9.3 and 29.1 rpm, the largest particles of 3.5 cm gave the lowest $U_o$ values, while at higher speeds (101 rpm) the lowest $U_o$ values were obtained from the smaller particles of 2.22 cm diameter. In all of these experiments, cans of size 303 × 406 were used at a particle volume fraction $\varepsilon = 0.293$. The data collected by Deniston (1987) showed that $h_{fp}$ was not greatly influenced by increasing particle size.

Lenz and Lund (1978) reported that increasing particle size resulted in higher overall heat transfer coefficients, but Sablani and Ramaswamy (1997) found that $U$ values decreased by about 9% in oil and 6% in water as the size of nylon particle increased from 19.05 to 25 mm diameter, and concluded that particle size influenced more $h_{fp}$ than it did $U$. In their experiments, $h_{fp}$ decreased by 13% in oil when the particle diameter increased from 19.05 to 22.25 mm while further increase in size to 25 mm diameter reduced $h_{fp}$ by 24%. Furthermore, $h_{fp}$ decreased by 9% in water when the particle diameter increased from 19.05 to 25 mm. Meng and Ramaswamy (2007b) reported no significant effect of particle size on $U_a$ and $h_{ap}$. Dwivedi (2008) reported that through analysis of variances, it can be seen that in free axial modes, particle size had a significant effect on $U$ and $h_{fp}$. The $U$ values in free axial mode for the three concentrations observed (20%, 30%, and 40%) were found to decrease by 33%, 25%, and 29% as the size of the particles increased from 19.05 to 25 mm. Under the same increasing size conditions, $h_{fp}$ also decreased, probably due to the thicker boundary layers associated with larger particle diameters. It can be seen that the increasing diameter had a greater influence on $h_{fp}$ than on $U$ ($h_{fp}$ decreased by 32%, 34%, and 41% as the size went from 19.05 to 25 mm in the three concentrations).

### 10.5.6 Particle Shape

To determine the effect of particle shape might have on $U$, Ramaswamy and Sablani (1997) used nylon particle of various shape such as cylinders, cubes, and spheres, and found that when there were multiple particles inside the can, the $U$ values for the cube-shaped particles were about 6% lower than those of the cylinder, and 6% lower than that of the spheres. When studied in oil, the average $U$ value of the cubes became 3% lower than the cylinders and 6% lower than the sphere-shaped particles. The authors explained this by saying that the different shapes created different void areas between the particles, allowing for different levels of mixing to take place inside the can. Particle shape was already known to have an influence on the heat transfer coefficients due to differences in liquid flow over the different shaped surfaces of the particles. $h_{fp}$ was determined to be affected by shape as well: in experiments conducted by Ramaswamy and Sablani (1997) using multiple particles submerged in oil, $h_{fp}$ increased by 6% with cubes compared to that of a cylinder which in turn increased by 20% compared to that of sphere, and the influence of shape was even more noticeable in water. They explained this by saying that the disturbances in the flow field near the surface of the cylinder and the cube made the effects of the particle shape more predominant. The cube-shaped particle, for example, behaved like an extreme case of a particle with a rough surface (Astrom et al., 1994). As for cases with more than one particle, the results were reversed with the sphere having the highest $h_{fp}$ values and the cubes having the lowest, and this was explained by way the different shapes interlock with each other and pack into the can, creating void spaces of different extents with each different shape.

### 10.5.7 Particle Density

Particle density can affect the particle fluid motion pattern inside the can, thereby affecting the heat transfer coefficient. When using a single particle inside a can, Sablani (1996) found a significant effect of density on the $h_{fp}$ value. He noted that particles of greater density settled in the can faster, resulting in more motion inside the can due to the higher particle/liquid relative velocity and creating a higher $h_{fp}$. When the particles with densities close to each other (nylon and acrylic) had similar $h_{fp}$ values in both oil and water, he concluded that particle thermal properties did not influence $h_{fp}$. However, Stoforos and Merson (1992a and b) reported that their Teflon spheres had a higher $h_{fp}$ value than their aluminum particles of the same size, and besides the particle density, the thermal properties of the particle matter were presumed to be the explanation. Meng and Ramaswamy (2007b) reported that the $h_{fp}$ values increased with the increase in the particle density. Dwivedi (2008) studied the influence of particle density on $U$ and $h_{fp}$ in free axial mode and found that at 20% particle concentration, polypropylene particles are shown to have the lowest $h_{fp}$ (495 W/m$^2$ °C) and Teflon particles the highest (922 W/m$^2$ °C), while nylon particles fell in between with $h_{fp}$ of 764 W/m$^2$ °C.

## 10.6 PREDICTION MODELS FOR HEAT TRANSFER COEFFICIENTS ($U$ AND $h_{fp}$)

Since the heat transfer coefficients in agitation processing are influenced by several outside factors, it is important and necessary to observe and catalog the influence of these factors. Empirically, dimensionless correlations are widely used in the prediction of heat transfer coefficients, although in recent times such models have also been made from artificial neural networks (ANNs), and these models have shown great promise (Sablani et al., 1997a; Meng et al., 2006; Dwivedi, 2008).

### 10.6.1 Dimensionless Correlations

Dimensional analysis is a preferred and widely used technique for generalizing data because it limits the number of variables that must be studied and permits the grouping of physical variables that affect the process of heat transfer (Ramaswamy and Zareifard, 2003). It is also meaningful to generalize these effects in order to broaden their scope for scale-up considerations. In the dimensional analysis of forced convection heat transfer, where fluid is forced over the solid by external means, the Nusselt number ($N_u$), a dimensionless measure of convective heat transfer coefficient, is related with other dimensionless numbers such as Reynolds number ($R_e$) and Prandtl number ($P_r$). In the case of free or natural convection, where fluid motion is determined by the buoyancy forces, Nusselt number ($N_u$) is correlated to the Rayleigh number (which is the product of two dimensionless numbers, Grashof ($G_r$) and Prandtl ($P_r$) numbers). The Grashof number plays the same role in free convection that the Reynolds number plays in forced convection. Where Reynolds number is used to measure the ratio of the inertial to viscous forces acting on a fluid element, the Grashof number indicates the ratio of the buoyancy force to the viscous forces acting on the fluid (Incropera and Dewitt, 1996). These dimensional numbers give a better understanding of the physical phenomenon and can also be easily used for scale-up purposes. Dimensionless correlations involve overall heat transfer coefficients ($U$) because earlier studies were focused on the convective heat transfer in liquid foods. Rao and Anantheswaran (1988) presented an insightful review of these studies; however, for canned liquid particulate food systems, there are scant few studies available.

Sablani et al. (1997a) established correlations for overall heat transfer coefficients, as well as the fluid to particle heat transfer coefficients for liquid/particle mixtures in EOE rotation. In his studies, he correlated $N_u$ with other dimensionless numbers, including Archimedes and Froude ($F_r$) numbers, the ratio of headspace to the length of the can ($H_s/H_c$), the ratio of the particle to liquid concentration [$e/(100 - e)$], the ratio of the equivalent particle diameter to the diameter of the can ($d_e/D_c$), particle sphericity ($\psi$), and the density simplex ($p_p - p_l/p_l$). In EOE rotation, Meng and Ramaswamy (2007a) developed the dimensionless correlations for high-viscosity fluid/particle mixtures using apparent heat transfer coefficients $h_{ap}$ and $U_a$. Lenz and Lund (1978) and Deniston et al. (1987) presented correlations in free axial rotation for $U$ relating $N_u$ to $R_e$, $P_r$, and other dimensionless numbers, although they did not develop the relation between fluid and particle heat transfer. In the following years, Fernandez et al. (1988) presented dimensionless correlations for $h_{fp}$ in cans with axial rotation, adding to their studies. Using the fluidized bed and packed bed approaches and using the modified Stanton number and Colburn $j$-factor in empirical correlations, they were able to model their correlations; however, he was unable to develop correlations for $U$. The time–temperature prediction at the particle center requires appropriate correlations for both $U$ and $h_{fp}$, and cannot be made with only one of these coefficients. Because of the difficulties in obtaining time–temperature measurements of the liquid and particle simultaneously, literature pertaining to free axial agitation lacks the correlations of $U$ and $h_{fp}$.

Anantheswaran and Rao (1985a) tried different variables to determine characteristic length, and found that using the sum of the diameter of rotation and the length of the can in Newtonian liquid resulted in a statistically acceptable $R^2$ (0.92) for $U$ values. In axial rotation, reel radius was used as the characteristic length, as demonstrated in studies by Lenz and Lund (1978), and in studies by Deniston et al. (1987), the diameter of the can was used in equations to describe the heating behavior of liquid in the presence of multiple particles. Fernandez et al. (1988) used the diameter of the particle in the development of a dimensionless correlation for fluid to particle heat transfer coefficient $h_{fp}$, whereas Sablani et al. (1997) tried different variables and came to the conclusion that for the most appropriate $U$ correlations it was best to use the sum of the rotation diameter and the diameter of the can as the characteristic length, and the particle's shortest dimension for $h_{fp}$ correlations.

Sablani et al. (1997) and Meng and Ramaswamy (2007a) found that for EOE rotation, forced convection was the dominant mechanism for heat transfer for the canned liquid/particle mixture, and modeled their particulate laden products on this basis. On the contrary, Fand and Keswani (1973) reported that in all forced convection situations, the natural convection phenomenon continued to operate since buoyant forces resulting from density differences exist.

While in Steritort, Rao et al. (1985) reported that because the can rotates only over one-third of the cycle, it can be expected that natural (free) convection is an important mode of heat transfer. Since fluid flow agitation during the upper two-third part of the retorts are low, the mechanisms involved in modeling the dimensionless number in case of Steritort are the complex combination of the natural and forced convections. In mixed convection, buoyant as well as inertial forces operate simultaneously within the fluid, and in some situations, may have the same order of magnitude (Fand and Keswani, 1973; Chapman, 1989). Marquis et al. (1982) acknowledged that in his research, natural convection was important during the heating of liquids in bottles subjected to axial rotation, and even Lenz and Lund (1978) correlated their data with a two-term model that can be interpreted as the sum of natural and forced convection contributions. Even though they had included both convection modes in their equations, they supplied only a fixed value to free convection and concentrated their work with forced convection. As discussed above, there is little or no information available in previously published literature where $U$ and $h_{fp}$ correlations are calculated simultaneously in free axial mode, while some correlations exist for both individually. Despite the fact that this concept of modeling the time–temperature profile shows promise in a wide range of industrial applications, little research has been done to explore its potential in free axial rotation. In his study, Lenz and Lund's study was taken a step further by Dwivedi (2008), where he applied variable free convection terms on the basis of Grashof and Prandtl numbers. He developed the dimensionless correlations for estimating heat transfer coefficients in canned Newtonian liquids, both with and without particles, under mixed and forced convection heat transfers, using multiple nonlinear regressions of statistically significant dimensionless groups. Data on overall heat transfer coefficient $U$ and fluid to particle heat transfer coefficients $h_{fp}$ were obtained for several processing conditions and were analyzed separately for particle and particleless conditions. The dimensionless correlation was formulated on the basis that heat transfer to the canned liquid in Steritort takes place by combined natural and forced convections, this theory having provided the best correlation between the estimated and experimental values of the form $N_u = A1(G_rP_r)^{A2} + A3(R_e)^{A4}(P_r)^{A5}(F_r)^{A6}(\rho_p/\rho_l)^{A7}\,[e/(100-e)]^{A8}\,(d_p/D_c)^{A9}\,(K_p/K_l)^{A10}$. The combination of the radius of the reel, can, and the particles were chosen as the characteristic lengths for the correlation. $U$ and $h_{fp}$ in free and fixed axial modes in mixed convection mode, while in the presence of particles, were correlated to Reynolds, Prandtl, Grashof, and Froude numbers, as well as the relative density of particle to liquid ($\rho_p/\rho_l$), particle concentration [$e/(100-e)$], diameter of particle to can ($d_p/D_c$), and particle/liquid thermal conductivity ratio ($K_p/K_l$). In the absence of particles, $U$ was correlated only to the Reynolds, Prandtl, Grashof, and Froude numbers. During the presence of particles, an excellent correlation of ($R^2 > 0.93$) was obtained between the Nusselt number and the dimensionless groups for all models. When the particles were absent, in EOE mode in a pure forced convection situation, $R^2$ was only 0.81. Introducing natural convection ($G_r \times P_r$) increased $R^2$ to 0.97, which shows that there was a marked improvement in the model due to free convection, even when forced convection was the dominant mode of heat transfer. The validity of the mixed convection equations for $U$ in free axial mode was limited to $R_e$ in the range from 61 to $2.75 \times 10^3$; $P_r$, $5.76 \times 10^2$ to $8.04 \times 10^3$; $F_r$, $2.1 \times 10^{-3}$ to $7.5 \times 10^{-1}$; $G_r$, $4.89 \times 10^4$ to $8.12 \times 10^6$; $\rho_p/\rho_l$, 0.65–1.75; $d_p/D_c$, 0.21–0.28; and [$e$/(100 – $e$)], 25%–66%, whereas $h_{fp}$ was in the range of $R_e$ from 57 to $2.60 \times 10^3$; $P_r$, $5.76 \times 10^2$ to $8.04 \times 10^3$; $F_r$, $2.0 \times 10^{-3}$ to $7.3 \times 10^{-2}$; $G_r$, $4.48 \times 10^4$ to $7.43 \times 10^6$; $\rho_p/\rho_l$, 0.65–1.75; $d_p/D_c$, 0.21–0.28; and [$e$/(100 – $e$)], 25%–66%.

In his study, Dwivedi (2008) reported a number of dimensionless correlations under mixed convection mode situations for canned liquid with and without particle situations that are presented in Appendix A.

### 10.6.2 Artificial Neural Network

The term "neural network" resulted from artificial intelligence research, which attempts to understand and model brain behavior. "Artificial intelligence is the part of computer science concerned with designing intelligent computer systems, that is, systems that exhibit characteristics were associated with intelligence in human behavior." Over the past few years, neural networks have shown great promise, and when solving nonlinear prediction problems prove to be more powerful than most other statistical methods (Bochereay et al., 1992). The goal of the neural network is to map a set of input patterns onto a set of corresponding output patterns, and Figure 10.2 shows a typical neural network formed by an interconnection of nodes. Normally, a network can have one to three hidden layers, but this figure has one hidden layer, an input layer, and an outer layer, and each layer is essential to the success of the network. The entire network can be viewed as black box into which a specific input to each node is sent in the input layer. Although the entire processing step is hidden, the network processes information through the interconnections between nodes, and finally gives us an output on the output layer.

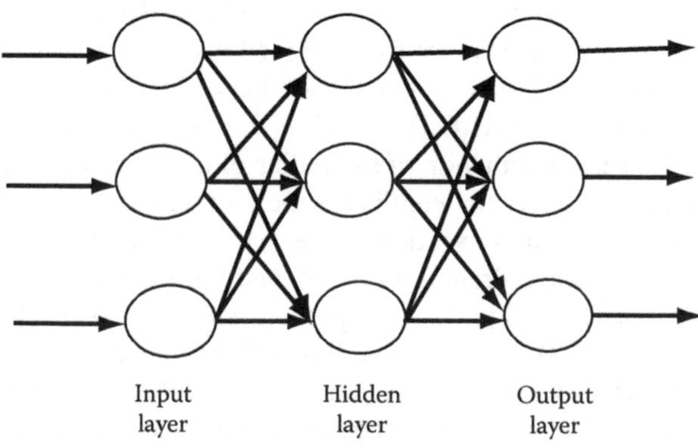


**FIGURE 10.2** Typical multilayer neural network with one hidden layer.

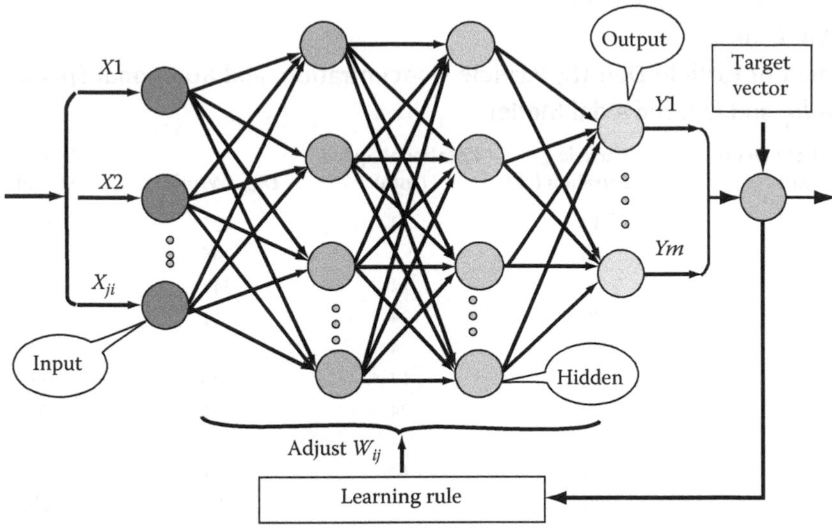


**FIGURE 10.3** Typical multilayer neural networks.

Figure 10.3 shows that the neural network is made up of neuron arranged in layers, and each neuron in a layer is linked to many other neurons in other layers with varying connection weights ($W_{ij}$) that represent the strength of these connections. Sometimes, the neurons are even connected to neurons in their own layer. Within each neuron threshold value, bias ($B_i$) is added to this weighted sum of connection weights and nonlinearity is transformed using an activation function to generate the output signals response ($O$) of each neuron ($i$) to the input signals ($I$) from connecting neurons ($J$), which can be mathematically expressed as

$$O_i = f\left(\sum_{j=1}^{m} I_j W_{ij} + B_i\right) \tag{10.11}$$

The architecture and the algorithmic method chosen for training is what will determine the learning ability of the neural network. Due to its relative simplicity and its stability, back propagation algorithm is very popular (Neural Ware, Inc., Carnegie, PA, 1996). In a back propagation algorithm, the learning process starts with random initialization of connection weights. The response of each output neuron during learning is compared with a corresponding desired response. Any errors associated with the output neurons are computed and transmitted from the output layer to the input layer through the hidden layer. The weights are adjusted at the end of the back propagation cycle in order to minimize the errors, and this procedure is repeated over the entire learning cycle for a set number of times, chosen by trail and error. In practice, if sufficient numbers of these input–output combinations are used for the learning of the neural network, it should be able to predict the output for new inputs.

An ANN model was developed by Meng and Ramaswamy (2006) for the apparent heat transfer coefficients associated with canned particulates in high-viscous Newtonian and non-Newtonian fluids during EOE rotation in a pilot scale rotary

**TABLE 10.1**
**Effect of Particle Density, Particle Concentration, and Rotational Speed on $h_{fp}$ and $U$ (Free Axial Mode)**

| Particle Density (kg/m³) | Particle Concentration (%) | Rotational Speed (rpm) | $U_{free}$ (W/m²K) | $h_{fp}$ Free (W/m²K) |
|---|---|---|---|---|
| 128 | 40 | 24 | 423 | 891 |
| 1128 | 40 | 4 | 245 | 411 |
| 1128 | 40 | 14 | 347 | 721 |
| 1128 | 20 | 24 | 411 | 764 |
| 1128 | 20 | 4 | 238 | 397 |
| 1128 | 20 | 14 | 340 | 607 |
| 1128 | 30 | 24 | 431 | 909 |
| 1128 | 30 | 4 | 250 | 427 |
| 1128 | 30 | 14 | 334 | 621 |
| 2210 | 40 | 24 | 494 | 1030 |
| 2210 | 40 | 4 | 307 | 503 |
| 2210 | 40 | 14 | 411 | 839 |
| 2210 | 20 | 24 | 502 | 922 |
| 2210 | 20 | 4 | 312 | 508 |
| 2210 | 20 | 14 | 417 | 731 |
| 2210 | 30 | 24 | 545 | 1140 |
| 2210 | 30 | 4 | 338 | 565 |
| 2210 | 30 | 14 | 504 | 921 |
| 830 | 40 | 24 | 310 | 499 |
| 830 | 40 | 4 | 158 | 225 |
| 830 | 40 | 14 | 256 | 406 |
| 830 | 20 | 24 | 320 | 495 |
| 830 | 20 | 4 | 164 | 228 |
| 830 | 20 | 14 | 263 | 391 |
| 830 | 30 | 24 | 338 | 634 |
| 830 | 30 | 4 | 175 | 266 |
| 830 | 30 | 14 | 320 | 528 |

retort. For the overall heat transfer coefficients as well as the fluid to particle heat transfer coefficients for liquid particulate systems in cans subjected to EOE mode, Sablani and Ramaswamy (1997) developed another ANN model. The multilayer neural network had six input neurons and two output layers corresponding to $U$ and $h_{fp}$, and was developed for the presence of multiple particles inside the can. Using the back propagation algorithm, the optimal neural network was obtained by varying numbers of hidden layers, number of neurons in each hidden layer, and the number of learning runs.

ANN was used to optimize the conduction of heated foods (Sablani et al., 1995) and a correlation was developed among optimal sterilization temperature, corresponding processing time and quality factor retention with the can dimension, food thermal diffusivity, and kinetic parameters of the quality factors. The models developed with ANN were able to predict the optimal sterilization temperatures with an accuracy of ±0.5°C and other responses with less than 5% associated errors.

Dwivedi (2008) also developed ANN models to predict overall heat transfer coefficient ($U$) and the fluid to particle heat transfer coefficient ($h_{fp}$) in canned Newtonian fluids both with and without particles in fixed, free, and EOE modes (Table 10.1). He further reported that the developed ANN models were found to predict the responses with the mean relative error (MRE) of 2.6% and 2.9% for $h_{fp}$ and $U$ in fixed axial mode, and 1.85% and 2.5% in free axial mode, respectively. Similarly, MREs associated with the overall heat transfer coefficient in Newtonian liquid without particulates in EOE mode were 3.8%, and was 2.06% in the case of free axial mode. Neural network-based equations for estimation of $U$ and $h_{fp}$ in canned Newtonian fluids with particles under free axial mode are presented in Appendix B.

### 10.6.3 Comparison between ANN and Dimensionless Correlation Models

Sablani et al. (1997b) developed ANN models and reported that the prediction errors were about 50% better than those associated with dimensionless correlations: less than 3% for $U$ and 5% for $h_{fp}$. Meng (2006) found that the ANN models were able to predict responses with MRE of 2.9%–3.9% in Newtonian fluids and 4.7%–5.9% in non-Newtonian fluids. These results were 27%–62% lower than those associated with the dimensionless correlations. Dwivedi (2008) reported in his study that MRE for ANN models were much lower compared with MREs associated with dimensionless correlations: 74.46% lower than dimensionless correlations for $h_{fp}$ and 66.35% lower for $U$ in fixed axial mode with particulate in liquid. Under the same conditions, they were 77.71% lower for $h_{fp}$ and 66.98% lower for $U$ in free axial mode. MRE for ANN models were again lower when compared to $U$ in EOE and free axial modes for fluids without particulate: 37.02% lower than dimensionless correlations for EOE mode and 75.53% lower for free axial mode. ANN models yielded much higher $R^2$ values than those obtained with dimensionless numbers (Tables 10.2 and 10.3).

**TABLE 10.2**
**Comparisons of Errors for ANN Models vs. DCs for Liquid with Particulates**

| | Fixed Axial Mode (with Particles) | | | | Free Axial Mode (with Particles) | | | |
|---|---|---|---|---|---|---|---|---|
| | $h_{fp}$ (W/m²K) | | $U$ (W/m²K) | | $h_{fp}$ (W/m²K) | | $U$ (W/m²K) | |
| | DC | ANN | DC | ANN | DC | ANN | DC | ANN |
| MRE (%) | 10.24 | 2.6 | 8.62 | 2.9 | 8.3 | 1.85 | 7.35 | 2.5 |
| $R^2$ | 0.92 | 0.98 | 0.92 | 0.97 | 0.95 | 0.99 | 0.95 | 0.98 |

**TABLE 10.3**
**Comparisons of Errors for ANN Models vs. Dimensionless Correlations for Liquid without Particulates**

| | EOE Mode | | Free Axial Mode | |
|---|---|---|---|---|
| | *U* (W/m²K) | | *U* (W/m²K) | |
| | DC | ANN | DC | ANN |
| MRE (%) | 6.05 | 3.81 | 8.34 | 2.06 |
| $R^2$ | 0.97 | 0.98 | 0.95 | 0.99 |

## 10.7 FLOW VISUALIZATION

To gain a better understanding of the associated heat transfer mechanism within the can, flow visualizations can be used to trace and study the motion of particles through liquid or the motion of liquid inside a can being processed (Rao and Ananthswaran, 1988). Since the motion inside the can has already been understood to affect the heat transfer rates, flow visualization becomes an invaluable tool in process calculations. Using flow visualizations, Meng and Ramaswamy (2006) were able to observe the relative movement of the particles in viscous fluids in containers subjected to flow conditions simulating the thermal process in EOE rotation. They indicated that fluid motion could easily affect heat transfer, and that differences observed between different particulate food systems could be related in some way to the motion inside the can. Sablani and Ramaswamy (1998) explained the effects of some parameters on heat transfer coefficients after having studied the particle mixing behavior in EOE rotation. Merson and Stoforos (1990) used cans in axial rotation and studied the motion of spherical particles in liquid, and they calculated their relative liquid to particle velocity from idealized particle motion and liquid velocities. Using these velocities, they predicted the heat transfer coefficients using dimensionless correlation. Subsequently, Stoforos and Merson (1992a and b) attempted to explain the decrease in fluid to particle heat transfer coefficient with increasing rotational speed and decreasing liquid viscosity by relating particle motion to heat transfer.

## NOMENCLATURE

| | |
|---|---|
| $a$ | radius of a sphere or cylinder or half length of the side of cube; or radius of a can, m |
| $A$ | surface area, m² |
| $Bi$, | Bias |
| $C/C_p$ | specific heat capacity |
| $d$ | diameter of the particle, m |
| $D$ | diameter of the can, m |
| $h_{ap}$ | apparent heat transfer coefficient between retort and particle, W/m²K |

| | |
|---|---|
| $H_c$, | Can head space, meter |
| $h_{fp}$ | fluid to particle heat transfer coefficient, $W/m^2K$ |
| $H_s$ | can headspace, m |
| $i$, | neurons |
| $I$, | Input signals |
| $J$ | Connecting neurons |
| $k$ | thermal conductivity, W/mK |
| $K_p$, | Conductivity of the particle, W/mC |
| $m$ | mass, kg |
| $O$, | Output signal response |
| $p$, | Density. $Kg/m^3$ (its mentioned in nomenclature under greek symbol) |
| $p_p$, | It can be read using subscript as density of the particle, no need to rewrite it |
| $p_l$, | Density of the liquid (no need to rewrite) |
| $r$ | radial coordinate (in spherical or cylindrical coordinates) |
| $R^2$ | coefficient of determination |
| $T$ | temperature, °C |
| $t$ | time, s |
| $U$ | overall heat transfer coefficient, $W/m^2K$ |
| $U_a$ | apparent overall heat transfer coefficient, $W/m^2K$ |
| $U_o$ | Overall heat transfer coefficient |
| $Wij$, | Connection weights |

*Subscripts*

| | |
|---|---|
| c | can |
| i | initial condition, space index |
| l | liquid |
| p | particle |
| r/R | retort |
| s | surface |

*Greek Symbols*

| | |
|---|---|
| $\alpha$ | thermal diffusivity, $m^2/s$ |
| $\beta$ | root of the equation: $\beta \cot \beta B_i - 1 = 0$ |
| $\rho$ | density, $kg/m^3$ |
| $\varepsilon$ | particle concentration, % (particle volume/can volume) |
| $\psi$ | sphericity of the particle ($A_{sphere}/A_p$) |
| $\tau_l$ | time constants for liquid |
| $\tau_p$ | time constants for particle |
| $\omega$ | can angular velocity |

*Dimensionless Parameters*

| | |
|---|---|
| $B_i$ | Biot number |
| $F_r$ | Froude number |
| $G_r$ | Grashof number |
| $N_u$ | Nusselt number |
| $P_r$ | Prandtl number |
| $R_e$ | Reynolds number |

# APPENDIX A

Dimensionless correlations for estimation of convective heat transfer coefficient under mixed convection mode situations for canned liquid with and without particle situation

## A.1 LIQUID WITH PARTICULATE

For $U$ in free axial mode

$$N_u = 29(G_r P_r)^{-0.02} + 0.03306(R_e)^{0.66}(P_r)^{0.49} \times (F_r)^{-0.1635} \times \left(\frac{\rho_p}{\rho_l}\right)^{0.42} \times \left(\frac{e}{100-e}\right)^{0.009} \times \left(\frac{d_p}{D_c}\right)^{-0.70}$$

For $U$ in fixed axial mode

$$N_u = 0.07(G_r P_r)^{0.37} + 0.0379(R_e)^{0.50}(P_r)^{0.43} \times (F_r)^{-0.06} \times \left(\frac{\rho_p}{\rho_l}\right)^{0.6342} \times \left(\frac{e}{100-e}\right)^{0.012} \times \left(\frac{d_p}{D_c}\right)^{-1.27}$$

For $h_{fp}$ in free axial mode

$$N_u = 20(G_r P_r)^{0.01} + 0.00955(R_e)^{0.72}(P_r)^{0.56} \times (F_r)^{-0.195} \times \left(\frac{\rho_p}{\rho_l}\right)^{1.165} \times \left(\frac{e}{100-e}\right)^{0.022} \times \left(\frac{d_p}{D_c}\right)^{-1.05} \left(\frac{K_p}{K_l}\right)^{2.26}$$

For $h_{fp}$ in fixed axial mode

$$N_u = 0.07(G_r P_r)^{0.37} + 0.078(R_e)^{0.50}(P_r)^{0.0.45} \times (F_r)^{-0.06} \times \left(\frac{\rho_p}{\rho_l}\right)^{0.619} \times \left(\frac{e}{100-e}\right)^{0.009} \times \left(\frac{d_p}{D_c}\right)^{-1.12} \times \left(\frac{K_p}{K_l}\right)^{0.48}$$

## A.2 LIQUID WITHOUT PARTICULATE

For $U$ in free axial mode

$$N_u = 58.06(G_r P_r)^{0.31} + 0.015(R_e)^{0.757}(P_r)^{0.708} \times (F_r)^{0.121}$$

For $U$ in EOE mode

$$N_u = 2.17(G_r P_r)^{0.10} + 0.016(R_e)^{0.695}(P_r)^{0.66} \times (F_r)^{-0.08}$$

## APPENDIX B

Neural network-based equations for estimation of overall heat transfer coefficient ($U$) and the fluid to particle heat transfer coefficient $h_{fp}$ in canned Newtonian fluids with particles under free axial mode:

$$U = -0.1192a_1 - 0.1659b_1 + 0.2741c_1 - 0.2004d_1 + 0.2129e_1 - 0.1863f_1 - 0.3840g_1 - 0.2793h_1 - 0.3508$$

$$h_{fp} = -0.2055a_1 - 0.2294b_1 + 0.4056c_1 + 0.1151d_1 + 0.0355e_1 - 0.1788f_1 - 0.2268g_1 - 0.2449h_1 - 0.4326$$

where

$a_1 = \tanh(-0.8985X_1 - 0.2482X_2 - 1.0276X_3 - 0.3915X_4 - 0.1569X_5 + 0.6981X_6 - 0.1100)$

$b_1 = \tanh(-0.0216X_1 + 0.2126X_2 - 1.2319X_3 + 1.2408X_4 - 0.1549X_5 + 0.1875X_6 - 0.7559)$

$c_1 = \tanh(3.5283X_1 - 0.1889X_2 - 0.1636X_3 - 1.1130X_4 + 1.0590X_5 - 0.0168X_6 + 2.5774)$

$d_1 = \tanh(-0.1155X_1 + 0.439X_2 + 0.0980X_3 + 0.5226X_4 - 0.1307X_5 + 1.2359X_6 - 0.9304)$

$e_1 = \tanh(0.0668X_1 - 0.1132X_2 + 0.3295X_3 + 0.0458X_4 - 0.1195X_5 - 0.2453X_6 + 0.1183)$

$f_1 = \tanh(-0.2191X_1 + 0.1217X_2 + 0.2566X_3 - 0.4450X_4 - 0.1066X_5 + 0.3127X_6 - 0.7145)$

$g_1 = \tanh(0.2252X_1 - 1.0693X_2 - 0.0950X_3 - 0.1022X_4 + 0.1485X_5 - 0.0415X_6 + 0.5558)$

$h_1 = \tanh(-0.0740X_1 + 0.4963X_2 - 0.2223X_3 - 0.2070X_4 - 0.0326X_5 + 0.1962X_6 - 0.4295)$

## REFERENCES

Abbatemarco C and Ramaswamy HS. 1994. End-over-end thermal-processing of canned vegetables—effect on texture and color. *Food Research International* 27(4):327–334.

Abbatemarco C and Ramaswamy HS. 1995. Retention of nutrients in canned buffered aqueous solutions as influenced by rotational sterilization in a steam/air retort. *Canadian Journal of Agricultural Engineering* 37:345–350.

Anantheswaran RC and Rao MA. 1985a. Heat transfer to model Newtonian liquid foods in cans during end-over-end rotation. *Journal of Food Engineering* 4(1):1–19.

Anantheswaran RC and Rao MA. 1985b. Heat transfer to model non-Newtonian liquid foods in cans during end-over-end rotation. *Journal of Food Engineering* 4(1):21–35.

Astrom A and Bark G. 1994. Heat transfer between fluid and particles in aseptic processing. *Journal of Food Engineering* 21:97–125.

Berry MR and Bradshaw JG. 1980. Heating characteristics of condensed cream of celery soup in a steritort: Heat penetration and spore count reduction. *Journal of Food Science* 45(4):869–874.

Berry MR and Bradshaw JG. 1982. Heat penetration for sliced mushrooms in brine processed in still and agitating retorts with comparisons to spore count reduction. *Journal of Food Science* 47(5):1698–1704.

Berry MR and Dickerson RW. 1981. Heating characteristic of whole kernel corn processed in a steritort. *Journal of Food Science* 46(2):889–895.

Berry MR and Kohnhorst AL. 1985. Heating characteristics of homogeneous milk-based formulas in cans processed in an agitating retort. *Journal of Food Science* 50(1):209–214.

Berry MR, Savage RA, and Pflug IJ. 1979. Heating characteristics of cream-style corn processed in a steritort: Effects of headspace, reel speed and consistency. *Journal of Food Science* 44(3):831–835.

Bochereau L, Bourgien P, and Palagos B. 1992. A method for prediction by combining data analysis and neural networks: Applications to prediction of apple quality near infra-red spectra. *Journal of Agricultural Engineering Research* 51(3):207–216.

Britt IJ. 1993. Thermal processing in batch-type rotational retorts. PhD dissertation, Technical University of Nova Scotia, Halifax, NS.

Chapman AJ. 1989. *Heat Transfer*, 4th edn. Macmillan Publishing Co., New York.

Clifcorn LE, Peterson GT, Boyd JM, and O'Neil JH. 1950. A new principle for agitating in processing of canned foods. *Food Technology* 4:450–460.

Conley W, Kapp L, and Schuhmann L. 1951. The application of "end-over-end" agitation to the heating and cooling of canned food products. *Food Technology* 5:457–460.

Deniston MF. 1984. Heat transfer coefficients to liquids with food particles in axially rotating cans. MSc thesis, University of California, Davis, CA.

Deniston MF, Hassan BH, and Merson RL. 1987. Heat transfer coefficients to liquids with food particles in axially rotating cans. *Journal of Food Science* 52(4):962–966.

Dwivedi M. 2008. Heat transfer to canned Newtonian liquid particulate mixture subjected to axial agitation processing. PhD thesis (February), Department of Food Science, McGill University, Montreal, Canada.

Fand RM and Keswani KK. 1973. Combine natural and forced convection heat transfer from horizontal cylinde to water. *International Journal of Heat and Mass Transfer* 16:1175–1191.

Fernandez CL, Rao MA, Rajavasireddi SP, and Sastry SK. 1988. Particulate heat transfer to canned snap beans in a steritort. *Journal of Food Process Engineering* 10(3):183–198.

Guiavarc'h YP, Deli V, Van Loey AM, and Hendrickx ME. 2002. Development of an enzymic time temperature integrator for sterilization processes based on Bacillus licheniformis α-amylase at reduced water content. *Journal of Food Science* 67:285–291.

Haentjens TH, Van Loey AM, Hendrickx ME, and Tobback PP. 1998. The SSE of [alpha]-amylase at reduced water content to develop time temperature integrators for sterilization processes. *Lebensmittel-Wissenschaft und-Technologie* 31(5):467–472.

Hassan BH. 1984. Heat transfer coefficients for particles in liquid in axially rotating cans. PhD thesis, Department of Agricultural Engineering, University of California, Davis, CA.

Hiddink J. 1975. Natural convection heating of liquids with reference to sterilization of canned food. Agricultural Research Report No. 839. Center for Agricultural Publishing and Documentation, Wageningen, the Netherlands.

Holdsworth SD. 1997. *Thermal Processing of Packaged Foods*, 1st edn. Blackie Academic and Professional, an imprint of Chapman and Hall, London, United Kingdom.

Hotani S and Mihori T. 1983. Some thermal engineering aspects of the rotation method in sterilization. In *Heat Sterilization of Food*. T. Motohito and K. I. Hayakawa (Eds.). Koseisha-Koseikaku Co., Ltd., Tokyo, Japan, pp. 121–129.

Houtzer RL and Hill RC. 1977. Effect of temperature deviation on process sterilization value with continuous agitating retorts. *Journal of Food Science* 42:775–777.

Hughes JP, Jones TER, and James PW. 2003. Numerical simulation and experimental visualization of the isothermal flow of liquid containing a headspace bubble inside a closed cylinder during off-axis rotation. *Food and Bioproducts Processing* 81(C2):119–128.

Incropera FP and Dewitt DP. 1996. *Introduction to Heat Transfer.* John Wiley & Sons, New york.

James PW, Hughes JP, Jones TER, and Tucker GS. 2006. Numerical simulations of non-isothermal flow in off-axis rotation of a can containing a headspace bubble. *Chemical Engineering Research and Design* 84(A4):311–318.

Lekwauwa AN and Hayakawa K-I. 1986. Computerized model for the prediction of thermal responses of packaged solid-liquid food mixture undergoing thermal processes. *Journal of Food Science* 51(4):1042–1049.

Lenz MK and Lund DB. 1978. The lethality-Fourier number method. Heating rate variations and lethality confidence intervals for forced-convection heated foods in containers. *Journal of Food Process Engineering* 2(3):227–271.

Maesmans G, Hendrickx M, DeCordt S, Fransis A, and Tobback P. 1992. Fluid-to-particle heat transfer coefficient determination of heterogeneous foods: A review. *Journal of Food Processing and Preservation* 16(1):29–69.

Marquis F, Bertch AJ, and Cerf O. 1982. Sterilisation des liquides dans des bouteilles d'axe horizontal. *Influence de la vitesse de rotation sur le transfert de chaleur.* Le Lait 62:220–231.

May NS. 2001. Retort technology. In *Thermal Technologies in Food Processing.* P. Richardson (Ed.). Woodhead Publishing, Cambridge, United Kingdom, pp. 7–28.

Meng Y. 2006. Heat transfer studies on canned particulate viscous fluids during end-over-end rotation. PhD thesis, McGill University, Montreal, Canada.

Meng Y and Ramaswamy HS. 2007a. Dimensionless heat transfer correlations for high viscosity fluid-particle mixtures in cans during end-over-end rotation. *Journal of Food Engineering* 80(2):528–535.

Meng Y and Ramaswamy HS. 2007b. Effect of system variables on heat transfer to canned particulate non-Newtonian fluids during end-over-end rotation. *Food and Bioproducts Processing* 85(C1):34–41.

Merson RL and Stoforos NG. 1990. Motion of spherical particles in axially rotating cans. Effect of liquid particle heat transfer. In *Engineering and Food.* Vol. 2. W.E.L. Spiess and H. Schubert (Eds.). Elsevier Science, New York, pp. 60–69.

Naveh D and Kopelman IJ. 1980. Effect of some processing parameters on the heat-transfer coefficients in a rotating autoclave. *Journal of Food Processing and Preservation* 4(1–2):67–77.

Neural Ware, Inc. 1996. *Using Neural Works.* Neural Ware, Inc. Technical Publication Group, Pittsburgh, PA.

Quast DG and Siozawa YY. 1974. Heat transfer rates during axially rotated cans. *Proceedings of the International Congress of Food Science Technology* 4:458.

Ramaswamy HS, Awuah GB, and Simpson BK. 1997. Heat transfer and lethality considerations in aseptic processing of liquid/particle mixtures: A review. *Critical Reviews in Food Science and Nutrition* 37(3):253–286.

Ramaswamy HS and Marcotte M. 2005. *Food Processing Principles and Applications,* 1st edn. CRC Press, Boca Raton, FL.

Ramaswamy HS and Sablani SS. 1997. Particle motion and heat transfer in cans during end over end processing. Influence of physical properties and rotational speed. *Journal of Food Processing and Preservation* 21:105–127.

Ramaswamy HS and Zareifard MR. 2003. Dimensionless correlations for forced convection heat transfer to spherical particles under tube-flow heating conditions. In *Transport Phenomena in Food Processing.* J. Welti-Chanes, J.F. Velez-Ruiz, and G. Barbosa-Canovas (Eds.). CRC Press, Boca Raton, FL, pp. 505–520.

Rao MA and Anantheswaran RC. 1988. Convective heat transfer to fluid foods in cans. *Advances in Food Research* 32:39–84.

Rao MA, Cooley HJ, Anantheswaran RC, and Ennis RW. 1985. Convective heat transfer to canned liquid foods in a Steritort. *Journal of Food Science* 50(1):150–154.

Ruyter PWD and Brunet R. 1973. Estimation of process conditions for continuous sterilization of foods containing particulates. *Food Technology* 27(7):44.

Sablani SS. 1996. Heat transfer studies of liquid particle mixtures in cans subjected to end over end processing. PhD thesis, Department of Food Science, McGill University, Montreal, Canada.

Sablani SS and Ramaswamy HS. 1995. Fluid-to-particle heat-transfer coefficients in cans during end-over-end processing. *Food Science and Technology [Lebensmittel-Wissenschaft & Technologie]* 28(1):56–61.

Sablani SS and Ramaswamy HS. 1996. Particle heat transfer coefficients under various retort operating conditions with end-over-end rotation. *Journal of Food Process Engineering* 19(4):403–424.

Sablani SS and Ramaswamy HS. 1997. Heat transfer to particles in cans with end-over-end rotation: Influence of particle size and concentration (%v/v). *Journal of Food Process Engineering* 20(4):265–283.

Sablani SS and Ramaswamy HS. 1998. Multi-particle mixing behavior and its role in heat transfer during end-over-end agitation of cans. *Journal of Food Engineering* 38(2):141–152.

Sablani SS and Ramaswamy HS. 1999. End-over-end agitation processing of cans containing liquid particle mixtures. Influence of continuous vs. oscillatory rotation. *Food Science and Technology International* 5(5):385–389.

Sablani SS, Ramaswamy HS, and Prasher, SO 1995. A neural-network approach for thermal-processing applications. *Journal of Food Processing and Preservation* 19(4):283–301.

Sablani SS, Ramaswamy HS, and Mujumdar AS. 1997a. Dimensionless correlations for convective heat transfer to liquid and particles in cans subjected to end-over-end rotation. *Journal of Food Engineering* 34(4):453–472.

Sablani SS, Ramaswamy HS, Sreekanth S, and Prasher, SO. 1997b. Neural network modeling of heat transfer to liquid particle mixtures in cans subjected to end-over-end processing. *Food Research International* 30(2):105–116.

Stoforos NG. 1988. Heat transfer in axially rotating canned liquid/particulate food system. PhD thesis, University of California, Davis, CA.

Stoforos NG and Merson RL. 1990. Estimating heat transfer coefficients in liquid/particulate canned foods using only liquid temperature data. *Journal of Food Science* 55(2):478–483.

Stoforos NG and Merson RL. 1991. Measurement of heat-transfer coefficients in rotating liquid/particulate systems. *Biotechnology Progress* 7(3):267–271.

Stoforos NG and Merson RL. 1992a. Physical property and rotational speed effects on heat-transfer in axially rotating liquid particulate canned foods. *Journal of Food Science* 57(3):749–754.

Stoforos NG and Merson RL. 1992b. Physical property and rotational speed effect on heat transfer in agitating liquid/particulate canned food. *Journal of Food Process Engineering* 18:165–185.

Stoforos NG and Merson RL. 1995. A solution to the equations governing heat-transfer in agitating liquid particulate canned foods. *Journal of Food Process Engineering* 18(2):165–185.

Stoforos NG and Reid DS. 1992. Factors influencing serum separation of tomato ketchup. *Journal of Food Science* 57(3):707–713.

Tattiyakul J, Rao MA, and Datta AK. 2002a. Heat transfer to a canned corn starch dispersion under intermittent agitation. *Journal of Food Engineering* 54(4):321–329.

Tattiyakul J, Rao MA, and Datta AK. 2002b. Heat transfer to three canned fluids of different thermo-rheological behaviour under intermittent agitation. *Food and Bioproducts Processing* 80(C1):20–27.

Teixeira AA, Dixon JR, Zahradnik WZ, and Zinsmeister GE. 1969. Computer optimization of nutrient retention in the thermal processing of conduction-heated foods. *Food Technology* 23:845–850.

Tsurkerman VC, Rogachev VI, Fromzel QG, Pogrebenco NM, Tsenkevich VI, Evstigneer GM, and Kolodyazhnyi CF. 1971. Preservation of tomato paste in large containers under rotation. *Konsewnaya i Ovashchesushi "naya promyshlennost"* 11:11–13 (abstract). (Cited by Quast and Siozawa (1974).)

Varma MN and Kannan A. 2006. CFD studies on natural convective heating of canned food in conical and cylindrical containers. *Journal of Food Engineering* 77(4):1024–1036.

Weng ZJ, Hendrickx M, Maesmans G, and Tobback P. 1992. The use of a time temperature-integrator in conjunction with mathematical-modeling for determining liquid particle heat-transfer coefficients. *Journal of Food Engineering* 16(3):197–214.

Willhoft EMA. 1993. *Aseptic Processing and Packaging of Particulate Foods*, 1st edn. Blackie Academic & Professional, London, United Kingdom.

# 11 Numerical Model for Ohmic Heating of Foods

*Michele Chiumenti, Cristian Maggiolo, and Julio García*

## CONTENTS



## 11.1 INTRODUCTION

During the last years, very complex phenomena have been successfully studied with the help of sophisticated numerical tools. Every year, the accuracy of the results coming from the numerical simulation of different industrial processes is increasing due to a better knowledge of both the governing laws, which describe the physics and the constitutive laws, which define the food behavior.

Nowadays, complex formulations are able to describe physical phenomena such as the heat transfer, fluids dynamics, chemical reactions, microbial growth, among others (Wappling-Raaholt et al. 2002).

In this chapter, both food durability and food preservation processes are studied within the framework provided by the numerical simulation, a powerful prediction tool able to quantify the effectiveness of a treatment in a food preservation process.

This way, it is possible to predict with a certain degree of accuracy the temperature field as well as the food humidity, or any other variable, which can be used to control and optimize both food quality and safety, reducing the number of experimental testing.

This chapter will focus on the ohmic heating as an emerging food preservation technology currently used by the food industry.

## 11.2 OHMIC TREATMENT

The ohmic heating (also called Joule heating or electric resistance heating) consists of an interesting alternative to generate the thermal field required in a food preservation process (Palaniappan and Sastry 2002, Fryer and Davies 2001, and Zhang and Fryer 1994). In this case, an electric current passing through the food generates an internal heat source induced by the thermal resistance (Joule effect) of the food (U.S. Food and Drug Administration, 2000).

The ohmic treatment requires the use of electrodes in contact with the food, a frequency range without restrictions (varying from radio to microwave frequencies), and a wave shape also without any restrictions (usually sinusoidal waves are used).

The possible range of applications for the ohmic heating goes from blanching processes, evaporation processes, to fermentation processes and extraction processes. As an example, the ohmic heating is successfully and largely used to process any kind of liquid foods such as the fruit juice, leading to safe products with a minimal distortion of the original taste.

The most important advantage of the ohmic treatment is the fast and uniform food heating required by the microbial control in a food preservation process. Another important advantage of such technique is the possibility of treating liquids containing particles (Sastry 1992, Sastry and Salengke 1998). In fact, through the ohmic heating, it is possible to warm up the particles as fast as the fluid due to the different ionic content (electrical conductivity) of the liquid phase if compared to the particles. As a result, the ohmic treatment is able to optimize the process when compared to more conventional techniques reducing the process time and increasing the final quality.

## 11.3 GOVERNING LAWS

In this section, the governing laws used to describe the ohmic process are presented.

### 11.3.1 Heat Transfer Model

The temperature field, $T$, induced by the electric current in the ohmic heating is computed solving the balance of energy equation given by

$$\frac{\mathrm{d}H}{\mathrm{d}t} = R - \nabla \cdot \mathbf{Q} \tag{11.1}$$

where

$H$ is the enthalpy function

$\mathbf{Q}$ is the heat flux

$R$ is the thermal source (per unit of volume) generated by the thermal resistance of the substance and induced by the electric field

The enthalpy function measures the amount of heat absorbed by the system and it can be computed as

$$\frac{\mathrm{d}H}{\mathrm{d}t} = \left(\frac{\mathrm{d}H}{\mathrm{d}T}\right)\left(\frac{\mathrm{d}T}{\mathrm{d}t}\right) = C_{\text{food}} \frac{\mathrm{d}T}{\mathrm{d}t} \tag{11.2}$$

where $C_{food}(T) = dH(T)/dT$ is the heat capacity of the food substance selected. The heat capacity can also be expressed in terms of density and specific heat, $\rho_{food}(T)$ and $c_{food}(T)$, respectively, as $C_{food}(T) = \rho_{food}(T)c_{food}(T)$ so that the balance of energy equation results in

$$\rho_{food}\, c_{food}\, \frac{\partial T}{\partial t} = R - \nabla \cdot \mathbf{Q} \tag{11.3}$$

The heat flux is computed using the Fourier's law as

$$\mathbf{Q} = -K_{food} \nabla T \tag{11.4}$$

where $K_{food}(T)$ is the (temperature dependent) thermal conductivity of the food product.

The next step is the integration of the balance of energy equation in whole domain of the problem, $V$. Applying the Galerkin method, the resulting weak form yields:

$$\int_V \delta T \cdot \left( \rho_{food}\, c_{food}\, \frac{\partial T}{\partial t} \right) dV + \int_V \nabla(\delta T) \cdot K_{food} \cdot \nabla T\, dV = \int_V \delta T \cdot R\, dV + \int_{S_q} \delta T \cdot \bar{q}\, dV \tag{11.5}$$

where $\bar{q} = \mathbf{Q} \cdot \mathbf{n}$ is a prescribed heat flux (Neumann's condition) normal to the boundary domain $S_q$. Finally, suitable initial and Dirichlet's boundary conditions must be introduced as

$$\begin{aligned} T(t=0) &= T_0 \quad \text{on} \quad V \\ T &= \bar{T} \quad \text{on} \quad S_T \end{aligned} \tag{11.6}$$

### 11.3.2 Thermophysical Properties

The thermophysical properties introduced above are a fundamental ingredient to solve heat transfer equation above presented. It is well known that material properties are strongly depending on food structure, composition, and temperature, among others. On the other hand, such temperature field is changing during the process operation affecting both the properties and the process itself, leading to a strong coupled problem to be solved. The solution proposed herewith computes the thermophysical properties as a function of the proximal composition (defined in terms of macronutrient fractions) according to the prediction laws proposed by Choi and Okos (1986) and Toledo (1999).

According to this work, once specified the original composition of the food product in terms of proteins, carbohydrates, fats, fibers, and minerals content, any material property is built up using the mixture theory as

$$P_{food}(T) = \sum_{j=1}^{nc} P_j(T) X_j \tag{11.7}$$

Hence, the generic food property $P_{food}(T)$ depends on the properties of each component, $P_j(T)$ weighted by its volume fraction (in dry basis), $X_j$.

### 11.3.3 Electric Field and the Ohm's Law

The heat generated (per unit of volume) during the ohmic treatment is introduced into the balance of energy equation through the source term, $R$ as

$$R = \mathbf{E} \cdot \mathbf{J} \tag{11.8}$$

where $\mathbf{J}$ and $\mathbf{E}$ are the current density and the electric field, respectively. The Ohm's law states the relationship between those two fields as

$$\mathbf{J} = \mathbf{E}\sigma_{\text{food}} \tag{11.9}$$

where $\sigma_{\text{food}}(T)$ is the electric conductivity. For most of the materials, including food products, and in our case liquid foods, the electric conductivity is strongly dependent on the temperature field. A possible relationship can be established introducing the following linear law:

$$\sigma_{\text{food}}(T) = \sigma_0(1 - mT) \tag{11.10}$$

where

$\sigma_0$ is the electric conductivity at the reference temperature
$m$ is the temperature coefficient as shown by Palaniappan and Sastry (1991)

On the other hand, the electric field, $\mathbf{E}$, can be expressed as the gradient of a scalar function called electrostatic potential, $\phi$ (also known as the voltage). An electric field, points from regions of high potential, to regions of low potential, expressed mathematically solving the following equation:

$$\mathbf{E} = -\nabla\phi \tag{11.11}$$

As a result, the heat generated in the ohmic treatment by means of an electric potential difference applied between two conducting electrodes immersed in the substance to be heated yields:

$$R = \sigma_{\text{food}} \left\|\nabla\phi\right\|^2 \tag{11.12}$$

Observe that the electrostatic potential associated with a conservative electric field, in the absence of unpaired electric charge can be computed using the following equation:

$$\nabla \cdot \mathbf{E} = 0 \tag{11.13}$$

which can be rewritten in terms of the electric potential as a Laplace's equation of the form:

$$\nabla^2\phi = 0 \quad \text{or} \quad \Delta\phi = 0 \tag{11.14}$$

This equation must be integrated within the entire domain of interest, $V$, and complemented by the corresponding boundary conditions in terms of prescribed potential (Dirichlet's conditions). Applying the Galerkin method, the weak form of the electrostatic potential equation results in (see De Alwis and Fryer 1990)

$$\int_V \nabla(\delta\phi)\cdot\nabla\phi\,\mathrm{d}V = 0 \quad \text{on} \quad V \tag{11.15}$$

$$\phi = \overline{\phi} \quad \text{on} \quad S_\phi$$

### 11.3.4 Microbial Lethality Criterion

To compute the thermal lethality of the microbial population, $F$, the general method of Bigelow has been implemented as follows (more details in Simpson et al., 2003):

$$F = D_{\text{ref}} \cdot x = \int_0^t 10^{\frac{T - T_{\text{ref}}}{z_{\text{ref}}}}\,\mathrm{d}t \tag{11.16}$$

where

$T_{\text{ref}}$ is the process temperature
$z_{\text{ref}}$ is the temperature gradient necessary for a decimal reduction of the initial microbial population
$D_{\text{ref}}$ is the time necessary for a decimal reduction of the microorganisms
$x$ is the number of decimal reductions required

The final concentration of microorganisms, $C_{\text{f}}$, can be expressed in a practical format as the initial concentration, $C_0$, reduced decimally ($x$ times) during the process, as

$$C_{\text{f}} = C_0 \times 10^{-x} = C_0 \times 10^{-\frac{F}{D_{\text{ref}}}} \tag{11.17}$$

## 11.4 EXPERIMENTAL SETTING

A number of experiments have been carried out to, first, calibrate the model according to the experimental evidence and, later, to validate the proposed numerical tool.

The experiments have been carried out in pilot ohmic heater consisting of a glass chamber, T-shaped, where the food can be processed (Figure 11.1). The heater is 18 cm long with a diameter of 4 cm. A thermocouple can be introduced from the

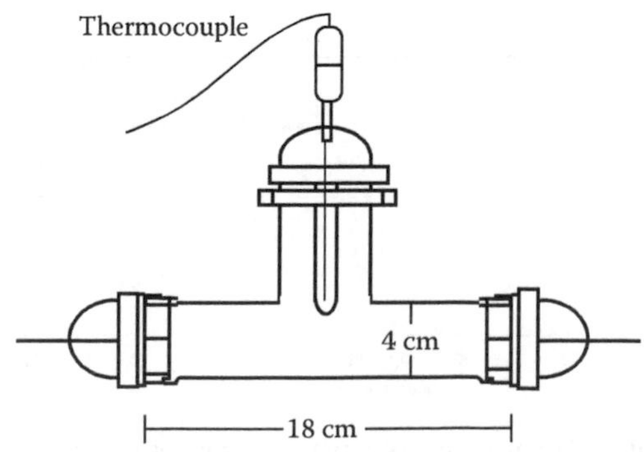


**FIGURE 11.1** Outline of the pilot ohmic heater.

**TABLE 11.1**
**Proximal Composition of Orange Juice**

| Macronutrient | Proximal Fraction |
|---|---|
| Water | 0.883 |
| Protein | 0.007 |
| Fat | 0.002 |
| Carbohydrate | 0.164 |
| Ash | 0.004 |

upper side while at the lateral extreme there are two stainless steel electrodes used to apply the electric potential difference.

The model has been validated processing orange juice with the ohmic treatment. The orange juice proximal composition has been obtained from the Chilean foods compositions table (Schmidt-Hebbel et al. 1992) and it is presented in Table 11.1.

Using the food characterization proposed by Choi and Okos (1986), it is possible to get the following temperature-dependent laws for the thermophysical properties of the orange juice: (Figures 11.2 through 11.4)

$$\rho_{\text{food}}(T) = (-0.0036)T^2 - (0.0106)T + 1042.1 \tag{11.18}$$

$$c_{\text{food}}(T) = (-0.0055)T^2 + (0.1428)T + 3871.1 \tag{11.19}$$

$$K_{\text{food}}(T) = (6 \times 10^{-6})T^2 + (0.0017)T + 0.5426 \tag{11.20}$$

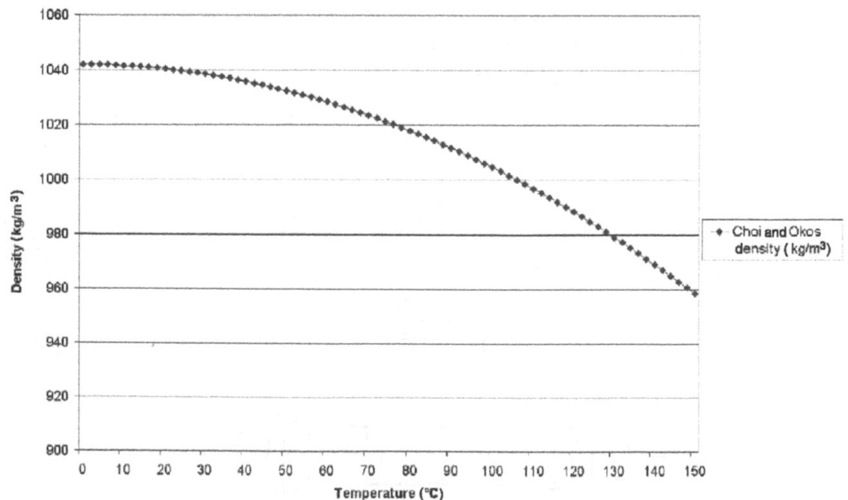


**FIGURE 11.2** Density of orange juice (kg/m³) vs. temperature (°C).

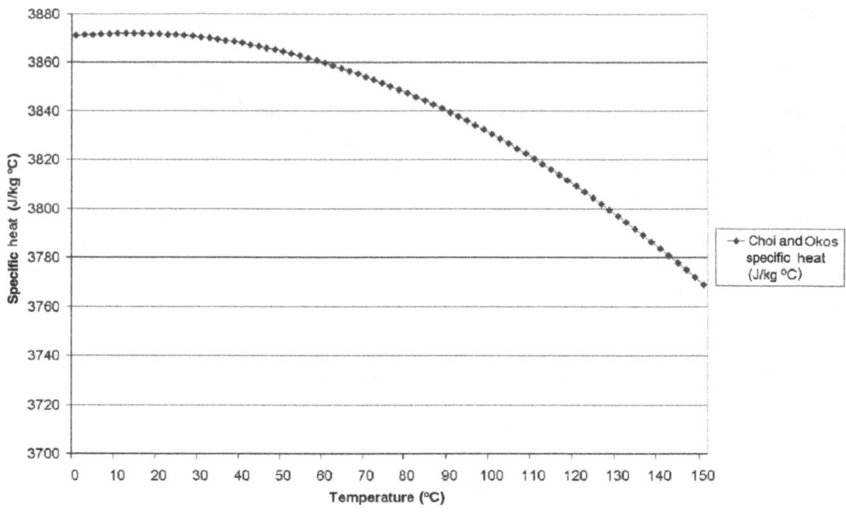


**FIGURE 11.3** Specific heat (J/kg °C) vs. temperature (°C).

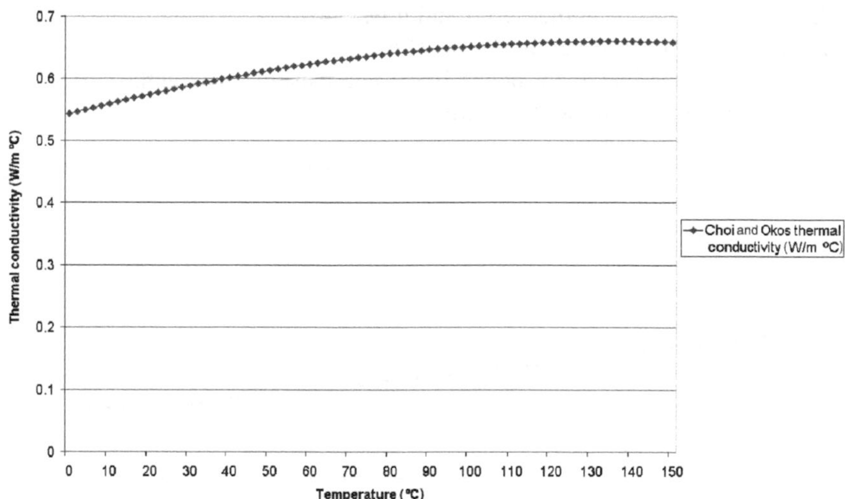


**FIGURE 11.4** Thermal conductivity of orange juice (W/m °C) vs. temperature (°C).

A lineal relationship between the electrical conductivity and the temperature field (see Palaniappan and Sastry 1991) has been assumed as follows (Figure 11.5):

$$\sigma_{\text{food}}(T) = (0.0057)T + 0.1279 \tag{11.21}$$

The electric potential difference imposed to the electrodes is $\phi = 230$ V. On the other side, during the experiment, the electric voltage effectively supplied to

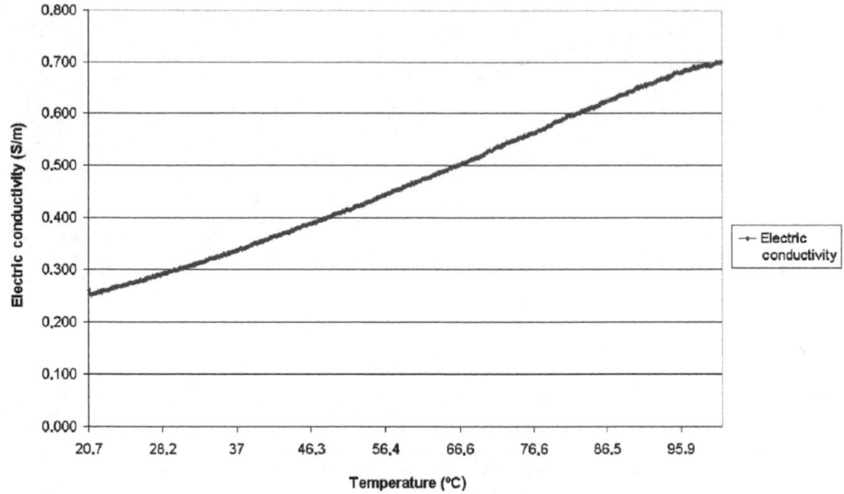


**FIGURE 11.5** Electrical conductivity (S/m) vs. temperature (°C).

the ohmic heather has been measured and it was possible to see a time-dependent evolution as it can be observed in Figure 11.6

Note that it is possible to consider a linear relationship between the current voltages with time, introduced into the model as

$$\phi = (-0.0341)t + 229.7 \tag{11.22}$$

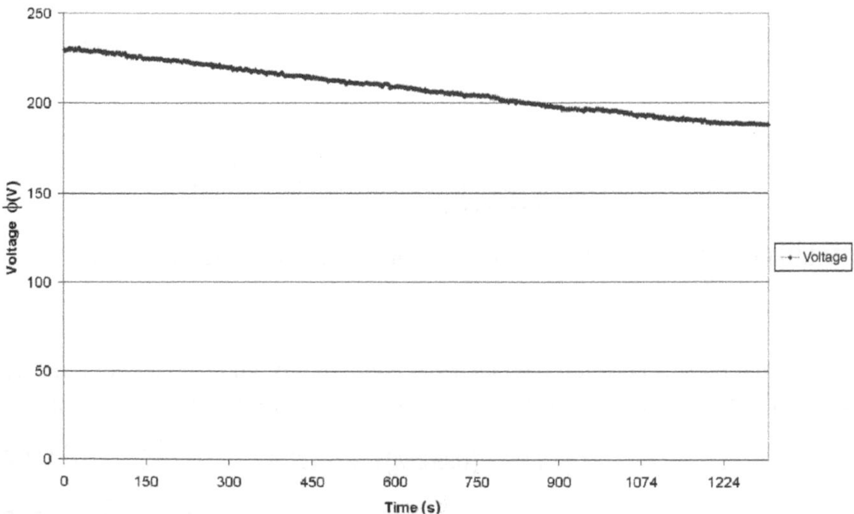


**FIGURE 11.6** Current voltage ϕ(v) vs. time (s).

**TABLE 11.2**
***E. coli* Thermal Resistance Parameters**

| | $T_{ref}$ (°C) | $z_{ref}$ (°C) | $D_{ref}$ (min) |
|---|---|---|---|
| *E. coli* | 62.8 | 5.3 | 0.47 |

Finally, the thermal lethality of the microbial population (*Escherichia coli*) for acidic fruit juice such as the orange juice, U.S. Food and Drug Administration (1999) (pH = 3.0), can be computed using the thermal resistance parameters defined in Table 11.2

The U.S. Food and Drug Administration (1999) suggests as a thermal lethality criterion for acidic juices $F = 5 \cdot D_{ref}$, which means that the food safety is ensured when $F \geq 2.35\,\text{min}$.

## 11.5 NUMERICAL RESULTS

The numerical simulations have been carried out using an in-house software package called *ProFood* (http://www.cimne.upc.es/profood/). *ProFood* is a finite element-based software, which allows the numerical simulation of coupled analyses such as coupled thermomechanical simulation, or heat–mass transfer (dehydration or drying processes) as well as thermoelectrical simulations as presented in this work and required for the ohmic preservation process optimization.

Both preprocessing tools (used for the meshing operation and problem definition) and postprocessing framework (to visualize the result obtained) are integrated into the *ProFood* software platform.

A preliminary numerical test has been performed considering a vertical section of the original geometry. The resulting 2D geometry has been meshed using 3200 triangular elements and 1700 nodes (see Figure 11.7).

Figure 11.8a and b shows the electric voltage contour fields at the beginning and at the end of the process, respectively. It is possible to observe how the electric

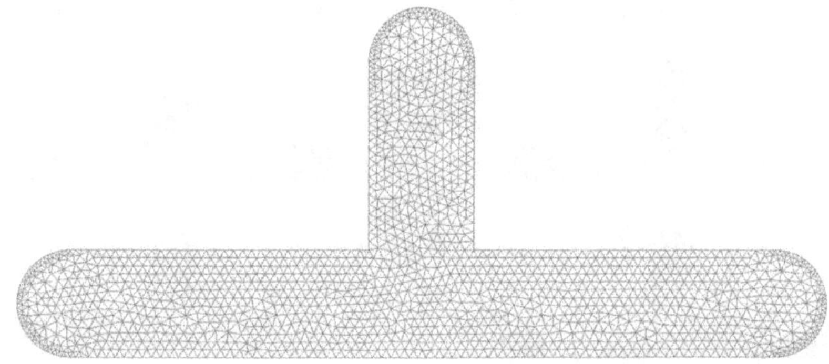

**FIGURE 11.7** 2D finite element mesh of ohmic heater.

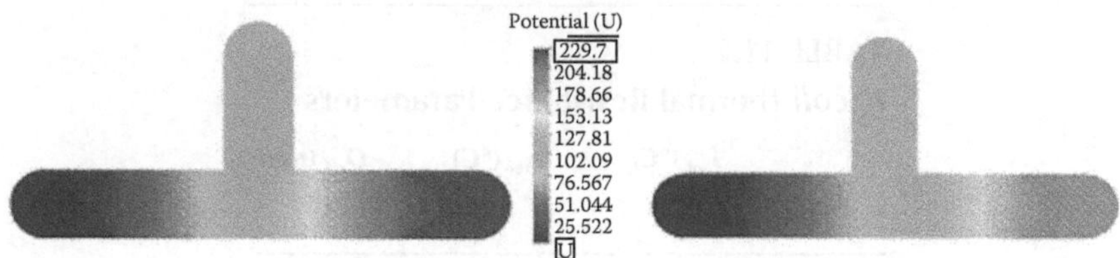


**FIGURE 11.8** Electric potential at the beginning and at the end of the process time.

potential difference is decreasing according to the experimental evidence. This result is used by the heat transfer solver as an input data to compute the thermal resistance (ohmic heat source). Note that the electric potential is mostly linear between the two electrodes and constant in the vertical tube as well as at the two extremes of the horizontal tube meaning that there will exist a heat source only between the two electrodes in the horizontal tube and this should be reflected in the temperature contour fields.

Figure 11.9 shows the evolution of the temperature field at different time steps. Observe how the temperature increases only where the electric field induces the thermal dissipation by ohmic resistance.

Figure 11.10 compares the temperature evolution registered by the thermocouple with the one computed at the same location. It is possible to observe how the numerical results agree very well with available experimental data.

Finally, Figure 11.11 compares the thermal lethality computed using the prediction model implemented in *ProFood* software with experimental data. Also in this case, the difference is negligible. The graph shows how the lethality required (2.35 min) for the sterilization process is achieved after 742 s.

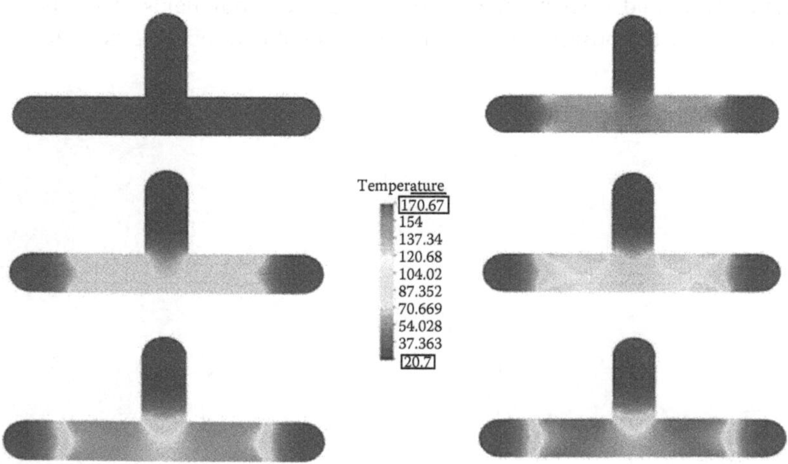


**FIGURE 11.9** Temperature contour-fill at different time steps.

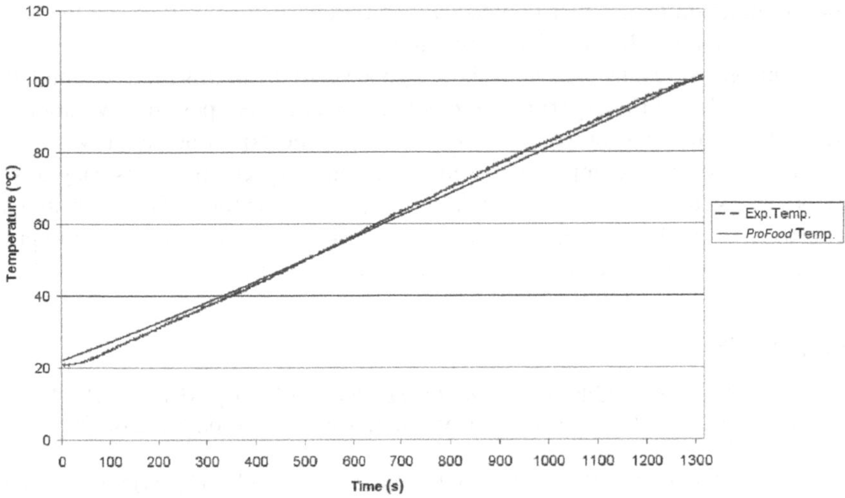


**FIGURE 11.10** Temperature evolution at the thermocouple location: *ProFood* software results vs. experimental data.

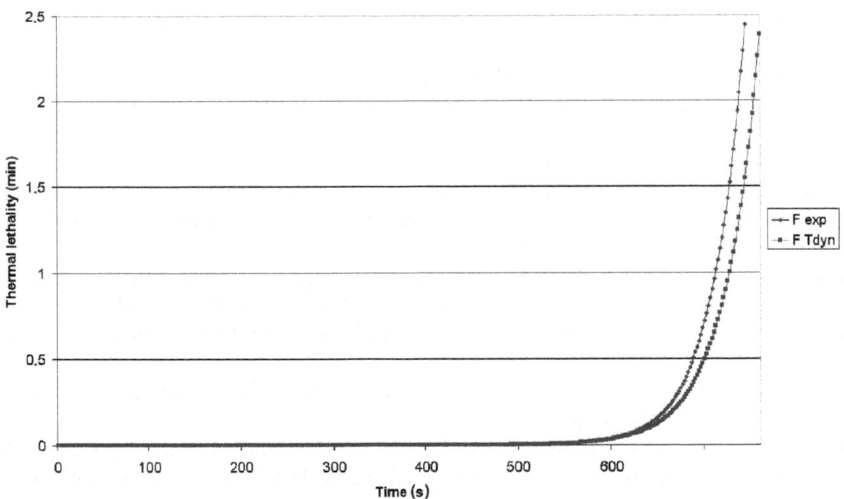


**FIGURE 11.11** Thermal lethality evolution: *ProFood* software results vs. experimental data.

## 11.6 CONCLUSIONS

The objective of the work of this chapter was the numerical simulation of the ohmic heating process as an interesting preservation technology now extensively used in the food industry. The governing equations which describe the physics of the problem have been detailed in terms of a heat transfer model coupled with the electric field generated by the electric potential difference imposed to the electrodes.

A characterization of the thermophysical food properties as well as the microbial lethality criterion has been added to the model.

The numerical results presented show a good accuracy of this process simulation when compared with the experimental evidence. Both the temperature evolution and the microbial concentration compare very well with the experimental curve.

Extension to more complex geometries or boundary conditions is easy to perform as well as the coupling with any kind of optimization tools. The software platform presented can be considered as a very interesting tool to optimize and improve the industrial use of this kind of food preservation process.

## REFERENCES

Choi, Y. and Okos, M.R. 1986, *Effects of Temperature and Composition on the Thermal Properties of Foods*, eds. M.L. Maguer, and P. Jelen. New York, NY, Elsevier Science, pp. 93–101.

De Alwis, A.A.P. and Fryer, P.J. 1990, A finite element analysis of heat generation and transfer during ohmic heating of food. *Chem. Eng. Sci.*, 45, 1547–1559.

Fryer, P.J. and Davies, L.J. 2001, Modelling electrical resistance ("ohmic") heating of foods. *Food Sci. Technol.*, 107, 225–264.

Palaniappan, S. and Sastry, S.K. 1991, Electrical conductivity of selected solid foods during ohmic heating. *J. Food Process Eng.*, 14, 221–236.

Palaniappan, S. and Sastry, S.K. 2002, Ohmic heating. *Food Sci. Technol.*, 114, 451–460.

Sastry, S.K. 1992, A model for heating of liquid–particle mixtures in a continuous flow ohmic heater. *J. Food Proc. Eng.*, 15, 263–278.

Sastry, S.K. and Salengke, S. 1998, Ohmic heating of solid–liquid mixtures: A comparison of mathematical models under worstcase heating conditions. *J. Food Proc. Eng.*, 21, 441–458.

Schmidt-Hebbel, H., Pennacchiotti, M.I., Masson, S.L., Vinagre, L.J., Mella, R.M.A., Zuccarelli, P.M.T., Carrasco, D.C., Jaña, M.W., and Oliver, A.H. 1992, *Tabla de Composición Química de Alimentos Chilenos.* 8ª Edición. Editorial Universitaria. Facultad de Ciencias Químicas y Farmacéuticas, Universidad de Chile.

Simpson, R., Almonacid, S., and Teixeira, A. 2003, Bigelow's general method revisited: Development of a new calculation technique. *J. Food Sci.*, 68(4), 1324–1333.

Toledo, R. 1999, *Fundamentals of Food Process Engineering.* Gaithersburg, MD, Aspen Publishers.

U.S. Food and Drug Administration (FDA) 1999, *Juice HACCP.* Available at: http://www.cfsan.fda.gov/~comm/haccpjui.html

U.S. Food and Drug Administration (FDA), *Kinetics of Microbial Inactivation for Alternative Food Processing Technologies 2000.* Available at: http://www.cfsan.fda.gov/~comm/ift-exec.html

Wappling-Raaholt, B., Scheerlinck, N., Galt, S., Banga, J.R., Alonso, A., Balsa-Canto, E., Van Impe, J., Ohlsson, T., and Nicolai, B.M., 2002, Combined electromagnetic and heat transfer model for heating of foods in microwave combination ovens. *J. Microwave Power EE.*, 37(2), 97–111.

Zhang, L. and Fryer, P.J. 1994, Food sterilization by electrical heating: Sensitivity to process parameters. *Am. Inst. Chem. Eng. J.*, 40, 888.

# 12 Computational Fluid Dynamics in Thermal Processing

*Tomás Norton and Da-Wen Sun*

## CONTENTS

## 12.1 INTRODUCTION

The application of the principles of fluid motion and heat transfer to design problems in the food industry has undergone remarkable development since the early 1990s. Problems involving heat and mass transfer, phase change, chemical reactions, and complex geometry, which once required either highly expensive or oversimplified computations, can now be efficiently completed with a high level of spatial and temporal accuracy on personal computers. This remarkable progression is owing to the development of a design and analysis tool, known as computational fluid dynamics (CFD), which can be used to tackle complex problems in fluid mechanics and heat transfer, and many other physical processes. CFD is based on numerical methods that solve for the governing transport mechanisms over a multidimensional domain of interest. Its physical basis is rooted in classical fluid mechanics. Since its first computer implementation in the 1950s, the technical formulation of CFD has been evolving contemporaneously with the digital computer (Norton and Sun, 2007). In its present-day form, CFD can be used to efficiently quantify many complex dynamic phenomena and as a result has developed into a multifaceted industry, generating billions of Euros worldwide within a vast range of specializations (Abbott and Basco, 1989; Xia and Sun, 2002).

Thermal processes are widely recognized as major preservation techniques, which can be applied by a processor to render a food product commercially sterile, as well as to modify sensory characteristics (Richardson, 2001). During the thermal processing of foods, thermodynamic and fluid-dynamic phenomena govern many and are involved in almost every process. Due to the complexity of the physical mechanisms involved in thermal processing, the ability to fully quantify the phenomena is dependent both on the physics and the level of precision associated with the analysis tools. Even though physical experimentation may allow for direct measurement, without the need for modeling assumptions, a comprehensive analysis would not only necessitate expensive equipment and require considerable time and expertise, but could also be highly intrusive (Verboven et al., 2004). Fortunately, ubiquitous processes such as blanching, cooking, drying, sterilization, and pasteurization, which rely on thermal exchange to raise a product thermal-center to a prespecified temperature, are highly amenable to CFD modeling, and have unarguably contributed to its exponential take-up witnessed in recent years. Coupled with other technological advancements, CFD has led to an improvement in both food quality and safety alongside reducing energy consumption, and has reduced the amount of experimentation and empiricism associated with a design process (Norton and Sun, 2007). Further evidence suggests that present-day CFD software, with its rich tapestries of mathematical physics, numerical methods, user interfaces, and state-of-the-art visualization techniques, is currently recognized as a formidable technology to permit comprehensive analyses of thermal processing (Verboven et al., 2004).

Recently, Sun (2007) has compiled a state-of-the-art reference of CFD applications in the food industry in which light was shed by experts on both the understanding of processing technologies via CFD and the advantages that CFD can offer in a commercial setting. General applications of CFD in the food industry have also been reviewed by some authors (Scott and Richardson, 1997; Xia and Sun, 2002; Norton and Sun, 2007). In a review, more specific to thermal processing, Verboven et al. (2004) have discussed the fundamentals of CFD applications and highlighted the existing limitations and challenges that face the current users of this technology. In light of the rapid development in both thermal processing and CFD technology that has taken place over recent years, this chapter gives a comprehensive account of the latest advances, made through successful applications in research. First, the need for CFD to model thermal processes will be discussed. The main equations governing the physical mechanisms encountered in thermal processing will then be discussed followed by an extensive overview of the conventional processes, which have been modeled with CFD. An account of the CFD modeling studies of emerging thermal processing technologies will then be highlighted. Finally, an overview of the challenges that are still encountered in the CFD modeling of thermal processes is given.

## 12.2 HETEROGENEITY IN THERMAL PROCESSING: THE NEED FOR DISTRIBUTED PARAMETER MODELS

Owing to their complex thermophysical and in some cases polymorphic properties, heat transfer to and from foods can stimulate complex chemical and physical alteration. Nowadays, with both food preservation and safety being equally important objectives of processing, it is necessary to promote quality characteristics of food whilst eradicating the threat of spoilage. For this to happen efficiently, not only must the appropriate temperature and duration of heating be known but other influential phenomena must also be understood. Mathematical models permit the representation of a physical process, from the analysis of measured data or physical properties over a range of experimental conditions, and allow process parameters to then be ascribed as constant or variable, lumped or distributed. Such a representation is necessary in order for model predictions to be generalized in a manner amenable to equipment design and optimization. Both the theoretical and descriptive aspects of food processing have long been regarded as highly important. Heldman and Singh (1981) have described food engineering as being composed of two distinct parts, (a) the descriptive part which deals with detailed descriptions of physical processes and equipment and (b) the theoretical part, which comprises a mathematical description of the processing equipment and the changes that may occur during processing and storage, etc. As these two parts supplement and support each other, mathematical modeling of unit processes has always been an integral part of industrial design.

During thermal processing, heat transfer occurs due to one or more of the three mechanisms: conduction, convection, and radiation, and depending on the thermal process one mechanism is usually dominant (Wang and Sun, 2003). As the thermal exchange phenomena are either distributed in space or in space and time, their fundamental representation is by means of partial differential equations (PDEs). However, the full solution of these equations requires complex time-consuming iterative techniques, which have only been attainable within the last number of years. Thus, their simplification has proved, in the past, to be necessary. For example, temperature evolution through a three-dimensional (3D) solid is accurately represented by a parabolic PDE. By assuming that linear conditions prevail, this equation can be integrated to describe 1D steady-state heat transfer between the hot and the cold surface of a solid, and provided that the proportionality constant (which is an intensive property of the conducting material) is adequately represented, a reasonably accurate solution can be obtained. Also in the past, when spatially dependent and transient solutions were sought to optimize the thermal processing of solids, the 1D Fourier equation was used in combination with finite difference methods (Hendrickx et al., 1993). Simplifying conduction to a 1D problem has proved to be a suitable means of describing conduction heat transfer where conditions allow, as heat flows at right angles to the temperature gradient. However, more involved methods are required when the temperature depends on more than one space variable and when heat is transferred through irregular geometry; as generally encountered in the food industry (Tewkesbury et al., 2000).

In contrast to conduction, convection permits heat transfer by mass motion of a fluid system when it is heated and caused to move, carrying energy away from the source of heat, thereby allowing cooler fluid to replace it. Convection above a hot surface occurs because hot fluid expands, becomes less dense, and rises. Depending on the flow velocity, convection can be free (i.e., natural) or forced. Newton's law of cooling describes the energy flow rate due to convective heat transfer at the interface between a solid object and the bulk fluid environment, and is a function of a proportionality constant known as the heat transfer coefficient. Unlike the proportionality constant for conduction, the heat transfer coefficient is an empirically determined parameter with units of W $m^{-2}$ $K^{-1}$, whose value depends on many relevant conditions, which are all process dependent. For convenience, this coefficient can be determined by optimization methods, which iterate through solutions until the heat transfer model for the solid arrives at a good fit to experimental data (Varga and Oliveira, 2000). An average or global heat transfer coefficient has been used in many simulations (Wang and Sun, 2003). However, as discussed by Nicolaï et al. (2001) and Verboven et al. (2004), allowing a global heat transfer coefficient to be representative of the convection process cannot serve as a basis of process improvement when it is inherently a local phenomenon dependent on surface geometry, flow characteristics, and other physical processes such as mass transfer, etc. Nicolaï and de Baerdemaeker (1996) showed that for a small global heat transfer coefficient, small changes in its local value may cause large temperature differences across a food product, thereby suggesting that the global value may find better use as an indicator of product sensitivity in such circumstances. Moreover, in the determination of the global heat transfer coefficient, correlations of the type $Nu = cRe^{m}Pr^{n}$ are widely used in practice. Zheng et al. (2007) noted that with $c$, $m$, and $n$ being constants, it is difficult to describe, to good level of accuracy, the change of Nusselt number with Reynolds and Prandtl numbers over a wide range of these parameters. Therefore, such correlations require a comprehensive experimental approach in their validation, as well as information regarding the specific range and conditions from which the correlations were developed. Unfortunately, this is not always the case.

Radiation heat transfer is entirely different to conduction and convection heat transfer as a medium is not required. Radiant exchange occurs at a rate proportional to the fourth power of the absolute temperature difference between two surfaces. Inclusion of radiant heat transfer in calculations is often required when modeling thermal exchange owing to high-temperature surfaces, e.g., in baking ovens. Microwave or radio frequency radiation in heating or drying operations may need modeling; however, in contrast to thermal radiation, these mechanisms permit the penetration of electromagnetic waves into a food product. Similar to the cases of heat transfer by conduction or convection, and provided conditions allow, the radiant heat transfer can be linearized so that it can be calculated as a function of the temperature difference, the area of transfer, and a proportionality constant of the surfaces (Wang and Sun, 2005). However, radiation can present many difficulties that cannot be modeled by simplified means. For example, bread baked in an oven is subject to radiant energy, which can be blocked by vapor boundary layers that develop during the baking process (Zhou and Therdthai, 2007). Furthermore, color change such as browning influences baking by increasing the absorption of infrared rays (Zhou and Therdthai, 2007). As a consequence, radiation causes localized temperature differentials of an exposed surface, particularly the darkened area (Matz, 1989). In other operations, electromagnetic and radio frequency radiation is used to volumetrically deposit energy into a product, the rate of which is a function of the dielectric properties of the food. To calculate the electric field inside a food, exponential decay has traditionally been used in a lot of oven investigations (Singh and Heldman, 2001). Unfortunately, for many situations, this is a qualitative assumption and can be wrong depending on size of the food and its dielectric properties, a more correct method would be to solve the Maxwell electromagnetic equations for the oven by numerical means, and then obtain the volumetric rate of heating (Verboven et al., 2007).

The transport equations for conduction, convection, and radiation can also be solved analytically for a small number of boundary, and initial conditions (Nicolaï et al., 2001). An analytical procedure has even been developed to solve a conjugate thermal problem associated with radiation, convection, and conduction (Nakayama et al., 2004). However, the localized contributions cannot be determined analytically and numerical analysis is required to establish the distribution of the variables, i.e., temperature, fluid velocity, and mass fraction, during thermal processing. With this in mind, CFD codes have been developed around numerical algorithms that solve the nonlinear PDEs governing all fluid flow, heat transfer, and many other physical phenomena. In other words, CFD techniques can be used to build a distributed parameter model that is spatially and temporally representative of the physical system, thereby permitting the achievement of a solution with a high level of physical realism. The accuracy of a CFD simulation is a function of many parameters including the level of empiricism involved, i.e., via turbulence models or additional physical models; the assumptions involved, i.e., the Buossinesq versus the ideal gas approximation for inclusion of buoyancy effects; the simplification of both geometry and boundary conditions made in order to reduce processing time; and whether process modifying physical mechanisms are included, i.e., chemical kinetics. The greater the amount of approximations, the less accurate the CFD solution will become (Verboven et al., 2004). However, if used correctly, CFD will provide understanding on the physics of a system in detail, and can do so through nonintrusive flow, thermal, and concentration field predictions. Furthermore, after many years of development, CFD codes can now solve advanced problems of combined convection, radiation, and conduction heat transfer flows in porous media, multiphase flows, and problems involving chemical kinetics.

Many applications of CFD in modern thermal processing exist, i.e., in oven, heat exchanger, spray dryer, and pasteurization technologies, amongst others. CFD simulations are taking giant leaps in realism each year as experience is gained in each specialization, alongside the continual development of commercial CFD codes. For example, recent oven analyses have used CFD in conjunction with other mathematical models of food quality attributes in order to optimize variations in the weight loss and crust color of bread products within the oven (Zhou et al., 2007). Also, modern operations like air impingement ovens have seen CFD being employed so that the interaction of the jet flow pattern with the product could be understood; spatially dependent heat and mass transfer coefficients could be predicted; and equipment design parameters could be optimized (Kocer et al., 2007; Olsson and Trägårdh, 2007). In recent times, 3D CFD modeling of plate heat exchangers (PHE) has allowed the milk fouling process to be simulated based on both hydrodynamic and thermodynamic principles. Such a fouling model permits the influence of corrugation shapes and orientations on the PHE performance to be assessed (Jun and Puri, 2007). The current standard in the CFD modeling of spray dryers has also raised the value of simulations by extending the basic models of flow patterns and particle trajectories with submodels such as drying models, kinetic models of thermal reactions, submodels describing the product stickiness, and agglomeration models, etc. (Straatsma et al., 2007). In other areas, CFD has recently been applied to the pasteurization of intact eggs, which was formally an application that had little information concerning the temperatures and process times that should be applied (Denys et al., 2007).

All of these applications are good examples of the propensity of CFD to promote comprehensive solutions of both novel and conventional thermal processes, and even for processes which had little prior knowledge. Up to now, CFDs remarkable success can be attributed to the active role mostly played by convection heat transfer during many thermal processes. It is now inevitable that modeling studies will exploit the current ability of CFD to model all modes of heat transfer alongside the convection-driven processes.

## 12.3 EQUATIONS GOVERNING THE MAIN THERMAL PROCESSES

### 12.3.1 Navier–Stokes Equations

The mathematical formulation of fluid motion has been complete for almost two centuries, since the emergence of three very important scientists in the field of fluid mechanics. The first one being the Swiss mathematician and physicist Leonhard Euler (1707–1783) who formulated the Euler equations, which describe the motion of an inviscid fluid based on the conservation laws of physics, now defined as classical physics; namely the conservation of mass, momentum, and energy. The French engineer and physicist, Claude-Louis Navier (1785–1836), and the Irish mathematician and physicist, George Gabriel Stokes (1819–1903) later introduced viscous transport into the Euler equations by relating the stress tensor to fluid motion. The resulting set of equations now termed the Navier–Stokes equations for Newtonian fluids have formed the basis of modern day CFD (Anon, 2007).

$$\nabla \cdot \vec{v} = 0 \tag{12.1}$$

$$\rho \frac{\partial v_i}{\partial t} + \rho \ \vec{v} \cdot \nabla \vec{v}_i = -\nabla p + \mu \nabla^2 \ \vec{v}_i + \rho g \tag{12.2}$$

Equation 12.1 is a mathematical formulation of the law of conservation of mass (which is also called the continuity equation) for a fluid element where $\vec{v}$ consists of the components of $\vec{v}_i$, the solution for each requires a separate equation. The conservation of mass states that the mass flows entering a fluid element must balance exactly with those leaving for an incompressible fluid. Equation 12.2 is the conservation of momentum for a fluid element, i.e., Newton's second law of motion, which states that the sum of the external forces acting on the fluid particle is equal to its rate of change of linear momentum.

#### 12.3.1.1 Relationship between Shear Stress and Shear Rate

Any fluid that does not obey the Newtonian relationship between the shear stress and shear rate is called a non-Newtonian fluid. The shear stress in a Newtonian fluid is represented by the second term on the right-hand side of Equation 12.2. Many food processing media have non-Newtonian characteristics and the shear thinning or shear thickening behavior of these fluids greatly affects their thermal–hydraulic performance (Fernandes et al., 2006). Over recent years, CFD has provided better understanding of the mixing, heating, cooling, and transport processes of non-Newtonian substances. Of the several constitutive formulas that describe the rheological behavior of substances, which include the Newtonian model, power law model, Bingham model and the Herschel–Bulkley model, the power law is the most commonly used in food engineering applications (Welti-Chanes et al., 2005). However, there are some circumstances where modeling the viscosity can be avoided as low velocities permit the non-Newtonian fluid to be considered Newtonian, e.g., as shown by Abdul Ghani et al. (2001).

### 12.3.2 Heat Transfer Equation

The modeling of thermal processes requires that the energy equation governing the heat transfer within a fluid system to be solved. This equation can be written as follows:

$$\rho\frac{\partial\left(c_p\,\vec{v}_i\right)}{\partial t}+\vec{v}\cdot\nabla T=\lambda\nabla^2 T+s_{\mathrm{T}} \tag{12.3}$$

The transport of heat in a solid structure can also be considered in CFD simulations, and becomes especially important when conjugate heat transfer is under investigation, in which case it is important to maintain continuity of thermal exchange across the fluid–solid interface (Verboven et al., 2004). The Fourier equation which governs heat transfer in an isotropic solid can be written as

$$\rho\frac{\partial\left(c_p\,\vec{v}_i\right)}{\partial t}=\lambda\nabla^2 T+s_{\mathrm{T}} \tag{12.4}$$

Evidently, the only difference between the two equations is that the Fourier equation lacks a convective mixing term for temperature, which is incorporated into Equation 12.3. For a conjugate heat transfer situation where evaporation at the food surface is considered, and where the heat transfer coefficient is known, the boundary condition for Equation 12.4 may be written as follows:

$$h\left(T_{\mathrm{bf}}-T_{\mathrm{s}}\right)+\varepsilon\sigma\left(T_{\mathrm{bf}}^4-T_{\mathrm{s}}^4\right)=-\lambda\frac{\partial T}{\partial n}-\alpha N_{\mathrm{s}} \tag{12.5}$$

where

- $\varepsilon$ is the emission factor coefficient
- $\sigma$ is the Stefan–Boltzmann constant

Heat transported by radiation and convection from air to food raises the sample temperature and also goes toward evaporating the free water at the surface (Aversa et al., 2007). The solution of Equation 12.3 can be used on the food surface to calculate the local heat transfer coefficients (Verboven et al., 2003),

$$h = \frac{-\lambda \left.\frac{\partial T}{\partial n}\right|_{\text{surface}}}{(T_{\text{bf}} - T_{\text{s}})} \tag{12.6}$$

Equation 12.6 can be used, provided the surface temperature is assumed independent of the coefficient during calculations.

#### 12.3.2.1 Equation of State

The equation of state relates the density of the fluid to its thermodynamic state, i.e., its temperature and pressure. The profiles of thermal processing variables, i.e., temperature, water concentration, and fluid velocity, are functions of density variations caused by the heating and cooling of fluids. Since fluid's flows encountered in thermal processing can be regarded as incompressible, there are two means of modeling the density variations that occur due to buoyancy. The first is the well-known Boussinesq approximation (Ferziger and Peric, 2002). This has been used successfully in many CFD applications (Abdul Ghani et al., 2001):

$$\rho = \rho_{\text{ref}} \left[1 - \beta \left(T - T_{\text{ref}}\right)\right] \tag{12.7}$$

The approximation assumes that the density differentials of the flow are only required in the buoyancy term of the momentum equations. In addition, a linear relationship between temperature and density, with all other extensive fluid properties being constant, is also assumed. This relationship only considers a single-component fluid medium; however, by using Taylor's expansion theorem the density variation for a multicomponent fluid medium can also be derived.

Unfortunately, the Boussinesq approximation is not sufficiently accurate at large temperature differentials (Ferziger and Peric, 2002). Therefore, in such cases another method of achieving the coupling of the temperature and velocity fields is necessary. This can be done by expressing the density difference by means of the ideal gas equation:

$$\rho = \frac{p_{\text{ref}} M}{RT} \tag{12.8}$$

This method can model density variations in weakly compressible flows, meaning that the density of the fluid is dependent on temperature and composition but small pressure fluctuations have no influence.

### 12.3.3 Turbulence Models

Thermal processes are usually associated with turbulent motion; primarily due to the involvement of high flow rates and heat transfer interactions. Presently, even though the Navier–Stokes equations can be solved directly for laminar flows, it is not possible to solve the exact fluid motion in the Kolmogorov microscales associated with engineering flow regimes and thus turbulence requires modeling

(Friedrich et al., 2001). For this, there are many turbulence models available, and their prediction capability has undergone great improvement over the last number of years. It should be noted, however, that none of the existing turbulence models are complete, i.e., their prediction performance is highly reliant on turbulent flow conditions and geometry. Without a complete turbulence model capable of predicting the average field of all turbulent flows, the present understanding of turbulence phenomena will reduce the generality of solutions. In the following, some of the best performing turbulence models are discussed. A brief synopsis of the advantages and disadvantages of various turbulence models is given in Table 12.1.

#### 12.3.3.1 Eddy Viscosity Models

In turbulent flow regimes, engineers are generally content with a statistical probability that processing variables (such as velocity, temperature, and concentration) will exhibit a particular value, in order to undertake suitable design strategies. Such information is afforded by the Reynolds averaged Navier–Stokes equations (RANS), which determine the effect of turbulence on the mean flow field through time averaging. By averaging in this way, the stochastic properties of turbulent flow are essentially

**TABLE 12.1**
**Turbulence Models Used when Modeling Thermal Processing and Their Attributes**

| Turbulence Model | Advantages | Disadvantages | References |
|---|---|---|---|
| $k$–ε variants | Reasonably good swirling flows with dense meshing can include buoyancy turbulence production. | Poor predictions of flows with adverse pressure gradients, impinging flows, and separated flows. | Verboven et al. (2001a,b) |
| $k$–ω variants | No use of empirical log-law correlation. Therefore, better prediction of impinging flows. | Predictions are very sensitive to the free-stream turbulence conditions. | Olsson et al. (2007) |
| RSM | Each Reynolds stress is modeled, therefore contributing to a more accurate solution. | Convergence can be difficult. Circumspect choices for the most applicable variant must be made. | Olsson et al. (2007) Moureh and Flick (2005) |
| LES | Modeling the large-scale turbulence with the Navier–Stokes equations permits the understanding of mechanisms that explain the effect of free-stream turbulence heat transfer. | Perfect experimental knowledge of the turbulent field in the form of statistical accounts of simulations of numerous solution runs. | Kondjoyan (2006) |

disregarded and six additional stresses (Reynolds stresses) result, which need to be modeled by a physically well-posed equation system to obtain closure that is consistent with the requirements of the study.

The eddy viscosity hypothesis (Boussinesq relationship) states that an increase in turbulence can be represented by an increase in effective fluid viscosity, and that the Reynolds stresses are proportional to the mean velocity gradients via this viscosity (Ferziger and Peric, 2002). For a $k$–$\varepsilon$ type turbulence model, the following representation of eddy viscosity can be written as

$$\mu_t = \rho C_\mu \frac{k^2}{\varepsilon} \tag{12.9}$$

For the $k$–$\omega$ type turbulence without the low-Reynolds number modifications, the eddy viscosity can be represented by Wilcox (1993)

$$\mu_t = \rho \frac{k}{\omega} \tag{12.10}$$

This hypothesis forms the foundation for many of today's most widely used turbulence models, ranging from simple models based on empirical relationships to variants of the two-equation $k$–$\varepsilon$ model, which describes eddy viscosity through turbulence production and destruction (Versteeg and Malaskeera, 1995). All eddy viscosity models have relative merits with respect to simulating thermal processes.

The standard $k$–$\varepsilon$ model (Launder and Spalding, 1974), which is based on the transport equations for the turbulent kinetic energy $k$ and its dissipation rate $\varepsilon$, is semiempirical and assumes isotropic turbulence. Although it has been successful in numerous applications and is still considered an industrial standard, the standard $k$–$\varepsilon$ model is limited in some respects. A major weakness of this model is that it assumes an equilibrium condition for turbulence, i.e., the turbulent energy generated by the large eddies is distributed equally throughout the energy spectrum. However, energy transfer in turbulent regimes is not automatic and a considerable length of time may exist between the production and the dissipation of turbulence.

The renormalization group (RNG) $k$–$\varepsilon$ model (Choundhury, 1993) is similar in form to the standard $k$–$\varepsilon$ model but owing to the RNG methods from which it has been analytically derived, it includes additional terms for dissipation rate development and different constants from those in the standard $k$–$\varepsilon$ model. As a result, the solution accuracy for highly strained flows has been significantly improved. The calculation of the turbulent viscosity also takes into account the low-Reynolds number if such a condition is encountered in a simulation. The effect of swirl on turbulence is included in the $k$–$\varepsilon$ RNG model, thereby enhancing accuracy for recirculating flows.

In the realizable $k$–$\varepsilon$ model (Shih et al., 1995), $C_\mu$ is expressed as a function of mean flow and turbulence properties, instead of being assumed constant, as in the case of the standard $k$–$\varepsilon$ model. As a result, it satisfies certain mathematical constraints on the Reynolds stress tensor that is consistent with the physics of turbulent flows (e.g., the normal Reynolds stress terms must always be positive). Also, a new model for the dissipation rate is used.

The $k$–$\omega$ model is based on modeled transport equations, which are solved for the turbulent kinetic energy $k$ and the specific dissipation rate $\omega$, i.e., the dissipation rate per unit turbulent kinetic. An advantage that the $k$–$\omega$ model has over the $k$–$\varepsilon$ model is that its performance is improved for boundary layers under adverse pressure gradients as the model can be applied to the wall boundary, without using empirical log-law wall functions. A modification was then made to the linear constitutive equation of $k$–$\omega$ model to account for the principal turbulence shear stress. This model is called the shear-stress transport (SST) $k$–$\omega$ model and provides enhanced resolution of boundary layer of viscous flows (Menter, 1994).

#### 12.3.3.2 Near-Wall Treatment

An important feature of $k$–$\varepsilon$ turbulence models is the near-wall treatment of turbulent flow. Low Reynolds number turbulence models solve the governing equations all the way to the wall. Consequently, a high degree of mesh refinement in the boundary layer is required to satisfactorily represent the flow regime, i.e., $y+\leq 1$. Conversely, high Reynolds number $k$–$\varepsilon$ models use empirical relationships arising from the log-law condition that describes the flow regime in the boundary layer of a wall. This means that the mesh does not have to extend into this region; thus the number of cells involved in a solution is reduced. The use of this method requires $30 < y+ < 500$ (Versteeg and Malalsakeera, 1995).

#### 12.3.3.3 Reynolds Stress Model

The Reynolds stress closure model (RSM) generally consists of six transport equations for the Reynolds stresses—three transport equations for the turbulent fluxes of each scalar property and one transport equation for the dissipation rate of turbulence energy. RSMs have exhibited far superior predictions for flows in confined spaces where adverse pressure gradients occur. Terms accounting for anisotropic turbulence, which are included in the transport equations for the Reynolds stresses, mean that these models provide a rigorous approach to solving complex engineering flows. However, storage and execution time can be expensive for 3D flows.

#### 12.3.3.4 Large Eddy Simulation and Direct Eddy Simulation

Large eddy simulation (LES) forms a solution given the fact that large turbulent eddies are highly anisotropic and dependent on both the mean velocity gradients and geometry of the flow domain. With the advent of more powerful computers, LES offers a way of alleviating the errors caused by the use of RANS turbulence models. However, the lengthy time involved in arriving at a solution means that it is an expensive technique of solving the flow (Turnbull and Thompson, 2005). LES provides a solution to large-scale eddy motion in methods akin to those employed for direct numerical simulation. It also acts as spatial filtering, thus only the turbulent fluctuations below the filter size are modeled. More recently a methodology has been proposed by which the user specifies a region where the LES should be performed, with RANS modeling completing the rest of the solution; this technique is known as DES and has found to increase the solution rate by up to four times (Turnbull and Thompson, 2005).

### 12.3.4 Equation for Mass Transfer

In general, mass transfer in food products depends on local water concentration and is governed by Fick's law of diffusion of the form

$$\frac{\partial C}{\partial t} = \vartheta \nabla^2 C + s_c \tag{12.11}$$

where

$C$ is the water concentration in food (ppm or mol $m^{-3}$)
$\vartheta$ is the effective diffusion coefficient of water in food ($m^2$ $s^{-1}$)

If the food is highly porous and dehydrated, then water vapor diffusion may be significant. However, for products with a void fraction lower than 0.3 (May and Perrè, 2002), vapor diffusion can be neglected. For mass transfer in the fluid medium, a passive scalar can represent the diffusion of mass and be written as

$$\frac{\partial C}{\partial t} + \nabla \cdot (\vec{v}C) = \vartheta \nabla^2 C + s_c \tag{12.12}$$

In flows involving gases mixing with air and chemicals in water, diffusion coefficients can be found in the literature, but for liquid foods a reasonable approach is to make an educated guess and conduct a sensitivity analysis (Verboven et al., 2004). Using a passive scalar is only valid in low concentrations, and becomes invalid when particulate sizes of about 1 μm are present, which can influence the flow properties.

Simulations of spray dryers, solid–gas flows, or nonhomogenous liquid foods exemplify applications where solid and fluid phases coexist and interact with each other. For this modeling, either of these two approaches are used: the Eulerian–Lagrangian or the Eulerian–Eulerian. Using the former, the bulk fluid is modeled as a continuum carrying discrete solid particles (Nijdam et al., 2006). The Lagrangian equations of mass and momentum are then solved to determine the trajectory of each particle within the continuum. As discussed by Verboven et al. (2004), in turbulent flow simulations, the turbulent velocity fluctuations have been averaged out by the RANS models and the turbulent dispersion of the particles requires modeling by mimicking these fluctuations. This is done by extracting random numbers from a Gaussian distribution with a computed mean and variance (Harral and Burfoot, 2005). When the particle passes through any arbitrary mesh element, the energy, mass, and momentum transferred to the fluid continuum are calculated and added to the source terms of that element. In this way, all trajectories are calculated one by one. In the next iteration, the fluid flow is solved using these source terms, and the calculation loop is repeated until sufficient convergence is achieved. Whilst the Eulerian–Lagrangian approach provides a direct physical interpretation of the particle–fluid, particle–particle, and particle–wall interactions, computational times can become excessive, depending on the number of particles to be solved in the system (Szfran and Kmiec, 2004). This means that applications of the Eulerian–Lagrangian approach are still limited to small-scale computations. In the Eulerian approach,

**TABLE 12.2**
**Features of CFD Software and the Relevant Effects on Model Development**

| Feature of CFD | Effect | Reference of Use |
|---|---|---|
| Eulerian–Lagrangian modeling | Good level of accuracy for modeling particle interactions, however storage demands increases with number of the simulated particles. | Langrish and Fletcher (2003) |
| Eulerian–Eulerian modeling | Bulk movement of particle and fluid phases modeled, particle interaction not primary focus. The model can be parallelized to increase efficiency. | Szafran and Kmiec (2005)<br>Zhonghua and Mujumdar (2007) |
| Scalar modeling | Can be used to track movement of mass, which depends on bulk fluid motion. Can model a process under destruction, i.e., microbial death. | Abdul Ghani and Farid, (2002a,b) |
| Coupled solver | Can increase solving efficiency in presence of large sources. Can be used to control initiation of unsteady phenomena. | Fletcher et al. (2006) |
| Dynamic meshing | Allows a long time-dependent batch process to be modeled. | Therdthai et al. (2004b)<br>Wong et al. (2007a,b) |
| User defined coding | Allows the user to integrate submodels in CFD simulations where other food-related process occur. | Straatsma et al. (2007) |

the fluid and particle phases are treated as interacting and interpenetrating continua (Nijdam et al., 2006). The governing equations for each phase are convection–diffusion equations, which contain extra source terms to account for the turbulent dispersion of the particles. Coupling of the phases is achieved through pressure and interphase exchange coefficients. The equations appropriate to the Eulerian approach are presented by Verboven et al. (2004) and Nijdam et al. (2006). Many of the features of CFD software are summarized in Table 12.2.

### 12.3.5 Radiation Models

In recent years, the number of radiation models incorporated into commercial CFD codes has increased, and some have been designed by the code developers themselves. The most common radiation models used in thermal processing simulations include discrete ordinate (DO) or surface-to-surface (S2S) models. The DO model takes into account media participation (Modest, 1992). The S2S model considers the radiation heat exchange between two surfaces only (Siegel and Howell, 1992), and the amount of radiation received and emitted by each surface is defined by the surface's view factors and the thermal boundary conditions. In contrast to an S2S solution, the solution for the DO model is coupled to the flow solution and energy is exchanged between the fluid and the radiation field. Therefore, solution times for S2S model can be almost half those of the DO model (Mistry et al., 2006). Radiation models have enjoyed much use in oven modeling, as to be shown later.

## 12.4 SOLVING THE TRANSPORT PHENOMENA: STATE OF THE ART IN CFD SOLUTIONS

### 12.4.1 Numerical Discretization

CFD code developers have a choice of many different numerical techniques to discretize the transport equations. The most important of these include finite difference, finite elements, and finite volume. The finite difference technique is the oldest one used, and many examples of its application in the food industry exist. However, due to difficulties in coping with irregular geometry, finite difference is not commercially implemented. Furthermore, the current trend of commercial CFD coding is aimed toward developing unstructured meshing technology capable of handling the complex 3D geometries encountered in industry. Therefore, the prospects of finite difference being used in industrial CFD applications seem limited.

Finite element methods (FEMs) have historically been used in structural analysis where the equilibrium of the solution must be satisfied at the node of each element. Nicolaï et al. (2001) provided a short introduction to the use of the finite elements method in conduction heat transfer modeling, and it will not be discussed here. Suffice to say that as a result of the weighting functions used by this method, obtaining a 3D CFD solution with a large number of cells is impractical at present. Therefore, finite elements are not generally used by commercial CFD developers, especially as many of these CFD codes are marketed toward solving aerodynamic problems. Nonetheless, finite elements methods have enjoyed the use in the modeling of electromagnetic heating in microwave ovens (Verboven et al., 2007; Geedipalli et al., 2007); vacuum microwave drying (Ressing et al., 2007); radio frequency heating of food (Marra et al., 2007); and conduction and mass transport during drying (Aversa et al., 2007). Therefore, it seems that the finite elements method is amenable to the modeling of novel thermal processes, once the details of fluid flow do not need explicit quantification.

With finite volume techniques, the integral transport equations governing the physical process are expressed in conservation form (divergence of fluxes) and the volume integrals are then converted to surface integrals using Gauss's divergence theorem. This is a direct extension of the control volume analysis that many engineers use in thermodynamics, heat transfer applications, etc., so it can be easily interpreted. Thus, expressing the equation system through finite volumes forms a physically intuitive method of achieving a systematic account of the changes in mass, momentum and energy, as fluid crosses the boundaries of discrete spatial volumes within the computational domain. Also, finite volume techniques yield algebraic equations that promote solver robustness, adding further reasons to why many commercial developers implement this technique.

### 12.4.2 Generic Equation and Its Numerical Approximation

The basic requirement of CFD is to obtain a solution to a set of governing PDEs of the transported variable. As discussed above, the transport phenomena are generally described by commercial CFD developers using the finite volume method. The generic conservation convection–diffusion equation, after the application of Gauss's theorem to obtain the surface integrals, can be described as follows:

$$\int_V \frac{\partial \rho\phi}{\partial t} + \int_A n \cdot (\rho U \phi) \mathrm{d}A = \int_A n \cdot \left(\Gamma_\phi \operatorname{grad}\phi\right) \mathrm{d}A + \int_V S_\phi \qquad (12.13)$$

The conservation principle is explicit in Equation 12.13, i.e., the rate of increase of $\phi$ in the control volume plus the net rate of decrease of $\phi$ due to convection is equal to the increase in $\phi$ due to diffusion and an increase in $\phi$ due to the sources (Versteeg and Malaskeera, 1995). For a numerical solution, both the surface and volume integrals need to be solved on a discrete level, which means numerical interpolation schemes are required. As the convection term, the only nonlinear term in the equation, needs to be approximated at each mesh element face, it presents the greatest challenge in allowing the numerical scheme to preserve properties such as the stability, transportiveness, boundedness, and accuracy of a solution. For the same reason, numerical schemes are often called convection schemes as the accurate and stable representation of the convection term is a major requirement (Patankar, 1980). The reader can refer to standard CFD textbooks, e.g., those of Patankar (1980) and Versteeg and Malaskeera (1995) for a complete discussion on the various properties of convection schemes.

The performance of a convection scheme is denoted by its ability to reduce the error once the mesh is refined (Ferziger and Peric, 2002). The first-order convection schemes are bounded and stable but are predisposed to numerical diffusion and exhibit a sluggish response to grid refinement. Owing to their favorable convergence attributes, these schemes are still prevalent in food engineering literature, which obviously casts serious doubts on the validity of some solutions especially when grid refinement studies have sometimes proved unattainable. To further this point, Harral and Boon (1997) showed that coarse grid predictions agreed more favorably with experimental measurements than the grid-independent solution. The higher-order scheme QUICK is more accurate and responsive to grid refinement but due to its unbounded nature, it often develops solutions with unphysical undershoots and overshoots when strong convection is present. Convergence may also be difficult, especially when nonlinear sources are present in the simulation.

Besides the above classical convection schemes, other higher order schemes are also available, and can be categorized as linear and nonlinear. Linear schemes, such as the cubic upwind scheme, offer good resolution but do not guarantee boundedness, meaning that unwanted oscillations or even negative values may be generated in regions of steep gradients. On the other hand, nonlinear schemes secure boundedness by means of nonlinear flux limiter which may, to a certain extent, reduce the numerical accuracy of the solution. Versteeg and Malaskeera (1995), Patankar (1980), and Ferziger and Peric (2002) provide more detail on the performance of classical numerical schemes, and CFD software user documentation should provide sufficient information on the use of other higher order schemes within a software package.

### 12.4.3 Meshing the Problem

The volume mesh in a simulation is a mathematical description of the space or the geometry to be solved. One of the major advances to occur in meshing technology over recent years was the ability for tetrahedral, hexahedral hybrid, and even polyhedral meshes to be incorporated into commercial codes. This has allowed mesh to be fit to any arbitrary geometry, thereby enhancing the attainment of solutions for many industrial applications. In addition, some modern commercial CFD codes promote very little interaction between the user and the individual mesh elements, with regard to specifying the physics of the problem. Through this decoupling of the physics from the mesh, the user is allowed to concentrate more on the details of the geometry, and the transfer of simulation properties and solutions from one mesh to another is easier, when mesh independence of a simulation needs to be studied or when extra resolution is required.

As a result of unstructured meshing, local mesh refinement without creating badly distorted cells is achievable. However, as with structured meshes, mesh quality is still an important consideration, as poor quality can affect the accuracy of the calculated convective and diffusive fluxes. A common measure of quality is the skewness angle, which determines whether the mesh elements permit the computation of bounded diffusion quantities. Code developers may use the actual angle or an index between 0 and 1 as the metric for skewness. When a skewness angle of 0° is obtained, the vector connecting the center of two adjacent elements is orthogonal to the face separating the elements; this is the optimum value. With skewness angles of 90° or greater problems in terms of accuracy are caused, the solver may even divide by zero. Other than this, the versatility of these meshes has led to an increased take-up by the CFD community and their uses are finding accurate solutions in many applications within the food industry.

### 12.4.4 Obtaining a Solution

#### 12.4.4.1 Algebraic Equation System

Concerning the discretization of Equation 12.13, the algebraic system for the transported variable $\phi$ at iteration $m + 1$, after the inclusion of an under-relaxation constant $\alpha$, can be written as

$$\frac{a_p}{\alpha}\phi_p^{m+1} + \sum_n a_n \phi_n^{m+1} = b + \frac{a_p}{\alpha}(1-\alpha)\phi_p^m \tag{12.14}$$

where the summation is over all the cells $n$ lying adjacent to the cell $p$. On the right-hand side of Equation 12.14, $b$ represents the contributions to the discretized equation evaluated from the iteration $m$. Under-relaxation is generally included to reduce the change in $\phi$ from iteration to iteration, so that stability can be enhanced throughout the solution process. Equation 12.14 can also be given in the following form:

$$\frac{a_p}{\alpha}\Delta\phi_p + \sum_n a_n \Delta\phi_n = b - a_p\phi_p^m - \sum_n a_n\phi_n^k \qquad (12.15)$$

where $\Delta\phi_p = \Delta\phi_p^{k+1} - \Delta\phi_p^k$. The right-hand side of this equation is termed residual, and represents the discretized form of Equation 12.13 at iteration $m$. Therefore, the residual will be zero when the discretized equation is satisfied exactly.

This system can be written in matrix notation as

$$Ax = b \qquad (12.16)$$

where

$A$ is a sparse matrix containing all the coefficients of the linear system
$x$ is a vector containing the unknown variable values ($\Delta\phi$ in Equation 12.15) at the grid nodes
$b$ is a residual of Equation 12.15

However for Equation 12.13, this system is still nonlinear, i.e., the coefficients in $A$ are a function of geometrical quantities, fluid properties, and the variable values themselves. Moreover, it may be possible that the source term or the transport coefficient may also be functions of the flow variable. In this case, an iterative solution is required, with two levels of iteration, i.e., an outer iteration loop controlling the solution update and an inner loop controlling the solution of the linear system. Since the outer iterations are repeated multiple times, the linear system need only be solved approximately at each iteration (Versteeg and Malasekeera, 1995; Peric and Ferziger, 2002).

#### 12.4.4.2 Inner Iteration Loop

Regarding inner iterations, an attempt is made to solve the equation system by making successive approximations to the solution starting from an initial guess, i.e., by finding an approximate solution vector $x^m$, a better solution is then sought giving $x^{m+1}$. The solution is then repeated until a convergence criterion is met. As discussed in depth by Ferziger and Peric (2002) and briefly by Nicolaï et al. (2001), iterating in this manner will not only reduce the residual but also, in a concomitant fashion, reduce the convergence error. Many of these basic iteration techniques exist from Gauss–Seidel iteration to alternating direction implicit, and are discussed in detail by Ferziger and Peric (2002). However, these basic iteration methods exhibit relatively sluggish convergence characteristics when grid resolution is increased, i.e., they are only effective at removing high-frequency error components.

Such features suggest that some of the work could be done on a coarse grid, since computations on coarse grids are much less costly and the Gauss–Seidel method converges four times faster on a grid half as fine (Ferry, 2002; Ferziger and Peric, 2002). The multigrid approach clusters the cells in the simulation to form coarse level of meshing. Then during the solution procedure, the residual from the fine level is transferred (called restricting) to the coarser level, the correction is then interpolated from the coarse level back to the fine level (called prolongation). Algebraic multigrid, as opposed to geometric multigrid, is the form commercially implemented. The reason being that unlike geometric multigrid, algebraic multigrid derives a coarse level system without reference to the underlying grid geometry or discrete equations. After the coarse level is derived, the solution process begins. In the meantime, each cell is divided into finer cells. Then, once a converged solution is found on the coarsest grid, it is interpolated to the next finer grid level to provide initialization. The process is repeated until a solution on finest grid level is obtained, with the short-wavelength error components being efficiently removed (Ferziger and Peric, 2002).

#### 12.4.4.3 Outer Iteration Loop

Solution update is controlled by the outer iteration loop, for which two types of solvers exist namely segregated and coupled solvers. Segregated solvers solve the fluid flow and energy equations for each component of velocity, and pressure and temperature sequentially, accounting for their coupling in a deferred manner. Also, owing to the inherent nonlinearity, the linkage between the momentum and continuity equations is often only achieved with a predictor–corrector approach like Semiimplicit method for pressure-linked equations, devised by Patankar and Spalding (1972), or its descendents, which are employed by many commercial packages to control solution update for the discretized equations. Albeit versatile and commonly used, a segregated solver, by its very design, presents some difficulties. First, as explained by Ferziger and Peric (2002) and Versteeg and Malaskeera (1995), the inception of this technique has been based around constant-density flow regimes, meaning that although a segregated solver is capable of handling mildly compressible flows and low Raleigh number natural convection, its suitability dwindles as Mach number and Raleigh number increase. Second, the segregated solver may require considerable under-relaxation before convergence is realized, and as a result solution times may be increased. This heightens the requirement to optimize the number of inner and outer iterations, alongside relaxation parameters, in order to achieve a converged solution efficiently and monotonically. An interesting point to note is that the roots of segregated solvers were formed in the staggered grid arrangement. However, segregated solvers have been developed for collocated grid arrangement since the 1980s, at which point their popularity gave rise to improved pressure–velocity coupling algorithms, e.g., Rhie–Chow techniques (Ferziger and Peric, 2002).

The coupled flow solver is another breed of solver, which has become more popular in recent times, thanks to more efficient and cheaper workstations. These solvers solve the conservation equations for mass and momentum simultaneously using a time-marching approach. The preconditioned form of the governing equations is used, ensuring that the eigenvalues of the system remain well conditioned with respect to the convective and diffusive timescales, even for incompressible and isothermal flows. This formulation experiences greater robustness when solving flows with large buoyancy or momentum sources, as these enhance the coupling between the equations within the system. An advantage of the coupled solver is that the number of iterations required for a solution is independent of mesh size.

## 12.5 OPTIMIZING CONVENTIONAL THERMAL PROCESSES WITH CFD

Over the years, many of the conventional process such as sterilization, drying, and cooking have been optimized with CFD. Some the studies conducted thus far are reviewed below.

### 12.5.1 Sterilization and Pasteurization

#### 12.5.1.1 Canned Foods

Sterilization is a conventional thermal process that can be modeled with CFD. In such a process, rapid and uniform heating is desirable to achieve a predetermined level of sterility with minimum destruction of the color, texture, and nutrients of food products (Tattiyakul et al., 2001). Siriwattanayotin et al. (2006) used CFD to investigate sterilization value ($F_0$) calculation methods, and concluded that when $F_0$ was determined using the "thermal death time" approach, the process time to achieve the desired temperature in a sugar solution was underestimated if the surrounding temperature was lower than the reference value.

Canned viscous liquid foods such as soup, carboxyl methyl cellulose (CMC), or corn starch undergoing sterilization have been simulated with CFD. In most cases where natural convection occurs, fluid velocities and shear rates are rather low and thus non-Newtonian fluids can be assumed as Newtonian (Abdul Ghani et al., 2003). Neglecting heat generation due to viscous dissipation is another assumption that is generally used. In such simulations, CFD has shown the transient nature of the slowest heated zone (SHZ), and has illustrated the large amount of time needed for heat to be transferred throughout food, as well as the sharp heterogeneity in temperature profile when the process is static (i.e., natural convection is dominant) (Abdul Ghani et al., 1999).

Abdul Ghani et al. (2002) did CFD studies of both natural and forced convection (via can rotation) sterilization processes of viscous soup, and showed that forced convection was about four times more efficient. Further simulation studies of a starch solution undergoing transient gelatinization showed that uniform heating could be obtained by rotating the can intermittently during the sterilization process (Tattiyakul et al., 2001). CFD simulations have also been used to generate the data required for the development of a correlation to predict the sterilization time (Farid and Abdul Ghani, 2004). A more recent study simulated the sterilization process of a solid–fluid mixture in a can, and showed that the position of the food in the can has a large influence on the sterilization times experienced (Abdul Ghani and Farid, 2006).

CFD has recently been used to study the effect of container shape on the efficiency of the sterilization process (Varma and Kannan, 2005, 2006). Conical-shaped vessels pointing upward were found to reach appropriate sterilization temperature the quickest (Varma and Kannan, 2005). Full cylindrical geometries performed best when sterilized in a horizontal position (Varma and Kannan, 2006).

#### 12.5.1.2 Foods in Pouches

In recent years, CFD has provided a rigorous analysis of the sterilization of 3D pouches containing liquid foods (Abdul Ghani and Farid, 2002a,b). Coupling first-order bacteria and vitamin inactivation models with the fluid flow has allowed transient temperature, velocity, and concentration profiles of both bacteria and ascorbic acid to be predicted during natural convection. The concentrations of bacteria and ascorbic acid after heat treatment of pouches filled with the liquid food were measured, and found close agreement with the numerical predictions. The SHZ was found to migrate during sterilization until eventually resting in a position about 30% from the top of the pouch. As expected, the bacterial and ascorbic acid destruction was seen to depend on both temperature distribution and flow pattern.

#### 12.5.1.3 Intact Eggs

CFD has been used to predict the transient temperature and velocity profiles during the pasteurization process of intact eggs (Denys et al., 2003–2005). Owing to its ability to account for complex geometries, heterogeneous initial temperature distributions, transient boundary conditions, and nonlinear thermophysical properties, CFD has permitted a comprehensive understanding of this thermal process (Denys, 2003). Such an analysis has allowed the gap in the knowledge of this area to be filled, as up to recently little information on the correct processing temperatures and times for safe pasteurization, without loss of functional properties, was available (Denys, 2004). In the series of studies published on this topic by Denys et al., a procedure to determine the surface heat transfer coefficient using CFD simulations of eggs filled with a conductive material of known thermal properties was first developed. After this, conductive and convective heating processes in the egg were modeled (Denys et al., 2004). From this, it was revealed that, similar to the phenomena noted by Abdul Ghani et al. (1999) in canned food, the cold spot was found to move during the process toward the bottom of the egg. Moreover, again similar to the findings of Ghani et al. (2002), a cold zone as opposed to a cold spot was predicted. The location of the cold zone in the yolk was predicted to lay below its geometrical center, even for the case where the yolk was positioned at the top of the egg. It was concluded that no convective heating takes place in the egg yolk during processing.

In the final paper of the series, Denys et al. (2005) coupled a first-order kinetic approach to predict the rate of thermal microbial inactivation for *Salmonella enteritidis* (*SE*) during the heating process. First, *SE* was observed to be less heat sensitive in the yolk when compared to egg-white. Then they found that processing at a temperature of 57°C enabled 5 log reductions of *SE* within 40 min, which was considered a reasonable process time (Figure 12.1) after balancing the desire for efficient processing with the desire to retain raw egg properties. The results of the CFD analysis were then presented graphically with an aim to help egg processors to optimize the pasteurizing process.

### 12.5.2 Aseptic Processing

#### 12.5.2.1 Plate Heat Exchangers for Milk Processing

In the dairy industry, it is essential to heat-treat milk products in a continuous process as, on one hand, it is necessary to promote microbial safety and increase the shelf life, whereas, on the other hand, efficient plant processes are also desirable. The adverse influences which heat has on the sensory and nutritional properties of

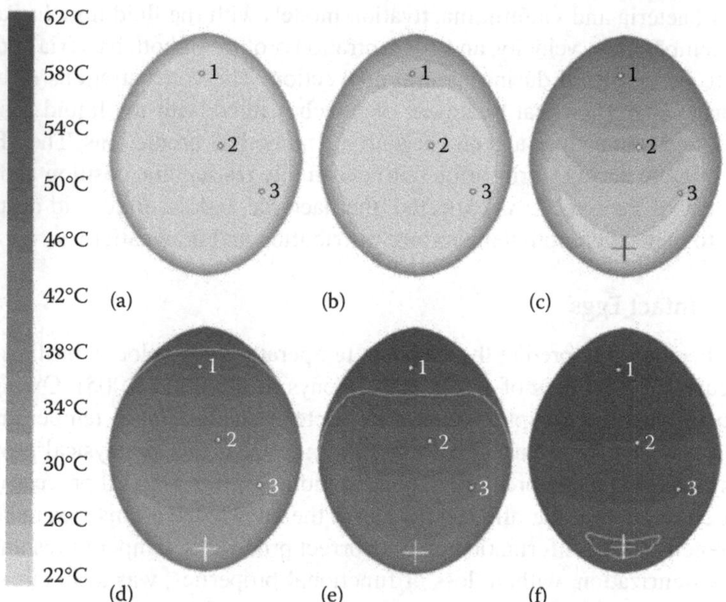


**FIGURE 12.1** CFD-predicted temperature contours in the cross section of an egg filled with CMC, which was heated in a water bath; (a) 5 s, (b) 10 s, (c) 30 s, (d) 80 s, (e) 150 s, and (f) 300 s. The white line shows the 53°C temperature contour, the cross is the coldest point. (From Denys, S., Pieters, J.G., and Dewettinck, K., *J. Food Eng.*, 63, 281, 2004. With permission.)

the final milk product act as a hindrance to efficient thermal processing. Therefore, a fine balance must be maintained between the residence time of the milk product and the heat flux applied throughout the process. Unfortunately, given that indirect, as opposed to direct, heat exchangers permit more control in this regard, fouling presents the greatest engineering and economic challenges to the dairy industry (Visser and Jeurnink, 1997).

The development of CFD models in recent years has contributed to the significant progress made in understanding of the thermal–hydraulics of heat exchangers (Grijspeerdt et al. 2005; Jun and Puri, 2005). It has been shown that the plate geometry can influence fouling rates, and so CFD models of PHE thermal–hydraulics can bring about significant benefits for system optimization (Park et al., 2004). Many CFD studies of PHEs exist, and have presented different techniques for geometry optimization, i.e., corrugation shape, or the optimization of other process parameters, i.e., inlet and outlet positions and PHE-product temperature differences (Kenneth, 2004; Grijspeerdt et al., 2007; Jun and Puri, 2005). Grijspeerdt et al. (2005) investigated the effect of large temperature differences between product and PHE, and noted that the larger the difference is, the greater the opportunity for fouling is. Whilst fouling has been studied by the same group of authors (Grijspeerdt et al., 2003), Jun and Puri (2005) were the first to couple a fouling model with a 3D thermal–hydraulic model within CFD simulations, as evidenced by Figure 12.2. In their study, Jun and Puri investigated the influence of various PHE designs on

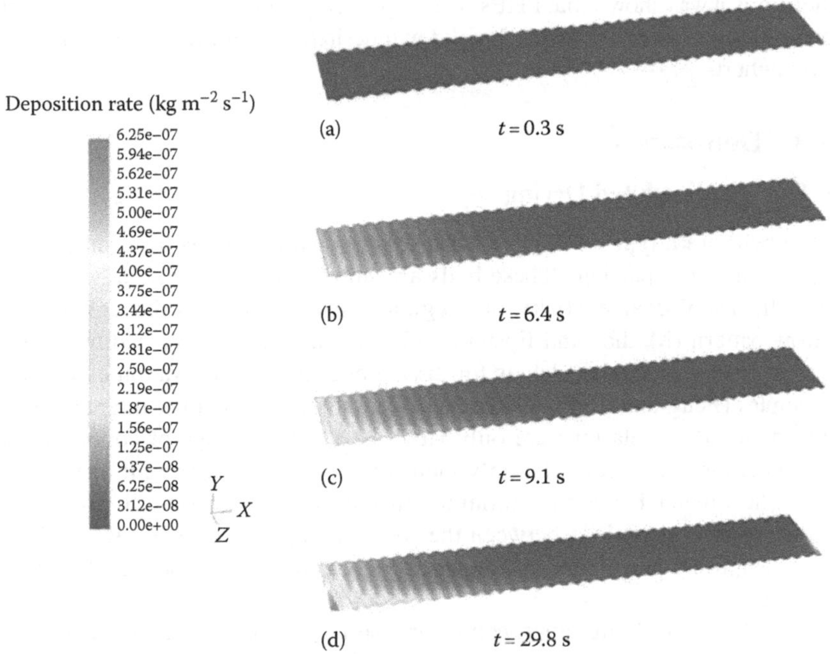


**FIGURE 12.2** Simulated fouling profiles of fluid milk under the unsteady-state conditions. (From Jun, S. and Puri, V.M. in *Computational Fluid Dynamics in Food Processing*, CRC Press, 417, 2007. With permission.)

fouling rates, from those typically used in the dairy industry to those used in the automobile industry. They concluded that employing the latter PHE design in milk processing could decrease fouling (i.e., deposited mass) by a factor of 10, compared to the current system under the identical energy basis. However, Grijspeerdt et al. (2007) noted a problem with this coupled model, namely that the deposition scheme did not include a recalculation area, plate dimensions, and the heat transfer characteristics, even though, strictly speaking, this would be necessary.

#### 12.5.2.2 Plate Heat Exchangers for Yoghurt Processing

Fernandes et al. (2005, 2006) studied the cooling of stirred yoghurt in PHEs with CFD simulations in order to investigate the thermal–hydraulic phenomena involved in the problem. They modeled the rheological behavior of yoghurt via a Herschel–Bulkley model, with temperature influence on viscosity being accounted for through Arrhenius-type behavior. As well as accounting for this rheological behavior, they also provided a high level of precision in the PHE geometrical design and the imposed boundary conditions. During the course of these studies, it was found that due to the higher Prandtl numbers and shear-thinning effects provided by the yoghurt, the Nusselt number of the fully developed flows was found to be more than 10 times higher than those of water. This result presented a substantial thermal–hydraulic performance enhancement in comparison with that from Newtonian fluids (Maia et al., 2007).

Furthermore, it was shown that PHEs with high corrugation angles may provide better opportunities for the gel structure breakdown desired during the production stage of stirred yoghurt.

### 12.5.3 Dehydration

#### 12.5.3.1 Fluidized Bed Drying

Spouted beds are a type of fluidized bed dryer in which heat is transferred from a gas to the fluidized particle. These beds are suitable for industrial unit-operations that handle heavy, coarse, sticky, or irregularly shaped particles through a circulatory flow pattern (Mathor and Epstein, 1974). Spouted beds have had many applications in drying of foods, such as for drying of grains (Madhiyanon et al., 2002), diced apples (Feng and Tang, 1998), etc. Because of the complex interactions that occur, empirical correlations are only valid for a certain range of conditions, and CFD simulations have been the only means of providing accurate information on the flow phenomena. However, similar to other drying applications, difficulties exist in modeling the interactions between the solid and liquid phases, as well as limitations in computing power, and consequently only a limited number of CFD simulations exist.

The Eulerian–Eulerian approach offers the most efficient way of representing the two phases in this type of system, when considering the present levels of computer power and the large number of granules (grain) in a typical system, and consequentially its use has been preferred over discrete methods by some researchers. Szafran and Kmiec (2004) used this approach, via the multifluid granular kinetic model of Gidaspow et al. (1992) and the $k$–$\varepsilon$ model, in their CFD simulations of the transport mechanisms in the spouted bed. Many physical mechanisms needed to be accounted for in the CFD model, as not only did the transport of the mass continuum, i.e., granular phase, need to be solved, but so did the interphase moisture transport. Using the appropriate mass flux equations, which were derived from the diffusion model of Crank (1975), the drying period was predicted in each mesh element without the use of the critical moisture content; this represented a considerable boost in accuracy. This accuracy was borne out in the mass flux computations; however, the heat-transfer rate was underpredicted, when compared to experimental results. An analysis of the transport mechanisms showed that fluctuations in airflow had an influence on the instantaneous distributions of the grain, air temperatures, and the local moisture content, but not on the mean mass-averaged values. In fact it was shown that the internal mass transfer resistance and the mass transfer depended on the zone of the dryer which the grain was positioned during the process.

A major difficulty encountered in modeling spouted bed dryers has been the excessive computing times required to simulate only a fraction of the drying process. For example, owing to the small time-steps required to resolve the instabilities in the flow regime, it would take a 2D CFD simulation almost 1 year to simulate 1 h of drying (Szafran and Kmiec, 2005). Thus at present CFD can only be used as a tool which permits a deeper understanding of the flow patterns and their effects on the drying kinetics rather than for design and optimization (Zhonghua and Mujumdar, 2007). That being said, a recent study has shown that the results from

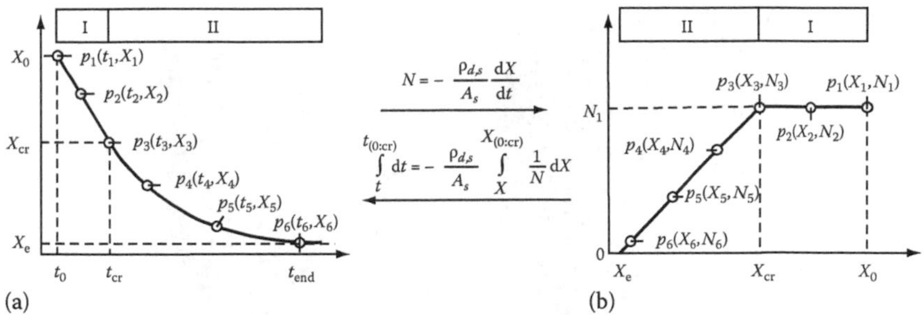


**FIGURE 12.3** Procedure adopted by Szafran and Kmeic (2005): (a) a theoretical drying curve and (b) theoretical drying rate curve. Region I denotes the first (constant) drying rate period, whereas region II denotes the second (falling) drying rate period.

CFD simulations can be transformed into time-independent data, extrapolated, and then transformed back to time-dependent data so that the mass flow rate over the full drying period can be quantified (Szafran and Kmiec, 2005). The means by which Szafran and Kmiec (2005) completed the transformation is illustrated in Figure 12.3. Overall, good agreement was found with the important phase of drying, which makes this technique an important consideration when undertaking such simulations, as system design or optimization may then be realized in reasonable time frames.

#### 12.5.3.2 Spray Drying

Spray drying is another traditional drying technique and is used to derive powders from products, with its main objective being to create a product that is easy to store, handle, and transport. As pointed out by Fletcher et al. (2006), an important requirement of the spray dryer operation is to avoid highly unsteady flows as these can cause deposition of partially dried product on the wall, resulting in a build up of crust, which is liable to catch fire due to overheating. As the phenomenological aspects of spray dryers are highly 3D and reliant on flow patterns, empirical techniques cannot provide the means of analyzing effects of chamber geometry or operating parameters. Therefore, CFD has been a necessary requisite for accurate spray dryer modeling, and has been employed for over 10 years now (Langrish and Fletcher, 2003). Figure 12.4 illustrates the effectiveness of CFD at capturing the different flow-field experienced as the inlet swirl angle changes.

The use of CFD in spray drying has been comprehensively reviewed in recent years, and for interested readers the articles of Langrish and Fletcher (2003) and Fletcher et al. (2006) provide a good understanding of the topic. It is important to note that most CFD of spray dryers use the Eulerian–Lagrangian approach, which, as discussed above, demands a large amount of storage, if the numbers of particles/droplets being modeled are great. Moreover, when large-scale systems need to be modeled, enhancing efficiency by parallel processing via domain decomposition would be difficult, as particles will not remain in one domain (Fletcher et al., 2006). Even though such difficulties are obviated with the Eulerian–Eulerian approach, many disadvantages exist, such as the loss of the time history of individual

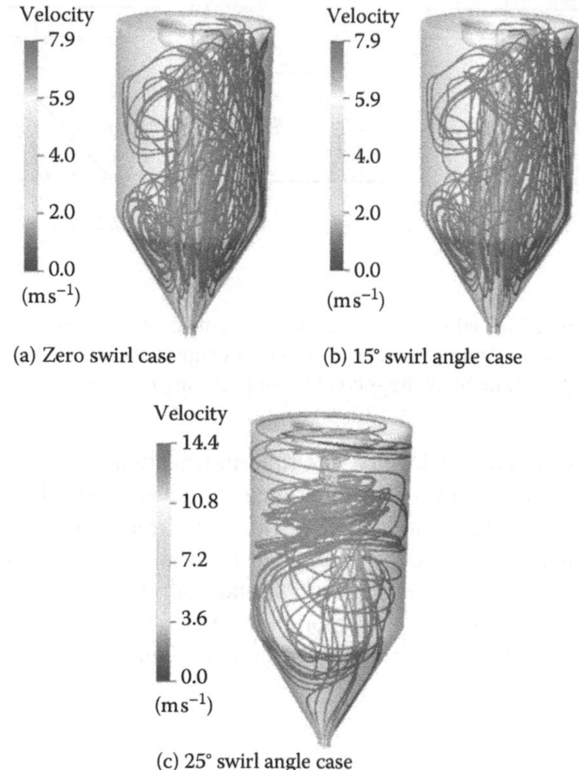


(a) Zero swirl case (b) 15° swirl angle case

(c) 25° swirl angle case

**FIGURE 12.4** Effect of swirl angle on the flow field behavior as visualized by streaklines starting from the inlet. (From Fletcher, D.F., Guo, B., Harvie, D.J.E., Langrish, T.A.G., Nijdam, J.J., and Williams, J., *Appl. Math. Model.*, 30, 1281, 2006. With permission.)

particles, the difficulty of modeling turbulent dispersion, and the inability to model interacting jets, with some of these problems sharing common ground with fluidized bed modeling. However, in phenomena unrelated to fluidized bed applications, the Eulerian–Lagrangian approach permits reasonable predictions of drying rate, particle–wall depositions, and particle agglomeration, whereas comparable applications using the Eulerian–Eulerian approach are limited. Therefore, Lagrangian particle tracking will remain a key feature of spray dryer modeling into the future.

One of the big difficulties when using CFD software packages in spray dryer modeling is that, owing to the presence of both solid and fluid, the mass transport limitations within a droplet cannot be easily taken into account, and therefore submodels must be included to do so; and for accurate solutions, these must be used alongside many other submodels that account for other phenomological aspects (Straatsma et al., 2007). In a recent study by Straatsma et al. (2007), submodels for mass transport, interparticle collision, agglomeration, thermal reactions, and stickiness were implemented with an Eulerian–Lagrangian model of an industrial dryer. The CFD simulations allowed the authors (Straatsma et al., 2007) to assess the agglomeration size of the particles and the stickiness of the particles colliding with each other and with the wall, and as a consequence allowed the fouling liability of the dryer to be evaluated.

#### 12.5.3.3 Forced-Convection Drying

Impinging jets are regularly used to dry food products in the food industry. Consequently, numerous investigations have numerically studied the heat and mass transfer characteristics of these applications. Many of these studies examine the influence of turbulence models (Seyedein et al., 1994; Ashfort-Frost et al., 1996), or the shape of impinging jet (Zhao et al., 2004), on Nusselt number distribution over the product surface. However, limited CFD studies that consider the coupled heat and mass transfer during drying exist.

Because of the complex geometry usually encountered in drying applications, theoretical studies are often not applicable, and to obtain the spatial distribution of transfer coefficients with reasonable accuracy it is necessary to solve the Navier–Stokes equations in the product surroundings. From the distribution of transfer coefficients, the correct temperature and moisture profiles within the product can be predicted, so that the drying process can be optimized. Kaya et al. (2007) used this approach with CFD to determine the transfer coefficients, and then the heat and mass transfer within the food was simulated with an external program. However, it is also possible to do this within the CFD package by determining the heat transfer coefficient with Equation 12.6, and the mass transfer with the Lewis relation, once a transient solution is performed. Such a technique was used by Suresh et al. (2001) in CFD simulations of the conjugate heat and mass transfer through a block immersed in a boundary layer flow. Conjugate heat and mass flux can also be determined with the help of empirically correlations of the transfer coefficients; however, these may not always be adequate (Zheng et al., 2007).

Fully coupled CFD solutions of heat and mass transfer in both solid and fluid phases also exist. De Bonis and Ruocco (2007) numerically modeled the heat and mass transfer in a food slab by an impinging heated air jet using time-dependent governing equations for a dilute mixture, via 2D CFD simulations. They described the evaporation kinetics as a first-order Arrhenius-type reaction and found strong process heterogeneity in moisture and temperature gradients. This heterogeneity was also found in solutions of surface water activity, which they attributed to the peculiar surface heat/mass transfer mechanism. A more advanced coupling procedure was presented by Erriguible et al. (2005, 2006), who explicitly accounted for the movement of bound water, free water, and vapor through the porous solid. They noted increase in the prediction accuracy when compared to those made via the use of heat transfer coefficients in the literature.

### 12.5.4 Cooking

#### 12.5.4.1 Natural Convection Ovens

Electric ovens are commonly used household appliances that rely on conjugate thermal exchange to produce the desired cooking effect in a foodstuff. For that reason, CFD is an appropriate tool to quantify the internal thermal field and mass transfer, both of which are important for robust design and performance. Natural convection is dominant in these ovens when it is produced by a source below the product (bake mode), with a source above the product driving radiative heat transfer (broil mode). Abraham and Sparrow (2003) used CFD to model the flow-field in an electric oven on baking mode with isothermal sidewalls, employing temperature boundary conditions consistent with experimental measurements. These simulations allowed insight into the relevant contributions of convective and radiative heat transfer to be gained, as well as determining the accuracy of predictions made by simple elementary surface correlations. Predictions were obtained with a steady-state solver and the solutions were then used as input into a quasisteady model to permit the time-wise temperature variation of the foodstuff to be analyzed, from which excellent agreement between experimental and numerical results was achieved. Abraham and Sparrow (2003) also found that for a blackened thermal load radiative heat transfer contributed to 72% of the total heat transfer, whereas this was 8% when the load was reflective. Heat flux distribution on the lower and upper surfaces varied as a function of contributions from temperature difference between the oven and the product, alongside that from the buoyant plume emanating from the heat source. From this, as shown in Figure 12.5, it was noted that steady-state heat transfer could be obtained at the top surface, whereas the undersurface was highly unsteady.

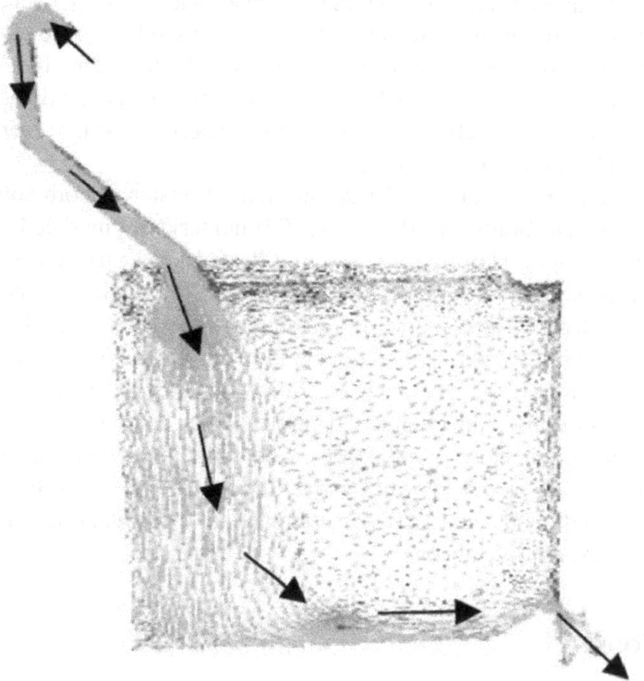

**FIGURE 12.5** Velocity field in the midplane of the oven with an atmospheric pressure boundary condition imposed at the boundaries, as simulated by CFD. CFD predicted an incorrect flow-field unless an artificial pressure drop was imposed. (From Mistry, H., Ganapathi-subbu, Dey, S., Bishnoi, P., and Castillo, J.L., *Appl. Thermal Eng.*, 26, 2448, 2006. With permission.)

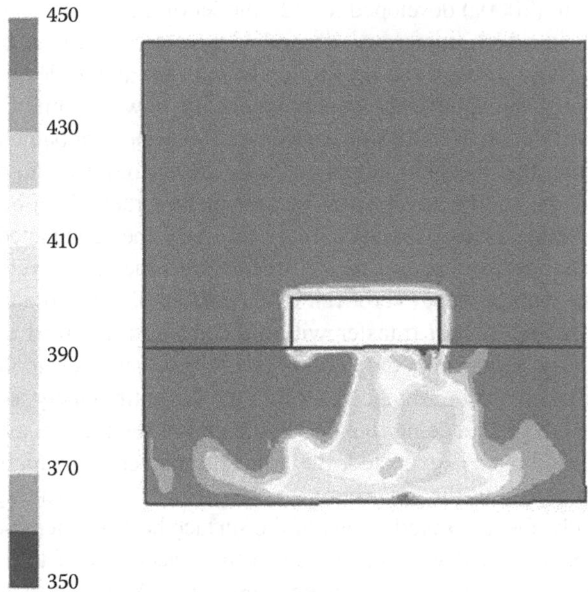


**FIGURE 12.6** CFD-predicted temperature contours of rectangular product in a natural convection oven with a temperature difference of 150K between product and enclosure. (From Abraham, J.P. and Sparrow, E.M., *Numer. Heat Tr. A-Appl.*, 44, 105, 2003. With permission.)

A thorough investigation into the thermal profile of an electrical oven, operated under both broil and bake modes, was completed by Mistry et al. (2006). The solution first obtained from the steady-state analysis yielded a flow-field, which opposed that evident from experimental observation, as illustrated in Figure 12.6. This was addressed by imposing an artificial, i.e., a "numerical," vent suction pressure, the value of which was tweaked until thermal field predictions corresponded with experimental measurements. Full cycling times, employing intermittent ON/OFF operation of heaters, were also simulated for both the broil and bake cycles. From the comparison of predictions, the broil cycle was confirmed to be less efficient, with a notable heterogeneity in temperature profile, owing to temperature stratification; this underscored the fact that the main thermal exchange in this cycle was due to radiation.

#### 12.5.4.2 Forced Convection Ovens

Over the years, forced convection ovens have been developed into two types of systems, which operate in very distinct and separate ways. The first type of conventional forced convection oven uses a fan to develop high-velocity convection currents around the enclosed foodstuffs, which enhances the rate of heat transfer, and permits faster cooking of the products. The second type of forced convection oven permits the use of impinging jets to provide the main thermal exchange. The physics associated with this type of oven has a direct relationship with air impingement drying; thus, it is easy to see why its correct utilization in the industry can provide high levels of efficiency.

Verboven et al. (2000a) developed a CFD model of an empty, isothermal forced-convection oven, in which numerous submodels were incorporated, including those describing the airflow through the heating coils and fan, alongside swirl in the flow regime added by the fan. Due to the complexity of the flow dynamics involved in the oven, a detailed validation exercise was performed. An important outcome of the study was the observation that the swirl model was a necessary requisite for accurate quantitative results, as the swirl actually improves the airflow rate whilst having no direct effect on the mechanical energy balance. In a later study, they conducted a CFD investigation into the temperature and airflow distribution in the same oven during operation cycle, loaded with product (Verboven et al., 2000b). Prior to the CFD modeling, a lumped FEM model of heat transfer within a food was designed to calculate the appropriate heat sources and boundary conditions, which were then applied in the CFD model. The prediction errors averaged to 4.6°C for the entire cavity at the end of the warm-up stage, with this reducing to 3.4°C at the end of the process as the variability in the system leveled out. As expected, the products subject to the highest heat transfer were situated closest to the fan where they experienced uniform high-velocity air flow. Unfortunately, the CFD predictions of the surface heat transfer coefficients were grossly underpredicted, and were attributed to the inaccuracy of the wall functions calculations. However, in order to circumvent these errors, Verboven et al. (2000b) used the local turbulence intensity and Reynolds number, obtained from the CFD model, to compute the heat transfer coefficients. The correlation derived alongside other published numerically derived correlations is summarized in Table 12.3.

Application of CFD in jet impingement oven systems provides detailed understanding of the effect of different oven geometries as well as object geometries on the system performance. A full 3D CFD model of a multiple jet impingement oven was developed by Kocer and Karwe (2005). Convection and conduction heat transfer were coupled. However, as the thermal exchange was a result of jet impingement only, radiation was ignored with a small compromise of accuracy. Moisture transfer was not considered. Kocer and Karwe (2005) then determined a correlation for the average Nusselt number in terms of Reynolds number for multiple jets impinging on the surface of a cylindrical model cookie, which indicated the strong dependence of surface heat transfer coefficient on velocity of the jet. In another CFD study of multiple jets flows (Olsson et al., 2005), the airflow and thermal exchange characteristics cylinders were found to be dependent on the distance and opening between the jets. At larger distances between the jets, more entrainment of air between the jets occurred, whereas for smaller spacing between the jets, the entrainment was suppressed, almost no air exited through the opening between the jets and the flow between the cylinders were almost stagnant.

#### 12.5.4.3 Baking Ovens

For the bakery industry, it is necessary to maintain consistency in product quality throughout processing. This requires quality-conscious control strategies to be implemented (De Vries et al., 1995), which should permit the adjustment of process parameters in response to disturbances, such as a change in the oven load. The successful implementation of such control strategies has been the long-term goal of process modeling via CFD for some time now (Zhou et al., 2007). The tone

**TABLE 12.3**
**Correlations for Surface Heat Transfer Coefficients Developed from CFD Models Predictions**

| Authors | Process | Turbulence Model | Correlation Developed | Comments |
|---|---|---|---|---|
| Verboven et al. (2000b) | Heating of food in forced convection oven | Std. $k$–$\varepsilon$ | $\frac{Nu}{Nu_0} = 1 + 0.088TuRe^{1/2}$ where, $Nu_0 = 0.245Re^{1/2}$ | Due to inaccuracies with wall functions, the correlation was not developed using turbulence values from the free-stream |
| Abraham and Sparrow (2003) | Oven heating of rectangular food in a natural convection oven | Spalart-Allmaras | Developed quasisteady model to predict temperature evolution of solid food | Reasonable agreement with surface heat transfer coefficients was found for surfaces where steady heat transfer was predicted |
| Ollson et al. (2004) | Jet impingement heating of cylindrical food placed on flat surface | SST, $k$–$\omega$ | $Nu_m = 0.14Re^{0.65} \times (H/d)^{-0.077} \times (d/D)^{0.32}$ | The heat transfer rate was found to be higher on the top of the product and in the wake, lower in the separation point and the back of the cylinder |
| Kocer and Karwe (2005) | Multiple jet impingement heating of cylindrical food (cookie) placed in center of oven | Std. $k$–$\varepsilon$ | $Nu = 0.58Re^{0.375}$ | Predictions showed a strong dependence between the surface heat transfer coefficients and air velocity, whereas they were almost independent of temperature |
| Nitin et al. (2006) | Jet impingement heating of cylindrical food placed on flat surface | Std. $k$–$\varepsilon$ | $Nu = 0.000348Re^{0.78} \times (d/D)^{0.65}$ | Predictions were found to depend on the geometry of the model object and jet flow field in the oven |

of the first studies with this aim was set out by Therdthai et al. (2003), who solved the 2D flow and temperature field in conventional “u-turn” bread-baking oven. They used the simulations to provide information for temperature control, as well as for control sensor position. However, a number of simplifications limited the veracity of the model, with regard to its representation of the physical process. For example, a steady-state flow regime was assumed, even though the process is inherently transient. Furthermore, the contribution from radiative heat transfer was ignored, despite the fact that this mode significantly influences thermal exchange. These assumptions were, however, addressed in a subsequent publication, in which a full 3D model was built, using a moving grid technique, to simulate the transient baking process (Therdthai et al., 2004a). Radiative heat transfer was also incorporated.

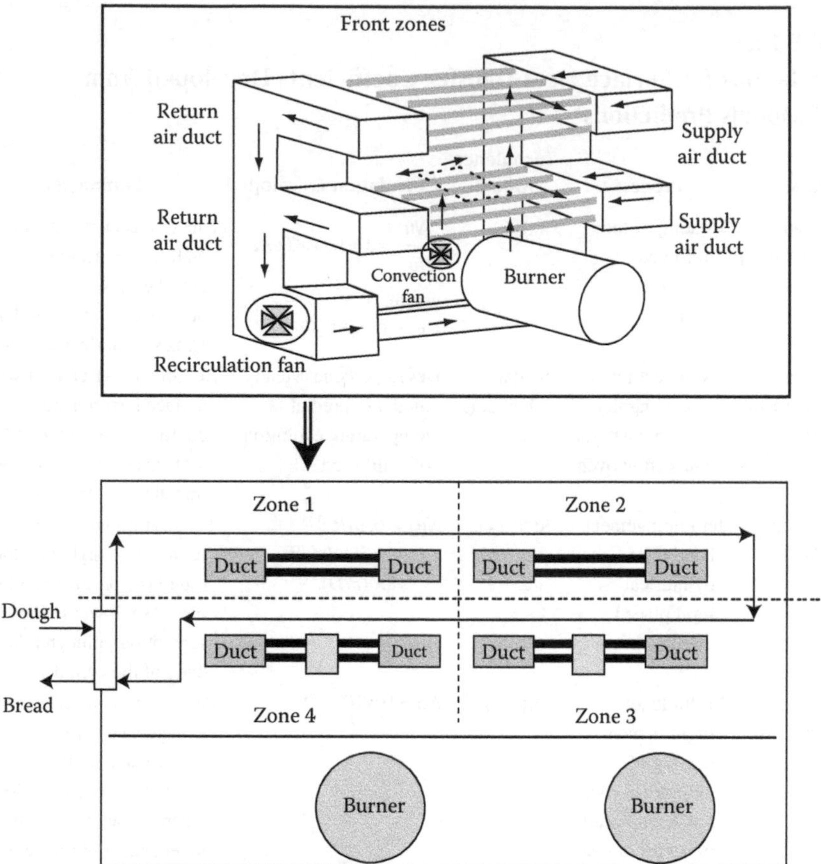


**FIGURE 12.7** The industrial bread baking oven. (As simulated by Therdthai, N., Zhou, W.B., and Adamczak, T.J., *Food Eng.*, 65, 599, 2004a. With permission.)

Therdthai et al. (2002) used mathematical models, which they had developed in a previous study, in conjunction with the CFD results to predict the quality attributes of bread. Using both sets of predictions, they provided recommendations to reduce the energy use during the process, whilst not compromising the bread quality. More specifically, with reference to Figure 12.7, the propositions were that the duct temperature in zones 1, 3, and 4 should be reduced by about 10°C so that the tin temperature in zones 1 and 3 would decrease, thereby reducing weight loss in this zone. At the same time, their observations showed that the flow rate of the convection fan in zone 3 should be increased by doubling it throughput, so that the tin temperatures in zones 2 and 4 could be maintained at high levels. This optimization was predicted to produce a weight loss of 7.95%, with the lightness values of the crust color on the top, bottom, and side of the loaf being around *L*-values of 50.68, 55.34, and 72.34, respectively. Later a starch gelatinization model, based on simple first-order kinetics, was incorporated to examine the effects of the optimized operation on dough gelatinization (Therdthai et al., 2004b). The rate of gelatinization was notably slower for the bread baked using the optimized conditions, especially

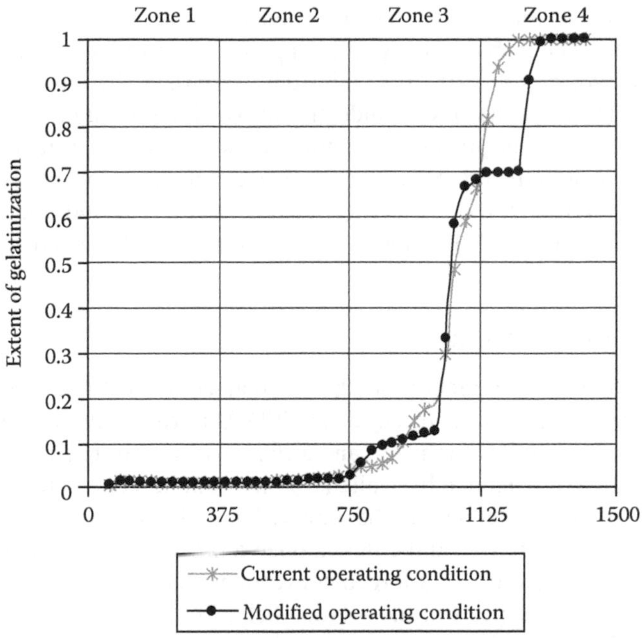


**FIGURE 12.8** Comparison of the starch gelatinization profiles during baking under the standard and optimized baking conditions. (From Therdthai, N., Zhou, W.B., and Adamczak, T., *J. Food Eng.*, 65, 543, 2004b. With permission.)

in zone 3, as a result of reduction in energy supplied to this zone. However, the gelatinization gradually picked up and reached maximum gelatinization extent at the end of the baking process, as shown in Figure 12.8. Other investigations have sought to assess and quantify the robustness of CFD models to changes in the physical properties of bread during the baking process (Wong et al., 2006). Wong et al. developed simple mathematical models, which described the changes in the temperature profiles as a function of the changes in the physical properties.

Recent CFD modeling studies of the bread baking process have looked at the 2D physical representation, coupling convection, and radiative heat transfer via DO model (Wong et al., 2007a). Moreover, the density, heat capacity, and thermal conductivity were allowed to vary with temperature. However, some discrepancies between predictions and measurements of the actual baking process were found, especially of those comparisons made at the dough center, which were probably caused by no modeling of the moisture transport in the dough and evaporation kinetics. Moreover, the confining effect afforded by the 2D model was seen to cause lack of correspondence in the validation study. Comparisons were also made with the 3D model of Therdthai et al. (2004a), from which it was summarized that the 2D model provided better predictions in the moving sensors. Most notably CFD predicted that at full oven condition the temperature readings from the stationary sensors oscillated in association with the movement of dough/bread along the traveling track owing to the reduced temperature gradient in the region around the sensors and bread surface. In a later study, proportional–integral (PI) controllers were incorporated to the existing 2D CFD model and the feasibility of establishing a framework that is capable of integrating a process controller with a CFD model was assessed (Wong et al., 2007b). This study comprises the long-term objective of process modeling in the bakery industry, in which a dual-mode controller was designed to suit different processing conditions, which were simulated to provide satisfactory control in the presence of disturbances and set point changes. From the study, it was shown how the preheating stage could be removed under controlled conditions, which retained the quality characteristics of the bread whilst saving energy.

#### 12.5.4.4 Microwave Ovens

Microwaves ovens are commonly used home appliances. Heat and mass transfer in these appliances have been modeled using CFD. Unfortunately, the coupled problem of electromagnetic, heat transfer, and airflow has not yet been solved with the CFD approach, as commercially available finite volume codes are not amenable for electromagnetic modeling. In the first CFD study of this nature, Verboven et al. (2003) numerically solved the flow around a product of uniform surface temperature, under both natural and forced convection conditions, in order to observe the phenomenological aspects in the solution field. The combination of natural and forced convection typically experienced in microwave ovens was modeled and the local heat transfer coefficients were computed. The study showed that, for the typical flow rates experienced in the oven, there was an overall increase in heat and mass transfer coefficient when compared to purely natural or forced convection situations. However, the mass transfer coefficient was not high enough to avoid the development of sogginess. To reduce sogginess, it was suggested that optimal placement of inlets and outlets alongside optimizing fan design would be required.

Even though commercial CFD codes may not be suitable for multiphysics problem including electromagnetic radiation, such coupling can be accomplished. Perre and Turner (1999) coupled heat and mass transfer with Maxwell's equations and the dielectric properties, which varied as a function of both temperature and moisture, so that the drying of wood could be quantified. As mentioned above, such coupling in food simulations has required synergy between different numerical approaches. For example, a weak coupling between a finite elements model for electromagnetic radiation and a CFD model has been achieved by Verboven et al. (2007), via surface heat transfer coefficients, in a jet impingement microwave oven. An example of full coupling between the calculations of heat and mass transfer and electromagnetic radiation has been achieved by Dincov et al. (2002), who mapped the (moisture and temperature dependent) porous media dielectric properties from the CFD mesh onto the electromagnetic finite difference mesh and then mapped the microwave heating function from the electromagnetic finite difference mesh onto the CFD mesh. Using this specialized coupling algorithm, Dincov et al. (2002) modeled the phase change that accompanies the temperature increase during microwave heating. The elevated internal temperatures coupled with the increase in internal vapor pressure were seen to drive the liquid from the medium quickly and efficiently. It was also seen that, because liquid water is the most active component in absorbing microwaves, the energy absorption is reduced during the process.

## 12.6 MODELING EMERGING THERMAL TECHNOLOGIES WITH CFD

### 12.6.1 High-Pressure Thermal Processing

High-pressure processing (HPP) is a novel food processing method, which has shown great potential in the food industry. Similar to heat treatment, HPP inactivates microorganisms, denatures proteins, and extends the shelf life of food products, but in the meantime, unlike heat treatments, high-pressure (HP) treatment can also maintain the quality of fresh foods, with little effects on flavor and nutritional value.

During compression/decompression phases, the internal energy of an HPP system changes, resulting in heat transfer between the internal system and its boundaries. These thermal–hydraulic characteristics were studied with CFD by Hartmann (2002). The technique was deemed necessary in gaining thorough understanding of the phenomena inherent in HPP, especially when the scale-up phenomena need to be analyzed (e.g., layout and design of HP devices, packages, etc.) (Hartman, 2002). Hartmann and Delgado (2002) used CFD and dimensional analyses to determine the timescales of convection, conduction, and bacterial inactivation, the relative values of which contribute to the efficiency and uniformity of conditions during HPP. Conductive and convective timescales were directly compared to the inactivation timescale in order to provide a picture of the thermal–hydraulic phenomena in the HP vessel during bacterial inactivation. The results showed that pilot scale systems exhibited a larger convection timescale than the inactivation timescale, and that the intensive fluid motion and convective heat transfer resulted in homogenous bacterial inactivation. Furthermore, the simulations of industrial scaled systems showed greater efficiency in bacterial inactivation as the compression heating subsisted for greater time periods when compared to smaller laboratory systems. Other CFD simulations showed that the thermal properties of the HP vessel boundaries have considerable influence on the uniformity of the process, and insulated material promoted the most effective conditions (Hartmann et al., 2004). As well as this, the insulated vessel was found to increase the efficiency of the HPP by 40%. A CFD and dimensionless analysis of the convective heat transfer mechanisms in liquid foods systems under pressure were also performed by Kowalczyk and Delgado (2007) who advised that HP systems with a characteristic dimension of 1 m alongside a low viscous medium should be used to avoid heterogeneous processing of the product.

CFD studies have also provided solutions to the thermal–hydraulic phenomena in HPP systems containing packaged ultrahigh temperature (UHT) milk (Hartmann et al., 2003), packaged enzyme mixture (Hartmann and Delgado, 2003), solid beef fat (Abdul Ghani and Farid, 2007), and solid food analog material (Otero et al., 2007) (e.g., tylose with similar properties to meat, and agar with similar properties to water). In both investigations (Hartmann et al., 2003; Hartmann and Delgado, 2003), the most significant results showed strong coupling between concentrations of the surviving microorganisms and the spatial distribution of the food package in the HP vessel, owing to the inhomogeneous temperature field. A low conductive package material was also found to improve the uniformity of processing by preserving the elevated temperature level within the package throughout the pressurization phase; an average difference of about two log reductions was found per 10-fold increase in the package thermal conductivity. The 2D CFD simulations of Otero et al. (2007) found that the filling ratio of the HP vessel played a major role in process uniformity, with convective currents having least effect on heat transfer when this ratio is large (Figure 12.3). Otero et al. (2007) also showed that by anticipating the temperature increase resulting from compression heating and by allowing the pressure transmitting medium to supply the appropriate quantity of heat, the uniformity of HPP was enhanced when both large and small sample ratios were used (Figure 12.4). More recently, Abdul Ghani and Farid (2007) used 3D CFD simulations to illustrate both convective and conductive heat transfer in an HPP system loaded with pieces of solid beef fat. The simulation showed a greater adiabatic heating in the beef fat than the pressure transmitting medium owing to the greater compression heating coefficient used in this case.

### 12.6.2 Ohmic Heating

The basic principle of ohmic heating is that electrical energy is converted to thermal energy within a conductor. In food processing, foodstuffs act as the conductor. The main advantage of ohmic heating is that, because heating occurs by internal energy generation within the conductor, this processing method leads itself to even distribution of temperatures within the food, and it does not depend on heat transfer mechanisms (Jun and Sastry, 2005a,b). Jun and Sastry (2005a,b) were the first to model the ohmic heating process, with the aim to enhance heating techniques for use by cabin crews during long-term space missions. They developed a 2D transient model, for chicken noodle soup (assumed single phase) and black beans, under the ohmic heating process, by solving the electric field via the Laplace equation. From this solution, the internal heat generation was obtained, which was added as a source to the Fourier equation, and was then numerically solved by the CFD code. Electrical conductivity was allowed to vary as a function of temperature. The CFD model was able to predict regions of electric field overshoot in the food, as well as the nonuniformities in the predicted thermal field. Moreover, they noted that as the electrode got wider, the cold zone area developed in the middle of the packaging diminished to a minimum and then appeared and grew at the corners of the packaging, clearly illustrating the existence of a threshold value for electrode size optimization. They later expanded the model so that a 3D representation of pouched tomato soup could be simulated (Jun and Sastry, 2005a,b), in which they found the electric field strength near the edges of electrodes to overshoot as it got close to the maximum value, as predicted by their first model. On the other hand, the food between the V-shaped electrodes experienced a weak electric field strength, which gave rise to cold zones in the food. Jun and Sastry (2005a,b) recognized that the presence of these cold zones merited further research on pouches via modeling and package redesign.

## 12.7 CHALLENGES FACE THE USE OF CFD IN THERMAL PROCESS MODELING

### 12.7.1 Improving the Efficiency of the Solution Process

As it can be seen above, the physical mechanisms that govern thermal processes generally include any combination of fluid flow, heat, mass, and scalar transport.

As CFD involves the modeling of such mechanisms, which mostly occur on different timescales, the temporal accuracy and stability of a solution is usually bounded by the ability of the model to capture the mechanism that is the quickest to occur. This is done through time stepping, which has therefore to be optimized during model development. The time-step must be small enough to resolve the frequencies of importance during a transient process. To do this, an appropriate characteristic length and velocity of the problem is necessary, which can be obtained from nondimensional numbers such as the Stroudal number, from experimental data, or from experience. For example, when Szafran and Kmiec (2005) developed a CFD model of a spouted bed dryer, they found that extremely small time-steps of the order of $1 \times 10^{-4}$ were required to resolve the instabilities in the flow regime as well as the circulation of phases. However, this meant that a year of computation would result in a solution for only 1 h of drying. Such a time frame is excessive by any length of the imagination, and overcoming this difficulty required considerable insight into the physical process. To do so, Szafran and Kmiec (2005) transformed the process from time dependent into time independent, from which drying curves could be developed, as described above (see Section 12.5).

Insight into the numerical abilities of CFD packages is also important if one needs to solve the problems of excessive computing times. Taking the parallelization features of commercial CFD codes as an example, these can allow a solution to be formed quicker, via domain decomposition, as long as the computing power is available and Lagrangian particle tracking is not employed. Alongside this, the solving techniques employed in commercial CFD codes have also been found to play a major role in efficiency. Fletcher et al. (2006) noted how segregated solvers and coupled solvers can bring different attributes to solution progression, and found that owing to the reduced levels of "random noise" introduced, the coupled solver permitted a high level of control over the solution process, allowing efficient and accurate predictions of the transient evolution of the flow instability in a spray dryer, when compared to the segregated solver.

Simplifying the geometrical representation of CFD models can also cut down on both preprocessing and solving time. The 2D modeling technique assumes that the length of a system is much greater than its other two dimensions, and that the process flow is normal to this length. As the effects of the confining geometry are essentially disregarded, accurate judgment of whether the process is amenable to the 2D assumption is required.

### 12.7.2 CFD and Controlling Thermal Processing

All thermal processes in the food industry are performed under controlled conditions. Unfortunately, due to the nonlinearity of the transport phenomena, CFD techniques are not yet amenable to the online control of thermal processes, and reduced order models which use statistical data to manipulate the process variables via controlled inputs are more appropriate. However, this does not mean that the actions of a control system cannot be modeled by CFD. Wong et al. (2007a,b) have been the first to implement a control system within a CFD model, in order to simulate its performance. Such abilities undoubtedly provide benefits during the predesign or optimization stages of system development.

However, research and technology have not yet reached the level where low-order and CFD models can be combined to effect control with accuracy during industrial thermal processing. This type of synergism would provide benefits in situations where accurate experiential readings, which act as the input to the control system, may not be representative of the total system, i.e., aseptic processing of nonhomogeneous food. In this instance, the use of CFD to generate the time series data could be a viable alternative. However, more research is needed before such approaches can be applied.

### 12.7.3 Turbulence

One of the main issues faced by the food industry over the last two decades is the fact that most turbulence models have shown to be application specific. Presently, there are many turbulence models available; however, until a complete turbulence model capable of predicting the average field of all turbulent flows is developed, the CFD optimization of many thermal processes will be hampered. The reason is that in every application many different turbulence models must be applied until the one that gives the best predictions is found. The closest to the complete turbulence model thus far is LES, which uses the instantaneous Navier–Stokes equations to model large-scale eddies, with smaller scales solved with a subgrid model. However, using this model demands large amounts of computer resources, which may not be presently achievable.

For many cases, the $k$–$\varepsilon$ model and its variants have proved to be successful in applications involving swirling flow regimes, once the mesh is considerably fine. This is because, as pointed out by Guo et al. (2003), when the mesh is very fine, the $k$–$\varepsilon$ model performs like the subgrid model of an LES, and handles small-scale turbulence, while the largest scales are solved by the transient treatment of the averaged equations. In practice, therefore, most of the important energy-carrying eddies can be solved by this means. However, $k$–$\varepsilon$ models have lacked performance when predicting impinging flows, or in flows with large adverse pressure gradients. In such instance, models like the $k$–$\omega$ or LES should be considered more closely.

### 12.7.4 Boundary Conditions

In CFD simulations, the boundary conditions must be adequately matched to the physical parameters of the process, with the precision of similarity being conditioned by the mechanism under study and the level of accuracy required. Even when this is done, CFD solution still may not be a correct physical representation of the physical system. This was shown by Mistry et al. (2006), who found that an artificial pressure differential was required to predict the correct flow patterns in an oven, which was heated by natural convection. Such results suggest the importance of sensitivity analysis studies alongside experimental measurements in the early stages of model development. Sensitivity analyses are also necessary for turbulence model specification, or turbulence model tuning via inlet conditions, and for CFD model simplification.

## 12.8 CONCLUSIONS

CFD has played an active part in the design of thermal processes for over a decade now. In recent years, simulations have reached higher levels of sophistication, as application specific models can be incorporated into the software with ease, via user defined files. The importance of maintaining a high level of accuracy via circumspect choices made during model development is evident from the reviewed studies, as many studies provide detailed validation exercises. Undoubtedly, with current computing power progressing unrelentingly, it is conceivable that CFD will continue to provide explanations for transport phenomena, leading to better design of thermal processes in the food industry.

## NOMENCLATURE

| | |
|---|---|
| $C$ | water concentration (mol m$^{-3}$) |
| $C_\mu$ | empirical turbulence model constant |
| $c_p$ | specific heat capacity (W kg$^{-1}$ K$^{-1}$) |
| $d$ | width of jet (m) |
| $D$ | width of cylinder (m) |
| $g$ | acceleration due to gravity (m s$^{-2}$) |
| $H/d$ | jet-to-cylinder distance ratio |
| $M$ | molecular weight (kg kmol$^{-1}$) |
| $Nu$ | Nusselt number |
| $p$ | pressure (Pa) |
| $R$ | gas constant (J kmol$^{-1}$ K$^{-1}$) |
| $Re$ | Reynolds number |
| $k$ | turbulent kinetic energy (m$^2$ s$^{-2}$) |
| $s_T$ | thermal sink or source (W m$^{-3}$) |
| $s_c$ | concentration sink or source (mol m$^{-3}$) |
| $T$ | temperature (K) |
| $Tu$ | turbulence intensity (%) |
| $t$ | time (s) |
| $U$ | velocity component (m s$^{-1}$) |
| $\vec{v}_i$ | velocity component (m s$^{-1}$) |
| $x$ | Cartesian coordinates (m) |

*Greek letters*

| | |
|---|---|
| $\rho$ | density (kg m$^{-3}$) |
| $\mu$ | dynamic viscosity (kg m$^{-1}$ s$^{-1}$) |
| $\beta$ | thermal expansion coefficient (K$^{-1}$) |
| $\lambda$ | thermal conductivity (W m$^{-1}$ K$^{-1}$) |
| $\alpha$ | water molar latent heat of vaporization (J mol$^{-1}$) |
| $\varepsilon$ | turbulent dissipation rate (m$^2$ s$^{-3}$) |
| $\phi$ | the transported quantity |
| $\Gamma$ | diffusion coefficient of transported variable |
| $\vartheta$ | the diffusivity of the mass component in the fluid (m$^2$ s$^{-1}$) |
| $\omega$ | specific dissipation (s$^{-1}$) |
| $\mu_t$ | turbulent viscosity (kg m$^{-1}$ s$^{-1}$) |

*Subscripts*

| | |
|---|---|
| i | Cartesian coordinate index |
| bf | bulk fluid |
| s | surface |
| $V$ | mesh element volume |
| $A$ | area of mesh element |
| $m$ | mean |
| 0 | with no turbulence |

## REFERENCES

Abbott M B and Basco B R. 1989. *Computational Fluid Dynamics—An Introduction for Engineers*. Longman Scientific & Technical, Harlow, UK, pp. 5–30.

Abdul Ghani A G and Farid M M. 2006. Numerical simulation of solid–liquid food mixture in a high pressure processing unit using computational fluid dynamics. *Journal of Food Engineering*, 80(4), 1031–1042.

Abdul Ghani A G and Farid M M. 2007. Numerical simulation of solid–liquid food mixture in a high pressure processing unit using computational fluid dynamics. *Journal of Food Engineering*, 80(4), 1031–1042.

Abdul Ghani A G, Farid M M, Chen X D, and Richards P. 1999. An investigation of deactivation of bacteria in a canned liquid food during sterilization using computational fluid dynamics (CFD). *Journal of Food Engineering*, 42, 207–214.

Abdul Ghani A G, Farid M M, Chen X D, and Richards P. 2001. Thermal sterilization of canned food in a 3-D pouch using computational fluid dynamics. *Journal of Food Engineering*, 48, 147–156.

Abdul Ghani A G, Farid M M, and Chen X D. 2002. Theoretical and experimental investigation of the thermal inactivation of *Bacillus stearothermophilus* in food pouches. *Journal of Food Engineering*, 51(3), 221–228.

Abdul Ghani A G, Farid M M, and Zarrouk S J. 2003. The effect of can rotation on sterilization of liquid food using computational fluid dynamics. *Journal of Food Engineering*, 57, 9–16.

Abraham J P and Sparrow E M. 2003. Three-dimensional laminar and turbulent natural convection in a continuously/discretely wall-heated enclosure containing a thermal load. *Numerical Heat Transfer Part A: Applications*, 44(2), 105–125.

Anon. 2007. A brief history of computational fluid dynamics. Internet article, www.fluent.com, Fluent Inc. 10 Cavendish Court, Lebanon, NH 03766, USA.

Aversa M, Curcio S, Calabro V, and Iorio G. 2007. An analysis of the transport phenomena occurring during food drying process. *Journal of Food Engineering*, 78(3), 922–932.

Choundhury D. 1993. Introduction to the renormalization group method and turbulence modelling. Fluent Inc. Technical Memorandum TM-107.

Crank J. 1975. *The Mathematics of Diffusion*, 2nd ed. Clarendon Press, Oxford, UK.

De Bonis M V and Ruocco G. 2007. Modelling local heat and mass transfer in food slabs due to air jet impingement. *Journal of Food Engineering*, 78(1), 230–237.

De Vries U, Velthuis H, and Koster K. 1995. Baking ovens and product quality—a computer model. *Food Science and Technology Today*, 9(4), 232–234.

Denys S, Pieters J G, and Dewettinck K. 2003. Combined CFD and experimental approach for determination of the surface heat transfer coefficient during thermal processing of eggs. *Journal of Food Science*, 68(3), 943–951.

Denys S, Pieters J G, and Dewettinck K. 2004. Computational fluid dynamics analysis of combined conductive and convective heat transfer in model eggs. *Journal of Food Engineering*, 63(3), 281–290.

Denys S, Pieters J G, and Dewettinck K. 2005. Computational fluid dynamics analysis for process impact assessment during thermal pasteurization of intact eggs. *Journal of Food Protection*, 68(2), 366–374.

Denys S, Pieters J G, and Dewettinck K. 2007. CFD analysis of thermal processing of eggs. Chapter 14 in *Computational Fluid Dynamics in Food Processing*, Da-Wen Sun (Editor), 28 chapters, CRC Press, Boca Raton, FL, pp. 347–381.

Dincov D D, Parrott K A, and Pericleous K A. 2002. Coupled 3-D finite difference time domain and finite volume methods for solving microwave heating in porous media. *Lecture Notes in Computer Science*, 2329, 813–822.

Erriguible A, Bernada P, Couture F, and Roques M A. 2005. Modeling of heat and mass transfer at the boundary between a porous medium and its surroundings. *Drying Technology*, 23(3), 455–472.

Erriguible A, Bernada P, Couture F, and Roques M A. 2006. Simulation of convective drying of a porous medium with boundary conditions provided by CFD. *Chemical Engineering Research and Design*, 84(A2), 113–123.

Farid M and Abdul Ghani G A. 2004. A new computational technique for the estimation of sterilization time in canned food. *Chemical Engineering and Processing*, 43(4), 523–531.

Feng H and Tang J. 1998. Microwave finish drying of diced apples in a spouted bed. *Journal of Food Science*, 63, 679–683.

Fernandes C S, Dias R, Nóbrega J M, Afonso I M, Melo L F, and Maia J M. 2005. Simulation of stirred yoghurt processing in plate heat exchangers. *Journal of Food Engineering*, 69, 281–290.

Fernandes C S, Dias R P, Nobrega J M, Afonso I M, Melo L F, and Maia J M. 2006. Thermal behaviour of stirred yoghurt during cooling in plate heat exchangers. *Journal of Food Engineering*, 69, 281–290.

Ferry M. 2002. New features of migal solver. *The Phoenics Journal*, 14(1), 88–96.

Ferziger J H and Peric M. 2002. *Computational Methods for Fluid Dynamics*. Berlin Heidleberg: Springer-Verlag, pp. 1–100.

Fletcher D F, Guo B, Harvie D J E, Langrish T A G, Nijdam J J, and Williams J. 2006. What is important in the simulation of spray dryer performance and how do current CFD models perform? *Applied Mathematical Modelling*, 30(11), 1281–1292.

Friedrich R, Huttl T J, Manhart M, and Wagner C. 2001. Direct numerical simulation of incompressible turbulent flows. *Computers and Fluids*, 30, 555–579.

Geedipalli S S R, Rakesh V, and Datta A K. 2007. Modelling the heating uniformity contributed by a rotating turntable in microwave ovens. *Journal of Food Engineering*, 82(3), 359–368.

Ghani A G A, Farid M M, and Chen X D. 2002. Theoretical and experimental investigation of the thermal destruction of vitamin C in food pouches. *Computers and Electronics in Agriculture*, 34(1–3), 129–143.

Gidaspow D, Bezburuah R, and Ding J. 1992. Hydrodynamics of circulating fluidized beds, kinetic theory approach. In Fluidization VII, *Proceedings of the 7th Engineering Foundation Conference on Fluidization*, Brisbane, Australia, pp. 75–82.

Grijspeerdt K, Vucinic D, and Lacor C. 2007. CFD modelling of the hydrodynamics of plate heat exchangers for milk processing. Chapter 19 in *Computational Fluid Dynamics in Food Processing*, Da-Wen Sun (Editor), 28 chapters, CRC Press, Boca Raton, FL, pp. 403–417.

Harral B B and Boon C R. 1997. Comparison of predicted and measured airflow patterns in a mechanically ventilated livestock building without animals. *Journal of Agricultural Engineering Research*, 66, 221–228.

Harral B and Burfoot D. 2005. A comparison of two models for predicting the movements of airborne particles from cleaning operations. *Journal of Food Engineering*, 69, 443–451.

Hartmann C. 2002. Numerical simulation of thermodynamic and fluid-dynamic processes during the high-pressure treatment of fluid food systems. *Innovative Food Science and Emerging Technologies*, 3(1), 11–18.

Hartmann C and Delgado A. 2002. Numerical simulation of convective and diffusive transport effects on a high-pressure-induced inactivation process. *Biotechnology and Bioengineering*, 79(1), 94–104.

Hartmann C and Delgado A. 2003. The influence of transport phenomena during high-pressure processing of packed food on the uniformity of enzyme inactivation. *Biotechnology and Bioengineering*, 82(6), 725–735.

Hartmann C, Delgado A, and Szymczyk J. 2003. Convective and diffusive transport effects in a high pressure induced inactivation process of packed food. *Journal of Food Engineering*, 59(1), 33–44.

Heldman D R and Singh R P. 1981. *Food Process Engineering*. AVI Publishing, Westport, CT, pp. 1–10.

Hendrickx M, Silva C, Oliveria F, and Tobback P. 1993. Generalised (semi-) empirical formulas for optimal sterilisation temperatures of conduction-heated foods with infinite surface heat transfer coefficients. *Journal of Food Engineering*, 19(2), 141–158.

Jun S and Puri V M. 2005. 3D milk-fouling model of plate heat exchangers using computational fluid dynamics, *International Journal of Dairy Technology*, 58(4), 214–224.

Jun S and Puri V M. 2007. Plate heat exchange thermal and fouling analysis. Chapter 19 in *Computational Fluid Dynamics in Food Processing*, Da-Wen Sun (Editor), 28 chapters, CRC Press, Boca Raton, FL, pp. 417–433.

Jun S and Sastry S. 2005a. Reusable pouch development for long term space missions: A 3D ohmic model for verification of sterilization efficacy. *Journal of Food Engineering*, 80(4), 1199–1205.

Jun S and Sastry S. 2005b. Modeling and optimization of ohmic heating of foods inside a flexible package. *Journal of Food Process Engineering*, 28(4), 417–436.

Kaya A, Aydin O, and Dincer I. 2007. Numerical modeling of forced-convection drying of cylindrical moist objects. *Numerical Heat Transfer Part A: Applications*, 51(9), 843–854.

Kenneth J B. 2004. Heat exchanger design for the process industries. *Journal of Heat Transfer*, 126(6), 877–885.

Kocer D and Karwe M V. 2005. Thermal transport in a multiple jet impingement oven. *Journal of Food Process Engineering*, 28(4), 378–396.

Kocer D, Nitin N, and Karwe M V. 2007. Applications of CFD in Jet impingement oven. Chapter 19 in *Computational Fluid Dynamics in Food Processing*, Da-Wen Sun (Editor), 28 chapters, CRC Press, Boca Raton, FL, pp. 469–487.

Kondjoyan A. 2006. A review on surface heat and mass transfer coefficients during air chilling and storage of food products. *International Journal of Refrigeration*, 29(6), 863–875.

Kowalczyk W and Delgado A. 2007. On convection phenomena during high pressure treatment of liquid media. *International Journal of High pressure Research,* 27(1): 85–92.

Langrish T A G and Fletcher D F. 2003. Prospects for the modelling and design of spray dryers in the 21st century. *Drying Technology*, 21(2), 197–215.

Launder B E and Spalding D B. 1974. The numerical computation of turbulent flows. *Computer Methods in Applied Mechanics and Engineering*, 3, 269–289.

Madhiyanon T, Soponronnarit S, and Tia W. 2002. A mathematical model for continuous drying of grains in a spouted bed dryer. *Drying Technology*, 20, 587–614.

Marra F, Lyng J, Romano V, and McKenna B. 2007. Radio-frequency heating of foodstuff: Solution and validation of a mathematical model. *Journal of Food Engineering*, 79(3), 998–1006.

Matz S. 1989. *Equipment for Bakers*. Elsevier Science, New York.

May B K and Perre P. 2002. The importance of considering exchange surface area reduction to exhibit a constant drying flux period in foodstuffs. *Journal of Food Engineering*, 54(4), 271–282.

Menter F R. 1994. Two-equation eddy-viscosity turbulence models for engineering applications. *AIAA Journal*, 32(8), 1598–1604.

Mistry H, Ganapathi-subbu, Dey S, Bishnoi P, and Castillo J L. 2006. Modeling of transient natural convection heat transfer in electric ovens. *Applied Thermal Engineering*, 26(17–18), 2448–2456.

Modest M. 1992. *Radiative Heat Transfer.* Academic Press, New York.

Moureh J and Flick D. 2005. Airflow characteristics within a slot-ventilated enclosure. *International Journal of Refrigeration*, 26, 12–24.

Nakayama A, Kuwahara F, Xu G, and Kato F. 2004. An analytical treatment for combined heat transfer by radiation, convection and conduction within a heat insulating wall structure. *Heat and Mass Transfer*, 40(8), 621–626.

Nicolaï B M and de Baerdemaeker J. 1996. Sensitivity analysis with respect to the surface heat transfer coefficient as applied to thermal process calculations. *Journal of Food Engineering*, 28, 21–33.

Nicolaï B M, Verboven P, and Scheerlinck N. 2001. Modelling and simulation of thermal processes. Chapter 6 in *Thermal Technologies in Food Processing*, Philip Richardson (Editor), 14 chapters, Woodhead Publishing Ltd, Cambridge, England, pp. 91–109.

Nijdam J J, Guo B Y, Fletcher D F, and Langrish T A G. 2006. Lagrangian and Eulerian models for simulating turbulent dispersion and coalescence of droplets within a spray. *Applied Mathematical Modelling*, 30(11), 1196–1211.

Nitin N, Gadiraju R P, and Karwe M V. 2006. Conjugate heat transfer associated with a turbulent hot air jet impinging on a cylindrical object. *Journal of Food Processing Engineering*, 29(4), 386–399.

Norton T and Sun D-W. 2007. An overview of CFD applications in the food industry. Chapter 1 in *Computational Fluid Dynamics in Food Processing*, Da-Wen Sun (Editor), 28 chapters, CRC Press, Boca Raton, FL, pp. 1–43.

Olsson E and Trägårdh C. 2007. CFD modelling of jet impingement during heating and cooling of foods. Chapter 20 in *Computational Fluid Dynamics in Food Processing*, Da-Wen Sun (Editor), 28 chapters, CRC Press, Boca Raton, FL, pp. 478–505.

Olsson E E M, Ahrne L M, and Tragardh A C. 2005. Flow and heat transfer from multiple slot air jets impinging on circular cylinders. *Journal of Food Engineering*, 67(3), 273–280.

Otero L, Ramos A M, de Elvira C, and Sanz P D. 2007. A model to design high-pressure processes towards an uniform temperature distribution. *Journal of Food Engineering*, 78(4), 1463–1470.

Park K, Choi D-H, and Lee K-S. 2004. Optimum design of plate heat exchanger with staggered pin arrays. *Numerical Heat Transfer Part A: Applications*, 45(4), 347–361.

Patankar S V. 1980. *Numerical Heat Transfer.* Hemisphere Publishing, Washington, DC.

Patankar S V and Spalding D B. 1972. A calculation procedure for heat, mass and momentum transfer in three-dimensional parabolic flows. *International Journal for Heat and Mass Transfer*, 15, 1787–1806.

Perre P and Turner I W. 1999. A 3-D version of TransPore: A comprehensive heat and mass transfer computational model for simulating the drying of porous media. *International Journal of Heat and Mass Transfer*, 42(24), 4501–4521.

Ressing H, Ressing M, and Durance T. 2007. Modelling the mechanisms of dough puffing during vacuum microwave drying using the finite element method. *Journal of Food Engineering*, 82(4), 498–508.

Richardson P. 2001. *Thermal Technologies in Food Processing*, Philip Richardson (Editor), 14 chapters, Woodhead Publishing Ltd, Cambridge, England, pp. 1–3.

Scott G and Richardson P. 1997. The application of computational fluid dynamics in the food industry. *Trends in Food Science and Technology*, 8, 119–124.

Seyedein S H, Hasan M, and Mujumdar A S. 1994. Modelling of a single confined turbulent slot jet impingement using various $k$–ε turbulence models. *Applied Mathematical Modelling*, 18, 526–537.

Shih T H, Liou W W, Shabbir A, and Zhu J. 1995. A new $k$–ε eddy viscosity model for high Reynolds number turbulent flows—model development and validation. *Computers and Fluids*, 24(3), 227–238.

Siegel R and Howell J R. 1992. *Thermal Radiation Heat Transfer*. Hemisphere Publishing, Washington, DC.

Singh R P and Heldman D R. 2001. *Introduction to Food Engineering*. Academic Press, A Harcourt Science and Technology Company, San Diego, CA.

Siriwattanayotin S, Yoovidhya T, Meepadung T, and Ruenglertpanyakul W. 2006. Simulation of sterilization of canned liquid food using sucrose degradation as an indicator. *Journal of Food Engineering*, 73, 307–312.

Straatsma J, Verdurmen R E M, Verscheuren M, Gunsing M, and de Jong P. 2007. CFD simulation of spray drying of food products. Chapter 10 in *Computational Fluid Dynamics in Food Processing*, Da-Wen Sun (Editor), 28 chapters, CRC Press, Boca Raton, FL, pp. 249–287.

Sun D-W. 2007. *Computational Fluid Dynamics in Food Processing*, Da-Wen Sun (Editor), 28 chapters, CRC Press, Boca Raton, FL.

Suresh H N, Narayana P A A, and Seetharamu K N. 2001, Conjugate mixed convection heat and mass transfer in brick drying. *Heat and Mass Transfer*, 37, 205–213.

Szafran R G and Kmiec A. 2004. CFD modeling of heat and mass transfer in a spouted bed dryer. *Industrial and Engineering Chemistry Research*, 43(4), 1113–1124.

Szafran R G and Kmiec A. 2005. Point-by-point solution procedure for the computational fluid dynamics modeling of long-time batch drying. *Industrial and Engineering Chemistry Research*, 44(20), 7892–7898.

Tattiyakul J, Rao M A, and Datta A K. 2001. Simulation of heat transfer to a canned corn starch dispersion subjected to axial rotation. *Chemical Engineering and Processing*, 40, 391–399.

Tewkesbury H, Stapley A G F, and Fryer P J. 2000. Modelling temperature distributions in cooling chocolate moulds. *Chemical Engineering Science*, 55(16), 3123–3132.

Therdthai N, Zhou W B, and Adamczak T. 2002. Optimisation of the temperature profile in bread baking. *Journal of Food Engineering*, 55(1), 41–48.

Therdthai N, Zhou W B, and Adamczak T. 2003. Two-dimensional CFD modelling and simulation of an industrial continuous bread baking oven. *Journal of Food Engineering*, 60(2), 211–217.

Therdthai N, Zhou W B, and Adamczak T. 2004a. Three-dimensional CFD modelling and simulation of the temperature profiles and airflow patterns during a continuous industrial baking process. *Journal of Food Engineering*, 65(4), 599–608.

Therdthai N, Zhou W B, and Adamczak T. 2004b. Simulation of starch gelatinisation during baking in a travelling-tray oven by integrating a three-dimensional CFD model with a kinetic model. *Journal of Food Engineering*, 65(4), 543–550.

Turnbull J and Thompson C P. 2005. Transient averaging to combine large eddy simulation with Reynolds averaged Navier–Stokes simulations. *Computers and Chemical Engineering*, 29, 379–392.

Varga S and Oliveira J C. 2000. Determination of the heat transfer coefficient between bulk medium and packed containers in a batch retort. *Journal of Food Engineering*, 44(4), 191–198.

Varma M N and Kannan A. 2005. Enhanced food sterilization through inclination of the container walls and geometry modifications. *International Journal of Heat and Mass Transfer*, 48, 3753–3762.

Varma M N and Kannan A. 2006. CFD studies on natural convective heating of canned food in conical and cylindrical containers. *Journal of Food Engineering*, 77(4), 1027–1036.

Verboven P, Scheerlinck N, De Baerdemaeker J, and Nicolaï B M. 2000a. Computational fluid dynamics modelling and validation of the isothermal airflow in a forced convection oven. *Journal of Food Engineering*, 43(1), 41–53.

Verboven P, Scheerlinck N, De Baerdemaeker J, and Nicolaï B M. 2000b. Computational fluid dynamics modelling and validation of the temperature distribution in a forced convection oven. *Journal of Food Engineering*, 43(2), 61–73.

Verboven P, Datta A K, Anh N T, Scheerlinck N, and Nicolaï B M. 2003. Computation of airflow effects on heat and mass transfer in a microwave oven. *Journal of Food Engineering*, 59(2–3), 181–190.

Verboven P, de Baerdemaeker J, and Nicolaï B M. 2004. Using computational fluid dynamics to optimise thermal processes. Chapter 4 in *Improving the Thermal Processing of Foods*, Philip Richardson (Editor), 23 chapters, Woodhead Publishing Ltd, Cambridge, England, pp. 82–102.

Verboven P, Datta A K, and Nicolaï B M. 2007. Computation of airflow effects in microwave and combination heating. Chapter 12 in *Computational Fluid Dynamics in Food Processing*, Da-Wen Sun (Editor), 28 chapters, CRC Press, Boca Raton, FL, pp. 313–331.

Versteeg H K and Malalsekeera W. 1995. *An Introduction to Computational Fluid Dynamics*. Longman Group Ltd, Harlow, UK, pp. 1–100.

Visser J and Jeurnink T J M. 1997. Fouling of heat exchangers in the dairy industry. *Experimental Thermal and Fluid Science*, 14(4), 407–424.

Wang L and Sun D-W. 2003. Recent developments in numerical modelling of heating and cooling processes in the food industry—a review. *Trends in Food Science and Technology*, 14, 408–423.

Wang L and Sun D-W. 2005. Heat and mass transfer in thermal food processing. Chapter 2 in *Thermal Food Processing*, Da Wen Sun (Editor), CRC Press, Taylor & Francis group, New York, pp. 35–71.

Welti-Chanes J, Vergara-Balderas F, and Bermúdez-Aguirre D. 2005. Transport phenomena in food engineering: Basic concepts and advances. *Journal of Food Engineering*, 67, 113–128.

Wilcox D C. 1993. Comparison of 2-equation turbulence models for boundary-layer flows with pressure gradient. *AIAA Journal*, 31(8), 1414–1421.

Wong S Y, Zhou W B, and Hua J S. 2006. Robustness analysis of a CFD model to the uncertainties in its physical properties for a bread baking process. *Journal of Food Engineering*, 77(4), 784–791.

Wong S Y, Zhou W B, and Hua J S. 2007a. CFD modeling of an industrial continuous bread-baking process involving U-movement. *Journal of Food Engineering*, 78(3), 888–896.

Wong S Y, Zhou W B, and Hua J S. 2007b. Designing process controller for a continuous bread baking process based on CFD modelling, *Journal of Food Engineering*, 81(3), 523–534.

Xia B and Sun D-W. 2002. Applications of computational fluid dynamics (CFD) in the food industry: A review. *Computers and Electronics in Agriculture*, 34, 5–24.

Zhao W N, Kumar K, and Mujumdar A S. 2004. Flow and heat transfer characteristics of confined noncircular turbulent impinging jets. *Drying Technology*, 22(9), 2027–2049.

Zheng L, Delgado A, and Sun D-W. 2007. Surface heat transfer coefficients with and without phase change. Chapter 23 in *Food Properties Handbook*, 2nd ed., Shafiur Rahman (Editor), CRC Press, Boca Raton, FL.

Zhonghua W and Mujumdar A S. 2007. Simulation of the hydrodynamics and drying in a spouted bed dryer. *Drying Technology*, 25, 59–74.

Zhou W and Therdthai N. 2007. Three-dimensional CFD modelling of a continuous baking process. Chapter 11 in *Computational Fluid Dynamics in Food Processing*, Da-Wen Sun (Editor), 28 chapters, CRC Press, Boca Raton, pp. 287–213.

# Part III

## Optimization

# 13 Global Optimization in Thermal Processing

*Julio R. Banga, Eva Balsa-Canto, and Antonio A. Alonso*

## CONTENTS



## 13.1 INTRODUCTION

Model-based simulation of food processing units and/or full plants has received great attention (Fryer, 1994; Datta, 1998; Nicolaï et al., 2001). Since most processes are operated in batch mode, these models are usually dynamic in nature, consisting of sets of ordinary and/or partial differential and algebraic equations (PDAEs). However, less work has been done on the full exploitation of these models, via mathematical optimization, in order to build powerful decision support systems.

In this contribution, we will consider the optimization of thermal processing, which seems to be the type of process that has received most attention regarding its optimization so far. The mathematical models of the different thermal food processing operations have certain characteristics, which will pose special difficulties for their mathematical optimization. In particular, the following attributes are especially relevant in this context: (1) nonlinear, dynamic models (i.e., batch or semibatch processes), (2) distributed systems (temperature is distributed in space), and (3) nonlinear constraints coming from (microbiological) safety and quality requirements.

Thus, these mathematical models usually consist of sets of algebraic, partial, and ordinary differential equations (PDAEs), or even integro-partial differential-algebraic equations (IPDAEs), with possible logic conditions (transitions, i.e., hybrid systems). These PDAE models are usually transformed into DAEs via suitable spatial discretization methods (e.g., finite differences, method of lines, finite elements, etc.).

## 13.2 DYNAMIC OPTIMIZATION

Since thermal process models have an inherent dynamic nature, we will need to use methods designed for the optimization of dynamic systems in order to arrive to optimal decisions. The type of problem especially relevant for us is the so-called dynamic optimization problem. Dynamic optimization (also called open loop optimal control) aims to find the optimal operating policies (controls) of the thermal process, modeled as a nonlinear dynamic system, in order to optimize a performance index (functional).

In more detail, the "dynamic optimization" problem mentioned above is usually formulated as

1. Find the controls, the process time, and a set of time-independent parameters
2. To minimize a performance index (functional expressing the criterion to optimize)
3. Subject to several sets of constraints
   a. A set of differential-algebraic equality constraints (dynamics of the system)
   b. Sets of path equality and inequality constraints
   c. Bounds (lower and upper limits) for the controls and/or the states

For the simpler case of no time-independent parameters and a fixed process time, the corresponding mathematical statement would be

Find $\mathbf{u}(t)$ over $t \in [t_o, t_f]$ to minimize (or maximize):

$$J[\mathbf{x},\mathbf{u}] = \theta[\boldsymbol{x}\{t_f\}] + \int_{t_0}^{t_f} \phi[\boldsymbol{x}\{t\}, \boldsymbol{u}\{t\}, t]\mathrm{d}t \tag{13.1}$$

subject to

$$\Psi\left(\mathbf{x}, \mathbf{x}_\xi, \mathbf{x}_{\xi\xi}, \ldots, \xi\right) = 0 \quad \xi \in \Omega, \boldsymbol{x}_\xi = \partial \boldsymbol{x}/\partial\xi \tag{13.2}$$

$$\Gamma\left(\mathbf{x}, \mathbf{x}_\xi, \ldots, \xi\right) = 0 \quad \xi \in \Omega \tag{13.3}$$

$$\frac{\mathrm{d}\boldsymbol{x}}{\mathrm{d}t} = \varphi[\boldsymbol{x}\{t\}, \boldsymbol{u}\{t\}, t] \quad \boldsymbol{x}(t_0) = \boldsymbol{x}_0 \tag{13.4}$$

$$\mathbf{h}\left[\mathbf{x}(t),\mathbf{u}(t)\right]=0 \tag{13.5}$$

$$\mathbf{g}\left[\mathbf{x}(t),\mathbf{u}(t)\right]\leq 0 \tag{13.6}$$

$$\mathbf{x}^{L}\leq\mathbf{x}(t)\leq\mathbf{x}^{U} \tag{13.7}$$

$$\mathbf{u}^{L}\leq\mathbf{u}(t)\leq\mathbf{u}^{U} \tag{13.8}$$

In the above statement, $J$ is the performance index, which includes a cost term at final time and also a cost term which takes into account the states and control histories. In addition, $\mathbf{x}$ is the vector of state variables, $\mathbf{u}$ is the vector of control variables, $\xi$ are the independent variables (including time $t$ and spatial position $\chi$). Equation 13.2 is the system of governing partial differential equations (PDEs) within the domain $\Omega$; Equation 13.3 is the auxiliary conditions of the PDEs (boundary and initial conditions) on the boundary $\delta\Omega$ of the domain $\Omega$; Equation 13.4 is the system of ordinary differential equality constraints with their initial conditions; Equations 13.5 and 13.6 are the equality and inequality algebraic constraints and Equations 13.7 and 13.8 are the upper and lower bounds on the state and control variables.

This dynamic optimization problem is usually solved using the so-called direct approach (Banga et al., 1994; Balsa-Canto et al., 2005a), where either the control variables, or both the controls and the states, are discretized using some basis functions (e.g., piecewise constant or piecewise-linear, Lagrange polynomials, etc.). In the case of control parametrization methods, the resulting problem is a nonlinear programming problem subject to differential algebraic equations (NLP-DAEs problem).

## 13.3 MULTIMODALITY: NEED OF GLOBAL OPTIMIZATION

Most state-of-the-art methods (and software) available for the solution of NLP are iterative, gradient-based methods. These methods are of local nature, i.e., they converge to local solutions, usually the nearest maximum (or minimum) to the starting point. This is fine for some problems of interest, which are known to be unimodal (i.e., with a single extreme).

However, it turns out that many NLP of practical interest are in fact multimodal, i.e., there are several extremes, with many local solutions and possibly a single global optimum. If a gradient method is not started close to the global maximum, it might arrive to one of the local (worse) solutions.

In fact, the NLP-DAEs discussed previously can be (and often are) multimodal. This is mainly due to the highly nonlinear and constrained nature of these problems. Thus, in order to ensure proper solution of these problems, we must use suitable global optimization (GO) methods.

## 13.4 GLOBAL OPTIMIZATION METHODS

GO is a research area, which has started to receive increased attention during the last decade. Existing GO methods can be roughly classified as deterministic

(Floudas, 2000) and stochastic strategies (Guus et al., 1995; Törn et al., 1999). It should be noted that, although deterministic methods can guarantee global optimality for certain GO problems, no algorithm can solve general GO problems with certainty in finite time. In fact, although several classes of deterministic methods (e.g., branch and bound [B&B]) have sound theoretical convergence properties, the associated computational effort increases very rapidly (often exponentially) with the problem size.

In contrast, many stochastic methods can locate the vicinity of global solutions with relative efficiency (Banga and Seider, 1996), but the cost to pay if that global optimality cannot be guaranteed. However, in practice the process designer can be satisfied if these methods provide him/her with a very good (often, the best available) solution in modest computation times. Furthermore, stochastic methods are usually quite simple to implement and use, and they do not require transformation of the original problem, which can be treated as a black box. This characteristic is especially interesting since very often the process engineer must link the optimizer with a third-party software package, where the process dynamic model has been implemented. Finally, many stochastic methods lend themselves to parallelization very easily, which means that medium-to-large-scale problems can be handled in reasonable wallclock time.

In the work performed by our group during the last decade, we have considered a set of selected stochastic and deterministic GO methods which can handle black box models. The selection has been made based on their published performance and on our own experiences considering their results for a set of GO benchmark problems. Although none of these methods can guarantee optimality, at least the user can solve a given problem with different methods and take a decision based on the set of solutions found. Usually, several of the methods will converge to essentially the same (best) solution. It should be noted that although this result cannot be regarded as a confirmation of global optimality (it might be the same local optimum), it does give the user some extra confidence. Further, it is usually possible to have estimates of lower bounds for the cost function and its different terms, so the goodness of the "global" solution can be evaluated (sometimes a "good enough" solution is sufficient).

## 13.5 GLOBAL OPTIMIZATION OF NONLINEAR DYNAMIC SYSTEMS

In this section, we will briefly review the state of the art regarding the GO of nonlinear dynamic systems. In the case of deterministic methods, Esposito and Floudas (2000) have provided what appears to be the first contribution applying this type of methods. Their elegant approach, based on B&B, can solve to global optimality if certain conditions hold. Its main drawbacks are the significant computational effort even for small problems and the differentiability conditions that must be met (which do not hold for many problems of interest). In any case, promising research continues in this direction.

In the case of stochastic methods, several approaches have been presented, which are able to escape from local optima (e.g., Luus, 1990; Banga and Seider, 1996; see also the review in Banga et al., 2003). Many of these methods are able to find approximate solutions found in reasonable CPU times. An additional advantage of high importance regarding their practical application is that arbitrary black box DAEs can be considered (including discontinuities, highly nonlinear terms, etc.). However, their main drawback is that global optimality cannot be guaranteed (although some of them have weak asymptotic convergence proofs).

In this domain, our research objectives during recent years have been

- To select the best stochastic methods, i.e., those with the best compromise between efficiency and robustness, and providing a suitable handling of path constraints (Banga et al., 1994, 1998)
- To design hybrid stochastic–deterministic methods: These new hybrid techniques attempt to combine the robustness and global convergence properties of stochastic methods with the efficiency of gradient-based local search methods when they are started close to the global solution (Balsa-Canto et al., 1998, 2005b)

We have tested these new methods with several types of thermal processes. In particular, we have attempted to devise methods which are (1) capable of solving NLP-DAEs of arbitrary complexity (treating them as "black box" problems), and (2) efficient for large-scale applications (i.e., NLP-DAEs with over 100 decision variables and over $10^4$ states). In Section 13.6, we will present a brief overview of the stochastic methods we have explored, highlighting the main conclusions that this research has produced. We believe that these results can be a helpful guideline for other researches, who are facing similar problems.

## 13.6 STOCHASTIC METHODS FOR GLOBAL OPTIMIZATION

A critical review of stochastic methods for GO of food processing was presented in Banga et al. (2003). Stochastic methods, in contrast with deterministic approaches, do not perform search in a systematic predefined way, but they rather introduce a certain degree of randomness during the iterative generation of new vectors of decision variables.

During more recent years, we have also evaluated a number of other (the so-called) metaheuristics (i.e., guided heuristics), which are mostly based on certain biological or physical optimization phenomena, and with combinatorial optimization as their original domain of application. Examples of these more recent methods are Taboo Search (TS), Ant Colony Optimization (ACO; Bonabeau et al., 2000), and Scatter Search (Egea et al., 2007).

Despite the huge popularity of genetic algorithms (Goldberg, 1989), our research has indicated that these are not the best methods for the GO of nonlinear dynamic systems. In fact, our main conclusion has been that the following methods usually provide the best compromise between robustness and efficiency:

- Differential evolution (Storn and Price, 1997)
- Stochastic ranking evolution strategy (SRES) (Runarsson and Yao, 2000)
- Globalm (Csendes et al., 2008)
- Scatter Search (Egea et al., 2007)

### 13.6.1 Software for Global Optimization

During our research, we have also developed a software toolbox (Garcia et al., 2005), which allows the dynamic optimization of arbitrary nonlinear dynamic processes, including distributed processes. This modular and flexible toolbox, named NDOT (nonlinear dynamic optimization toolbox), combines the control vector parameterization approach with a number of local and global nonlinear programming solvers and suitable dynamic simulation methods. NDOT is able to solve dynamic optimization problems for both lumped and distributed nonlinear processes, so it is an ideal tool for the optimization of thermal food processing. Moreover, using this tool it is very easy to compare the performance of different optimization methods, so we can evaluate if a given problem is multimodal, and then select the most suitable optimizer.

#### Examples

Here we will summarize the results obtained using GO for three relevant thermal processes:

- Thermal sterilization and pasteurization
- Contact cooking
- Microwave heating

### 13.6.2 Dynamic Optimization of Thermal Sterilization

This class of problems has been studied by many authors, as reviewed by, e.g., Durance (1997), Banga et al. (2003), and Awuah et al. (2007), and it is still receiving great attention. Examples of recent works are those of Chen and Ramaswamy (2004), Simpson et al. (2004), and Erdogdu (2003). Here, we will consider a particular example from Banga et al. (1991), who studied the formulation and solution of several problems with different objective functions and constraints. In this contribution, the formulation of the maximization of the final retention of a nutrient is considered, with constraints on the microbiological lethality at final time and on the final temperature in the hottest point.

We will consider the particular case of a cylindrical container of volume $V_T$ (radius $R$ and height $2L$) filled with a conduction-heated canned food. The objective is to find the optimal retort temperature, $T_{\text{ret}}(t)$, between the given lower and upper bounds, which maximizes the final retention of a nutrient ($J$, normalized between 0 and 1):

$$J = \frac{1}{V_T}\int_0^{V_T} \exp\left(\frac{-\ln 10}{D_{N,\text{ref}}}\int_0^{t_f} \exp\left(\frac{T(r,z,t)-T_{N,\text{ref}}}{Z_{N,\text{ref}}}\ln 10\right)\mathrm{d}t\right)\mathrm{d}V \tag{13.9}$$

with the heat transfer dynamics given by the conduction (Fourier) equation:

$$\frac{\partial T}{\partial t} = \alpha\left(\frac{\partial^2 T}{\partial r^2} + \frac{1}{r}\frac{\partial T}{\partial r} + \frac{\partial^2 T}{\partial z^2}\right) \tag{13.10}$$

with the following boundary and initial conditions:

$$T(R,z,t) = T_{\text{ret}}(t) \tag{13.11}$$

$$T(r,L,t) = T_{\text{ret}}(t) \tag{13.12}$$

$$\frac{\partial T}{\partial r}(0,z,t) = 0 \tag{13.13}$$

$$\frac{\partial T}{\partial z}(r,0,t) = 0 \tag{13.14}$$

$$T(r,z,0) = T_0 \tag{13.15}$$

plus the following constraints at final time:

1. On the final temperature in the hottest point:

$$T(r,z,t_{\text{f}}) \leq T_0 \quad \forall r,z \in V_T \tag{13.16}$$

2. On the microbiological lethality at final time:

$$F_S(t_{\text{f}}) \geq F_{S,D} \tag{13.17}$$

where

$$F_S(t_{\text{f}}) = \log\left(\frac{1}{V_T}\int_0^{V_T} \exp\left(\frac{-\ln 10}{D_{M,\text{ref}}}\int_0^{t_{\text{f}}} \exp\left(\frac{T(r,z,t)-T_{M,\text{ref}}}{Z_{M,\text{ref}}}\ln 10\right)\text{d}t\right)\text{d}V\right) \tag{13.18}$$

These expressions are derived from the assumption of pseudo-first-order kinetics for the thermal degradation of nutrients and microorganisms:

$$\frac{\text{d}C_M(t)}{\text{d}t} = \frac{-\ln 10}{D_{M,\text{ref}}} C_M(t) \exp\left(\frac{T(r,z,t)-T_{M,\text{ref}}}{Z_{M,\text{ref}}}\right) \tag{13.19}$$

$$\frac{\text{d}C_N(t)}{\text{d}t} = \frac{-\ln 10}{D_{N,\text{ref}}} C_N(t) \exp\left(\frac{T(r,z,t)-T_{N,\text{ref}}}{Z_{N,\text{ref}}}\right) \tag{13.20}$$

In the particular case considered, taken from Banga et al. (1991), the food is a pork puree. The bounds for the retort temperature were taken as $20°\text{C} \leq T_{\text{ret}} \leq 140°\text{C}$, and the final time considered was 7830 s. Banga et al. (1991) used integrated controlled random search for dynamic systems (ICRS/DS), a simple GO method based on an adaptive random search sequential strategy. These authors were able to reach values of nutrient retention of about 47.9%, but at a rather large computational cost.

When this problem was solved in our group with the NDOT (Garcia et al., 2005) comparing different local and global solvers, we found that, as expected, most gradient (local) methods failed to converge or converged to local solutions. However, good results were obtained if a control mesh refinement scheme was used in successive reoptimizations (Garcia et al., 2006). This indicates that the problem is only mildly multimodal, and that such multimodality can be reduced by the successive reoptimization approach. For the case of global methods, both DE and SRES obtained values similar to those of ICRS in reasonable computation times, although DE seems to be more robust than SRES.

In order to reduce the computational cost associated with GO, Balsa-Canto et al. (2002a,b) presented a new method, which combines stochastic GO with a reduced-order model of the original system. With this novel approach, this problem could be solved to global optimality in just a few seconds using a standard PC. The capabilities and possibilities of this approach for other complex processes were illustrated by Balsa-Canto et al. (2005a).

### 13.6.3 Dynamic Optimization of Contact Cooking

This research was performed in collaboration with the group of Professor R. Paul Singh (UC Davis). The objective was to obtain the optimal operating procedures for contact cooking of nonhomogeneous foods. The particular case study considered was one of high economical and societal importance: hamburger patties. In this process, there are a number of important safety issues due to several outbreaks of *Escherichia coli* O157:H7 in recent years. Besides, there also are several quality issues to consider, namely the final patty yield and its sensorial and nutritional quality.

One of the dynamic optimization problems considered was to find the optimal heating surface temperature to maximize the final patty yield subject to

- Microbial and temperature final-time constraints
- The process dynamic model

The process dynamic model was rather complex (PDAEs with moving fronts and phase change) due to the different coupled phenomena involved:

- Heat and mass transfer involving dimensional changes and phase change (patties are frozen initially)
- Kinetics for *E. coli* O157:H7 destruction
- Kinetics for textural modifications, yield losses, etc.
- Detailed phase change and moving front: ice and fat melting, water evaporation and moving crust, time-dependent heat transfer coefficient, etc.

Thus, when the solution of this problem was attempted by using standard (local) optimization solvers (e.g., Sequential Quadratic Programming, SQP), it was not

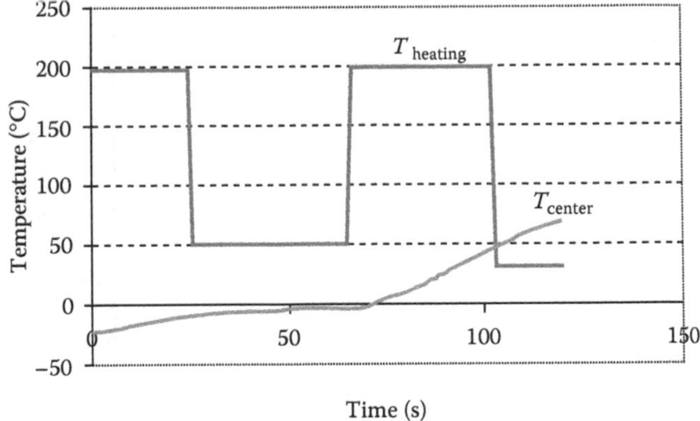


**FIGURE 13.1** Example of optimal control (heating temperature) for contact cooking.

surprising to find convergence failures or premature stops at local solutions. In contrast, the problem was easily solved by means of stochastic GO methods, with a very acceptable computational effort (minutes of CPU in a simple PC). An example of an optimal control profile is given in Figure 13.1 (more details are provided in Banga et al. (2001)). The computed optimal operating policies improved the final yield significantly. Thus, the above statement and its corresponding software implementation can be used as the kernel of a decision support system for the optimal operation and/or the design of new grills.

### 13.6.4 Optimal Control of Microwave Heating

This research was performed in the framework of a project funded by the European Union, with partners from K. U. Leuven (Belgium), SIK (Sweden), and Whirlpool (Sweden). The main objective was to improve the operation of microwave multimode heating ovens in order to optimize the final product temperature and quality uniformity while minimizing quality losses due to overheating and maintaining microbiological safety.

A typical realistic optimal control problem had 2 control variables and over 5000 states (after discretization using finite elements). Due to the nonlinearity of the models, state-of-the-art gradient methods (like SQP) failed to converge or converged to local solutions. Again, simple stochastic methods presented good convergence properties, arriving to satisfactory solutions, although in this case the computation times were 2–4 h (PC PIII) due to the large dimensionality of the simulation problem.

It was found that the computed optimal control policies lead to more uniform heating of problematic loads, like bricks or cylinders (Saa et al., 1998; Banga et al., 1999; Nicolaï et al., 2000; Sánchez et al., 2000). Experimental validation of these optimal polices confirmed the goodness of these results. A typical set of optimal control policies is presented in Figure 13.2. Once more, the resulting software can be used as the key element of a decision support system for deriving guidelines for optimal oven operation, or for the design of new ovens.

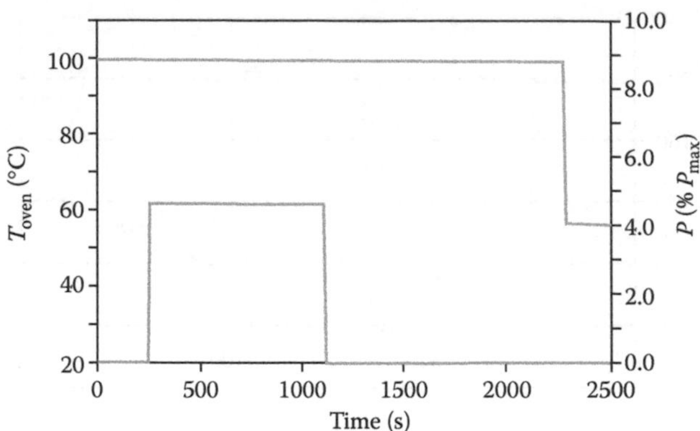


**FIGURE 13.2** Example of optimal operating policies (oven temperature and magnetron power) for a brick-shaped food load in a microwave combination oven.

## 13.7 CONCLUSIONS

The optimization of thermal processing of foods can be formulated as a dynamic optimization problem. Frequently, these problems turn out to be multimodal (several local solutions, and possibly a single global solution). The objective of this paper was to present an overview of the possibilities of GO techniques applied to thermal food processing.

For several food processes like thermal sterilization, contact cooking, and microwave processing, we have found that the optimal operating procedures computed via GO methods provide significant improvements over nominal processes. Thus, we believe that GO methods should be the key element behind decision support systems for food thermal processing.

## REFERENCES


Awuah, G.B., H.S. Ramaswamy, and A. Economides. 2007. Thermal processing and quality: Principles and overview. *Chemical Engineering and Processing*, 46(6): 584–602.

Balsa-Canto, E., A.A. Alonso, and J.R. Banga. 1998. Dynamic optimization of bioprocesses: Deterministic and stochastic strategies. Presented at ACoFop IV (Automatic Control of Food and Biological Processes). Göteborg, Sweden.

Balsa-Canto, E., A.A. Alonso, and J.R. Banga. 2002a. A novel, efficient and reliable method for thermal process design and optimization. Part I: Theory. *Journal of Food Engineering*, 52: 227–234.

Balsa-Canto, E., A.A. Alonso, and J.R. Banga. 2002b. A novel, efficient and reliable method for thermal process design and optimization. Part II: Applications. *Journal of Food Engineering*, 52: 235–247.

Balsa-Canto, E., A.A. Alonso, and J.R. Banga. 2005a. Dynamic optimization of complex distributed process systems. *Chemical Engineering Research and Design*, 83(A8): 1–6.

Balsa-Canto, E., V.S. Vassiliadis, and J.R. Banga. 2005b. Dynamic optimization of single- and multi-stage systems using a hybrid stochastic–deterministic method. *Industrial and Engineering Chemistry Research*, 44(5): 1514–1523.

Banga, J.R. and W.D. Seider. 1996. Global optimization of chemical processes using stochastic algorithms. In: *State of the Art in Global Optimization*, C.A. Floudas and P.M. Pardalos (Eds.), Kluwer Academic, Dordrecht, the Netherlands.

Banga, J.R., R.P. Martin, J.M. Gallardo, and J.J. Casares. 1991. Optimization of thermal processing of conduction-heated canned foods: Study of several objective functions. *Journal of Food Engineering*, 14(1): 25–51.

Banga, J.R., A.A. Alonso, R.I.P. Martin, and R.P. Singh. 1994. Optimal control of heat and mass transfer in food and bioproducts processing. *Computers and Chemical Engineering*, 18: S699–S705.

Banga, J.R., R. Irizarry, and W.D. Seider. 1998. Stochastic optimization for optimal and model-predictive control. *Computers and Chemical Engineering*, 22(4–5): 603–612.

Banga, J.R., J. Saa, and A.A. Alonso. 1999. Model-based optimization of microwave heating of bioproducts. In: *Proceedings of 7th International Conference on Microwave and High Frequency Heating*. Valencia, Spain.

Banga, J.R., Z. Pan, and R.P. Singh. 2001. On the optimal control of contact-cooking processes. *Food and Bioproducts Processing*, 79(C3): 145–151.

Banga, J.R., E. Balsa-Canto, C.G. Moles, and A.A. Alonso. 2003. Improving food processing using modern optimization methods. *Trends in Food Science and Technology*, 14(4): 131–144.

Bonabeau, E., M. Dorigo, and G. Theraulaz. 2000. Inspiration for optimization from social insect behaviour. *Nature*, 406: 39–42.

Chen, C.R. and H.S. Ramaswamy. 2004. Multiple ramp-variable retort temperature control for optimal thermal processing. *Food and Bioproducts Processing*, 82(C1): 78–88.

Csendes, T., L. Pál, J.O.H. Sendín, and J.R. Banga. 2008. The GLOBAL optimization method revisited. *Optimization Letters*, 2(4): 445–454.

Datta, A.K. 1998. Computer-aided engineering in food process and product design. *Food Technology*, 52(10): 44.

Durance, T.D. 1997. Improving canned food quality with variable retort temperature processes. *Trends in food science & Technology*, 8: 113–118.

Egea, J.A., M. Rodriguez-Fernandez, J.R. Banga, and R. Martí. 2007. Scatter search for chemical and bioprocess optimization. *Journal of Global Optimization*, 37(3): 481–503.

Erdogdu, F. 2003. Complex method for nonlinear constrained multi-criteria (multi-objective function) optimization of thermal processing. *Journal of Food Process Engineering*, 26(4): 357–375.

Esposito, W.R. and C.A. Floudas. 2000. Deterministic global optimization in nonlinear optimal control problems. *Journal of Global Optimization*, 17: 97–126.

Floudas, C.A. 2000. Deterministic global optimization: Theory, methods, and applications. Nonconvex optimization and its applications, vol. 37. Kluwer Academic, Dordrecht, Boston, MA, pp. xvii, 739.

Fryer, P. 1994. Mathematical models in food processing. *Chemistry and Industry (London)*, 13: 515–518.

García, M.S.G., E. Balsa-Canto, A.A. Alonso, and J.R. Banga. 2005. NDOT: A software toolbox for the dynamic optimization of nonlinear processes. *European Symposium on Computer Aided Process Engineering, ESCAPE-15*, May 29 to June 1, Barcelona, Spain.

García, M.S.G., E. Balsa-Canto, A.A. Alonso, and J.R. Banga. 2006. Computing optimal operating policies for the food industry. *Journal of Food Engineering*, 74(1): 13–23.

Goldberg, D.E. 1989. *Genetic Algorithms in Search, Optimization and Machine Learning*. Addison Wesley Longman, Boston, MA.

Guus, C., E. Boender, and H.E. Romeijn. 1995. Stochastic methods. In: *Handbook of Global Optimization*, R.H.a.P.M. Pardalos (Ed.) Kluwer Academic, Dordrecht, The Netherlands.

Luus, R. 1990. Optimal control by dynamic programming using systematic reduction in grid size. *International Journal of Control*, 51(5): 995–1013.

Nicolaï, B.M., A. Alonso, J.R. Banga, E.B. Canto, S. Galt, M. Idebro, T. Ohlsson, J. Saa, I. Sánchez, N. Scheerlinck, H. Stigter, J.V. Impe, and B. Wäppling-Raaholt. 2000. Optimal control of microwave combination ovens for food heating. In: *Proceedings of ICEF-8*, Puebla, Mexico.

Nicolaï, B.M., N. Scheerlinck, P. Verboven, and J. DeBaerdemaeker. 2001. Stochastic finite-element analysis of thermal food processes. *Food Science and Technology*, 107: 265–304.

Runarsson, T.P. and X. Yao. 2000. Stochastic ranking for constrained evolutionary optimization, *IEEE Transactions Evolutionary Computation*, 4: 284–294.

Saa, J., A.A. Alonso, and J.R. Banga. 1998. Optimal control of microwave heating using mathematical models of medium complexity. In: *Proceedings of ACoFoP IV (Automatic Control of Good and Biological Processes)*, Göteborg, Sweden.

Sánchez, I., A.A. Alonso, and J.R. Banga. 2000. Temperature control in microwave combination ovens. *Journal of Food Engineering*, 46: 21–29.

Simpson R., S. Almonacid, and M. Mitchell. 2004. Mathematical model development, experimental validation and process optimization: retortable pouches packed with seafood in cone frustum shape. *Journal of Food Engineering*, 63(2): 153–162.

Storn, R. and K. Price. 1997. Differential evolution—a simple and efficient heuristic for global optimization over continuous spaces. *Journal of Global Optimization*, 11: 341–359.

Törn, A., M. Ali, and S. Viitanen. 1999. Stochastic global optimization: Problem classes and solution techniques. *Journal of Global Optimization*, 14: 437.

# 14 Optimum Design and Operating Conditions of Multiple Effect Evaporators: Tomato Paste

*Ricardo Simpson and Danilo López*

CONTENTS



## 14.1 INTRODUCTION

Process optimization has always been a noble objective of engineers entrusted with the responsibility for process development and improvement throughout the food industry. Examples of sophisticated mathematical approaches to process optimization, in which some objective function is maximized or minimized subject to chosen constraints, are widely published in literature (Douglas, 1988). On the other hand, the chemical industry has used cost analysis in several cases in relation to design and process optimization. A classical example in the chemical industry is the determination of the optimal number of effects in a evaporation system, where the optimum is found when there is an economic balance between energy saving and added investment, this is, a minimization of the total cost (Kern, 1999). In this vision, although correctly, quality is not considered as a parameter in the determination of the optimum number of effects, so the process specifications and operating conditions (OC) are assumed independent of both product quality and its sale price. The purpose of this manuscript is to suggest that the extrapolation of optimization problems from chemical industry to the food industry may often be restricted to an unnecessarily narrow or local domain, and that a more global perspective may reap greater rewards. Questions as to just what should be maximized or minimized or what are the real constraints, as opposed to only those that are immediately apparent, are questions often posed without a broad enough view of the big picture.

The production of tomato paste is highly seasonal, and then, maximizing production levels in this industry is of vital importance. The process is generally done in multiple evaporation systems, with a different number of effects, through which the content of water is diminished until a final concentration from 30°Bx to 32°Bx is acquired, and where temperatures generally do not exceed 70°C.

Lycopene is the main carotenoid found in tomatoes and many studies have showed its inhibiting effect on carcinogenic cell growth (Shi et al., 2007). It is also the component that generates the red characteristic color in tomatoes, among other fruits and vegetables (Goula and Adamopoulus, 2006). A study developed by the University of Harvard revealed that the consumption of lycopene reduced the probabilities of generating prostate cancer by 45%, in a population of 48,000 subjects who had at least 10 rations of tomatoes or subproducts in their weekly diet. Other research discovered that lycopene also reduces cholesterol levels in the form of a lipoprotein of low density, which produces atherosclerosis; this means that the consumption of tomatoes reduces the effects produced by cardiovascular diseases.

Lycopene as the main organic compound presents a denaturalization reaction rate that is time and temperature dependent. Then, for the mathematical model of the behavior of multieffect evaporators, it is very important to have a good overview of the general fluctuation of lycopene retention or loss under different system designs and OC.

As aforementioned, most food processes have been adapted and extrapolated from the chemical engineering industry without an adequate consideration of product quality during system design and process optimization. That is certainly a good start, but maybe somewhat limited and might have inhibited us to take a more global view. For example, it appears of extreme relevance to consider quality more frequently as an intrinsic and integral part of process design. In the food industry, the main effort is commonly related to the maximization of the quality of the product, which is not necessarily the case in the chemical industry. Generally, the optimization of food processes have been restricted to determine the optimal OC of an allegedly, well-designed food process. Nevertheless, if quality is considered as a parameter in the system design, it is very probable that the new design will differ from the original design.

For example, in the case of a multiple effect evaporator system for the processing of tomatoes, the optimization of the design is only focused on an economic analysis, which combines the investment (number of effects) and the operating costs (steam consumption) (Kern, 1999). This strategy does not include quality as an integral part of the economic evaluation, even though previous studies have demonstrated the dependence of the final product-price toward quality of the final product (Schoorl and Holt, 1983).

The purpose of this chapter is to propose a new economic evaluation procedure to optimize the system design and operation of tomato juice, multiple effect evaporator, and compare it to the traditional chemical engineering approach of total cost minimization. The proposed strategy incorporates a quality factor that is expressed as a function of lycopene concentration on the final product to find the optimal number of effects and OC through the maximization of the net present value (NPV).

## 14.2 METHODOLOGY

### 14.2.1 Problem Description

Cost analysis has been extensively and correctly utilized in finding the best process design in several chemical engineering plants. A classical example is multi effect distillation. In this case, cost analysis should aim to determine the optimum number of effects in multiple-stage equipment. According to the literature in chemical engineering "The optimum number of effects must be found from an economic balance between the savings in steam obtained by multiple effect operation and added investment." It is important to elucidate whether the aforementioned approach is recommendable in the optimization of food processes. From a microeconomics point of view, this approach is correct, but it is important to consider that the different equipment configurations are producing exactly the same quality of the end product. Meaning that, independent of the number of effects, not only will we be able to reach the same degree of concentration, but also the same quality. In addition, by changing the equipment configuration, it is possible to attain the same degree of concentration but with different product quality.

At least, for multi effect evaporation in food processing, the referred approach is not necessarily the right microeconomic tool to find the optimum number of effects. For this kind of application, a correct microeconomic analysis should consider not only all costs but also the expected benefits. According to the relevant technical literature, an adequate microeconomic procedure is to maximize the NPV.

In the following, we compare two different economic approaches: (a) determination of the optimal number of effects by minimizing the total cost and (b) maximization of the NPV, considering quality as an intrinsic parameter of the modeled system.

### 14.2.2 Product Quality

To reach the objective of the present research work, a quality parameter must be considered to the mathematical model of the evaporation system. The chosen parameter is lycopene, because, as mentioned before, this carotenoid pigment is what gives tomatoes their characteristic color, and, in addition, it has some medical benefits.

Usually, degradation rates in sensitive food components are modeled as a first-order kinetic, as follows:

$$r = -kY \quad \text{or} \quad \frac{dY}{dt} = -kY \tag{14.1}$$

The Arrhenius equation relates specific reaction rate constant to temperature according to

$$k = k_0 \cdot \exp\left(-\frac{E}{R \cdot T}\right) \tag{14.2}$$

The first-order kinetic for lycopene degradation has been confirmed by Goula and Adamopoulus (2006). In the same research study, an equation was obtained to determine the reaction rate in the lycopene degradation, as a function of temperature and soluble solids concentration $X$ expressed in degree Brix.

$$\text{For } X \geq 55, \quad k = 0.121238\,e^{0.0188X} \exp\left(-\frac{2317}{T + 273.15}\right) [\text{min}^{-1}] \tag{14.3}$$

$$\text{For } X \leq 55, \quad k = 0.275271\,e^{0.00241X} \exp\left(-\frac{2207}{T + 273.15}\right) [\text{min}^{-1}] \tag{14.4}$$

In our research study, the system to be modeled should consider tomato concentration in the range of 5°Bx–35°Bx, so only Equation 14.4 is required.

### 14.2.3 Model Development

The evaporation process involves mass and heat transfer (Himmelblau and Bischoff, 1968). The tomato juice was considered as a binary solution of water and soluble solids, both considered inert in a chemical sense. Under these considerations, one effect of the industrial evaporator can be shown in the manuscript by Miranda and Simpson (2005).

So the macroscopic model is of the knowledge type based on conservation laws and also empirical relationships that describe the equilibrium phases. These relationships have been rearranged from nonlinear algebraic equations from literature, with the experience taken from the experimental site. Only the juice phase is considered for modeling.

The modeling assumptions are

- Homogenous composition and temperature inside each evaporator
- Constant juice level in each evaporator
- Thermodynamic equilibrium (liquid–vapor) for the whole modeled system

The mathematical model developed in this research study included specific relationships for lycopene degradation. The general system that must be solved (see Figure 14.1

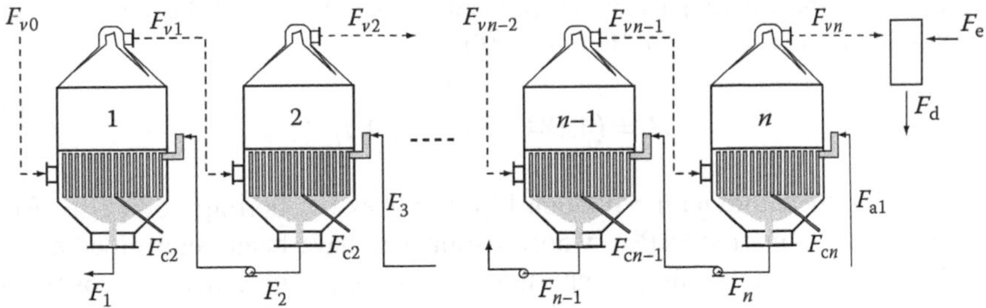


**FIGURE 14.1** Evaporation system of $n$ effects operated in countercurrent. (From Simpson, R., Almonacid, S., Lopez, D., and Abakarov, A., *J. Food Eng.*, 89, 488, 2008. With permission.)

for a schematic representation of the system) operates on countercurrent and the total number of effects varies from 1 to $n$. The value of $n$ and the OC are determined at the end of this chapter through the maximization of the NPV.

The total mass balance in evaporator effect $i$ is

$$\frac{dM_i}{dt} = F_{i+1} - Fv_i - F_i \tag{14.5}$$

If the mass within the evaporator effect is controlled, then, under steady state, Equation 14.5 can be written as

$$0 = F_{i+1} - Fv_i - F_i \tag{14.6}$$

In the same way, a mass balance for soluble solids at effect $i$ can be written as

$$\frac{d(M_i X_i)}{dt} = F_{i+1} \cdot X_{i+1} - F_i \cdot X_i \tag{14.7}$$

Under steady-state condition,

$$0 = F_{i+1} \cdot X_{i+1} - F_i \cdot X_i \tag{14.8}$$

The corresponding energy balance for the evaporator effect $i$ is

$$\begin{aligned} \frac{d(H_i M_i)}{dt} = {} & F_{i+1} \cdot H_{i+1} + Fv_{i-1} \cdot Hv_{i-1} - F_i \cdot H_i - Fv_i \cdot Hv_i \\ & - Fc_i \cdot Hc_i + Q_p \end{aligned} \tag{14.9}$$

Under steady-state condition,

$$\begin{aligned} 0 = {} & F_{i+1} \cdot H_{i+1} + Fv_{i-1} \cdot Hv_{i-1} - F_i \cdot H_i - Fv_i \cdot Hv_i \\ & - Fc_i \cdot Hc_i + Q_p \end{aligned} \tag{14.10}$$

The enthalpy of the tomato paste was estimated through the specific heat ($C_p$), utilizing the following expression (Tonelli et al., 1990):

$$H_i = (4.184 - 2.9337X_i)\,T_i \tag{14.11}$$

The following thermodynamic relationship describes the boiling point rise (BPR) or boiling point elevation (BPE), whose parameters have been determined experimentally. It is one of the three important properties (specific heat, viscosity, and BPR) that must be specified in a multiple effect evaporator (Rizvi and Mittal, 1992). This property (BPR) is significant at high soluble solids concentration. On a multiple effect equipment, the effective temperature differences decrease for the combination of boiling point. The following correlation reported by Miranda and Simpson (2005) was utilized:

$$\Delta eb = 0.175X^{1.11}e^{3.86X}P^{0.43} \tag{14.12}$$

Vapor was considered saturated within the evaporator. Correlations were obtained from Perry and Chilton (1973), and were allowed for the estimation of vapor properties with an error of less than 1% (see Miranda and Simpson (2005)).

To estimate lycopene degradation (or retention) in each evaporator effects, a mass balance at effect $i$ was carried out as follows:

$$F_{i+1}\cdot X_{i+1}\cdot Y_{i+1} - F_i\cdot X_i\cdot Y_i + M_i\left(\frac{\mathrm{d}(X_iY_i)}{\mathrm{d}t}\right) = \frac{\mathrm{d}(M_iX_iY_i)}{\mathrm{d}t} \tag{14.13}$$

Assuming steady state, perfect mixing, and first-order lycopene degradation rate, the following expression is obtained:

$$F_{i+1}\cdot X_{i+1}\cdot Y_{i+1} - F_i\cdot X_i\cdot Y_i - M_i\cdot X_i\cdot k_i\cdot Y_i = 0 \tag{14.14}$$

Solving for $Y_i$

$$Y_i = \frac{F_{i+1}\cdot X_{i+1}\cdot Y_{i+1}}{F_i\cdot X_i + M_iX_ik_i} \tag{14.15}$$

Combining Equations 14.4 and 14.15, the following equation is obtained:

$$Y_i = \frac{F_{i+1}\cdot X_{i+1}\cdot Y_{i+1}}{F_i\cdot X_i + M_i\cdot X_i\cdot\left[0.275271\cdot\exp(0.00241\cdot X_i)\cdot\exp\left(-\frac{2207}{T_i+273.15}\right)\right]} \tag{14.16}$$

With Equation 14.16, lycopene concentration for the output flowrate in each evaporator effect can be estimated knowing the steady-state values, the mass inside each evaporator effect, and the lycopene concentration in the input flowrate of each evaporator effect. It is important to mention that when trying to estimate lycopene concentration, a degree of freedom is added to the system, which is satisfied with the data of lycopene concentration in the evaporator system input (feeding) flowrate.

### 14.2.4 Economic Evaluation

First, we do a simple and preliminary analysis to find an optimum number of effects when processing a food product. The idea is to do a direct comparison between total cost minimization and NPV maximization. As stated earlier, the focus of the economic analysis is the inclusion of quality as an intrinsic parameter of the process design.

Generally, in the chemical industry, the quality of the product to be concentrated on is associated with the final concentration of the product independent of equipment design and OC. It is for this reason that the economic evaluation and optimization of these processes have been based on a total cost minimization. However, clearly, for food processes, product quality is highly dependent on equipment design and OC.

In the particular case of tomato paste, the most important component together with the compliance of concentration and consistency is the final content of lycopene in the product.

### 14.2.5 Maximization of NPV and Minimization of Total Cost

A criterion to optimize process design is to determine the number of effects that maximize the NPV of the invested capital for the new process line. This can be approached on the basis of microeconomics. The following equation is the expression for NPV:

$$\text{NPV} = -I + \sum_{j=1}^{n} \frac{\beta_j}{(1+i)^j} \tag{14.17}$$

From Equation 14.17, two significant terms can be distinguished: total investment ($-I$) and annual benefits ($\beta_j$), where the investment—for a given capacity—can be expressed as a function of the number of effects ($N_E$), and benefits are related to the product's unit-price ($P_u$) and costs per unit ($C_u$):

$$\beta_j = Q_j^* (P_u - C_u) \tag{14.18}$$

Clearly, in food products, unit-price is not constant and it is directly related to the final quality. As aforementioned, product quality is related to the number of effects ($N_E$), in general with process arrangement (PA) and OC. In addition, the incidence of energy in unit-cost can be expressed as a function of the number of effects ($N_E$) and OC too. Therefore, Equation 14.17 can be expressed as a function of the numbers of effects and OC as follows:

$$\text{NPV} = -I(N_E) + \sum_{j=1}^{n} \frac{Q_j^*[P_u(N_E, \text{OC}) - C_u(N_E, \text{OC})]}{(1+i)^j} \quad (14.19)$$

For fixed OC, to find the critical value for the number of effects ($N_E$ for maximum NPV), it is necessary to derive Equation 14.19 and then equalize it to zero, so $N_E^*$ can be obtained (critical value). On the other hand, critical value ($N_E^*$) represents a maximum for NPV if the second derivative is smaller than zero:

$$\frac{d(\text{NPV})}{dN_E} = -\frac{d[I(N_E)]}{dN_E} + \frac{d}{dN_E}\left\{\sum_{j=1}^{n} \frac{Q_j^*[P_u(N_E) - C_u(N_E)]}{(1+i)^j}\right\} \quad (14.20)$$

Considering

$$\beta_1 = \beta_2 = \ldots = \beta_n = \beta = Q^*[P_u(N_E) - C_u(N_E)] \quad (14.21)$$

and expressing the annual benefits as a present value, the second term of the right-hand side of Equation 14.17 can be reduced to

$$\sum_{j=1}^{n} \frac{\beta_j}{(1+i)^j} = \sum_{j=1}^{n} \frac{Q^*[P_u(N_E) - C_u(N_E)]}{(1+i)^j} = K'[P_u(N_E) - C_u(N_E)] \quad (14.22)$$

Therefore, replacing it in Equation 14.20, we obtain

$$\frac{d(\text{NPV})}{dN_E} = -\frac{d[I(N_E)]}{dN_E} + K'\frac{d[P_u(N_E) - C_u(N_E)]}{dN_E} \quad (14.23)$$

$$\frac{d(\text{NPV})}{dN_E} = -\frac{d[I(N_E)]}{dN_E} + K'\frac{d[P_u(N_E)]}{dN_E} - K'\frac{d[C_u(N_E)]}{dN_E} \quad (14.24)$$

In the case of a typical chemical engineering analyses, quality is considered to be independent of the number of effects ($N_E$). Meaning that the unit-price ($P_u$) is constant (independent of $N_E$); therefore, Equation 14.24 can be expressed as

$$\frac{d(\text{NPV})}{dN_E} = -\frac{d[I(N_E)]}{dN_E} - K'\frac{d[C_u(N_E)]}{dN_E} \quad (14.25)$$

By inspection of Equation 14.25, it is clear that the maximum NPV value is the same as the minimum total cost. This is the reason why in most chemical engineering analyses the search for the optimum process is reduced to find the minimum total cost. Quantification of NPV and total cost was mainly due by literature references and some direct quotations (Peters et al., 2003; Maroulis and Maroulis, 2005).

In food processing, quality has a strong effect on product price and, in addition, has a long-term effect on the consumer's perception of the food company. In this case, but also in general, the optimum process for a specific technology can be obtained by implementing and managing Equation 14.19.

## 14.3 CASE STUDIES

The mathematical model was developed for evaporator systems from 1 to 7 effects, operated under countercurrent. In each system, the restriction was equal areas for each evaporator effect. The input values to the model, shown in Table 14.1, were the same for all of the systems and obtained from an actual industrial plant, complemented with available online information from manufacturers.

### 14.3.1 Steady-State Conditions

From the mass and energy balance equations, liquid–vapor equilibrium equation, and specific relations for the tomato paste, a steady-state model for the evaporator system was developed, considering 1 up to 7 effects. From this information, it is possible to verify the decrease in vapor flowrate necessary for the operational process and an increase in the total system area, when augmenting the number of effects (Figure 14.2).

To have a more precise view of the product behavior in the evaporation system, residence time and their respective temperatures are presented in Table 14.2 for each effect in different systems.

### 14.3.2 Lycopene Retention

Lycopene retention in the final product was estimated for each one of the alternative systems from the data obtained under steady-state operation. From the results shown in Figure 14.3, it is clear that lycopene concentration in the final product has a linear decay when augmenting the number of effects in the evaporation system.

**TABLE 14.1**
**Input Data for Mathematical Model Implementation**

| Name | Variable | Value |
|---|---|---|
| Input flowrate | $F_{AI}$ (kg/h) | 50,000 |
| Input temperature | $T_{AI}$ (°C) | 98 |
| Initial soluble solids concentration | $X_{AI}$ (kg ss/kg) | 0.05 |
| Input concentration of lycopene | $Y_{AI}$ (kg Lic/kg ss) | 0.01 |
| Final soluble solids concentration | $X_1$ (kg ss/kg) | 0.3 |
| Steam inlet pressure | $Pv_0$ (kPa) | 143.4 |
| Temperature change in condensator | $Tv_n - T_d$ (°C) | 2 |
| Operation pressure in evaporator $n$ | $P_n$ (kPa) | 16.5 |

*Source:* Simpson, R., Almonacid, S., Lopez, D., and Abakarov, A., *J. Food Eng.*, 89, 488, 2008. With permission.

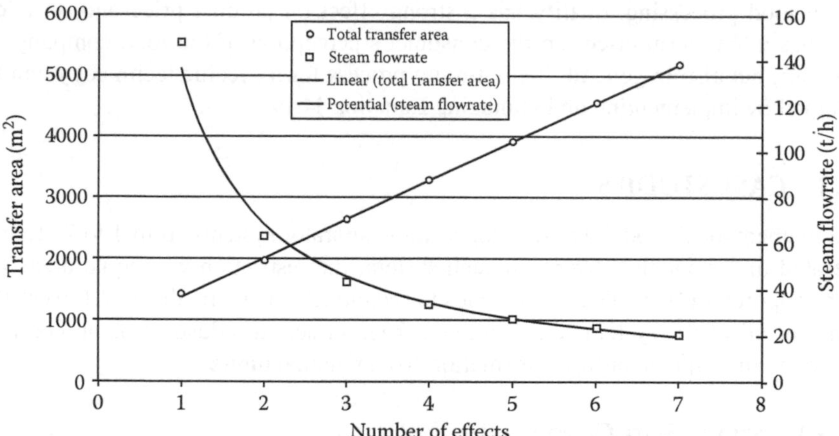


**FIGURE 14.2** Total transfer area ($m^2$) and steam inlet flowrate (t/h) vs. number of effects. (From Simpson, R., Almonacid, S., Lopez, D., and Abakarov, A., *J. Food Eng.*, 89, 488, 2008. With permission.)

The previous result gives a clue of how the content of lycopene in the final product is affected as a function of the residence time in the evaporation system. While the number of effects increases in the evaporation system, the total residence time increases as well as the temperature at which the product is exposed, and, therefore, there is also an increase of the lycopene degradation.

Lycopene retention for the theoretical evaporation system for tomato paste is shown in Figure 14.4. Clearly, lycopene retention decreases as the number of effects increase, which is justified by the augmentation of the total residence time of the system and also because of the temperature rise at which the tomato paste is exposed to. In regard to the supply flowrate, there is an increase in the lycopene retention when augmenting the flowrate. The increase in lycopene retention is less abrupt when the supply flowrate is over 100 t/h.

### 14.3.3 Economic Evaluation

The economic evaluation consists of determining the optimum number of effects and OC of the system. The economic evaluation of the system was done in two different ways. First, an economic evaluation with the concept of minimizing the total costs and second, an economic evaluation to maximize the NPV taking into account the impact of the process design and OC on product quality.

### 14.3.4 Optimum Number of Effects

The economic evaluation was carried out by simple inspection. This is where the steady-state conditions for systems with 1–7 effects were found, and then total cost minimization and NPV maximization methodologies were used. The search was focused to find the number of effects that minimize the total cost and, in addition, to find the number of effects that maximize the NPV.

**TABLE 14.2**
**Temperature Data (°C) and Residence Time (h) for Systems from 1 to 7 Effects with an Input Flowrate of 50 t/h**

| | Number of Effects in the System | | | | | | | | | | | | | |
|---|---|---|---|---|---|---|---|---|---|---|---|---|---|---|
| | 1 | | 2 | | 3 | | 4 | | 5 | | 6 | | 7 | |
| Effect Number | Residence Time | Temperature | Residence Time | Temperature | Residence Time | Temperature | Residence Time | Temperature | Residence Time | Temperature | Residence Time | Temperature | Residence Time | Temperature |
| 1 | 1.10 | 55.9 | 0.23 | 55.6 | 0.17 | 55.6 | 0.15 | 55.6 | 0.14 | 55.6 | 0.13 | 55.6 | 0.13 | 55.6 |
| 2 | | | 0.84 | 76.8 | 0.28 | 67.5 | 0.20 | 63.8 | 0.17 | 61.8 | 0.15 | 60.5 | 0.14 | 59.6 |
| 3 | | | | | 0.78 | 84.7 | 0.31 | 74.3 | 0.22 | 69.3 | 0.19 | 66.3 | 0.17 | 64.3 |
| 4 | | | | | | | 0.75 | 89.3 | 0.34 | 78.9 | 0.24 | 73.3 | 0.20 | 69.8 |
| 5 | | | | | | | | | 0.73 | 92.3 | 0.36 | 82.3 | 0.26 | 76.4 |
| 6 | | | | | | | | | | | 0.71 | 94.4 | 0.38 | 84.8 |
| 7 | | | | | | | | | | | | | 0.70 | 96.1 |

*Note*: Feed enters effect 1 and fresh vapor pressure is 143.4 kPa.

Output pressure at each effect

| Pressure (kPa) | 1 Effect | 2 Effects | 3 Effects | 4 Effects | 5 Effects | 6 Effects | 7 Effects |
|---|---|---|---|---|---|---|---|
| P1 | 16.5 | 40.18 | 56.4 | 67.7 | 76.05 | 82.41 | 87.5 |
| P2 | | 16.5 | 27.7 | 36.5 | 44.76 | 51.6 | 57.31 |
| P3 | | | 16.5 | 23.8 | 29.75 | 34.91 | 40.15 |
| P4 | | | | 16.5 | 21.82 | 26.38 | 30.32 |
| P5 | | | | | 16.5 | 20.65 | 24.28 |
| P6 | | | | | | 16.5 | 19.86 |
| P7 | | | | | | | 16.5 |

*Source:* Simpson, R., Almonacid, S., Lopez, D., and Abakarov, A., *J. Food Eng.*, 89, 488, 2008. With permission.

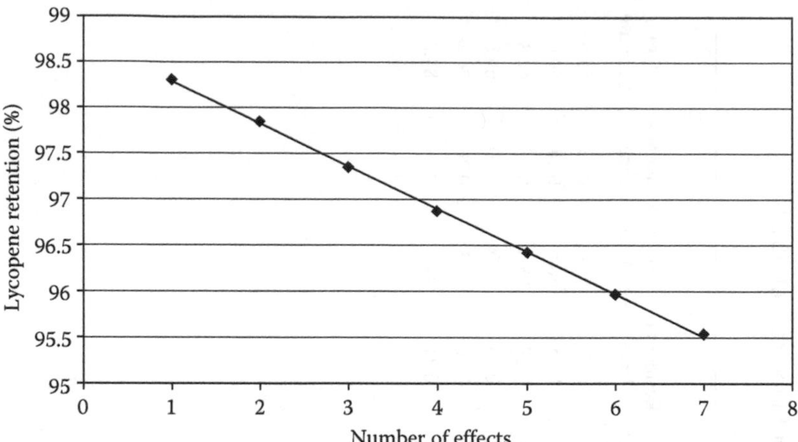


**FIGURE 14.3** Lycopene retention (%) vs. number of effects for an input of 50 t/h. (From Simpson, R., Almonacid, S., Lopez, D., and Abakarov, A., *J. Food Eng.*, 89, 488, 2008. With permission.)

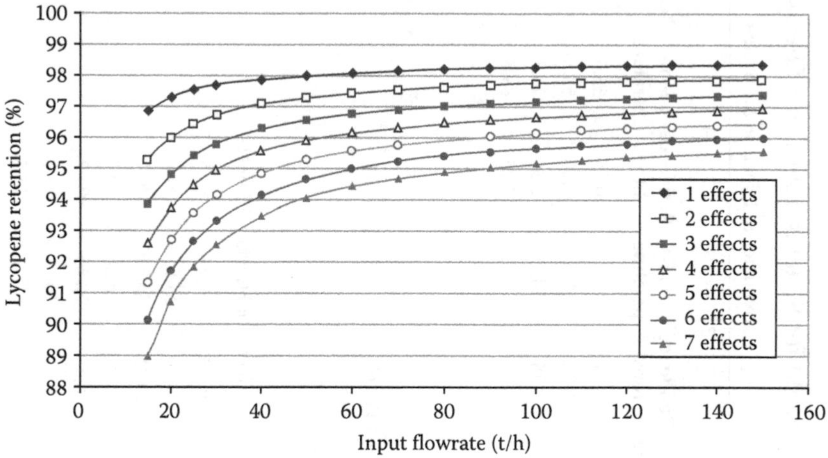


**FIGURE 14.4** Lycopene retention (%) for a system from 1 to 7 effects. (From Simpson, R., Almonacid, S., Lopez, D., and Abakarov, A., *J. Food Eng.*, 89, 488, 2008. With permission.)

The results for each evaluation systems are shown in Figures 14.5 and 14.6. The total cost minimization (Figure 14.5) shows an optimum of 4 effects. Nevertheless, when doing NPV maximization (Figure 14.6), the number of optimum effects was 3 due to the inclusion of the quality parameter on the evaluation procedure. Naturally, for different processing capacities, the optimum number of effects varies for both evaluation procedures. This is why differences are encountered in the optimum number of effects in some operation ranges. In Figure 14.7, the optimum number of effects is presented for different operation ranges. As it is observed in Figure 14.7, when evaluating the evaporation system, including the quality parameter, in the range of 25–50 t/h, the optimum number of effects decreases, in comparison to the evaluation done based on total costs only. This is explained with previous results where a decrease of lycopene retention was a result of the increase of the number of

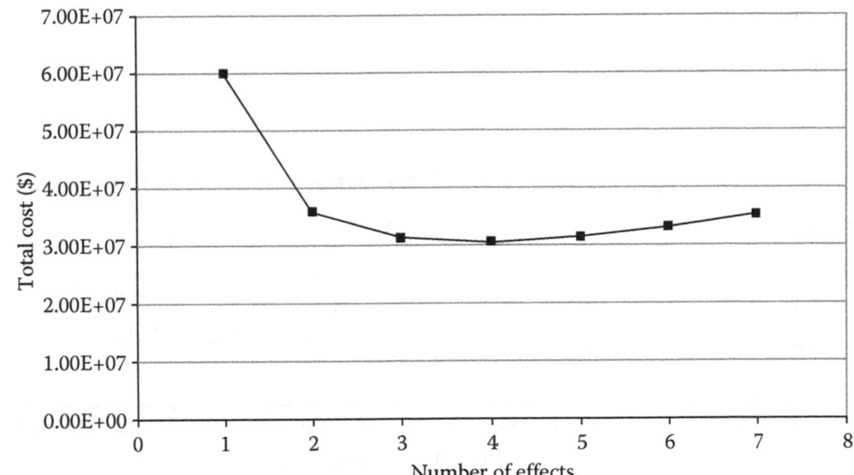


**FIGURE 14.5** Cost evaluation for an evaporator system with an input flowrate of 50 t/h. (From Simpson, R., Almonacid, S., Lopez, D., and Abakarov, A., *J. Food Eng.*, 89, 488, 2008. With permission.)

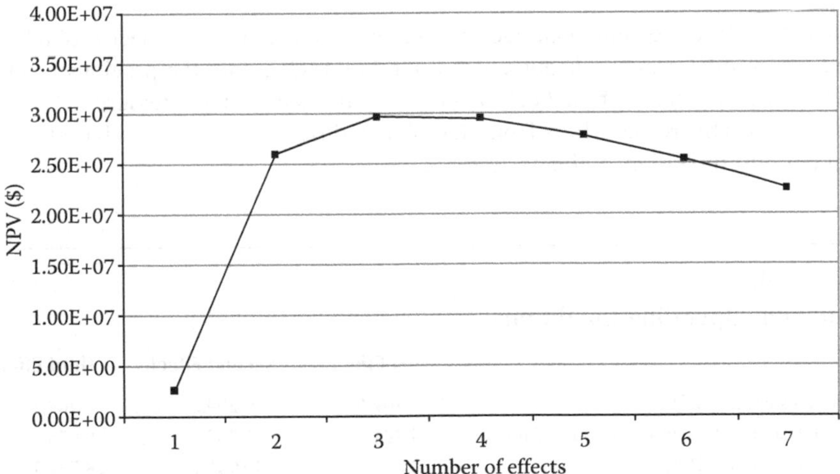


**FIGURE 14.6** NPV evaluation for evaporator systems with input flow of 50 t/h. (From Simpson, R., Almonacid, S., Lopez, D., and Abakarov, A., *J. Food Eng.*, 89, 488, 2008. With permission.)

effects. It is for this reason that the NPV maximization, in this particular case, tends to be a lower number of effects.

### 14.3.5 Optimum Operating Conditions

In the search for the optimum OC of evaporation system, the system was economically evaluated under a variable steam inlet pressure ($Pv_0$) where the inclusion of lycopene as a quality parameter was considered. As a constraint to the problem, it was estimated that the output temperature of the tomato paste (highest temperature), with the optimum number of effects for a defined vapor pressure, could not be

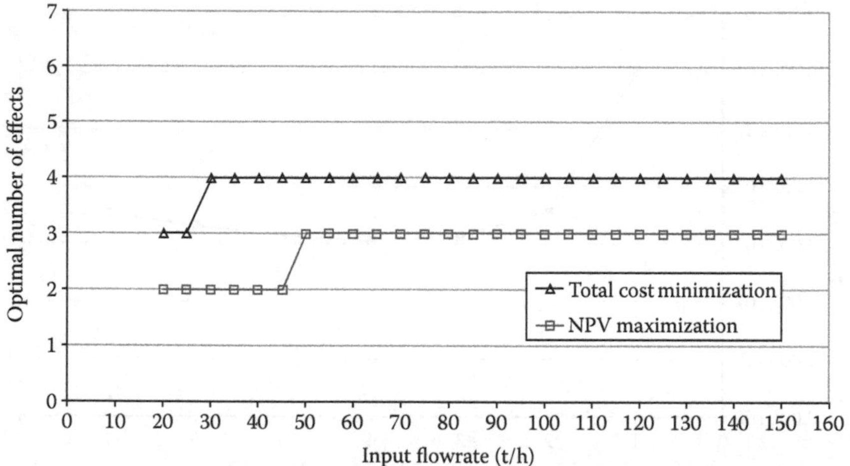


**FIGURE 14.7** Optimum number of effects for different input flowrates (t/h) according to total cost minimization and NPV maximization. (From Simpson, R., Almonacid, S., Lopez, D., and Abakarov, A., *J. Food Eng.*, 89, 488, 2008. With permission.)

higher than 95°C. Results obtained showed an increase in the project profitability when steam inlet pressure is augmented. Optimum OC, with the previously stated constraints, are found to be a 3 effects system with a steam inlet pressure of 260 kPa (Table 14.3). This result differs from that obtained from the evaluation based on the total cost minimization (5 effects system).

**TABLE 14.3**
**Optimum Operating Conditions**

| | First Effect | Second Effect | Third Effect |
|---|---|---|---|
| Heat transfer area ($m^2$) | 209.6 | 209.2 | 209.1 |
| Global heat transfer coefficient (kJ/°C/$m^2$/h) | 4,494.5 | 6,767.2 | 8,475.4 |
| Heat transferred (MJ/h) | 32,803.8 | 30,941.1 | 28,727.7 |
| BPR (°C) | 0.53 | 0.07 | 0.03 |
| Holdup (kg) | 5,320 | 4,907 | 4,815 |
| Residence time (h) | 0.639 | 0.224 | 0.141 |
| $\Delta T$ Nukiyama | 34.8 | 21.9 | 16.2 |
| **Steady-State Values** | | | |
| Steam inlet flowrate (kg/h) | 15,012.9 | | |
| Steam inlet pressure (kPa) | 260 | | |
| Steam inlet temperature (°C) | 129.1 | | |
| Steam inlet enthalpy (kJ/kg) | 2,725.45 | | |
| Output flowrate (kg/h) | 8,322.5 | 21,930.0 | 34,265.4 |
| Temperature (°C) | 94.26 | 71.88 | 55.60 |
| Concentration (kg ss/kg) | 0.300 | 0.114 | 0.073 |

**TABLE 14.3 (continued)**
**Optimum Operating Conditions**

| | First Effect | Second Effect | Third Effect |
|---|---|---|---|
| Lycopene concentration (kg/kg ss) | 0.0096790 | 0.0099130 | 0.0099724 |
| Lycopene retention (%) | 96.79 | 99.13 | 99.72 |
| Concentrate enthalpy (kJ/kg) | 311.3 | 276.7 | 220.7 |
| Vapor flowrate (kg/h) | 13,607.5 | 12,335.4 | 15,734.6 |
| Vapor pressure (kPa) | 81.84 | 32.9 | 16.5 |
| Vapor temperature (°C) | 93.73 | 71.81 | 55.57 |
| Vapor Enthalpy (kJ/kg) | 2,666.3 | 2,629.6 | 2,602.4 |
| Condensed flowrate (kg/h) | 15,012.9 | 13,607.46 | 12,335.4 |
| Condensation temperature (°C) | 129.08 | 93.7 | 71.81 |
| Condensed enthalpy (kJ/kg) | 540.4 | 392.4 | 300.7 |

*Source:* Simpson, R., Almonacid, S., Lopez, D., and Abakarov, A., *J. Food Eng.*, 89, 488, 2008. With permission.

**TABLE 14.4**
**Optimum Number of Effects for Different Steam Inlet Pressures**

| | Number of Optimum Effects | |
|---|---|---|
| $Pv_0$ (kPa) | Minimum Total Cost | Maximum NPV |
| 110 | 4 | 2 |
| 120 | 4 | 2 |
| 130 | 4 | 3 |
| 140 | 4 | 3 |
| 150 | 4 | 3 |
| 160 | 4 | 3 |
| 170 | 4 | 3 |
| 200 | 4 | 3 |
| 230 | 4 | 3 |
| 260 | 5 | 3 |

*Source:* Simpson, R., Almonacid, S., Lopez, D., and Abakarov, A., *J. Food Eng.*, 89, 488, 2008. With permission.

Table 14.4 depicts the results obtained by both economics ways: total cost minimization and NPV maximization.

## 14.4 CONCLUSIONS

The steady-state values of the evaporator system were able to be linked to the reaction kinetics of the target attribute, lycopene. A mathematical model was successfully developed, and then an economic evaluation of the optimum design and OC of the evaporation system (1–7 effects operated under countercurrent) was carried out.

It was possible to determine that the lycopene retention has a linear decay respect to the number of effects used in the evaporation system. When analyzing the behavior of a 5 effects evaporator system, an increase in the processing capacity from 50 to 75 t/h augments the lycopene retention in the final product from 95.25% to 96.27%.

The previous result is due to the decrease in residence time, and independent of an increase in the evaporator's temperature. This result is important because it indicates that the increment in tomato paste production, in this particular case, is restricted only by mechanical factors like available vapor pressure, maximum specified holdup, pump power, etc.

The total cost minimization allows the determination of the best equipment design (optimum number of effects), but no information on OC and product quality is obtained. On the other hand, with the NPV approach, it is possible to optimize the system design and OC simultaneously. In addition, the NPV approach considers the final product quality as an intrinsic parameter of the system.

With the inclusion of lycopene as a quality parameter (NPV), the optimum number of effects decreases from 4 to 3 when compared with total cost analysis. In addition, it was also possible to determine the optimum OC of the 3 effects system at 260 kPa.

It appears of extreme relevance to consider quality as an intrinsic and integral part of the process design, as it is then possible to identify several potential improvements in different food processes.

## NOMENCLATURE

| | |
|---|---|
| $C_u$ | cost per unit, \$/kg |
| $C_p$ | specific heat of concentrate, kJ/kg °C |
| $\Delta eb$ | BPE or BPR, °C |
| $E$ | activation energy, kJ/mol |
| $F$ | mass flowrate, kg/h |
| $H$ | enthalpy, kJ/kg |
| $i$ | annual interest rate |
| $I$ | total investment, \$ |
| $k$ | reaction rate constant, 1/h |
| $k_0$ | frequency factor, 1/h |
| $K'$ | constant |
| $M$ | mass in the evaporator, kg |
| $N_E$ | number of effects |
| NPV | net present value, \$ |
| $OC$ | operating conditions |
| $P$ | pressure, kPa |
| $PA$ | process arrangements |
| $P_u$ | unit sales price, \$/kg |
| $PV_0$ | inlet pressure |
| $Q$ | heat flow, kJ |
| $Q_j^*$ | annual production at period $j$, units per year |
| $R$ | ideal gases constant, °C kJ/mol |

| | |
|---|---|
| ® | registered mark |
| $r$ | reaction rate |
| $T$ | temperature, °C |
| $t$ | time, h |
| $X$ | concentration of soluble solids, kg SS/kg |
| $Y$ | lycopene concentration kg L/kg SS |

*Subscripts*

| | |
|---|---|
| $c$ | condensing |
| $d$ | download |
| $e$ | cooling |
| $i$ | evaporator effect $i$ |
| $j$ | evaluation period $j$ |
| $p$ | losses |
| $v$ | steam phase |

## ACKNOWLEDGMENTS

Ricardo Simpson is grateful for the financial support provided by CONICYT through the FONDECYT project number 1070946.

## REFERENCES

Douglas, J. M. 1988. *Conceptual Design of Chemical Processes*. McGraw-Hill International Edition, Chemical Engineering Series, New York.

Goula, A. and Adamopoulos, K. 2006. Prediction of lycopene degradation during a drying process of tomato pulp. *Journal of Food Engineering* 74(1): 37–46.

Himmelblau, D. M. and Bischoff, K. B. 1968. *Process Analysis and Simulation: Deterministic System*. Wiley, New York.

Kern, D. 1999. *Procesos de Transferencia de Calor*. Editorial Continental S.A., México.

Maroulis, A. Z. and Maroulis, Z. B. 2005. Cost data analysis for the food industry. *Journal of Food Engineering* 67(3): 289–299.

Miranda, V. and Simpson, R. 2005. Modeling and simulation of an industrial multiple effect evaporator: Tomato concentrate. *Journal of Food Engineering* 66(2): 203–210.

Perry, J. and Chilton, C. 1973. *Chemical Engineers Handbook*. McGraw-Hill, New York.

Peters, M., Timmerhaus, K., and West, R. 2003. *Plant Design and Economics for Chemical Engineers*. McGraw-Hill, New York.

Rizvi S. H. and Mittal G. S. 1992. *Experimental Methods in Food Engineering*. Van Nostrand Reinhold, New Delhi.

Schoorl, D. and Holt, J. E. 1983. An analysis of the effect of quality on prices of horticultural procedure. *Agricultural Systems* 12(2): 75–99.

Shi, J., Qu, Q., Kakuda, Y., Jun, S., Jiang, Y., Koide, S., and Shim, Y. 2007. Investigation of the antioxidant and synergistic activity of lycopene and other natural antioxidants using LAME and AMVN model systems. *Journal of Food Composition and Analysis* 20(7): 603–608.

Tonelli, M., Romagnoli, J., and Porras, J. 1990. Computer package for transient analysis of industrial multiple-effect evaporators. *Journal of Food Engineering* 12(4): 267–281.

# 15 Optimizing the Thermal Processing of Liquids Containing Solid Particulates

*Jasper D.H. Kelder, Pablo M. Coronel, and Peter M.M. Bongers*

## CONTENTS

## 15.1 INTRODUCTION

Optimization of in-line aseptic processing is essential for a successful commercial exploitation (Richardson, 2004). On one hand, in-line aseptic products boost economies of scale and the consumer is willing to pay a price premium for improved organoleptic and nutritional quality. On other hand, compared to traditional retorting processes, capital investment, complexity of operation, and the required level of operator skills are generally higher. Obviously constrained by safety considerations, these two considerations of product quality and cost suggest two main dimensions for optimization.

First, the final quality of the product as perceived by the consumer should be as high as possible for the price position, which is especially relevant for heterogeneous products (Sastry and Cornelius, 2002). Optimization of the product quality at the moment of consumption can be achieved at all stages of the supply chain: ingredient selection, equipment design, optimizing operational parameters such as flow rates and temperatures, controlling storage and transportation conditions, operator training, etc.

As the impact of temperature and time on quality is so important, this chapter starts with a fundamental discussion of thermal optimization. From the theoretical case of perfect heat transfer to a particle, recommendations are formulated that can be applied to all stages of an aseptic process.

Second, aseptic production can be optimized from a cost perspective. Global sourcing of nonperishable ingredients may bring benefits here, as well as the use of frozen ingredients to smoothen the seasonal availability (e.g., for vegetables). For manufacturers of a sufficiently global scale, production may shift between different production locations to follow local supply of fresh and less-expensive ingredients, and warehousing and transportation of finished goods may buffer demand.

On the scale of a single production site, there is also considerable scope to obtain cost benefits. Generally, line efficiencies are not high, which carries penalties in capital utilization as well as product losses during production. Extending the production run length (e.g., by reducing fouling or cleaning downtime), batch scheduling strategies, and optimal process control all offer opportunities for greater efficiency without compromising quality.

However, commercial production of aseptic products will always be a trade-off between final product quality and cost. This chapter discusses factors driving both quality and cost optimization, and aims to indicate ways to improve upon both.

## 15.2 FUNDAMENTALS OF THERMAL OPTIMIZATION

Thermal optimization of aseptic lines is a crucial factor driving the quality of final products. However, to optimize quality, its definition should be discussed more extensively. Second, the definition of the safety criterion in aseptic processing is revised to arrive at less conservative expressions amenable to heterogeneous foods. Third, the relevance of the high-temperature short-time (HTST) axiom to aseptic processing is discussed. Next, this section closes with a fundamental optimization exercise that highlights the factors driving optimal thermal sterilization of heterogeneous products.

### 15.2.1 Quality Definition and Assessment for Aseptic Products

Quality of a product is essentially defined by the satisfaction of the consumer upon use. Though factors such as packaging, marketing, and context (e.g., setting of the consumption and other meal components) play important roles in the eventual consumer appraisal, in this chapter, the scope of quality is limited to factors intrinsic to heterogeneous products that are also to some degree quantitatively measurable:

- Appearance (e.g., as measured by color, particle size, and shape)
- Texture and (liquid) consistency (e.g., through compression or extrusion tests and rheometry)
- Flavor and taste (e.g., artificial nose and composition)
- Micronutrients (e.g., vitamin C or thiamine levels)

Linking measured data to consumer perception and appraisal is still very challenging and product evaluation tests therefore involve extensive sensory analyses. In that way, the multisensory experience of product consumption can be fully accounted for, as this often cannot be broken down in data obtained by simple measurements. It should be emphasized that even superior scores on different sensorial dimensions do not necessarily imply consumer preference. However, the different methodologies to perform consumer trials and linking these to measurable product characteristics are outside the scope of this chapter.

Instead, quality as used in this chapter is limited to quality factors whose dependence on the thermal history can be readily integrated. Such factors include color, micronutrient levels, or textural changes. These changes are often expressed as the general effect of the heat treatment in terms of cooking, or the cook value (Equation 15.1):

$$C(t) = \int_0^t 10^{[T(t')-T_r]/Z} \mathrm{d}t' \tag{15.1}$$

where

$T_r$ is the reference temperature for quality (usually set to 100°C)
$T$ is the (variable) process temperature
$Z$ is the temperature difference required for a tenfold change in reaction rate

Interpreting the cook value in terms of a rough quality indication is straightforward. When vegetable particulates receive an insufficient cooking treatment ($C<5$ min), they can be considered raw with insufficient flavor development, whereas for cook values exceeding 20 min, virtually all vegetables loose their bite, turning into puree for very severe heat treatments ($C>40$). Similar arguments apply to the cooking of beef (raw at insufficient cook values versus tough or disintegrated at very high $C$), the activation and disintegration of thickeners such as starch, or the breakdown of valuable nutrients (e.g., vitamin C and thiamin). Many sources (e.g., Toledo, 1980; Holdsworth, 1992) give $D$ and $Z$ values (or rate constant and activation energies) for the quality factors depending on temperature.

### 15.2.2 Sterility Calculations in Mixed and Particulate Flows

In Chapter 3, the thermal profile was integrated in time to yield the cumulative lethality $F_0$ as applicable to a single fluid element or particle that was uniform in temperature at each instant in time. However, in real liquid and heterogeneous flows, significant radial velocity and temperature profiles exist in the cross section of the heat exchangers, which are manifest as a fast moving cold- (in heaters and holding tubes) or hot-core (in the cooling section). Assuming a uniform liquid velocity and temperature would therefore be incorrect, even in turbulent flows.

When particles are present, $F_0$ is customarily calculated based on a phantom particle having both the largest thermal path and the shortest residence time, and which has experienced the lowest external liquid temperature.

However, a particle will not always be the fastest, as it may radially migrate to streamlines of lower velocity and higher temperature (e.g., move closer to the wall of the heaters). Both phenomena may be related to mixing actions induced by the geometry (e.g., centrifugal forces in bends), or by particles interacting with the fluid or each other.

Also, sizing the aseptic system based on the lethal contribution of the holding tube alone neglects the often considerable lethality accumulated in the heating and cooling sections. For the liquid part of a product, $F_0$ achieved in the heating section is often considerable (Jung and Fryer, 1999; Kelder et al., 2002; Awuah et al., 2004), whereas the critical center of particles continues building sterility in the first part of the cooling section (Sandeep et al., 1999).

The most straightforward way to obtain more accurate predictions of lethality would be to model the high-temperature section of the aseptic system in three dimensions using a full computational fluid dynamics (CFD) analysis. For heterogeneous flows there would be a need to account for the varying degrees of particle loading and particle thermal development, and the two-way coupling between liquid and particles. In principle, this type of model would not rely on experimental/analytical heat transfer coefficients as these would follow from the calculations. The impact of bends, free convection, and time-dependent rheology could be included in such a model. Using a full geometric representation of all particulates in transient calculations, no further assumptions regarding fluid-particle heat transfer would need to be included. Based on the transient flow and temperature fields and microbial destruction kinetics, the actual microbial and spore concentration could be calculated throughout the system. As a final step, this concentration could be volume averaged at the end of the cooling section and used to calculate the total $F_0$ achieved.

Unfortunately, the computational burden of such modeling significantly overstresses software and computational capacity, both now and in the foreseeable future. Several researchers have therefore attempted to build simpler models allowing a more accurate calculation of $F_0$, and to include the contribution of the heating and cooling sections.

For liquid flows, Simpson and Williams (1974) presented the first non-isothermal results for lethality for a pseudoplastic fluid in a sterilizer consisting of a heater and a cooler. Kumar and Bhattacharya (1991) analyzed lethality in the developing flow of a pseudoplastic food with a temperature-dependent rheology. Jung and Fryer (1999) simulated a pseudoplastic model food with a temperature-dependent rheology in the heater, holder, and cooler of a sterilizer. Liao et al. (2000) studied lethality in a heater and cooler by solving the flow and temperature fields of a gelatinizing starch suspension. Bhamidipati and Singh (1994) applied the concept of thermal time distribution to shear thinning tomato juice with a temperature-dependent rheology in a tubular heat exchanger.

However, all studies regard the flow to be axisymmetric. When radial mixing is involved, for example, as a result of bends or free convection, fluid elements are no longer following streamlines parallel to the tube axis. The approach taken by Jung and Fryer (1999) for calculating the $F_0$ along such streamlines, and averaging appropriately at the end of the system (i.e., calculate $F_0$ into concentration, average and calculate the concentration back into $F_0$ again), is then difficult to implement.

A more accurate solution to lethality calculation in liquid flows is summation of the minimum lethal rate in the cross section of the tube along the length of a tubular section (Equation 15.2; Kelder et al., 2002). For this to be successful, information is needed on the radial velocity and temperature profiles:

$$F_0 = \sum_L \Delta F_{0_{\min}} = \sum_L \left( \Delta z \cdot \frac{dF_0}{dz} \right)_{\min} = \sum_L \left( \Delta z \cdot \frac{10^{\frac{T-T_r}{z}}}{w} \right)_{\min} \tag{15.2}$$

where

$\Delta z$ is an axial section of the tubular sterilizer
$w$ is the local axial velocity

This approach is more accurate than the cold- or hot-core assumption, as it accounts for temperature and velocity simultaneously. Though (mathematically) proven to be safe and more realistic than the cold-core assumption, this approach is still too conservative as it assumes a fluid element that receives the minimum treatment in each section of the unit.

For particulate flows, modeling efforts have been conducted with different degrees of approximation, older attempts that have been reviewed by Barigou et al. (1998). Among the simplest are heat conduction models describing the temperature evolution throughout particles as a function of varying temperature boundaries (Cacace et al., 1994; Mateu et al., 1997; Palazoglu and Sandeep, 2002). As a second step, lethality and quality were integrated and optimized as a function of, for example, heat transfer coefficient, residence time, or particle shape and size.

More extensive models also consider the thermal evolution of the liquid along the line, ideally coupled to the particles by heat balances and calculated simultaneously (Skjöldebrand and Ohlsson, 1993a,b; Mankad et al., 1995). Some of these models can, to a limited extent, account for the different residence time between the average fluid and the average particle in the system through the concept of a slip velocity (Mankad et al., 1995). Other models describe the aseptic process with greater flexibility by providing the option for multiple heating, holding, and cooling sections, some of which are marketed commercially (e.g., Weng, 1999). Most major suppliers of aseptic equipment and food manufacturers now seem to use this type of modeling (either developed in-house or licensed) to design and set their aseptic processes to safe and high-quality standards.

Models of a higher resolution have been scarce, both in terms of particle residence time and radial temperature distribution. Zhang and Fryer (1995) presented a moving mesh model that resolved the temperature distribution around and inside a two-dimensional particle sliding along a heated wall in the absence of convection. Mankad and Fryer (1997) presented a model of a stratified flow in a heater and holder allowing for sedimentation of particles and a differential residence time between liquid and particles. Sandeep et al. (2000) calculated residence time distributions of particles in a holding tube based on simulation of the flow and Lagrangian particle tracking. Particle–flow and particle–particle interactions were included and liquid and particle fields were iteratively solved. A second program was used to calculate the temperature distribution in the axial direction of the liquid and radially within the particles in a heat, hold, and cool section. Next, particle residence time distribution (RTDs) and temperature histories were integrated into $F_0$ values and the sensitivities of the process to particle concentration, RTD, and liquid-particle heat transfer coefficient were explored.

Though the models reviewed above have greatly improved understanding of the key factors driving aseptic sterilization, particle residence time distribution and radial temperature nonuniformities have still not fully been accounted for. An economic way to do so may be to reconstruct the radial temperature gradients based on the average fluid and wall temperature. This implies sufficiently developed flow and temperature profiles, and it requires a good estimate of the local heat transfer coefficients.

Next, the residence time distribution of particles can be expressed as a probability function of the radial position and hence velocity. A particle could for example be assumed to follow one streamline in one section of the heat exchanger, and be given a random new radial position in the next. This approach may be used to account for the mixing in the bends, particle–particle interactions, or free convective currents in the tube cross section. Also, the variability in size and thermal properties of particles may be included using the concept of size classes, improved accuracy of the temperature profiles in high particle fraction heterogeneous flows being the main benefit.

Finally, a risk-based approach using Monte Carlo simulations (Braud et al., 2000) could be adopted to include uncertainty in input data such as thermal properties (e.g., conductivity and heat capacity), formulation (e.g., particle mass fraction and size distribution), operating conditions (e.g., flow rate and medium temperatures), or microbial and quality kinetics. This type of modeling potentially yields a more realistic estimate of the lethality and quality factors of aseptic processes while avoiding accumulation of the worst-case assumption for each of the input parameters.

Whichever the course taken for future increases in model resolution, this will come at the expense of the experimental effort to obtain data on residence time distribution, heat transfer correlations, or rheological and thermal properties of liquids and particles. In the end, modeling and supporting experimental effort may yield hugely improved predictions, but these may never fully replace experimental validation of the actual formulation and thermal process.

### 15.2.3 HTST Treatments for Heterogeneous Products

At high temperatures, the inactivation of microbes and their spores exhibit higher rates than the destruction of quality factors. High temperatures applied for short times produce the same aseptic effect as longer process times at lower temperatures, whereas the destruction of quality factors is dramatically reduced. This can be easily shown by writing the ratio of the relative reactions rates $R_r$ (refer to Chapter 3) for kinetic processes 1 and 2 in terms of the Bigelow model (Equation 15.3):

$$\frac{R_{r1}}{R_{r2}} = \frac{10^{\frac{T-T_{r1}}{Z_1}}}{10^{\frac{T-T_{r2}}{Z_2}}} = 10^{\frac{(Z_2-Z_1)T+(Z_1T_{r2}-Z_1T_{r2})}{Z_1Z_2}} \tag{15.3}$$

When $Z_2>Z_1$, an increase in temperature speeds up the kinetics of process 1 relative to that of process 2. Generally, $Z$ is larger for quality factors than it is for microbial inactivation, and therefore employing the highest possible process temperature would be beneficial from a quality point of view. Expressed in Arrhenius terms, the process having the higher activation energy $E_a$ would respond more strongly to increases in temperature. This approach has become know as HTST concept (Holdsworth, 1992).

In well-mixed liquids of a uniform temperature, the HTST concept is constrained by several practical limitations. First, at very high wall and product temperatures, nonlinear fouling and browning reactions may render the operation uneconomical and product quality unacceptable (Simpson and Williams, 1974). Second, the maximum achievable medium temperature may be limited by the equipment and factory utilities (e.g., steam pressure available). Third, high-product temperatures necessitate a sufficient pressure in the aseptic system to avoid boiling that may exceed design specification. Fourth, high temperatures imply extremely short holding times, and accurate control of these may no longer be guaranteed. Finally, the validity of microbial destruction kinetics beyond the temperature range for which these were originally obtained is unreliable and product safety can no longer be guaranteed.

Aside from such practical considerations, the HTST approach should also be used with caution where residence time distributions and temperature nonuniformities exist, either in homogeneous (Jung and Fryer, 1999; Kelder et al., 2002) or heterogeneous flows (Sandeep et al., 2000). In such cases, the final product may exhibit an unacceptable distribution in quality factors.

To derive clues for temperature time combination for optimal product quality (i.e., a minimized and uniform cook value), a simple model was created to reconstruct the temperature evolution of both liquid and particles of a heterogeneous food along a tubular aseptic system. The underlying equations follow the heat transfer cascade as shown in Figure 3.1 and allow for an axial development of the liquid temperature (Equation 15.4), two-way energy transfer between the liquid and the fraction of representative particles (Equation 15.5), and a one-dimensional temperature distribution in cylindrical particles of infinite length (Equation 15.6):

$$\rho_f Cp_f A_f \frac{DT_f}{Dt} = U_{mf} l_f (T_m - T_f) + n_p h_{fp} A_p (T_p^s - T_f) \tag{15.4}$$

$$-\lambda_p \frac{\partial T_p^s}{\partial n} = h_{fp} (T_f - T_p^s) \tag{15.5}$$

$$\rho_p Cp_p \frac{\partial T_p(r)}{\partial t} = \lambda_p \nabla^2 T_p(r) \tag{15.6}$$

In Equations 15.4 through 15.6, $\rho$, $Cp$, and $\lambda$ are constant and have the usual meaning of density, heat capacity, and thermal conductivity; $t$ indicates time and $T$ temperature. Subscripts m, f, and p indicate the transfer medium, the liquid product fraction, and the particles, respectively, and superscript s the surface of the particle. In Equation 15.4, the substantial derivative (Bird et al., 1960) is used to calculate the energy flow to or from the liquid, where $A_f$ is the tube cross section, $l_f$ the tube perimeter, $n_p$ the specific number density of the particle, and $A_p$ the surface area of the representative particle. Finally, the overall heat transfer coefficients between the heating medium and the fluid, and between the fluid and the particles are given by $U_{mf}$ and $h_{fp}$, respectively.

To allow for a phase transition in the particles (either melting from the frozen state or crystallization of fats) an artificial heat capacity was included such that the enthalpy of phase transition $\Delta H$ was accounted for over the temperature range of transition spanning $T_s$ and $T_e$ (Equation 15.7):

$$\begin{aligned} Cp_p(T) &= Cp_{p1} \text{ for } T < T_s \\ Cp_p(T) &= Cp_{p2} \text{ for } T > T_e \\ \Delta H &= \int_{T_s}^{T_e} Cp_p(T)\mathrm{d}T \end{aligned} \tag{15.7}$$

A tubular heater and cooler were modeled to establish the boundaries of the cook values attainable using different flow and heat transfer scenarios. For illustrative purposes, the case of perfect heating is presented here to highlight the key factors impacting on thermal processing of heterogeneous products.

The scenario of perfect heat transfer applies to particulate flows subject to steam injection where the liquid base is heated up rapidly by the latent heat of condensing steam. It also presents a reasonable approximation for low-viscosity turbulent flows. In both cases, heat transfer into and through the liquid is very rapid, and is followed by a much slower process of heat conduction into and out of the core of the particles. This scenario was implemented in the model by setting both heat transfer coefficients $U_{mf}$ and $h_{fp}$ to very high values, effectively making conduction throughout the particle the limiting factor.

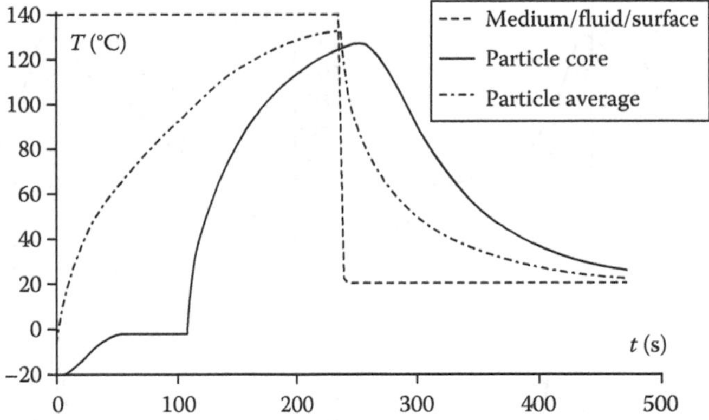


**FIGURE 15.1** Temperature time traces during a perfect heating–cooling process.

To evaluate industrially relevant processing conditions, a 10% mass fraction of infinite frozen cylinders of four different diameters was chosen, at a product mass flow of 5 t/h in 50 mm internal diameter tubes. In the heater, a constant wall temperature of 140°C was maintained (condensing steam), and in the cooler a wall temperature of 20°C was assumed. Heating and cooling sections were dimensioned such that the core $F_0$ value reached at least 2.5 min.

As an example, Figure 15.1 shows the average and core temperature traces for a particle of 15 mm diameter. As the heat transfer between medium and fluid is very good, the fluid temperature follows the medium temperature almost immediately, both in heating and cooling. Heat transfer between fluid and particle surfaces is also rapid and the surface temperature of the particle equals that of the fluid, both in the heater and in the cooler.

Even though the surface heats instantaneously, the particle core needs to go through the ice-water phase transition and for the greater part of the heating stage, large temperature differences throughout the particle persist. Looking at Equation 15.1, it becomes clear that the same must then be true of the cook value (or any other quality factor with similar kinetics).

This very nonuniform sterilization process can be strongly improved by giving the particles time to equilibrate at intermediate temperature levels. Figure 15.2a through d contains the temperature traces for a 1-, 2-, 3-, and 4-step equilibration processes.

By allowing equilibration, the particle residence time increases, but the temperature nonuniformity inside particles and between liquid and particles decreases. As a result, the distribution of the thermally affected quality factor becomes narrower. This exercise was repeated for particle sizes of 5, 10, and 20 mm diameter. The temperature evolution throughout the particles was integrated for the four particle sizes and the five processing scenarios (0–4 step equilibrations) in three cook values (Figure 15.3).

In Figure 15.3, the five upper traces represent the cook value of the surface of the particle, the middle five the volume average, and the lower five the core cook value.

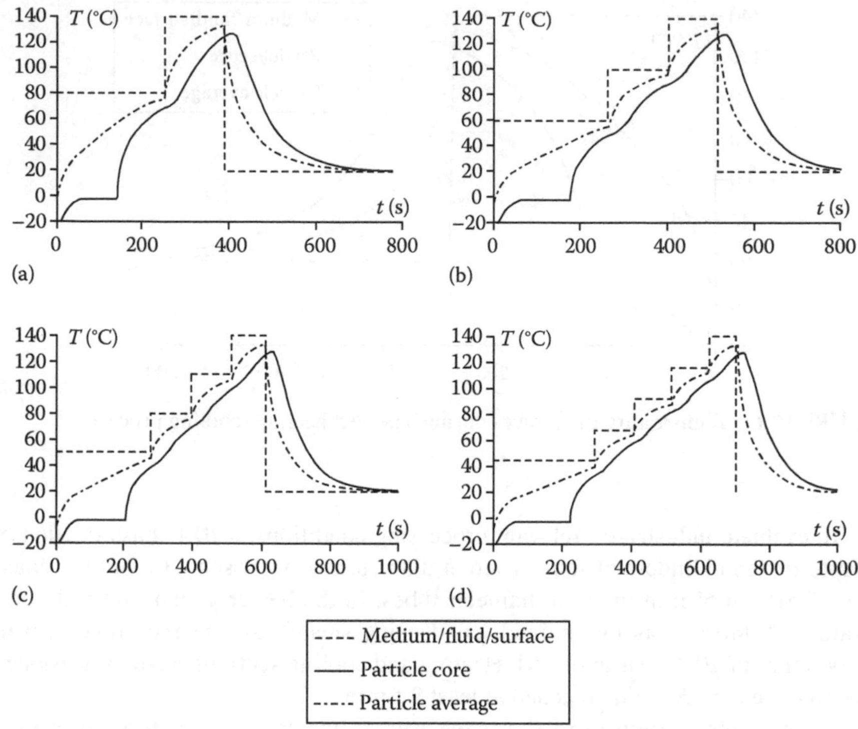


**FIGURE 15.2a–d** A 1–4 step equilibration perfect heating processes for 15 mm particles.

The numbers indicate the equilibration steps from 0 to 4. The reference line shows the 20 min value, beyond which most vegetables have softened to the point of losing textural integrity. Surface cook of particles in a process without equilibration surpasses this value for particles larger than about 7 mm in diameter. If equilibration is employed

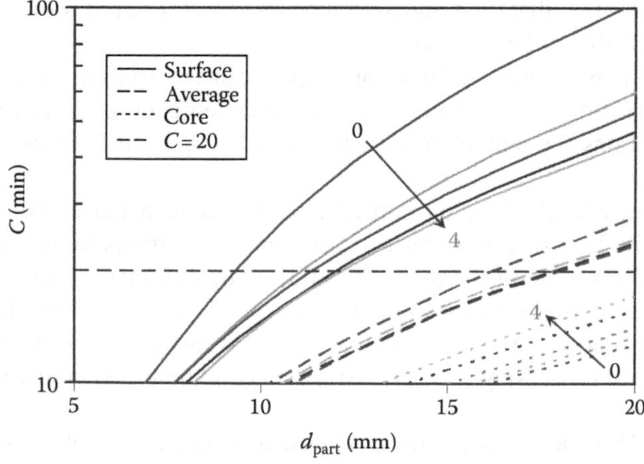


**FIGURE 15.3** Cook values in perfect heating processes as a function of particle size.

properly (e.g., in the 4-step process), particles could be processed up to 12 mm in diameter. This value is consistent with practically observed maximum particle sizes in tubular aseptic processing.

Though this perfect heat transfer case is a strong simplification of sterilization of heterogeneous products, it represents the lower bound to the severity of the heat treatment, and hence the upper bound to achievable quality or particle size. First, for a real system, a holding tube must be included, and allowance should be made for the residence time distribution of the particles. Both would lead to a prolongation of the heat treatment, and hence increase of the cook value. Second, particles normally have a size distribution, and smaller particles would heat up more rapidly to accumulate higher cook values. Third, appreciable heat transfer resistances are present in real systems between the medium and the liquid, and between liquid and particles, prolonging the product residence time and intensifying the heat treatment. Fourth, in sufficient mass fraction, the particles act as heat sinks in the heater and holding tube, and as heat sources in the cooler, again prolonging the heat treatment. Finally, as noted before, the liquid temperature is nonuniform in the cross section, which has to be allowed for by prolonging the process.

Despite the simplified nature of the perfect heat transfer case, a number of guidelines can be derived that are generally valid to improve product quality in aseptic processing of heterogeneous products:

1. Tempering of frozen ingredients, or the use of fresh ingredients at chilled or ambient temperature, reduces the thermal degradation of quality factors during preparation and reduces the thermal duty of the heaters. It also improves the temperature uniformity and spread in quality factors in particulates.
2. Batch cook vessel temperature can be optimized to achieve the same potential benefits as indicated for tempering.
3. Large temperature differentials between the medium and the product should be avoided throughout the heating stage. Several heating stages operated at increasingly higher temperatures allow for the particles to equilibrate thermally. The limit of an infinite number of heating stages is equivalent to a countercurrent heat transfer scenario, where the temperature differential between the medium and the particles is optimized.
4. Maximum sterilization temperature to be used is a function of the optimal driving force between the medium and the particulates.
5. Any decrease in residence time distribution allows for a less intense heating process.
6. Cooling rates must be maximized as far as practically and economically feasible.
7. Thermal path of the largest particles is the main barrier to obtain an acceptable quality in conduction limited thermal processes.

In Section 15.3, these guidelines drive the practical improvements possible for each step of the thermal process. In addition, the role of ingredients and auxiliary equipment (e.g., pumps) is discussed from the point of view of product quality.

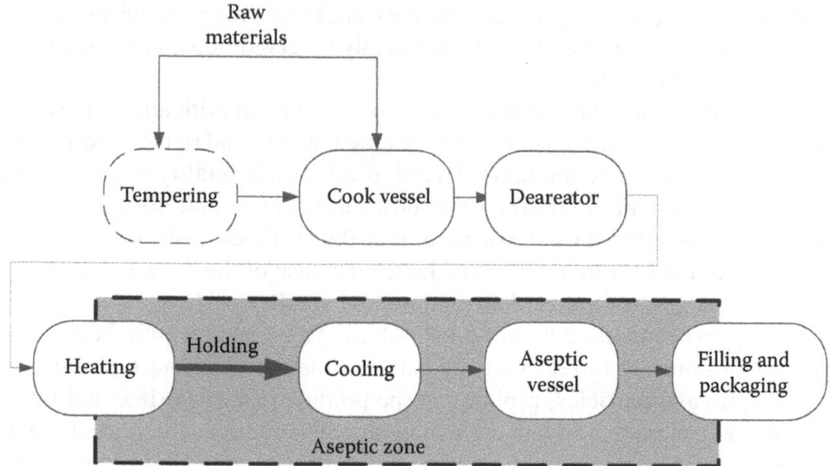


**FIGURE 15.4** Fundamental operations in optimized aseptic processing.

## 15.3 ASEPTIC QUALITY OPTIMIZATION

As a case study, the steps sketched in Section 3.4 were optimized from a quality point of view (Figure 15.4). As quality obviously strongly depends on the ingredients employed, this section sets out discussing the choice of ingredients for aseptic processing.

### 15.3.1 Role of Ingredients

Thermal processing induces changes in quality attributes of foods such as texture, nutritional content, and color. In continuous flow processes, further textural degradation and particle breakage may occur due to shear, contact with the wall, or interparticle collisions. Apart from an unfavorable appearance, shear damage promotes leaching of particulate material to the carrier liquid, thus inadvertently blending flavors and clouding the broth.

To make products of a high final quality, ingredients must be sufficiently robust to survive the harsh thermal treatment posed by any aseptic process, however optimized. Dairy components should be compatible with the ultra high temperature (UHT) process and may have to be pretreated (e.g., pasteurized) to render fouling rates acceptable. The cook profile of meats must match that of the aseptic process to yield the required succulence, and this may require precooking or frying. Starches should yield the right consistency throughout the process and in the finished product, and this often means the use of annealed or modified high-temperature starches. Also, blends of starches may be employed that are activated in different temperature ranges.

Amongst the most delicate ingredients to include in aseptic heterogeneous products are vegetables. For optimal nutritional retention, vegetables should be harvested at a predetermined ripeness and be processed immediately. Optimum textural properties can be obtained when vegetables are harvested before peak ripeness while maintaining flavor and color. In addition to the moment of harvest, the geographic location where the vegetable was grown, or the variety itself, can be optimized for either nutritional or organoleptic properties. Several treatments have been investigated to increase the firmness of vegetables, such as immersion in $Ca^{2+}$ salts (calcification), which reinforced the cellular structure of tomatoes and other delicate vegetables (Floros et al., 1992).

Though fresh ingredients generally perform better in aseptic processing from a quality point of view, these are not always available due to the seasonality of crops. Frozen vegetables are therefore widely used in aseptic processing to enable year round processing. However, in many cases, the quality of frozen vegetables is not on par with fresh vegetables. Prior to freezing, vegetables undergo a blanching step that inactivates microbes and enzymes and prevents color and nutrient degradation. At the same time, this process softens vegetable tissue that is aggravated by the water expansion in the subsequent freezing process. Upon thawing, the cellular damage sustained may cause significant loss of firmness, drip-losses occur, and the vegetables are more vulnerable to the detrimental impact of the heat treatment (Fellows, 1988). Vegetables can be conventionally frozen or of the higher individually quick-frozen quality grade, the latter being more expensive.

Finally, the overall quality of the product can be greatly improved by a careful dimensioning of the particles. Heating and cooling rate of particles and hence the minimum achievable intensity of the heat treatment strongly depend on the thermal path of the particle. If the thermal path is kept small but the particulates are long in the other dimensions (parallelepipeds or cylinders), the perceived size by the consumers can still remain quite large.

### 15.3.2 Raw Material Storage and Handling

Frozen raw materials may benefit from a tempering step to improve process throughput and product quality. When the thermal particle size or mass fraction is large, tempering and thawing may be required to meet safety and quality standards. Clumping of frozen particles and flotation in the cook vessel is a well-known phenomenon. Even when the ingredients are dumped slowly, large particle loadings (e.g., for rich formulations) will likely overstretch the heating duty of the cook vessels. Essentially, three methods are available for tempering and thawing:

1. Atmospheric tempering or exposure to (humid) air
2. Steam tempering
3. Volumetric tempering

Both atmospheric and steam temperings suffer from the conduction limitation for heat transfer, especially for large thermal path particles. In atmospheric tempering, the relatively small temperature difference between ingredient and environment further increases the required tempering time, which necessitates prohibitively large tempering areas, and cause potential microbiological issues. Steam tempering can be performed at higher driving forces, but care should be taken not to degrade the surface of the particles.

Volumetric tempering allows higher tempering rates and is available as microwave (915 MHz) or radio frequent (RF) technology (27.1 MHz). Both microwave and RF technology are readily available industrially and operational and cost aspects will be of decisive importance in selecting either.

### 15.3.3 Cook Vessel Operation

To minimize quality losses, the operating temperature of the cook vessels should be as low as possible. However, to allow activation of thickeners such as starch and to prevent growth of thermophilic spoilage organisms, the operating temperature should be above 70°C. This temperature must be well controlled, and the heating medium temperature must be minimized in order to limit heat damage. Using multiple small cook vessels and preventing stoppages also reduce the thermal load during preparation. To avoid burning of material around the free surface of the vessel's contents at partial loading or when draining, the heating jacket may consist of concentric surfaces that can be switched on or off according to demand.

### 15.3.4 Heating Stage

To maximize the heating rate of the product and minimize product holdup in the system, it is important to decrease the thermal path of the product flow. The heating rate approximately scales with the inverse of the square of the thermal path for viscous products. For viscous flows in tubular heaters, this can be achieved by decreasing the tube diameter to the smallest possible size (Section 3.4.3). However, the lower bound to the tube diameter is in practice given by the particle size, as tube blockage or particle damage due to shear must be avoided. As discussed in Section 3.2.1, conscious use of bends or employing continuously coiled heaters may also serve to reduce the thermal path.

A second method to improve heat transfer rates is the application of tube inserts or static mixers. Though static mixers are common in the chemical processing industry, they are less frequently used in the food processing industry. For heterogeneous flows, inserts or static mixers are unpractical as they obstruct the flow and may damage the particles. An exception to this may be the concept of the annulator. Annulators are inserts having channels of sufficient dimension to allow for the smooth passage of particles, while at the same time guiding the flow close to the tube axis to the wall, and vice versa (Figure 15.5). Annulators have been successfully applied to improve cooling and temperature uniformity in ice-cream processing, and they are also suitable for viscous particulate products with poor heat transfer characteristics.

Rapid heating of the liquid is only beneficial when heat transfer throughout the particles is not limited, a condition only satisfied for small particles. For large particles, heating rates should be limited to avoid a wide distribution of quality factors (Section 15.2.3). Improvement of (overall) heat transfer coefficients can still be beneficial; however, as the temperature differential between product and medium may be smaller at equal heating rate, which may reduce fouling and lower cost of energy.

From the above, it may have become clear that rapid heat transfer solutions such as steam injection or scraped surface heat exchangers (SSHEs) are not optimal in

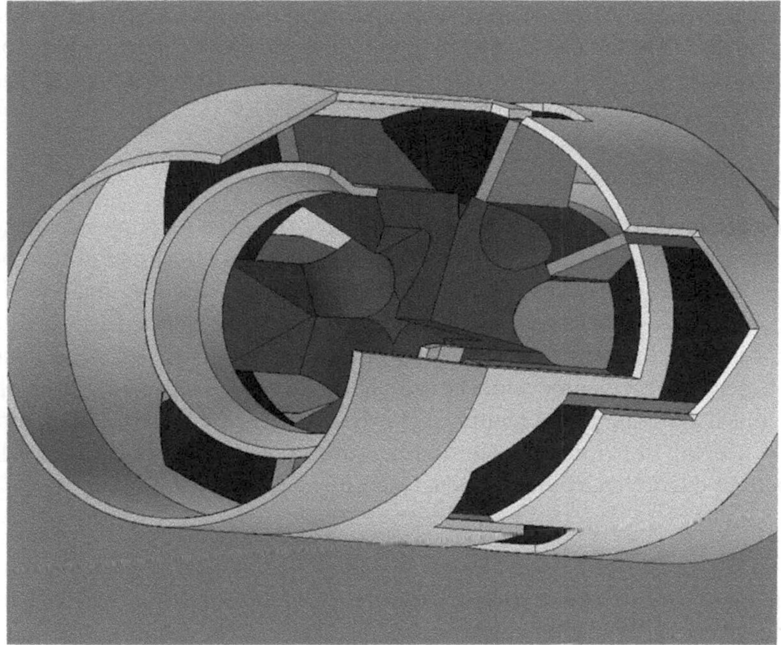

**FIGURE 15.5** Schematics of the annulator.

heterogeneous flows with particles exceeding certain dimensions. Of course such equipment may still be used to successfully heat (viscous) liquids, but in case of large particles these should have the opportunity to equilibrate at different heating temperatures. Ideally, the temperature differential should be constant, but Figure 15.3 shows that for the perfect heating case beyond three equilibration steps no significant improvements in textural quality are expected. Heat transfer scenarios closer to factory operating conditions (including finite heat transfer and particle slip) showed no qualitative differences to this scenario. Heating sections at three different heating medium temperatures would be equivalent to having three SSHEs at different temperatures followed by intermediate equilibration tubes.

### 15.3.5 Holding Tube

As discussed in Section 15.2.3, the maximum product temperature in the holding section is limited by several practical limitations. In addition, heat transfer resistances in the liquid and the particles necessitate optimization of the quality along the whole temperature treatment, while ensuring a sufficient $F_0$ in the holding tube (or in the high temperature section of the process if this can be justified). In practice, optimal liquid and particle core temperatures range between 125°C and 135°C, yielding significant quality benefits over conventional retorting.

Second, achieving a small residence time distribution of especially the particles in the heater and the holder enables a less conservative (that is smaller) sizing of the system. A previously proposed solution to give particles a sufficient and well-defined residence time was the Rota-Hold concept developed by Stork Food & Dairy systems (Holdsworth, 1992), but using coiled holding tubes (Sandeep et al., 2000) or holding tubes consisting of many bends to promote chaotic cross-sectional mixing (Kumar, 2007) may be employed to a similar effect.

### 15.3.6 Cooling Stage

To optimize product quality in the cooling stage, the same recommendations hold as in case of the heater (Section 15.3.4). As heat transfer at the end of the cooling stage is lowest as the viscosity of the liquid is generally highest, final cooling is critical, especially since the product liquid structure has fully formed and particulates have softened. Contrary to the heater, both heat transfer coefficients and temperature difference between product and medium should be maximized, for example, by using chilled water or glycol facilities. However, some caution should be exercised to avoid freezing or fat crystallization at the walls, or very nonuniform axial velocity distributions due to large differences in viscosities (Section 3.2.1).

### 15.3.7 Aseptic Tank

Aseptic tanks can be jacket cooled when the cooling capacity of the final in-line cooler is limited. In that case, incoming relatively warm product is mixed with cooled product in the tank, and the heat is transferred into the medium through the wall of the vessel. When the enthalpy entering with the product stream is removed by the jacket, the average temperature of the tank content is constant.

However, the thermal duty of jacketed aseptic vessels is generally less than that of cooking vessels, as the temperature between medium and vessel is smaller. To achieve a sufficient cooling capacity, it may be necessary to increase transfer area by adding cooling coils or internals to similar effect to the vessel.

### 15.3.8 Filler Operation and Packaging

The filler may impact on product quality by any shear damage occurring in its valves and flow passages. To avoid this, passages should allow for low shear conditions having smooth passages and sufficiently large cross sections.

Quality of finished heterogeneous product does not only hinge on particle integrity alone, but consumer appraisal is also important to achieve a well-defined and repeatable ratio of particulates to liquid in the pack. However, some limitations exist on the pack formats available for the aseptic filling of heterogeneous products: aseptic pouch filling for such products is currently unavailable.

Finally, the packaging impacts on the product quality throughout the shelf life. For a good and constant quality, it is essential that all packs are well sealed and prevent any quality deterioration by blocking light and oxygen ingress sufficiently.

### 15.3.9 Minimizing Shear Damage

Shear in continuous flows may affect particulates through a process of attrition and breakage, leading to a strong reduction in particle size and a clouding of otherwise clear broth. The rate of shear damage is maximal at the end of the heating, in the holding, and in the cooling sections when particulates have softened due to the thermal treatment. Components of a line need to be analyzed systematically to minimize the shear exerted on the particulates and the liquid structure. In practice, this means using short tubes of large diameters, avoiding sharp transitions in flow cross sections, and providing gentle agitation when required (e.g., in the aseptic tank). This is especially relevant when the particles have softened toward the end of the process.

Pumps and back-pressure devices require special attention in heterogeneous flows. Pressures in heterogeneous products may run as high as 30 bars or more and positive displacement pumps are required to deliver such heads. Rotary, reciprocal, screw, and lobe pumps can be used with different suppliers offering pumps for delicate particulates. Pump selection should also account for particle thermal history and the viscosity of the continuous phase.

Back-pressure devices are needed to maintain the pressure in the system and prevent boiling. For homogeneous products, spring valves are used that essentially generate a pressure drop over a narrow slit. For particulate products, this is obviously unsuitable, and back pressure can be generated having a second positive displacement pump at lower speed. Other options are especially designed using back-pressure devices (Cartwright, 2004) or using multiple aseptic tanks.

Aside from being part of a thermal optimization of the process, the aseptic tank can be used as a back-pressure device to avoid particle damage. In such operation, at least two aseptic tanks are operated in alternating mode under a sterile gas blanket (e.g., nitrogen). The first tank is filled up to the target level, upon which the flow is diverted to the next (pressurized) aseptic tank. Next, the pressure is released from the first tank and its content is fed into the filler. Though this mode of operation is more complicated in terms of control and hygiene, it eliminates the need for a back-pressure device that is incompatible with heterogeneous products.

### 15.3.10 Minimizing Product Oxidation

Oxidation is a well-known cause of quality deterioration, for example, vitamin C losses or the development of rancidity in products containing fats or oils. During cooking, oxygen may be entrained into the product, which may cause quality losses over the shelf life. A deaerator can be included in the design to remove most (dissolved) gasses.

A deaerator generally consists of a vessel in the 50–100 L range equipped with a level control and operated under a subatmospheric pressure (Figure 15.6). A feed pump following the cook vessel lowers the pressure prior to the deaerator. When a preheater is employed to enable a higher operating temperature in the deaerator, this feed pump is also used to overcome flow resistance in the preheater. A vacuum pump maintains a low pressure in the branch discharging dissolved gasses. Following the deaerator, the main pump increases the pressure to the level dictated by downstream flow resistance and system back-pressure. The product level in the deaerator is controlled by the feed pump.

The pressure–temperature combination in the deaerator is such that the liquid in it is essentially boiling. For temperature-equilibrated liquid–particle mixtures, this implies that boiling may occur both in the liquid as well in the particles and it should be

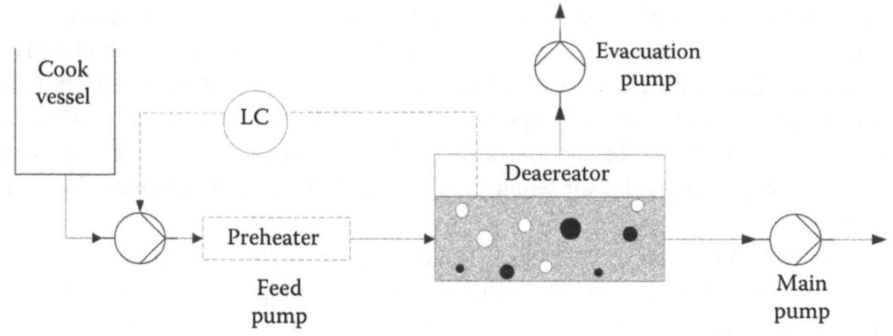


**FIGURE 15.6** Flow and control setup for the deaerator.

carefully checked not to interfere with particle integrity. Additionally, highly viscous products may impose limits to the temperature and pressure window due to minimal net positive suction head requirements for the main pump driving the flow.

Including a deaerator in the system also has implications for the downstream design. For deaeration to be meaningful, the aseptic vessels and the filling operation have to operate under inert gas. Though a deaerator adds complexity to the design of aseptic processes, it can improve quality upon storage of ambient stable products.

## 15.4 ASEPTIC COST OPTIMIZATION

Whereas Section 15.3 aimed to optimize from a quality perspective, this section is concerned with obtaining cost benefits by optimizing ingredients and production. Starting at global scale, we can optimize sourcing of some ingredients, whereas the availability of others may drive time and location of production. Second, at factory scale production scheduling may be of decisive importance for sound economics. Finally, proper control is instrumental to ensure safe processing, while at the same time creating conditions for uninterrupted production. Such run-length extension unlocks the full potential of the economies of scale provided by aseptic processing.

### 15.4.1 Optimized Ingredient Sourcing

Given the ingredient specifications, these may be obtained from various sources to obtain cost benefits. Nowadays, ambient stable ingredients like oils, flavors, powders, thickeners, and vegetable pastes are sourced on the global market. The same is true for frozen ingredients such as meats and vegetables, although sourcing may be more local due to the costs incurred by the frozen supply chain.

However, fresh ingredients such as dairy components and vegetables have a very limited shelf life even when kept refrigerated, and their use is generally constricted by geographic proximity. Though fresh produce is considerably cheaper than in the frozen state, seasonality of crops would limit production in factories to a few weeks per year with very low capital utilization and high product storage cost.

Clearly, this is a balance between capital expenditure, workforce utilization, cost of ingredients, and warehousing, so that production may (partially) be steered by availability of seasonal ingredients. However, for most aseptic factories, the main vegetable ingredients are used in the frozen state.

### 15.4.2 Factory-Wide Optimization

Aseptic processes may deliver a wide range of products in a variety of product formats. Products vary in color, viscosity, pH, and allergens. Especially allergens pose a challenge, as these are classified in groups that cannot be processed simultaneously. The time and effort it takes to change from one product to another depend on the difference in the specific characteristics (e.g., color or allergen). Especially after products containing allergens, a complete cleaned-in-place (CIP) cycle (including additional sterilization) may be required. In addition to changes in product, changes in product format can be time consuming as the aseptic filler and packing machine will require reassembly.

Generally, multiple manufacturing lines are located inside one factory, implying that lines share auxiliary resources. Typical examples of shared resources are the CIP/sterilize in place (SIP) system, waste treatment, and heating/cooling utilities. To reduce investment in auxiliaries, sizing is not based on peak demand: for example, the CIP system is not designed to provide CIP to all lines at the same time.

Optimizing the sequence of product and format on each line, and sharing auxiliary resources between lines can be achieved in two ways. The first option is to create sufficient slack on the production lines, at the cost of low capital utilization. The second option is to apply multistage scheduling to maximize the overall factory capacity. Bongers and Bakker (2006) applied the latter methodology to ice-cream manufacturing lines to the effect of significantly higher utilization and less unexpected shutdowns and product waste. This approach is expected to have a positive impact on industrial food manufacturing in general and will benefit most aspects of the product costs.

### 15.4.3 Optimal Control Strategies

The purpose of an aseptic control system is to maintain the specified time/temperature profiles despite disturbances and to ensure safe products (Hasting, 1992). A general controlled process setup is depicted in Figure 15.7. Disturbances acting upon the thermal process are

- Flow variations
- Fluctuations of the heating/cooling utilities
- Wearing of the control actuators and control valves (Ruel, 2000)
- Poor controller tuning (Buckbee, 2002)
- Fouling of the heat exchanger (both for product and medium side)

Apart from disturbances, bias in the temperature or flow sensors may result into lower lethal values. A measured temperature history is shown in Figure 15.8. Due to disturbances discussed before, considerable fluctuations are apparent.

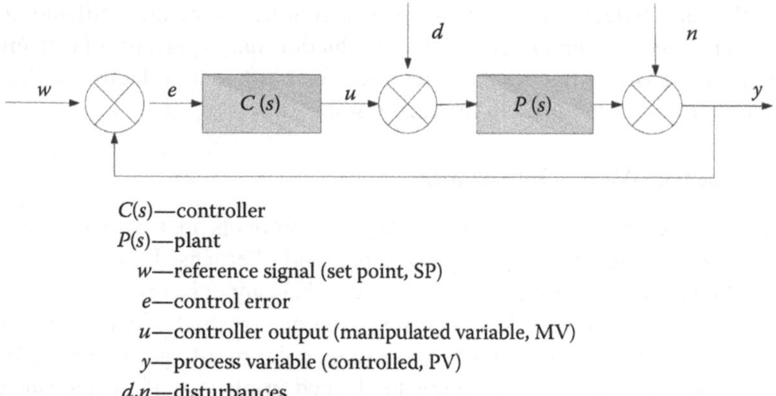


**FIGURE 15.7** Controlled process setup.

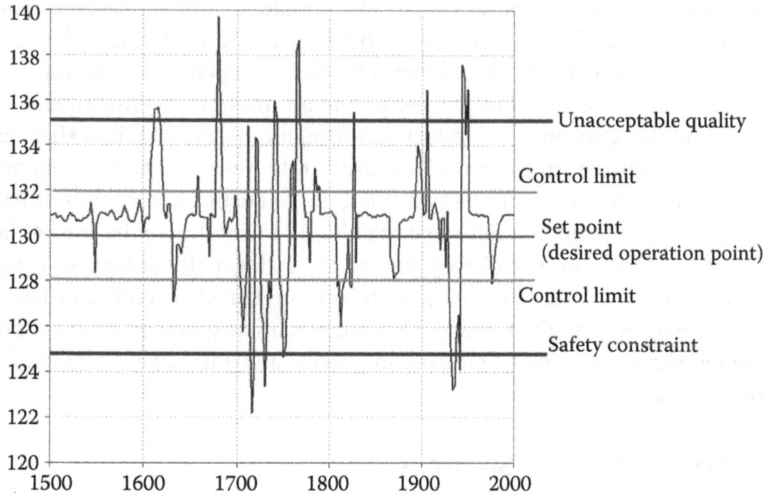


**FIGURE 15.8** Poorly controlled process before tuning.

Depending on these fluctuations, the temperature set point must be well above the safety constraint. As a consequence, the maximum temperatures occurring in the exchanger are also higher, leading to product deterioration and more rapid fouling.

To reduce temperature fluctuations, control is required during processing. A proportional–integral–derivative (PID) controller is the most popular feedback controller used in the process industries (Stephanopoulos, 1984). PID-controller parameters are

- P = proportional gain: feedback is proportional to the error measured
- I = integral time: eliminate offset between set point and process variable
- D = derivative time: reduce the time to reach the set point

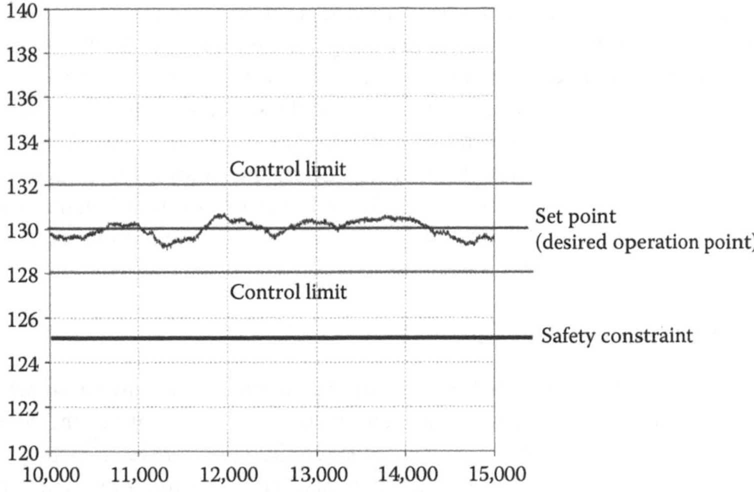


**FIGURE 15.9** Well-controlled process after tuning.

Tuning a control loop is the adjustment of its control parameters to values for the desired control response. This can be achieved most effectively using a process model (in some shape or form), and setting P, I, and D based on the model parameters. Manual tuning has proven inefficient, inaccurate, and often dangerous, and most modern industrial facilities use PID tuning and loop optimization software. These software packages will gather the data, apply process models, and suggest optimal tuning. Tuning the PID parameters for the process shown in Figure 15.8 resulted in stable operation (Figure 15.9).

One of the drawbacks of feedback control is that the deviations need to be substantial before the controller can respond. When a disturbance can be measured, feed-forward control (Figure 15.10) can compensate more rapidly and can thus be more effective.

Building on both concepts of feedback and feed-forward control, more advanced control strategies may be devised. Temporary failures downstream from the continuous part of the aseptic system (filler, packaging machines, etc.) cannot always be buffered by the aseptic tanks and the desired flow through the heat exchangers

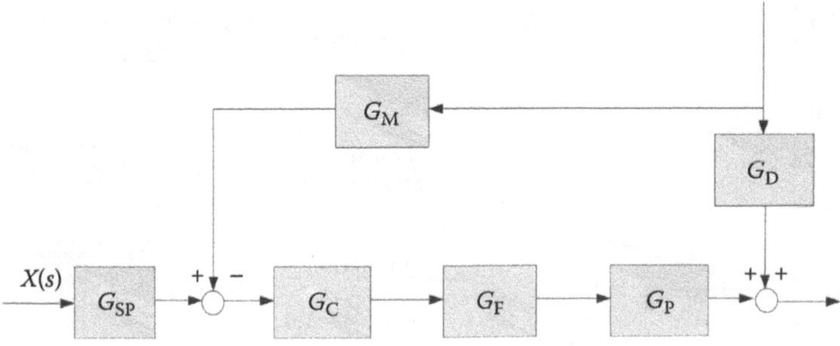


**FIGURE 15.10** Feed-forward control configuration.

may vary to prevent stoppage and product wastage. In modern aseptic systems, flow variations are possible within a limited range. However, when the temperature set points are not modified accordingly, under- or overprocessing results.

To overcome this, Negiz et al. (1996) and Schlesser et al. (1997) introduced the concept of multivariable control in HTST pasteurization systems to simultaneously control product flow (and hence residence time in the holding tube) and product temperature. This concept can be extended to maintain a constant lethality rate and optimize product quality attributes, and accommodate for production requirements (e.g., intermediate cleaning, filler availability).

### 15.4.4 Run-Length Extension

Fouling is unavoidable in thermal processing, but it should be minimized wherever possible. Deposits on the walls of the heat exchangers either affect the hydraulic performance of the equipment by increasing the required pressure head, or they diminish the efficiency of heat transfer. Fouling may also present a microbiological hazard in the accumulation of materials in nonhygienic parts of the line, or spoil product when large deposits are removed from the wall and are carried with the product.

Fouling is the result of a combination of several mechanisms such as crystallization, corrosion, caramelization, polymerization, Maillard reactions, and protein denaturation, all greatly accelerated at higher temperatures. Given the variety and complexity of the food materials, fouling can be a complex residue composed of carbohydrates, proteins, fats, and minerals, and the composition and properties of the deposits are very different from those of the food material.

A fouling process generally has the following stages: induction, transport, attachment, buildup, and aging. These steps can be monitored through the pressure drop on the product side, or the temperature (or pressure) of steam as a result of the reduced heat transfer. During induction and attachment very little changes can be observed, while during buildup and aging these changes can be observed easily (Figure 15.11).

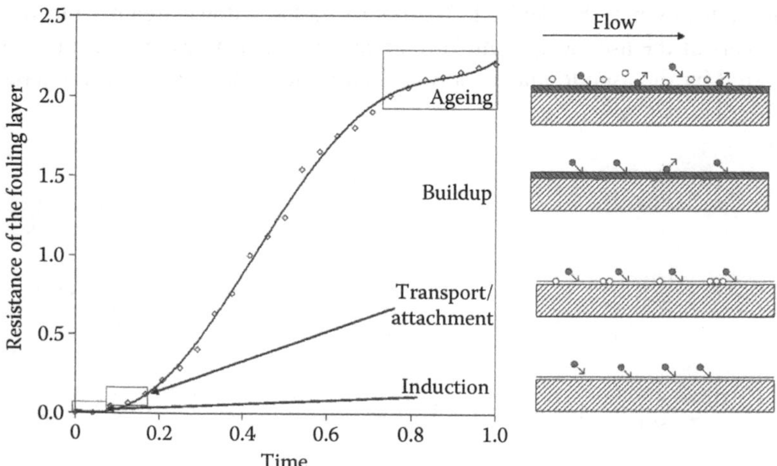


**FIGURE 15.11** Resistance of the fouling layer over time.

Regular cleaning of aseptic lines is required to remove the fouling that has accumulated and return the system to the initial state. Normally, limits on pressure drop and steam temperature trigger the shutdown of production and the start of a cleaning cycle. Modern plant equipment can be CIP where equipment is cleaned either by flow of the cleaning solutions or spraying such chemicals onto walls (e.g., cook vessel). However, some equipment requires opening and (manual) cleaning. CIP requires a parallel system to convey cleaning solutions and rinsing water.

Cleaning liquids include caustic, acid, and detergent solutions applied in different order depending on the fouling deposits. Milk protein fouling has been studied extensively and the cleaning of such deposits is performed by water rinse, caustic cleaner, acid rinse, and water rinse. Cleaning is performed in that order since the proteins are more susceptible to an alkali environment, whereas the salt deposits dissolve in acids. Other products may require different cleaning regimes since their deposits are composed of different elements (Gillham et al., 1999; Bansal and Chen, 2006).

Cleaning cycles are optimized in terms of utilization of cleaning solutions, energy usage, and water consumption, but the interruption of production remains costly both in terms of lost production time, waste, and energy. Though cleaning chemicals may be reused to save cost and protect the environment, they represent a potentially hazardous waste stream. Cleaning in principle solves the problem of fouling, but preventing it should have a high priority.

Fouling can be reduced by adequate setting of the operating parameters. This includes a moderate and approximately constant temperature differential between the product and the medium side. For this reason, countercurrent pressurized water is mostly used as the heating medium where direct application of steam would lead to burn on of the product. It is also vitally important that the process is well controlled, as large temperature fluctuations imply high maximum temperatures (Section 15.4.3), and can cause severe fouling. In some cases, direct (steam) heating may be applicable, as the heat is transferred directly into the bulk of the product by condensation. However, steam-injection points should be carefully designed and well insulated to avoid direct contact of the product with very hot surfaces associated with the steam supply.

Flow velocity is also an important factor, as an increase in velocity generally decreases the rate of fouling by increasing the turbulence and preventing that the fouling deposits stick to the wall. Second, minimizing the heat transfer resistance in the liquid as compared to the tube wall decreases the temperature at the metal–product interface, which decreases fouling rates. When fouling is severe, and other solutions do not apply, use of an agitated heat exchanger (SSHE) may be considered. Such heat exchangers permanently remove deposits from the walls, and can keep the heat transfer rate sufficiently high for longer times, provided the deposits do not impair final product quality.

Careful formulation, selection, and pretreatments of the ingredients may also mitigate fouling. Dairy ingredients may be replaced by oils and proteins having a vegetal origin, or by denatured versions of those ingredients such as pretreated (e.g., pasteurized) cream to inactivate the main fouling component β-lactoglobulin (Bansal and Chen, 2006). Protein-rich foods can be pretreated to prevent fouling, and starches and gums can be prehydrated and precooked to the desired values of viscosity (Lelieveld et al., 2005).

An aseptic process line can also be designed to minimize the severity or impact of fouling. An example of the first approach is to include intermediate holding tubes where fouling components (such as certain dairy proteins) are given a residence time to inactivate or denature (De Jong, 1997). The latter approach can be embodied in the design of a heat exchanger having a sufficiently large surface area. Deposits will now be spread over a larger surface area, and the run length can be extended before cleaning is required.

Finally, networks of heat exchangers are used in the sugar industry, where fouling is so severe that in intermediate, in-production cleaning is scheduled to certain portions of the heat exchanger network (Smaili and Angadi, 1999).

## 15.5 FUTURE TRENDS

Future trends in aseptic processing of (heterogeneous) aseptic liquid food products can also be grouped according to quality and cost optimization. To improve product quality, novel technologies such as volumetric heating (refer to Chapter 4) may bring significant benefits. Fruitful exploitation of such technologies may depend on a rigorous redesign of the current once-through processes, which should be driven by the required properties of the final product as received by the consumer. Examples could include split-stream processes, where the heat treatment could be optimized for the different product components, which could be aseptically assembled in the final pack.

From a cost optimization perspective, it is expected that production will be increasingly global and may shift between production locations to follow seasonal supply of raw materials. Alternatively, production facilities may acquire mobility to maximize capital utilization. Finally, though economy of scale is one of the selling points of aseptic processing over retorting, current trends indicate a demand for smaller economic production run-lengths to cater for shifting consumer preferences. Ideal aseptic systems should therefore combine the product quality of aseptic processing with the flexibility of the traditional retorting.

## 15.6 CONCLUDING REMARKS

Considerable challenges remain in exploiting the full potential of aseptic processing of heterogeneous foods. More accurate modeling tools are required to optimize the thermal profiles imposed upon products, and these should blend in seamlessly with the design and operation of aseptic lines. For mild sterilization, it is essential that process design is not based on the most conservative assumptions. Lethality in heater and cooler should be accounted for where appropriate, and $F_0$ should be based on realistic combinations of residence time and temperature distribution. The use of more robust ingredients, emerging heating technologies, and more complex process flow-sheets is expected to further boost organoleptic and nutritional quality.

At the same time, product cost can be minimized by global sourcing strategies and the use of ambient stable aseptic ingredients. Production can be scheduled on a global scale to leverage cost and quality benefits of fresh produce, or alternatively a portable factory that could follow seasonality. In all cases, it is important to put emphasis on flexibility of the aseptic plant early on in the design stage, as current trends toward smaller product batch sizes require aseptic quality at the ease of retorting.

However, industrial aseptic processing will always be a compromise between cost and quality, but the future has never looked brighter to push the envelope further toward quality.

## ACKNOWLEDGMENTS

The authors would like to thank Pierre Debias, Gary Mycock, Hans Hoogland, David Dearden, Roger Jordan, and Annemarie Elberse.

## NOMENCLATURE

| | | |
|---|---|---|
| $A$ | m$^2$ | surface area |
| $C$ | min | cook value |
| $Cp$ | J/kg °C | specific heat capacity |
| $D$ | min | $D$-value |
| $E_a$ | J/mol | activation energy |
| $F_0$ | min | equivalent process time at 121.1°C |
| $h$ | W/m$^2$ °C | heat transfer coefficient |
| $L_m$ | | length of section |
| $l_f$ | M | tube perimeter |
| $n_p$ | | number density of particles |
| $R_r$ | | relative reaction rate |
| $r_m$ | | radial position in Particle |
| $t$ | s | time |
| $T$ | °C | temperature |
| $T_r$ | °C | reference temperature |
| $T_s$ | °C | starting temperature of phase transition |
| $T_e$ | °C | end temperature of phase transition |
| $U$ | W/m$^2$ °C | overall heat transfer coefficient |
| $w$ | m/s | local axial velocity |
| $z$ | m | streamwise distance |
| $Z$ | °C | $Z$-value |
| $\Delta H$ | J/kg | enthalpy of melting |
| $\lambda$ | W/m °C | thermal conductivity |
| $\rho$ | kg/m$^3$ | density |

*Subscripts*

| | | |
|---|---|---|
| F | | of the fluid |
| M | | of the medium |
| P | | of the product/particle |

*Superscripts*

| | | |
|---|---|---|
| s | | surface of the particle |

## REFERENCES


G.B. Awuah, A. Economides, B.D. Shafer, and J. Weng. Lethality contribution from the tubular heat exchanger during high-temperature short-time processing of a model liquid food. *J. Food. Proc. Eng.*, 27:246–266, 2004.

B. Bansal and X.D. Chen. Effect of temperature and power frequency on milk fouling in an ohmic heater. *Food Bioprod. Process.*, 84(C4):286–291, 2006.

M. Barigou, S. Mankad, and P.J. Fryer. Heat transfer in two-phase solid-liquid food flows: A review. *Trans. IChemE*, Part C, 76:3–29, 1998.

S. Bhamidipati and R.K. Singh. Thermal time distributions in tubular heat exchangers during aseptic processing of fluid foods. *Biotechnol. Prog.*, 10:230–236, 1994.

R.B. Bird, W.E. Stewart, and E.N. Lightfoot. *Transport Phenomena*. John Wiley & Sons, New York, 1960.

P.M.M. Bongers and B.H. Bakker. Application of multi-stage scheduling. *ESCAPE 16 Proceedings*, Garmish, Germany, 2006.

L.M. Braud, M.E. Castell-Perez, and M.D. Matlock. Risk-based design of aseptic processing of heterogeneous food products. *Risk Anal.*, 20(4):405–412, 2000.

G. Buckbee. Poor controller tuning drives up valve costs. *IEEE Contr. Syst. Mag.*, 15:47–51, 2002.

D. Cacace, L. Palmieri, G. Pirone, G. Dipollina, P. Masi, and S. Cavella. Biological validation of mathematical modeling of the thermal-processing of particulate foods—The influence of heat-transfer coefficient determination. *J. Food Sci.*, 23(1):51–68, 1994.

G.D. Cartwright. Apparatus and method for controlling flow of process materials. Patent No. US2004/001335, 2004.

P.J. Fellows. *Food Processing Technology, Principles and Practice*. Woodhead Publishing Ltd., Cambridge, UK, 1988.

J.D. Floros, A. Ekanayake, G.P. Abide, and P.E. Nelson. Optimization of a diced tomato calcification process. *J. Food Sci.*, 57(5):1144–1148, 1992.

C.R. Gillham, P.J. Fryer, A.P.M. Hasting, and D.I. Wilson. Cleaning in place of whey protein deposits: Mechanisms controlling cleaning. *Trans. IChemE*, Part C, 77:127–136, 1999.

A.P.M. Hasting. Practical considerations in the design, operation and control of food pasteurization processes. *Food Control*, 3:7–32, 1992.

S.D. Holdsworth. *Aseptic Processing and Packaging of Food Products*. Elsevier Applied Science, Amsterdam, The Netherlands, 1992.

P. de Jong. Impact and control of fouling in milk processing, *Trends Food Sci. Tech.*, 8(12):401–405, 1997.

A. Jung and P.J. Fryer. Optimising the quality of safe foods: Computational modelling of a continuous sterilisation process. *Chem. Eng. Sci.*, 54:717–730, 1999.

J.D.H. Kelder, K.J. Ptasinski, and P.J.A.M. Kerkhof. Power-law foods in continuous coiled sterilisers. *Chem. Eng. Sci.*, 57:4605–4615, 2002.

A. Kumar and M. Bhattacharya. Numerical analysis of aseptic processing of a non-Newtonian liquid food in a tubular heat exchanger. *Chem. Eng. Commn.*, 103:27–51, 1991.

V. Kumar and K.D.P. Nigam. Laminar convective heat transfer in chaotic configuration. *Int. J. Heat Mass Transfer*, 50(13–14):2469–2479, 2007.

H.L.M. Lelieveld, M.A. Mostert, and J. Holah (Eds.). *Handbook of Hygiene Control in the Food Industry*. Woodhead Publishing Ltd., Cambridge, UK, 2005.

H.-J. Liao, M.A. Rao, and A.K. Datta. Role of thermo-rheological behaviour in simulation of continuous sterilization of starch dispersion. *Trans. IChemE*, Part C., 78:48–56, 2000.

S. Mankad and P.J. Fryer. A heterogeneous flow model for the effect of slip and flow velocities on food steriliser design. *Chem. Eng. Sci.*, 52(12):1835–1843, 1997.

S. Mankad, C.A. Branch, and P.J. Fryer. The effect of particle slip on the sterilisation of solid-liquid food mixtures. *Chem. Eng. Sci.*, 50(8):1323–1336, 1995.

A. Mateu, F. Chinesta, M.J. Ocio, M. Garcia, and A. Martinez. Development and validation of a mathematical model for HTST processing of foods containing large particles. *J. Food Protect.*, 60(10):1224–1229, 1997.

A. Negiz, A. Cinar, J.E. Schlesser, P. Ramanauskas, D.J. Armstrong, and W. Stroup. Automated control of high temperature short time pasteurization, *Food Control*, 7(6):309–315, 1996.

T.K. Palazoglu and K.P. Sandeep. Assessment of the effect of fluid-to-particle heat transfer coefficient on microbial and nutrient destruction during aseptic processing of particulate foods. *J. Food Sci.*, 67:3359–3364, 2002.

P. Richardson, *Improving the Thermal Processing of Foods.* Woodhead Publishing Ltd., Cambridge, UK, 2004.

M. Ruel. How valve performance affects the control loop. *Chem. Eng. Mag.*, 107(10):13–18, 2000.

K.P. Sandeep, C.A. Zuritz, and V.M. Puri. Determination of lethality during aseptic processing of particulate foods. *Trans. IChemE*, Part C, 77:11–17, 1999.

K.P. Sandeep, C.A. Zuritz, and V.M. Puri. Modelling non-Newtonian two-phase flow in conventional and helical-holding tubes. *Int. J. Food Sci. Technol.*, 35:511–522, 2000.

S.K. Sastry and B.D. Cornelius. *Aseptic Processing of Foods Containing Solid Particulates.* John Wiley & Sons, New York, 2002.

J.E. Schlesser, D.J. Armstrong, A. Cinar, P. Ramanauskas, and A. Negiz. Automated control and monitoring of thermal processing using high temperature, short time pasteurisation. *J. Diary Sci.*, 80:2291–2296, 1997.

S.G. Simpson and M.C. Williams. An analysis of high temperature/short time sterilisation during laminar flow. *J. Food. Sci.*, 39:1047–1054, 1974.

C. Skjöldebrand and T. Ohlsson. A computer simulation program for evaluation of the continuous heat treatment of particulate food products. Part 1: Design. *J. Food Eng.*, 20(2):149–165, 1993a.

C. Skjöldebrand and T. Ohlsson. A computer simulation program for evaluation of the continuous heat treatment of particulate food products. Part 2: Utilization. *J. Food Eng.*, 20(2):167–181, 1993b.

F. Smaili, D.K. Angadi, C.M. hatch, O. Herbert, V.S. Vassiliadis, and D.I. Wilson. Optimization of scheduling of cleaning in heat exchanger networks subject to fouling: Sugar industry case study. *Food Bioprod. Process.*, 77(C2):159–164, 1999.

G. Stephanopoulos. *Chemical Process Control, an Introduction to Theory and Practice.* Prentice-Hall, New York, 1984.

R. Toledo. *Fundamentals of Food Process Engineering*, AVI Publishing, New York, 1980.

Z. Weng, Aseptical™ software—a mathematical modeling package for multiphase foods in aseptic processing systems. *Advanced Aseptic Processing and Packaging*, Department of Food Science and Technology, UC Davis, Davis, CA, 1999.

L. Zhang and P.J. Fryer. A model for conduction heat transfer to particles in a hold tube using a moving mesh finite element method. *J. Food Eng.*, 26:193–208, 1995.

# 16 Optimizing Plant Production in Batch Thermal Processing: Case Study

*Ricardo Simpson and Alik Abakarov*

## CONTENTS



## 16.1 INTRODUCTION

Thermal processing is an important method for food preservation in the manufacture of shelf-stable canned foods, and it has been the cornerstone of the food processing industry for more than a century (Teixeira, 1992). The basic function of a thermal process is to inactivate food spoilage microorganisms in sealed containers of food using heat treatments at temperatures well above the ambient boiling point of water in pressurized steam retorts (autoclaves). Excessive heat treatment should be avoided because it means a detrimental effect to food quality and underutilization of plant capacity (Simpson et al., 2003a).

Thermal process calculations in which process times at specified retort temperatures are calculated in order to achieve safe levels of microbial inactivation (lethality) should be carefully carried out to ensure public health safety. Therefore, the accuracy of methods used for this purpose is highly relevant for food science and engineering professionals working in this field (Holdsworth, 1997).

The first procedure to calculate thermal processes was developed by W.D. Bigelow early in the twentieth century, and it is usually known as the General Method (Bigelow et al., 1920). The General Method makes direct use of the time–temperature history at the coldest point in order to obtain the lethality value of a process.

On the other hand, Formula Method (F.M.) works differently from the General Method. It makes use of the fact that the difference between retort and cold spot temperature decreases exponentially over process time after an initial lag period. Therefore, a semilogarithmic plot of this temperature difference over time (beyond the initial lag) appears as a straight line that can be mathematically described by a simple formula as well as related to lethality requirements by a set of tables that should be used in conjunction with the formula.

The referred procedures have the limitation providing safe information (process time and temperature) only to the recorded data at the experimental conditions. According to Simpson et al. (2003b), the Revisited General Method (RGM) was rather accurate when used for thermal process adjustment. The processing time was always estimated (overestimated) with an error under 5% (in all cases under study). For the F.M., it was common to find errors around 10%–20% or over. The new procedure safely predicted shorter process times than those predicted by the F.M. These shorter process times may have an important impact on product quality, energy consumption, plant production capacity, and adequate corrections for online control (process deviations).

According to Simpson et al. (2003b), the RGM has the capability to generate the isolethal processes from a single heat-penetration test. As described in Simpson et al. (2003a) when the isolethal processes have been generated, then utilizing the cubic spline interpolation procedure, it is possible to attain a continuous function. Natural cubic splines are used for creating a model that can fill in the holes between data, in effect, approximating a trend. They are, therefore, useful for making observations and inferences about an existing pattern in the data (Atkinson, 1985), a piecewise technique which is very popular as an interpolation tool.

Batch processing with a battery of individual retorts is a common mode of operation in many food-canning plants (canneries). Although high speed processing with continuous rotary or hydrostatic retort systems can be found in very large canning factories (where they are cost-justified by high volume throughput), such systems are not economically feasible in the majority of small- to medium-sized canneries (Norback and Rattunde, 1991). In such smaller canneries, retort operations are carried out as batch processes in a cook room in which the retort battery is located. Although the unloading and reloading operations for each retort are labor intensive, a well-designed and managed cook room can operate with surprising efficiency if it has the optimum number of retorts and the optimum schedule of retort operation.

This type of optimization in the use of scheduling to maximize efficiency of batch processing plants has become well known, and it is commonly practiced in many process industries. Several models, methods, and implementation issues related to this topic have been published in the process engineering literature (Barbosa and Macchietto, 1993; Kondili et al., 1993; Rippin, 1993; Lee and Reklaitis, 1995a,b;

Reklaitis, 1996). However, specific application to retort batteries in food-canning plants has not been addressed in the food process engineering literature. Food canneries with batch retort operations are somewhat unique, in that the cannery process line as a whole is usually a continuous process in that unit operations both upstream and downstream from the retort cook room are normally continuous (product preparation, filling, closing, labeling, case packing, etc.). Although retorting is carried out as a batch process within the cook room, unprocessed cans enter and processed cans exit the cook room continuously at the same rate. Since the entire process line operates continuously, food canneries are often overlooked as batch process industries.

Recent studies have proposed to operate autoclaves with several products at the same time (Simpson, 2005). This operation mode has been called as simultaneous sterilization. According to Simpson (2005), simultaneous sterilization applies, mainly, to the case of small canneries with few retorts that are frequently required to process small lots of different products in various container sizes that normally require different process times and retort temperatures. In these situations, retorts often operate with only partial loads because of the small lot sizes, and they are severely underutilized.

Over the last few years, many papers have been published where the mixed integer linear programming (MILP) have been successfully applied to find optimal scheduling. Examples of sequential MILP short-term scheduling models can be found in Méndez and Cerda (2000, 2002), Harjunkoski and Grossmann (2002), and Castro and Grossmann (2006). In the following research papers, sequence-based MILP models for multiproduct batch plants were presented: Jung et al. (1994), Moon et al. (1996), Castro and Grossmann (2005), Floudas and Lin (2004), Gupta and Karimi (2003), Maravelias (2006), Mendez et al. (2001, 2006), Ha et al. (2006), and Liu and Karimi (2008). An interesting application of MILP planning for a petrochemical plant has been described in Hui and Natori (1996). Erdirik-Dogan and Grossmann (2007) presented a multiperiod MILP model for the simultaneous planning and scheduling of single-stage multiproduct continuous plants with parallel units.

It is well known that in the MILP problem, it is necessary to restrict the decision variables of linear programming models to integer or binary values, for example, the decision variables represent a nonfractional entity such as people. As an example in the case of the sterilization problem, the MILP model fractional variables correspond to the amount of sterilizing products on some batch and nonfractional (binary) variable to simultaneous sterilization vectors.

Usually the MILP problems are much harder to solve than the pure linear programming problems, because the feasible region of MILP problems does not allow the application of the well-known simplex algorithm, except the combination of simplex algorithm (to solve the linear relaxation of MILP problem which is also called subproblem) with other algorithms, for example, branch-and-bound techniques, until the linear programming (LP) relaxation will not compute necessary solutions from the initial problem.

In this chapter, the main focus will be to solve two optimal scheduling sterilization problems, where, for the first one, a given amount of different canned food products, with specific quality requirement, should be sterilized for a minimal processing time, and for the second one, for a given processing time, the sterilization of different canned food products should be maximized. The two MILP problems were solved by "lp_solve" version 5.5.0.10* based on the revised simplex method and the branch-and-bound method for the nonfractional (binary) variables, and also by using online the Web Service of COIN-OR† (COmputational INfrastructure for Operations Research): Coin Branch and Cut Solver on NEOS Server version 5.0 using mathematical programming system (MPS) input.

## 16.2 METHODOLOGY

Like any another process, food sterilization of different products can be optimized. The optimization in this case means that for a given amount of each product and plant equipment (number and capacity of retorts), it is necessary to find the best scheduling which allows sterilizing all of given products in a minimum processing time.

The developed mathematical model, based on the possibility of simultaneous sterilization, has great potential in the case of small canneries with few retorts which are frequently required to process small amounts of different products in various container sizes which normally require different processing times and retort temperatures (Simpson, 2005).

In this situation, retorts often operate with only partial loads because of the small lot sizes, and they are normally underutilized. The simultaneous sterilization possibility has the advantage of the fact that, for any given product and container size, there is a number of alternative combinations for retort temperature (above the lethal range) and corresponding processing time that will deliver the same lethality ($F_0$ value) (Simpson, 2005). These can be called isolethal processes (Holdsworth and Simpson, 2007).

Important to this study is the fact (Simpson, 2005) that the differences found in the absolute level of quality retention were relatively small over a practical range of isolethal process conditions. This relative insensitivity of quality over a range of different isolethal process conditions opens the door to maximizing output from a fixed number of retorts for different products and container sizes. Isolethal processes can be identified for each of the various products, from which a common set of processes and conditions can be chosen for simultaneous sterilization of different product lots in the same retort.

### 16.2.1 MATERIALS

The following 16 canned food products were selected for heat-penetration tests and development of isolethal process conditions (Table 16.1):

- Peas (*Pisum sativum*)
- Corn (*Zea mays*)
- Asparagus (*Asparagus officinalis*)
- Green beans (*Phaseolus vulgaris*)

The referred vegetable products were selected for solving the simultaneous sterilization problem utilizing the developed MILP model.

* http://lpsolve.sourceforge.net/5.5/

† http://neos.mcs.anl.gov/neos/solvers/milp:Cbc/MPS.html

**TABLE 16.1**
**Products and Can Sizes Selected for this Research Study**

| | Can Size | | | | |
|---|---|---|---|---|---|
| **Product** | **211 × 400** | **300 × 407** | **307 × .409** | **307 × 113** | **401 × .411** |
| Asparagus | √ | √ | √ | √ | √ |
| Corn | √ | √ | √ | | √ |
| Green beans | √ | √ | √ | | √ |
| Peas | √ | | √ | | √ |

*Source:* Simpson, R. and Abakarov, A., *J. Food Eng.*, 90, 53, 2009. With permission.

### 16.2.2 Equipment

- Vertical Autoclave (static), Eclipse Lookout
- Digital Balance, Hispanic Precision
- Vertical Igniotubular Boiler, Eclipse Lookout
- Thermocouples, Ecklund-Harrison, model CNS and C-5
- Data logger, Ellab model CTF-84 (1984)

### 16.2.3 Heat-Penetration Tests for Products under Study

Heat-penetration experiments at different retort temperatures were conducted to examine the nature of the heat-penetration curves (center temperature histories). Experiments were carried out in triplicate, and thermocouples were located at the slowest heating point. In all cases under study, the center of the can was a representative location of the slowest heating point. The retort heating profile used consisted of an initial equilibrium phase at 20°C, followed by a linear coming-up-time (C.U.T.) of 7–9 min to accomplish venting, and holding phase at the specific retort temperature (TRT) over the calculated process time required to achieve the target lethality for the specific product under thermal processing.

### 16.2.4 Methods

#### 16.2.4.1 Simultaneous Sterilization Characterization

In terms of analysis, a range of isolethal processes for selected products and container sizes should be obtained from experimental work. Heat-penetration tests should be conducted on each product in order to establish process time at a reference retort temperature to achieve target lethality ($F_0$ values). A computer program was utilized to obtain the equivalent lethality processes according to the following specifications:

- Two $F_0$ values should be considered for each product ($F_{0\,\min}$ and $F_{0\,\max}$). The referred values are product-related, but in general, $F_{0\,\min}$ is chosen according to a safety criterion and $F_{0\,\max}$ according to a quality criterion.
- For each $F_0$ value ($F_{0\,\min j}$ and $F_{0\,\max j}$), isolethal processes at retort temperatures of $TRT_1$, $TRT_2$, $TRT_3$, …, $TRT_N$ should be obtained for each product.

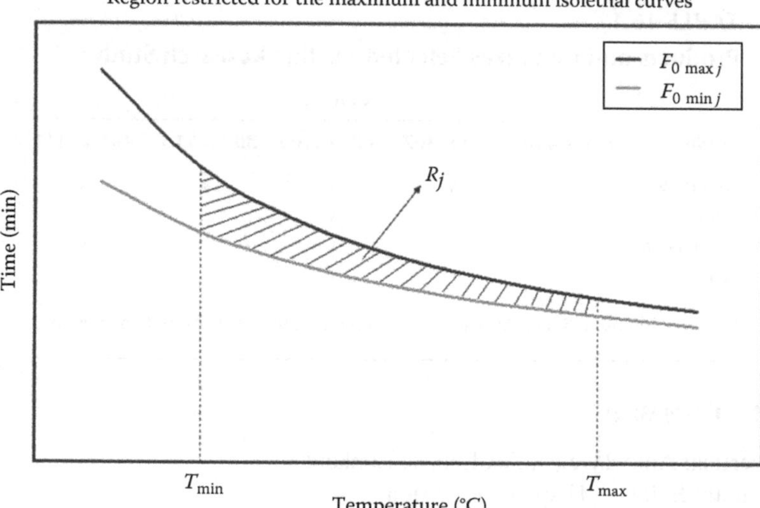


**FIGURE 16.1** Region restricted for the maximum and minimum isolethal curves ($F_{0\ \min j}$ and $F_{0\ \max j}$) for the *j–n*th product. (From Simpson, R. and Abakarov, A., *J. Food Eng.*, 90, 53, 2009. With permission.)

- Discrete values that define each process per product at different temperatures will be transformed as a continuous function through the cubic spline procedure (for both $F_{0\ \min}$ and $F_{0\ \max}$, per product), obtaining a set of two continuous curves per product (Figure 16.1).

In addition, the following criteria were established for choosing the optimum set of process conditions for simultaneous sterilization of more than one product:

- Total lethality achieved for each product must be equal or higher than the preestablished $F_{0\ \min}$ value for that specific product.
- Total lethality for each product must not exceed a preestablished maximum value ($F_{0\ \max}$) to avoid excessive overprocessing.

### 16.2.4.2 Cubic Splines

Natural cubic splines are used for creating a smooth function that can fill in the gaps between data points in a stepwise discrete function to approximate a smooth trend. They are useful for making observations and inferences about a pattern existing in the data (Atkinson, 1985). The philosophy in splining is to use low order polynomials to interpolate from grid point to grid point. This is ideally suited when one has control of the grid locations and the values of the data being interpolated, as in the work presented here. The method can be described briefly as follows.

Given a data set $\{(x_i, y_i)\}_{i=0}^{n}$, where $x_0 < x_1 < \cdots < x_n$, the function $S(x)$ is called a cubic spline if there exists $n$ cubic polynomials $s_i(x)$ with coefficients $s_{i,0}(x)$, $s_{i,1}(x)$, $s_{i,2}(x)$, $s_{i,3}(x)$ that satisfy the following properties:

1. $S(x) = S_i(x) = s_{i,0}(x) + s_{i,1}(x - x_i) + s_{i,2}(x - x_i)^2 + s_{i,3}(x - x_i)^3, x \in [x_i, x_{i+1}]$, $i = 0,1,\ldots,n - 1$.
2. $S(x) = y_i$, $i = 0,1,\ldots,n$. The spline passes through all given data points with a unique function between each set of data points.
3. $S_i(x_{i+1}) = S_{i+1}(x_{i+1})$, $i = 0,1,\ldots,n - 2$. The spline forms a continuous function over $[x_0, x_n]$.
4. $S_i'(x_{i+1}) = S_{i+1}'(x_{i+1})$, $i = 0,1,\ldots,n - 2$. The spline forms a smooth function.
5. $S_i''(x_{i+1}) = S_{i+1}''(x_{i+1})$, $i = 0,1,\ldots,n - 2$. The second derivative is continuous.

The function $S(x)$ called "natural cubic spline" must follow $S''(x_0) = S''(x_n) = 0$.

#### 16.2.4.3 Mathematical Formulation for Simultaneous Sterilization

Let us assume we have $n \geq 2$ products, say $P_1,\ldots,P_n$, which are processed at the plant location. Let us consider the index set $X = \{1,2,\ldots,n\}$.

Considering the temperature interval $[T_{\min}, T_{\max}]$, which denotes the temperature capabilities of the process.

Each product $P_j$, for $j \subset X$, has attached two strictly decreasing continuous functions, say

$$m_j, M_j : [T_{\min}, T_{\max}] \to (0, +\infty),$$

where $m_j(T) \leq M_j(T)$, for each $T \in [T_{\min}, T_{\max}]$. The meaning of $m_j(T)$(respectively, $M_j(T)$) is the minimum time (respectively, maximum time) needed to process the product $P_j$ at temperature $T$.

Defining the region

$$R_j : R_j = \{(T,t) : T_{\min} \leq T \leq T_{\max}, m_j(T) \leq t \leq M_j(T)\}.$$

The interpretation of $R_j$ is that the product $P_j$ can be processed at temperature $T$ with time $t$ if and only if $(T, t) \in R_j$. It is clear that a subcollection of products will be $P_{j1},\ldots, P_{jr}$, where $1 \leq j_1 \leq j_2 \leq \cdots \leq j_r \leq n$ can be simultaneously processed at temperature $T$ and time $t$ if and only if $(T, t) \in R_{j1} \cap R_{j2} \cap \cdots \cap R_{jr}$.

Then, obtaining all possible subcollection of products, which can be simultaneously processed, is equivalent to finding all possible subsets $Q = \{j_1,\ldots, j_r\} \subset X$, $r > 0, 1 \leq j_1 \leq j_2 \leq \cdots \leq j_r \leq n$. For which it holds that $I_Q = R_{j1} \cap R_{j2} \cap \cdots \cap R_{jr} \neq \emptyset$.

#### 16.2.4.4 Computational Procedure

In the practical sense, we have the products $P_1,\ldots, P_n$ and the temperature interval $[T_{\min}, T_{\max}]$.

**Computational procedure A**

1. We choose a positive integer $l \in \{1, 2, 3, \ldots\}$ and a partition $P = \{T_0 = T_{\min}, T_1, \ldots, T_l = T_{\max}\}$, where $T_m < T_{m+1}$ for $m = 0, \ldots, l-1$.
2. For each product $P_j$ we compute the values $m_j(T_m)$, $M_j(T_m)$, $m = 0, \ldots, l$.

3. For each $m \in \{0, 1, \ldots, l\}$, we define the values $m_p(T_m) = \text{Maximum}\ \{m_p(T_m)$: $j \in P\}$ $M_p(T_m) = \text{Minimum}\{M_p(T_m): j \in P\}$.
4. If for some $m \in \{0, 1, \ldots, l\}$ we have $m_p(T_m) \leq M_p(T_m)$, then we observe that the products $P_1, \ldots, P_n$ can be simultaneously processed at temperature $T_m$ with time $t \in [m_p(T_m), M_p(T_m)]$.

### 16.2.4.5 Mixed Integer Linear Programming Model

For food sterilization problem the following data is given and generated:

- Number of sterilization products: $N$
- Amount for each product: $a_j, j \in 1:N$
- Number of sterilization vectors: $M$
- Set of all possible sterilization vectors: $V_A = \{v^i\}, v_j^i \in \{0,1\},\ i \in 1{:}M, j \in 1:N,$
- Set of sterilization time: $T = \{t_i\}, i \in 1:M$
- Capacity of autoclaves: $C$

It should be noted that until the following condition (Equation 16.1) holds

$$a_j \geq C;\quad \forall j \in 1:N, \tag{16.1}$$

the algorithm to solve the above-mentioned sterilization problems is very simple: nonsimultaneous sterilization vectors should be applied (in case of maximization, nonsimultaneous sterilization vectors should be applied in order to increase its respective sterilization time), because (1) on any sterilization batch, the autoclave(s) can be completely filled with chosen products and (2) any other simultaneous sterilization vectors will let us sterilize a single batch of the same amount of products, but for higher or equal processing time (Table 16.2).

So, the case of food sterilization problems, when the following condition holds

$$a_j \leq C;\quad \forall j \in 1:N, \tag{16.2}$$

should be considered. To reduce number of decision variables of the MILP problem, the following procedure was utilized.

**Computational procedure B**

1. By using the procedure A, compute set $V_A$.
2. Compute such subset $V \subseteq V_A$ that for $\forall v^k \in V$ such that $j \in 1:N$ exists that if $\forall v^i \in V/v^k$ that $t_i \leq t_k$, then $(v_j^k = 1)$ and $(v_j^i = 0)$.

Procedure B generates such subset of simultaneous sterilization vectors $V$ that each vector of this subset is unique, i.e., any other vector of $V$ does not contain such combination of subcollection of products and sterilization time.

Let $V$ be a set of all existence sterilization vectors $v^i, i \in 1:M$, and $T$ be a set of sterilization time for each vector $t_i$, $i \in 1:M$. Before we consider the MILP problem statements, some definitions will be introduced.

**TABLE 16.2**
**Calculated Sterilization Vectors for Process Temperature 109°C**

| | Products | | | | | | | | | | | | | | | | |
|---|---|---|---|---|---|---|---|---|---|---|---|---|---|---|---|---|---|
| **Vector** | **1** | **2** | **3** | **4** | **5** | **6** | **7** | **8** | **9** | **10** | **11** | **12** | **13** | **14** | **15** | **16** | **Time (min)** |
| 1 | 0 | 0 | 1 | 0 | 0 | 0 | 0 | 0 | 0 | 1 | 0 | 0 | 0 | 0 | 0 | 0 | 107.63 |
| 2 | 1 | 0 | 1 | 0 | 0 | 1 | 0 | 1 | 0 | 1 | 0 | 0 | 0 | 0 | 1 | 0 | 110.91 |
| 3 | 1 | 0 | 1 | 1 | 0 | 1 | 0 | 1 | 0 | 1 | 0 | 0 | 0 | 0 | 1 | 0 | 111.05 |
| 4 | 1 | 0 | 1 | 1 | 0 | 1 | 0 | 1 | 0 | 1 | 0 | 0 | 0 | 1 | 1 | 0 | 112.56 |
| 5 | 1 | 0 | 1 | 1 | 0 | 1 | 0 | 1 | 0 | 1 | 0 | 1 | 0 | 1 | 1 | 0 | 114.71 |
| 6 | 1 | 1 | 1 | 1 | 0 | 1 | 0 | 1 | 0 | 1 | 0 | 1 | 0 | 1 | 1 | 0 | 118.12 |
| 7 | 0 | 0 | 0 | 0 | 1 | 0 | 1 | 0 | 1 | 0 | 1 | 0 | 1 | 0 | 0 | 1 | 152.25 |
| 8 | 0 | 1 | 0 | 0 | 1 | 0 | 1 | 0 | 1 | 0 | 0 | 1 | 1 | 0 | 0 | 1 | 144.38 |
| 9 | 1 | 1 | 1 | 1 | 0 | 1 | 0 | 1 | 0 | 1 | 0 | 1 | 0 | 1 | 1 | 1 | 121.17 |
| 10 | 1 | 1 | 1 | 1 | 0 | 1 | 0 | 1 | 1 | 1 | 0 | 1 | 0 | 1 | 1 | 1 | 123.36 |
| 11 | 1 | 1 | 1 | 1 | 0 | 1 | 1 | 1 | 1 | 1 | 0 | 1 | 0 | 1 | 1 | 1 | 127.89 |
| 12 | 1 | 1 | 1 | 1 | 1 | 1 | 1 | 1 | 1 | 1 | 0 | 1 | 0 | 1 | 1 | 1 | 133.08 |

*Source:* Simpson, R. and Abakarov, A., *J. Food Eng.*, 90, 53, 2009. With permission.

Let us to introduce computational procedure C to obtain, through sets $V$ and $T$, others expanded sets of sterilization vectors $\overline{V}$ and set of sterilization time $\overline{T}$.

**Computational procedure C**

(1) Beginning or starting point
(2) $\overline{V} := \emptyset$, $\overline{T} := \emptyset$, $k := 1$ ; $\overline{T} := \emptyset$, $k := 1$;
(3) $k := k + 1$;
(4) $s := \sum_{j=1}^{N} v_j^k a_j$;
(5) $n := \lfloor C/s \rfloor + 1$ ;
(6) for $l = 1 : n$
(7) $\overline{V} := \overline{V} \cup \{v^k\}$;
(8) $\overline{T} := \overline{T} \cup \{t_k\}$;
(9) end for;
(10) if $k < M$ then goto step 3;
(11) End

Obtained by procedure C, expanded sets $\overline{V}$ and $\overline{T}$ will be used in the developed MILP model instead of sets $V$ and $T$. We need to use the expanded set $\overline{V}$ because each sterilization vector $v^i$, $i \in 1 : M$, can be applied in the optimal scheduling more than once and so set $T$ should be expanded according to set $\overline{V}$. The maximum possible number of the vector $v^i$ implementation for the sterilizing process is calculated in lines (4) and (5) of the procedure C.

Let $\overline{M}$ be a power of obtained set $\overline{V}$, $\overline{M} = |\overline{V}|$.

Two types of decision variables will be used in the MILP models:

- Integer decision variables:

$$u_i = \begin{cases} 1, & \text{if vector } \bar{v}^i \in \bar{V}, \text{ is used for sterilization,} \\ 0, & \text{otherwise.} \end{cases},$$

- Continuous decision variables: $x_{ij}$, $i \in 1 : \bar{M}, j \in 1 : N$, corresponding to an amount of loaded product.

**MILP problem statements**

*Case 1. The minimization of the sterilization time for given amounts of product*

The objective function can be written by solving a problem such as

$$\sum_{i=1}^{\bar{M}} u_i \bar{t}_i \rightarrow \min, \tag{16.3}$$

where $\bar{t}_i \in \bar{T}$. It is obvious that the optimal value of a vector of integer variables $u^*$ equates as a minimum processing time.

Because all given products should be completely sterilized, we therefore have the following constraint:

$$\sum_{i=1}^{\bar{M}} x_{ij} = a_j, \quad \forall j \in 1 : N. \tag{16.4}$$

For all chosen simultaneous sterilization vectors $v^i$, $i \in 1 : \bar{M}$, the amount of products loaded in each batch should be less than the given capacity of autoclave $C$, and consequently we can write this constraint as

$$\sum_{j=1}^{N} x_{ij} \leq u_i C, \quad \forall i \in 1 : \bar{M}. \tag{16.5}$$

So, the minimization of processing time by MILP problem can be written as

Objective function:

$$\sum_{i=1}^{\bar{M}} u_i \, \bar{t}_i \rightarrow \min,$$

subject to:

$$\sum_{j=1}^{N} x_{ij} \leq u_i C, \quad \forall i \in 1 : \bar{M},$$

$$\sum_{i=1}^{\bar{M}} x_{ij} = a_j, \quad \forall j \in 1 : N.$$

*Case 2. Maximization of product amount for a given processing time t*
It is obvious, that the amount of sterilized product for some chosen set of the values $\{x_{ij}^*\}$, $i \in 1:\bar{M}, j \in 1:N$, is equal to $\sum_{i=1}^{\bar{M}} \sum_{j=1}^{N} x_{ij}^*$; therefore, the objective function for case 2 can be written as

$$\sum_{i=1}^{\bar{M}} \sum_{j=1}^{N} x_{ij} \rightarrow \max. \tag{16.6}$$

For all chosen simultaneous sterilization vectors $v^i$, $i \in 1:\bar{M}$, the amount of products loaded in each batch should be less or equal than the given capacity of autoclave $C$, and consequently we can write this constraint as

$$\sum_{j=1}^{N} x_{ij} \leq u_i C, \quad \forall i \in 1:\bar{M}. \tag{16.7}$$

Because the given amounts of each product to be sterilized should not be exceeded, the following constraint is included:

$$\sum_{i=1}^{\bar{M}} x_{ij} \leq a_j, \quad \forall j \in 1:N, \tag{16.8}$$

and because the processing time is fixed, the following constraint is necessary to use:

$$\sum_{i=1}^{M} u_i \bar{t}_i \leq T, \tag{16.9}$$

where $\bar{t}_i \in \bar{T}$.

## 16.3 ANALYSIS AND DISCUSSIONS

Figure 16.2 shows three isolethal curves for Peas (211 × 400) at $F_0$ values of 6, 8, and 10 min. Isolethal processes were obtained with RGM procedure for six different retort temperatures and fitted with cubic spline interpolation method. Those curves, at three $F_0$ levels (6, 8, and 10 min) were obtained for all 16 products under study. The information given in those curves is the basis to analyze and implement simultaneous sterilization at any given retort temperature in the range of 110°C–125°C (Simpson, 2005).

Figures 16.3 through 16.5 show process time per each product at different $F_0$ values (6, 8, and 10 min) at TRT = 110°C, 117°C, and 125°C, respectively. Similar information can be obtained at any given temperature from 110°C to 125°C. Due to the continuous data attained with the cubic spline method, simultaneous sterilization could be analyzed at any given retort temperature. Clearly, the possibility of implementing simultaneous sterilization is higher at lower retort temperature

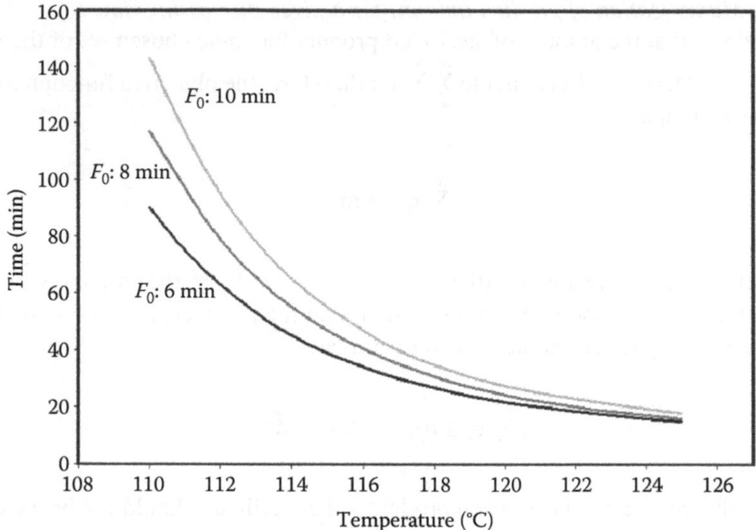


**FIGURE 16.2** Three isolethal curves for Peas (211 × 400) at $F_0$ values of 6, 8, and 10 min obtained with RGM procedure and fitted with cubic spline interpolation method. (From Simpson, R. and Abakarov, A., *J. Food Eng.*, 90, 53, 2009. With permission.)

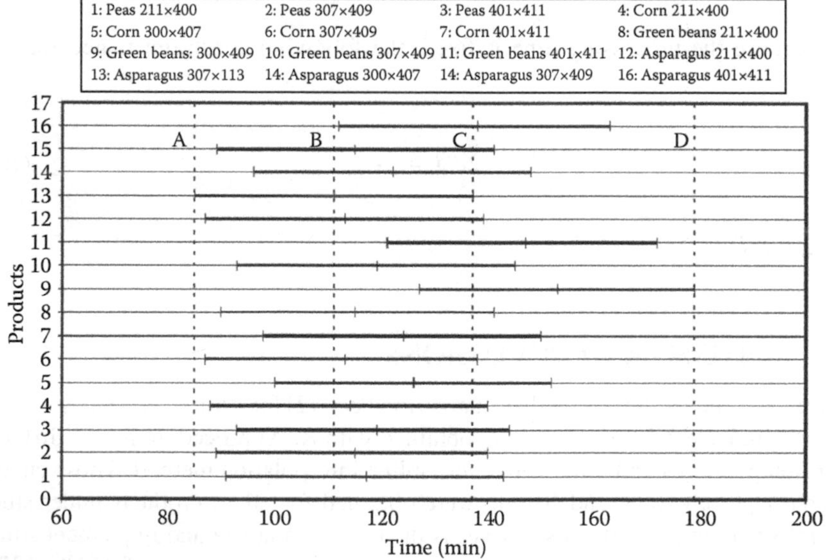


**FIGURE 16.3** Process time per each product at different $F_0$ values (6, 8, and 10 min) at TRT = 110°C obtained with RGM procedure. (From Simpson, R. and Abakarov, A., *J. Food Eng.*, 90, 53, 2009. With permission.)

(Figures 16.3 through 16.5). In terms of further analysis, two retort temperatures (110°C and 117°C) were chosen (Simpson, 2005).

Figure 16.3 shows the wide range of opportunities to implement simultaneous sterilization at 110°C. Independent of the maximum selected $F_0$ value (8 or 10 min), almost all products, under study, can be simultaneously processed at the same time. On the other hand, at retort temperature of 117°C, although several choices are

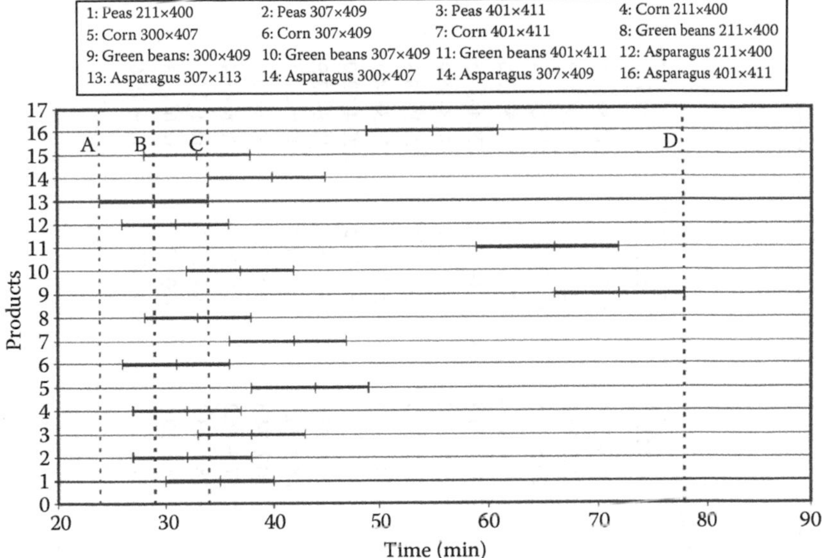


**FIGURE 16.4** Process time per each product at different $F_0$ values (6, 8, and 10 min) at TRT = 117°C obtained with RGM procedure. (From Simpson, R. and Abakarov, A., *J. Food Eng.*, 90, 53, 2009. With permission.)

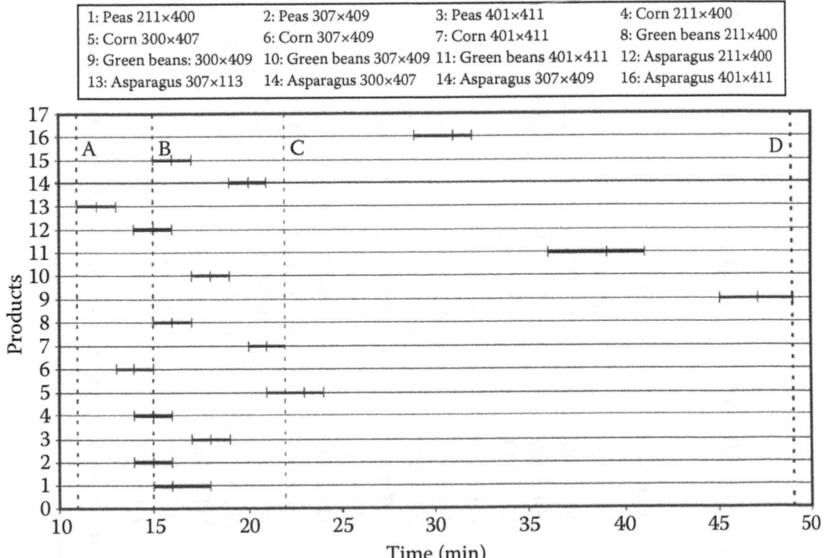


**FIGURE 16.5** Process time per each product at different $F_0$ values (6, 8, and 10 min) at TRT = 125°C obtained with RGM procedure. (From Simpson, R. and Abakarov, A., *J. Food Eng.*, 90, 53, 2009. With permission.)

available, it is hardly possible to attain simultaneous sterilization for all or most of the products at the same time (Simpson, 2005).

Table 16.3 shows process time per each product at two different $F_0$ values (6 and 10 min) at TRT = 109°C, 112°C, 116°C, 118°C, 123°C, and 125°C, respectively.

**TABLE 16.3**
**Processing Times for All 16 Products at $F_0$ = 6 min and $F_0$ = 8 min**

| Product | Can Size | Temperature (°C) | | | | | |
|---|---|---|---|---|---|---|---|
| | | 109 | 112 | 116 | 118 | 123 | 125 |
| Asparagus | 211 × 400 | 110.9–142.6 | 63.3–79.2 | 34.6–41.6 | 26.8–31.6 | 17.5–19.1 | 15.5–16.7 |
| | 300 × 407 | 118.1–149.8 | 70.8–86.5 | 40.9–48.3 | 30.2–37.7 | 22.3–24.3 | 19.9–21.4 |
| | 307 × 409 | 111.05–142.8 | 63.6–79.4 | 34.8–41.8 | 26.5–31.8 | 17.2–18.9 | 15.2–16.4 |
| | 307 × 113 | 106.5–138.2 | 59.0–74.8 | 30.2–37.2 | 22.4–27–2 | 13.4–15.0 | 11.5–12.6 |
| | 401 × 411 | 134.1–164–8 | 86.1–101.8 | 54.8–62.6 | 46.5–51.4 | 32.6–35.3 | 29.4–31.5 |
| Corn | 211 × 400 | 109.5–141.2 | 61.9–77.8 | 31.2–40.1 | 25.5–30.2 | 16.6–18.1 | 14.8–15.9 |
| | 300 × 407 | 127.9–159.6 | 79.8–96.8 | 47.3–55.0 | 38.4–43.9 | 26.0–28.4 | 22.9–24.9 |
| | 307 × 409 | 109.5–141.2 | 62.2–77.8 | 32.4–40.0 | 24.4–29.3 | 15.3–17.0 | 13.3–14.5 |
| | 401 × 411 | 123.4–155.1 | 73.3–89.9 | 42.7–49.9 | 34.7–39.5 | 23.0–25.2 | 20.4–22.0 |
| Green beans | 211 × 400 | 107.6–139.3 | 60.0–75.9 | 31.3–37–7 | 25.5–27.8 | 14.6–16.1 | 12.8–13.9 |
| | 300 × 407 | 152.3–184.0 | 104.9–120.9 | 74.4–81.8 | 65.3–70.5 | 51.2–54.2 | 47.2–49.9 |
| | 307 × 409 | 114.7–146.4 | 67.2–83.1 | 38.4–44.9 | 30.7–34.8 | 20.2–22.1 | 17.8–19.3 |
| | 401 × 411 | 144.4–176.1 | 97.1–112.7 | 68.5–75.4 | 57.3–64.2 | 42.8–46.0 | 38.8–41.4 |
| Peas | 211 × 400 | 112.6–144.3 | 65.1–80.9 | 36.3–43.3 | 28.5–33.3 | 18.4–20.3 | 16.1–17.5 |
| | 307 × 409 | 109.5–141.4 | 62.0–77.8 | 37.1–39.6 | 25.3–29.6 | 16.0–17.6 | 14.0–15.3 |
| | 401 × 411 | 121.17–152.9 | 71.3–88.7 | 40.2–47.5 | 31.1–36.9 | 20.5–22.7 | 17.9–19.6 |

*Source:* Simpson, R. and Abakarov, A., *J. Food Eng.*, 90, 53, 2009. With permission.

**TABLE 16.4**
**Amount of Each Product to be Processed with a Total Autoclaves Capacity of 10,000 L**

| **Product** | **1** | **2** | **3** | **4** | **5** | **6** | **7** | **8** |
|---|---|---|---|---|---|---|---|---|
| Amount (L) | 43,000 | 34,000 | 76,000 | 72,000 | 55,000 | 19,000 | 81,000 | 25,000 |
| Product | 9 | 10 | 11 | 12 | 13 | 14 | 15 | 16 |
| Amount (L) | 34,000 | 38,000 | 67,000 | 27,000 | 45,000 | 91,000 | 18,000 | 52,000 |

*Source:* Simpson, R. and Abakarov, A., *J. Food Eng.*, 90, 53, 2009. With permission.

Smooth curves were fitted utilizing cubic spline procedure. Due to the continuous data attained with the cubic spline method, simultaneous sterilization could be analyzed at any given retort temperature.

The developed MILP models were tested on 16 products presented in Table 16.1. Amounts for each given product were generated randomly in accordance with the condition (Equation 16.2) (Table 16.4). Autoclave(s) with capacity of 10,000L (available capacity for cans) were chosen.

For all 16 products, the isolethal processes were generated (calculated). By using the software developed by the authors, the subset $V$ containing 82 sterilization vectors for temperatures from $T_{min} = 109°C$ to $T_{max} = 124°C$ was obtained.

So, the optimum scheduling using simultaneous sterilization was solved with the following data:

1. Amount of each 16 sterilization products, presented in the Table 16.4
2. Number of sterilization vectors: $\bar{M} = 82$
3. Capacity of autoclaves: $C = 10{,}000$ L (available capacity for cans)

And, in addition, for case 2

4. Processing time: $T = 150$ min

Two software tools were utilized to solve the examples:

- "lp_solve" version 5.5.0.10
- Online Web Service of COIN-OR (COmputational INfrastructure for Operations Research): NEOS Server Version 5.0

*Case 1. Minimization of the sterilization time for a given amount of product*
In case of the nonsimultaneous sterilization possibility, the processing time for the referred example is equal to 332.5 min, the simultaneous sterilization possibility with MILP implementation reduced this processing time to 249.05 min (Table 16.5). Proposed MILP model always will give less or equal processing time in comparison with the nonsimultaneous sterilization, and the difference between both times will depend on the given data: simultaneous sterilization possibility, amounts of product, and autoclave capacity.

**TABLE 16.5**
**Results Obtained by Using the Developed MILP Mathematical Model (Case 1)**

| Solver Name | Computation Time (s) | Objective Function Value (min) | Nonsimultaneous Sterilization Case (min) |
|---|---|---|---|
| LP_Solver IDE v5.5.0.10 | 3114.062 (Pentium IV, 2.4 GHz, 2 Gb RAM) | 249.34 | 332.5 |
| Coin Branch and Cut Solver on NEOS Server version 5.0 using MPS input | 7.38 | 249.05 | |

*Source:* Simpson, R. and Abakarov, A., *J. Food Eng.*, 90, 53, 2009. With permission.

**TABLE 16.6**
**Results Obtained by Using the Developed MILP Mathematical Model (Case 2)**

| Solver Name | Computation Time (s) | Objective Function Value | Nonsimultaneous Sterilization Case |
|---|---|---|---|
| LP_Solver IDE v5.5.0.10 | 1377.69 (Pentium IV, 2.4 GHz, 2 Gb RAM) | 63,100 | 51,200 |
| Coin Branch and Cut Solver on NEOS Server version 5.0 using MPS input | 3.5 | 63,100 | |

*Source:* Simpson, R. and Abakarov, A., *J. Food Eng.*, 90, 53, 2009. With permission.

*Case 2. Maximization of product amount for a given processing time t*
In case of the nonsimultaneous sterilization possibility, the maximum amount of the products to be sterilized is equal to 51,200, the simultaneous sterilization possibility with MILP implementation increased this amount to 63,100 (Table 16.6). Proposed MILP model always will give more or equal amount of sterilized products in comparison with the nonsimultaneous sterilization, and the difference between both times will depend on the given data: simultaneous sterilization possibility, amounts of product, and autoclave capacity.

It should be noted that only for several products, the two MILP models has so many constraints, which are laborious to formulate in the required format, for example, by using lp_format, mps-format, or AMPL language. So the computer program was set to automatically formulate the problems for both cases in the lp or mps-format, for the given data of simultaneous sterilization.

## 16.4 CONCLUSIONS

The proposed MILP models based on the simultaneous sterilization possibility provide flexibility to optimize battery retort utilization and it is of special relevance for small and medium size canneries that normally work with many different products at the same time. The proposed MILP models are of special relevance for small and medium size canneries that normally work with many different products at the same time.

Furthermore, the proposed way for developing the MILP models allow, depending on practical situations, to consider other types of optimization problems with other type of constraints or objective functions. The implementation of the MILP model allows significant improvements in comparison with nonsimultaneous sterilization (normal practice in the canning industry).

## REFERENCES


Atkinson, K. 1985. *Elementary Numerical Analysis*. John Wiley & Sons, Inc., New York.

Barbosa, A. P. and Macchietto, S. 1993. Optimal design of multipurpose batch plants 1. Problem formulation. *Computers and Chemical Engineering*, 17, S33–S38.

Bigelow, W. D., Bohart, G. S., Richardson, A. C., and Ball, C. O. 1920. Heat penetration in processing canned foods. Bull. No. 16-L, Research Laboratory of the National Canners Association, Washington, DC.

Castro, P. M. and Grossmann, I. E. 2005. New continuous-time MILP model for the short-term scheduling of multi-stage batch plants. *Industrial and Engineering Chemistry Research*, 44(24), 9175–9190.

Castro, P. M. and Grossmann, I. E. 2006. An efficient MILP model for the short-term scheduling of single stage batch plants. *Computers and Chemical Engineering*, 30, 1003–1018.

Erdirik-Dogan, M. and Grossmann, I. 2007. Simultaneous planning and scheduling of single stage multi-product continuous plants with parallel lines. *Computers and Chemical Engineering*, 32, 2664–2683.

Floudas, C. A. and Lin, X. 2004. Continuous-time versus discrete-time approaches for scheduling of chemical processes: A review. *Computers and Chemical Engineering*, 28(11), 2109–2129.

Gupta, S. and Karimi, I. A. 2003. An improved MILP formulation for scheduling multi-product, multi-stage batch plants. *Industrial and Engineering Chemistry Research*, 42(11), 2365–2380.

Ha, J. H., Chang, H. K., Lee, E. S., Lee, I. B., Lee, B. S., and Yi, G. 2006. Inter-stage storage tank operation strategies in the production scheduling of multi-product batch processes. *Computers and Chemical Engineering*, 24, 1633–1640.

Harjunkoski, I. and Grossmann, I. E. 2002. Decomposition techniques for multistage scheduling problems using mixed-integer and constraint programming methods. *Computers and Chemical Engineering*, 26, 1533.

Holdsworth, S. D. 1997. *Thermal Processing of Packaged Foods*. Blackie Academic & Professional, London.

Holdsworth, S. D. and Simpson, R. 2007. *Thermal Processing of Packaged Foods*. 2nd Edtion. Springer, New York.

Hui, C. -W. and Natori, Y. 1996. An industrial application using mixed-integer programming technique: A multi-period utility system model. *Computers and Chemical Engineering*, S20, S1577–S1582.

Jung, J. H., Lee, H. K., and Lee, I. B. 1994. Completion times and optimal scheduling for serial multi-product processes with transfer and set-up times in zero-wait policy. *Computers and Chemical Engineering*, 18, 537–544.

Kondili, C., Pantelides, R., and Sargent, H. 1993. A general algorithm for short-term scheduling of batch operations. I. MILP formulation. *Computers and Chemical Engineering*, 17, 211–227.

Lee, B. and Reklaitis, G. V. 1995a. Optimal scheduling of cyclic batch processes for heat integration. I. Basic formulation. *Computers and Chemical Engineering*, 19(8), 883–905.

Lee, B. and Reklaitis, G. V. 1995b. Optimal scheduling of cyclic batch processes for heat integration. II. Extended problems. *Computers and Chemical Engineering*, 19(8), 907–931.

Liu, Y. and Karimi, I. 2008. Scheduling multistage batch plants with parallel units and no interstage storage. *Computers and Chemical Engineering*, 32, 671–693.

Maravelias, C. T. 2006. A decomposition framework for the scheduling of single- and multi-stage processes. *Computers and Chemical Engineering*, 30, 407–420.

Méndez, C. and Cerda, J. 2000. Optimal scheduling of a resource-constrained multiproduct batch plant supplying intermediates to nearby end-product facilities. *Computers and Chemical Engineering*, 24, 369.

Méndez, C. and Cerda, J. 2002. An efficient MILP continuous-time formulation for short-term scheduling of multiproduct continuous facilities. *Computers and Chemical Engineering*, 26, 687.

Mendez, C. A., Henning, G. P., and Cerda, J. 2001. An MILP continuous time approach to short-term scheduling of resource-constrained multi-stage flowshop batch facilities. *Computers and Chemical Engineering*, 25, 701–711.

Mendez, C. A., Cerda, J., Grossmann, I. E., Harjunkoski, I., and Fahl, M. 2006. State-of-the-art review of optimization methods for short-term scheduling of batch processes. *Computers and Chemical Engineering*, 30(6–7), 913–946.

Moon, S., Park, S., and Lee, W. K. 1996. New MILP models for scheduling of multi-product batch plants under zero-wait policy. *Industrial and Engineering Chemistry Research*, 35, 3458–3469.

Norback, J. and Rattunde, M. 1991. Production planning when batching is part of the manufacturing sequence. *Journal of Food Process Engineering*, 14, 107–123.

Reklaitis, G. V. 1996. Overview of scheduling and planning of batch process operations. In: *Batch Processing System Engineering*, G.V. Reklaitis, A.K. Sunol, D.W. Rippin, and O. Hortacsu (Eds.). Springer, Berlin, Germany, pp. 660–705.

Rippin, D. W. 1993. Batch process system engineering: A retrospective and prospective review. *Computer and Chemical Engineering*, 17, S1–S13.

Simpson, R. 2005. Generation of isolethal processes and implementation of simultaneous sterilization utilising the revisited general method. *Journal of Food Engineering*, 67(1–2), March, 71–79.

Simpson, R., Almonacid, S., and Teixeira, A. 2003a. Optimization criteria for batch retort battery design and operation in food canning-plants. *Journal of Food Process Engineering*, 25(6), 515–538.

Simpson, R., Almonacid, S., and Teixeira, A. 2003b. Bigelow's general method revisited: Development of a new calculation technique. *Journal of Food Science*, 68(4), 1324–1333.

Teixeira, A.A. 1992. Thermal process calculations. In: *Handbook of food Engineering,* Denis Heldman and Daryl Lund (Eds). CRC Press, Boca Raton, FL.

# Part IV

## Online Control and Automation

# 17 Online Control Strategies: Batch Processing

*Ricardo Simpson and Arthur Teixeira*

## CONTENTS



## 17.1 INTRODUCTION

Thermal processing is an important method of food preservation in the manufacture of shelf stable canned foods. Most of the works done in thermal processes deal with the microbiological and biochemical aspects of the process and are rarely related to the engineering aspects or practical industrial operations of the process. The basic function of a thermal process is to inactivate pathogenic and food spoilage causing bacteria in sealed containers of food using heat treatments at temperatures well above the ambient boiling point of water in pressurized steam retorts (autoclaves).

Excessive heat treatment should be avoided because it is detrimental to food quality, wastes energy, and underutilizes plant capacity. Thermal process calculations, in which process times at specified retort temperatures are calculated in order to achieve safe levels of microbial inactivation (lethality), must be carried out carefully to assure public health safety (Bigelow et al., 1920; Ball, 1928; Stumbo, 1973; Pham, 1987; Teixeira, 1992; Holdsworth, 1997). However, overprocessing must be avoided because thermal processes also have a detrimental effect on the quality (nutritional and sensorial factors) of foods. Therefore, the accuracy of the methods used for this purpose is of importance to food science and engineering professionals working in this field.

Control of thermal process operations in food canning factories consisted of maintaining specified operating conditions that have been predetermined from product and process heat penetration tests, such as the process calculations for the time and temperature of a batch cook. Sometimes unexpected changes can occur during the course of the process operation such that the prespecified processing conditions are no longer valid or appropriate. These types of situations are known as process deviations. Because of the important emphasis placed on the public safety of canned foods, processors must operate in strict compliance with the U.S. Food and Drug Administration's Low-Acid Canned Food regulations. A succinct summary and brief discussion of these regulations can be found in Teixeira (1992). Among other things, these regulations require strict documentation and record-keeping of all critical control points in the processing of each retort load or batch of canned product. Particular emphasis is placed on product batches that experience an unscheduled process deviation, such as when a drop in retort temperature occurs during the course of the process, which may result from the loss of steam pressure. In such a case, the product will not have received the established scheduled process, and must be either fully reprocessed, destroyed, or set aside for evaluation by a competent processing authority. If the product is judged to be safe then batch records must contain documentation showing how that judgment was reached. If judged unsafe, then the product must be fully reprocessed or destroyed. Such practices are costly.

In one very traditional method, online correction is accomplished by extending process time to that which would be needed had the entire process been carried out at the retort temperature reached at the lowest point in the deviation (Larkin, 2002). This method of correction will always assure sufficient food safety (process lethality), but often results in significant unnecessary overprocessing with concomitant deterioration in product quality. Alternative methods proposed initially by Teixeira and Manson (1982) and Datta et al. (1986), and later summarized in Teixeira and Tucker (1997) and demonstrated and validated by Teixeira et al. (1999) make use of heat transfer simulation software to automatically extend process time at the recovered retort temperature to reach precisely the original target lethality required of the process. An alternative approach to online correction of process deviations was described in the work of Akterian (1996, 1999), who calculated the correction factors using mathematical sensitivity functions.

The aim of this research study was the development of a safe, simple, efficient, and easy-to-use procedure to manage online corrections of unexpected process deviations in any canning plant facility. Specific objectives were to

1. Develop strategy to correct the process deviation by an alternative "proportional-corrected" process that delivers no less than final target lethality, but with near minimum extended process time at the recovered retort temperature
2. Demonstrate strategy performance by comparing proportional-corrected with "commercial-corrected" and "exact-corrected" process times
3. Demonstrate consistent safety of the strategy by exhaustive search over an extensive domain of product and process conditions to find a case in which safety is compromised
4. Quantify the economic and quality impact of online correction strategy
5. Develop and preliminary validate a strategy for online correction without extending process time for any low acid canned food

## 17.2 METHODOLOGY

To reach the objectives stated above, the approach to this work was carried out in four tasks, one in support of each objective. Task 1 consisted of developing the strategy for online correction of process deviations with minimum extended process time using the method of "proportional correction." Task 2 consisted of choosing appropriate mathematical heat transfer models for construction of the equivalent lethality curves or "look-up tables" needed for use with each respective strategy, and for determining the final lethality and quality retention for each of the thousands of cases simulated in the study. Task 3 included the complex optimization search routine that was carried out to demonstrate validity and consistent safety of the strategy. Task 4 quantifies the economic and quality impact of online correction strategy. And finally, Task 5 consisted of developing and validating the strategy for online correction of process deviations without extending process time. Methodology employed in carrying out each of these tasks is described in greater detail below.

### 17.2.1 Task 1: Proportional Correction Strategy Development

The objective for the strategy required in this task was to accomplish an online correction of an unexpected retort temperature deviation by an alternative process that delivers final target lethality, but with minimum extended process time at the recovered retort temperature. This would be accomplished with use of the same alternative process look-up tables that would normally be used with currently accepted methods of online correction of process deviations, but with a proportional correction applied to the alternative process time that would reduce it to a minimum without compromising safety. In order to fully understand this strategy, it will be helpful to first review the currently accepted method that is in common practice throughout the industry. Commercial systems currently in use for online correction of process deviations do so by extending process time to that which would be needed to deliver the same final lethality had the entire process been carried out with an alternative lower constant retort temperature equal to that reached at the lowest point in the deviation. These alternative retort temperature–time combinations that deliver the same final process lethality ($F_0$) are called equivalent lethality processes.

When these equivalent time–temperature combinations are plotted on a graph of process time versus retort temperature, they fall along a smooth curve called an equivalent lethality curve. These curves are predetermined for each product from heat penetration tests and thermal process calculations carried out for different retort temperatures.

The strategy will calculate the corrected process time ($t_D$) as a function of the temperature drop experienced during the deviation, but also the time duration of the deviation. The following expression illustrates mathematically how this proportional-corrected process time would be calculated for any number ($n$) of deviations occurring throughout the course of a single process:

$$t_D = t_{TRT} + \sum_{i=1}^{n}(t_D - t_{TRT})\left(\frac{\Delta t_i}{t_{TRT}}\right), \quad t_D \geq t_{TRT} \tag{17.1}$$

If we considered the commercial correction as a valid correction, that is,

$$t_{D\,\text{commercial}} = t_{TRT} + (t_D - t_{TRT})\frac{t_{TRTi}}{t_{TRT}} \tag{17.2}$$

Then it is intuitive that a safe correction for a process deviated for a short time is a correction proportional to the time the deviation occurs:

$$t_{D\,\text{proportional}} = t_{TRT} + (t_D - t_{TRT})\frac{\Delta t_i}{t_{TRT}} \tag{17.3}$$

### 17.2.2 Task 2: Performance Demonstration

This task consisted of demonstrating the performance of these strategies by simulating the occurrence of process deviations happening at different times during the process (early, late, and randomly) to both solid and liquid canned food products, calculating the alternative corrected process times, and predicting the outcomes of each corrected process in terms of final lethality and quality retention. For each deviation, three different alternative corrected process times were calculated:

- "Exact correction," giving corrected process time to reach precisely the final target lethality specified for the scheduled process, using computer simulation with heat transfer models
- Proportional correction, using the strategy described in this chapter with look-up tables
- "Commercial correction," using current industry practice with look-up tables (manually or computerized)

The heat transfer models were explicitly chosen to simulate the two extreme heat transfer cases encountered in thermal processing of canned foods. The rationale behind this decision was that canned foods possess heating characteristics between these two extreme situations. Conclusions extracted from these simulations will be extended to all canned foods.

The $F$-value is calculated by

$$F = \int_0^t 10^{\frac{T_{cp}(t) - T_{ref}}{z}} dt \tag{17.4}$$

First was the case of pure conduction heating of a solid product under a still-cook retort process (Equation 17.5 is the differential equation for heat conduction, Biot > 40). The second was the case of forced convection heating of a liquid product under mechanical agitation (Equation 17.6 is the differential equation for heat convection, Biot < 1). In both cases, the container shape of a finite cylinder was assumed, typical of a metal can or wide-mouthed glass jar. However, suitable models appropriate for a true container shapes can be used as required for this purpose. Examples of such models can be found in the literature (Teixeira et al., 1969; Manson et al., 1970, 1974; Datta et al., 1986; Simpson et al., 1989, Simpson, 2004). The product and process conditions chosen to carry out the demonstrated simulations for each case are given in Table 17.1.

$$\frac{\partial T}{\partial t} = \alpha \left[ \frac{1}{r} \frac{\partial}{\partial r} \left( r \frac{\partial T}{\partial r} \right) + \frac{\partial^2 T}{\partial h^2} \right] \tag{17.5}$$

$$\frac{\partial T(t)}{\partial t} = \frac{U_{ht} \cdot A}{\rho \cdot \text{Vol} \cdot Cp} [\text{TRT\_corr}(t) - T(t)] \tag{17.6}$$

### 17.2.3 Task 3: Demonstration of Safety Assurance by Complex Optimization Search Routine

This online correction strategy was validated and tested for safety assurance by executing a strict and exhaustive search routine with the use of the heat transfer models selected in Task 2 on high-speed computer. The problem to be solved by the

**TABLE 17.1**
**Product and Process Conditions Used for Online Correction Strategy Simulations**

| | Dimensions (cm) | | | Properties | | Normal Process | |
|---|---|---|---|---|---|---|---|
| **Product Simulated** | **Major** | **Intermedium** | **Minor** | **$\alpha$ ($m^2$/s)** | **$f_h$ (min)** | **Time (min)** | **TRT (°C)** |
| Pure conduction can, Biot > 40 | 11.3 | — | 7.3 | 1.70E–07 | 44.4 | 64.1 | 120 |
| Forced convection can, Biot < 1 | 11.3 | — | 7.3 | — | 4.4 | 15.6 | 120 |

*Source:* Simpson, R., Figueroa, I., Teixeira, A., *Food Control,* 18, 458, 2007. With permission.

search routine was to determine if the minimum final lethality delivered by all the corrected processes that could be found among all the various types of deviations and process conditions considered in the problem domain met the criterion that it had to be greater than or equal to the lethality specified for the original scheduled process. This criterion can be expressed mathematically:

$$\underset{U}{\text{Min}}\left[F_{\text{proportional}} - F_{\text{Tol}}\right] \geq 0 \tag{17.7}$$

Table 17.2 identifies the various types of deviations and process conditions that were explored and evaluated in the search routine (problem domain). The search routine was designed as an attempt to find a set of conditions under which the required search constraint was not met. The table gives the symbol used to represent each variable and a description of that variable, along with the minimum and maximum values limiting the range over which the search was conducted.

### 17.2.4 Task 4: Economic and Quality Impact of Online Correction Strategy

The purpose of this task was to estimate the impact on plant production capacity of alternative corrected process times used in response to the same frequency and type of deviations occurring annually in a typical cannery. The rationale behind this task stems from the realization that every time a process deviation is corrected in this way with any given retort in the cook room of a canning factory, the number of batches processed that day will be less than normal capacity. This can be translated into cost of lost productivity over the course of a canning season. This can be approached mathematically with the algorithm described below.

The following expression allows the calculation of the number of batch per autoclave at the processing plant:

$$N_{Bi} = \frac{H}{t_{ci} + t_{oj} + t_{di}} \tag{17.8}$$

**TABLE 17.2**
**Problem Domain for Search Routine**

| Process Variable | Description of Process Variable | Minimum Value | Maximum Value |
|---|---|---|---|
| TRT | Scheduled retort temperature | 110°C | 135°C |
| $TRT_i$ | Lowest retort temperature reached during deviation *i* | 100°C | TRT − 0.5°C |
| $t_{CUT}$ | Initial CUT of retort to reach TRT | 5 (min) | 15 (min) |
| $t_{dev-i}$ | Time during the process at which the deviation *i* begins | $t_{cut}$ | $t_{TRT}$ |
| $t_i$ | Time duration of the deviation *i* | 0.5 (min) | $t_{TRT} - t_{dev-i}$ |
| $T_{ini}$ | Initial product temperature | 20°C | 70°C |

*Source:* Simpson, R., Figueroa, I., Teixeira, A., *Food Control,* 18, 458, 2007. With permission.

where

$N_{Bi}$ is the number of batches processed per retort ***i*** during the season
$H$ is the operating time of the plant during the season (h)
$t_{ci}$ is the time to load retort ***i*** with product ***j*** (h)
$t_{oj}$ is the time to operate retort ***i*** (process cycle time) with product ***j*** (h)
$t_{di}$ is the time to download retort ***i*** with product ***j*** (h)

To simplify the analysis and to be able to have an estimate of the impact of operating time on plant production capacity, the following assumptions were made:

1. The plant has $N_A$ retorts and all of them are of equal size.
2. The number of containers (units) processed in each retort is the same ($N_{CBi}=N_{CB}$).
3. The plant is processing a single product ($t_{oj}=t_o$, $t_{ci}=t_c$, and $t_{di}=t_d$).

Therefore, the total number of units that can be processed in the whole season ($N_t$) can be expressed by thc following equation:

$$N_t = \frac{N_A * N_{CB} * H}{t_c + t_0 + t_d} \tag{17.9}$$

According to Equation 17.9, an extension in process time ($t_0$) will decrease plant production capacity ($N_t$). In addition, considering processing time as a variable and utilizing Equation 17.9 it is possible to quantify the impact of processing time in terms of plant production capacity.

Another way to assess the impact of the adopted strategy will be to consider that processors are operating at much higher retort temperature ($TRT_H$) to avoid deviant processes. For the purposes of analysis it was considered that processors operate each batch process at a temperature that is 2°C–3°C (common practice in the United States, although some plants even operate at higher temperatures) higher than the registered process and with this, practice processors are completely avoiding deviant processes. As mentioned earlier, the product and process conditions chosen to carry out the simulations to evaluate the impact on product quality are given in Table 17.2.

This research has described a practical, simple, and efficient strategy for online correction of thermal process deviations during retort sterilization of canned foods. The strategy is intended for easy implementation in any cannery around the world. This strategy takes into account the duration of the deviation in addition to the magnitude of the temperature drop. It calculates a proportional extended process time at the recovered retort temperature that will deliver the final specified target lethality with very little overprocessing in comparison to current industry practice. Results from an exhaustive search routine using the complex method support the logic and rationale behind the strategy by showing that the proposed strategy will always result in a corrected process that delivers no less than the final target lethality specified for the originally scheduled process. Economic impact of adopting this strategy over that currently used in industry practice can be a significant increase in production capacity for a typical cannery. In addition, utilizing this novel strategy canned products will attain a much higher quality.

### 17.2.5 Task 5: Online Correction without Extending Process Time

For the development of this strategy, we assume the retort control system to include a computer that is running the software containing the appropriate mathematical heat transfer model, and it reads the actual retort temperature from a temperature-sensing probe through an analogue/digital data acquisition system. This continual reading of retort temperature would be used as a real-time input of dynamic boundary condition for the mathematical heat transfer model. The model, in turn, would be accurately predicting the internal product cold spot temperature profile as it develops in response to the actual dynamic boundary condition (retort temperature). As the predicted cold spot temperature profile develops over time, the accumulating lethality ($F_0$) would be calculated by the General method, and would be known at any time during the process. Should a deviation occur during the process a simulated search routine would be carried out on the computer to find the combination of process conditions for the remainder of the process that would result in meeting the final target lethality without over extending processing time. The key in this strategy was to identify the retort temperature as the control variable to be manipulated during the remainder of the process (rather than process time). Therefore, upon recovery of the deviation, the search routine would find the new $TRT_H$ to be used for the remainder of the process, and send the appropriate signals through the data acquisition system to readjust the retort temperature accordingly.

Increasing retort temperature cannot be accomplished without increasing steam pressure correspondingly, which dictates a practical upper limit to choice of $TRT_H$. This upper limit comes into play when the deviation occurs near the end of the process, when the little time remaining forces the simulation search routine to choose the upper limit for retort temperature. In these cases, the safety requirement for reaching the final target lethality ($F_0$) must take priority over compromising process time. This will inevitably require some extension in process time, but it will be an absolute minimum, that would not likely upset scheduling routines. Validation of this method was confirmed by demonstrating consistent safety of the strategy by exhaustive optimization search over an extensive domain of product and process conditions in an attempt to find a case in which safety was compromised. No such case could be found.

This control strategy finds the lowest process temperature, $TRT_H$, at which the process temperature must be reestablishing after the deviation $i$ for the rest of the processing time, so the original process time remains the same and the required lethality value ($F_{obj}$) is reached.

$$TRT_H = \text{Min}\left\{TRT / F_p \geq F_{obj} \wedge t'_p = t_p\right\} \tag{17.10}$$

If $TRT_H$ surpass the practical limits, the final target lethality ($F_0$) is more important that the process time that must be extended according to

$$\begin{aligned}&\text{If}\quad \text{TRT}_{\text{H}} > \text{TRT}_{\max}\\ &\Rightarrow\quad \text{TRT}_{\text{H}} = \text{TRT}_{\max} \wedge t'_{\text{p}} = \text{Min}\left\{t_{\text{p}} / F_{\text{p}} \geq F_{\text{obj}}\right\}\end{aligned} \tag{17.11}$$

For a process that experiment more than one deviation, the temperature at which the retort temperature must be corrected after the deviation $i$ is

$$\text{TRT}_{\text{H}i} = \text{Min}\left\{\text{TRT} / F_{\text{p}} \geq F_{\text{obj}} \wedge t'_{\text{p}} = t_{\text{p}}\right\} \tag{17.12}$$

If $\text{TRT}_{\text{H}i}$ surpass the practical limit, then

$$\begin{aligned}&\text{If}\quad \text{TRT}_{\text{H}i} > \text{TRT}_{\max}\\ &\Rightarrow\quad TRT_{\text{H}i} = \text{TRT}_{\max} \wedge t'_{\text{p}} = \text{Min}\left\{t_{\text{p}} / F_{\text{p}} \geq F_{\text{obj}}\right\}\end{aligned} \tag{17.13}$$

This higher temperature, $\text{TRT}_{\text{H}}$, needs to be calculated using a look-up table or curve on a graph showing alternative retort temperature–time combinations that were predetermined to deliver the same target lethality (isolethality curves) for each product. Mathematically, this equivalent process time can be calculated from the anatomy of the recovered process deviation and original process conditions as follows:

$$t_{\text{H}} = \frac{t_{\text{TRT}}^2}{t_{\text{LDT}}} \tag{17.14}$$

Equation 17.14 can be obtained as follow: consider a process in which two deviations occur in sequence. The normal retort temperature is TRT, the first deviation occurs over a time interval $\Delta t$ at lower than normal retort temperature $\text{TRT}_{\text{D}}$ and the second occurs later over an equal time interval $\Delta t$ at a higher than normal retort temperature $\text{TRT}_{\text{H}}$. If a "proportional-corrected process" is applied to each one of the deviations as described in the previous tasks, the mathematical expressions for each correction will be as follows:

$$\text{Correction1} = \frac{t_{\text{TRT}} - t_{\text{H}}}{t_{\text{TRT}}}\Delta t \tag{17.15}$$

$$\text{Correction2} = \frac{t_{\text{LDT}} - t_{\text{TRT}}}{t_{\text{LDT}}}\Delta t \tag{17.16}$$

If we supposed now that $\text{TRT}_{\text{H}}$ is selected so that both corrections are equivalent, we can equate both terms:

$$\frac{t_{\text{TRT}} - t_{\text{H}}}{t_{\text{TRT}}}\Delta t = \frac{t_{\text{LDT}} - t_{\text{TRT}}}{t_{\text{LDT}}}\Delta t \tag{17.17}$$

$$\Rightarrow t_{\text{H}} = \frac{t_{\text{TRT}}^2}{t_{\text{LDT}}} \tag{17.18}$$

Equations 17.14 and 17.18 are identities.

**Validation of Task 5**

The utility of this approach to online correction of process deviations was demonstrated experimentally as a means of preliminary validation. Cylindrical cans (0.075 m diameter, 0.113 m height) containing a commercially prepared food product (Centauri Ravioli, 350 g) were thermally processed in a vertical still-cook retort under saturated steam with maximum working pressure of 40 psig at 140°C (Loveless, Model 177). Both retort temperature and internal product cold spot temperature were monitored with K-type thermocouples, and recorded with an Omega 220 data logger and modem with COM1 connection port. Cans were processed under different combinations of retort temperature and process time, with the temperatures recorded every 2 s. Each normal process was defined with come-up-time (CUT) of 7 min, during which the retort temperature increased linearly, followed by a period of constant retort temperature and a cooling cycle. Deviations during the process were deliberately perpetrated by manually shutting off the steam supply to the retort control system. Experiments were carried out in the Food Laboratory pilot plant of the Universidad Técnica Federico Santa Maria in Valparaiso, Chile.

## 17.3 ANALYSIS AND DISCUSSIONS

### 17.3.1 Equivalent Lethality Curves

Look-up tables are used to find the alternative process time for the corrected process, and can be presented graphically as "equivalent process lethality curves" for each scheduled product/process. Therefore, equivalent process lethality curves were constructed for each of the two simulated products used in this study, and are shown in Figures 17.1 and 17.2 for the case of solid (pure conduction) and liquid (forced convection), respectively.

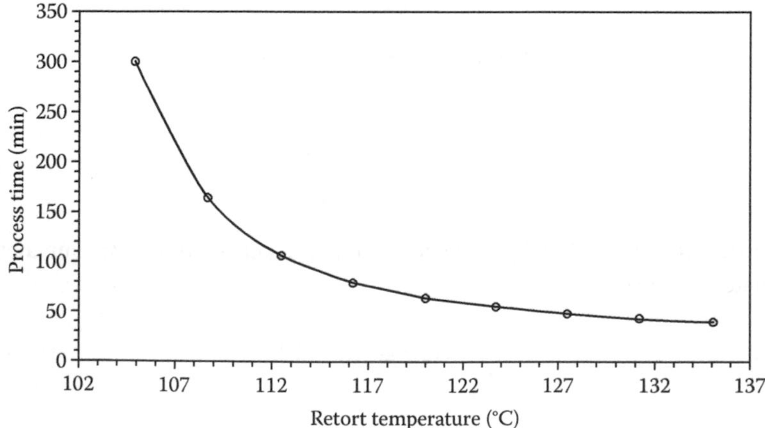


**FIGURE 17.1** Equivalent process lethality curve for simulated solid product under pure conduction heating, showing retort temperature/process time combinations that deliver the same final target lethality. (From Simpson, R., Figueroa, I., Teixeira, A., *Food Control,* 18, 458, 2007. With permission.)

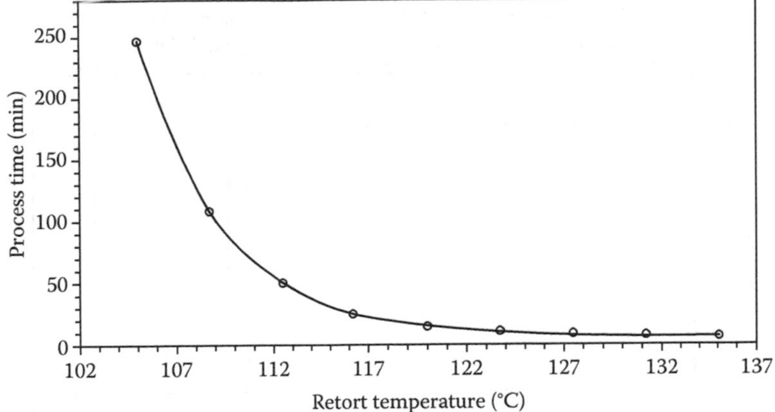


**FIGURE 17.2** Equivalent process lethality curve for simulated liquid product under forced convection heating, showing retort temperature/process time combinations that deliver the same final target lethality. (From Simpson, R., Figueroa, I., Teixeira, A., *Food Control,* 18, 458, 2007. With permission.)

### 17.3.2 Performance Demonstration

Figures 17.3 through 17.6 show results from the four product/process simulations carried out to demonstrate the performance of these strategies. The figures contain retort temperature profiles resulting from online correction of process deviations happening at different times during the process (early and late) to both solid and liquid

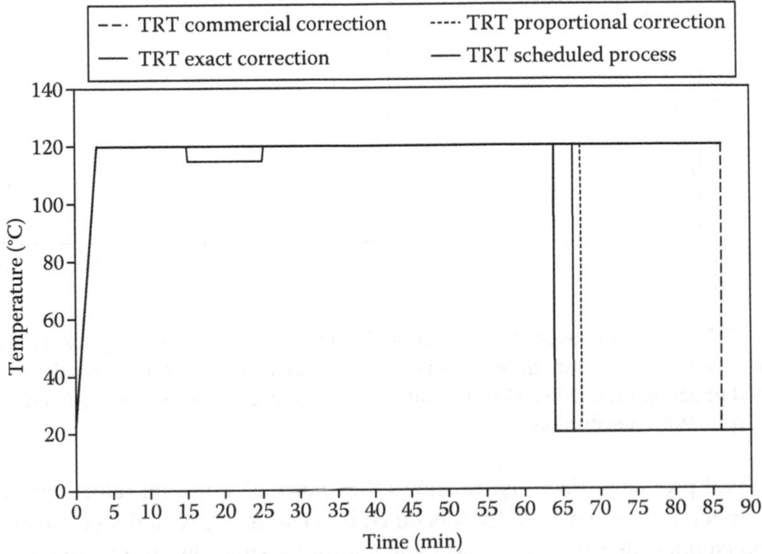


**FIGURE 17.3** Pure conduction simulation for online correction of an unexpected retort temperature deviation occurring early into the scheduled process for a cylindrical can of solid food under still cook. (From Simpson, R., Figueroa, I., Teixeira, A., *Food Control,* 18, 458, 2007. With permission.)

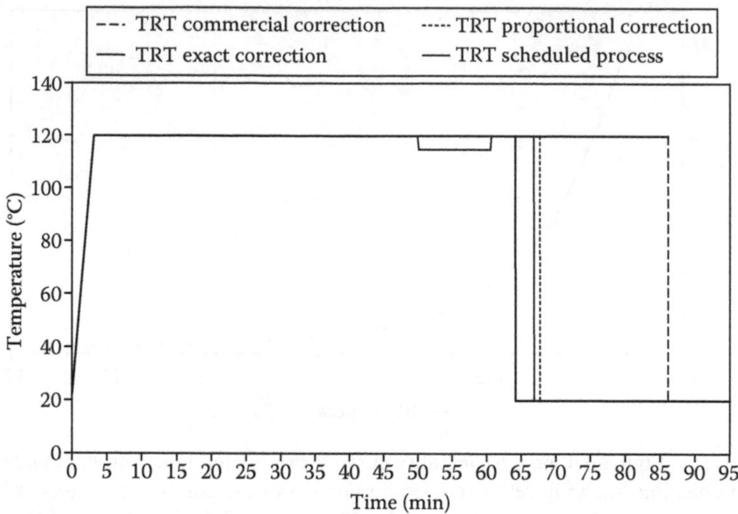


**FIGURE 17.4** Pure conduction simulation for online correction of an unexpected retort temperature deviation occurring late into the scheduled process for a cylindrical can of solid food under still cook. (From Simpson, R., Figueroa, I., Teixeira, A., *Food Control,* 18, 458, 2007. With permission.)

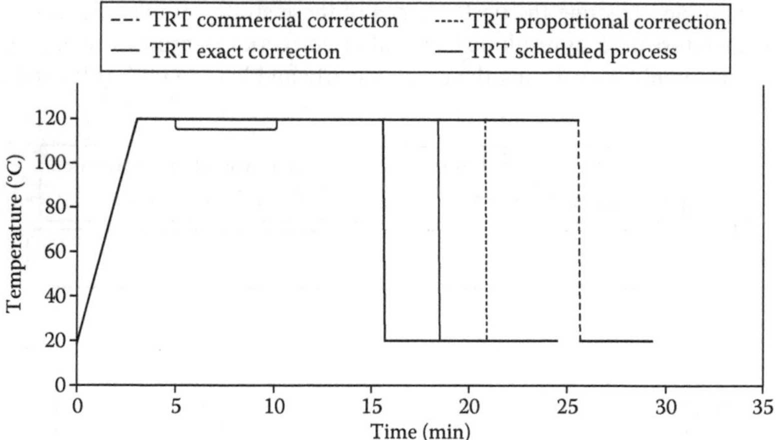


**FIGURE 17.5** Forced convection simulation for online correction of an unexpected retort temperature deviation occurring early into the scheduled process for a cylindrical can of liquid food under agitated cook. (From Simpson, R., Figueroa, I., Teixeira, A., *Food Control,* 18, 458, 2007. With permission.)

canned food products. Each figure shows the "normal" constant retort temperature profile expected for the originally scheduled process, along with the occurrence of a deviation (sudden step-drop in retort temperature for short duration) either relatively early or late into the process. In addition, for each deviation (one in each figure), three different alternative corrected process times are shown resulting from different strategies: exact correction, proportional correction, and commercial correction.

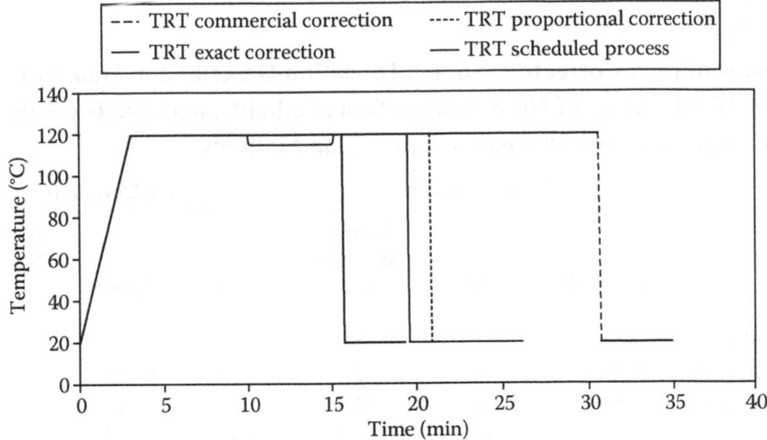


**FIGURE 17.6** Forced convection simulation for online correction of an unexpected retort temperature deviation occurring late into the scheduled process for a cylindrical can of liquid food under agitated cook. (From Simpson, R., Figueroa, I., Teixeira, A., *Food Control,* 18, 458, 2007. With permission.)

In all cases, the extended process time required by the commercial correction strategy is far in excess of the extended times called for by the other two strategies. Moreover, the new proportional correction strategy results in extending process time only slightly beyond that required for an exact correction, and will always do so. These results are summarized in Table 17.3, along with results from predicting the outcomes of each corrected process in terms of final process times required, and final lethality and quality retention achieved (using product/process data presented in Table 17.1). It is most interesting to note the dramatic improvement in nutrient (quality) retention between that resulting from the commercial correction (current industry practice) and from either of the other two strategies.

### 17.3.3 Demonstration of Safety Assurance by Complex Search Routine

This online correction strategy was validated and tested for safety assurance by executing an exhaustive search routine with the use of the heat transfer models. Recall, the problem to be solved by the search routine was to determine if the minimum final lethality delivered by all the corrected processes that could be found among all the various types of deviations and process conditions considered in the problem domain met the criterion that it had to be greater than or equal to the lethality specified for the original scheduled process. Table 17.2 identifies the problem domain by specifying the various types of deviations and process conditions that were explored and evaluated in the search routine. The search routine was designed to find a set of conditions under which the required search constraint was not met. No such conditions could be found.

### 17.3.4 Economic and Quality Impact of Online Correction Strategy

To analyze and assess the economic and quality impact of the novel strategy, two simulation procedures were executed. In both cases, 300 random simulations were carried out.

**TABLE 17.3**
**Outcomes of Each Corrected Process Deviation Described in Figures 17.3 through 17.6 in Terms of Final Process Time, Lethality, and Quality Retention for the Three Different Alternative Correction Methods**

| | Early Deviation | | | Late Deviation | | |
|---|---|---|---|---|---|---|
| | Time (min) | $F_0$ (min) | Nutrient Retention (%) | Time (min) | $F_0$ (min) | Nutrient Retention (%) |
| **Pure Conduction** | | | | | | |
| Scheduled process | 64.1 | 6.0 | 72.7 | 64.1 | 6.0 | 72.7 |
| Exact correction | 66.3 | 6.0 | 72.9 | 66.8 | 6.0 | 72.7 |
| Proportional correction | 67.5 | 6.5 | 72.2 | 67.5 | 6.2 | 72.3 |
| Commercial correction | 86.2 | 16.3 | 62.3 | 86.2 | 14.4 | 62.8 |
| **Forced Convection** | | | | | | |
| Scheduled process | 15.6 | 6.0 | 92.4 | 15.6 | 6.0 | 92.4 |
| Exact correction | 18.4 | 6.1 | 91.5 | 19.6 | 6.0 | 90.9 |
| Proportional correction | 20.8 | 8.0 | 89.7 | 20.8 | 7.0 | 89.9 |
| Commercial correction | 25.6 | 11.8 | 86.2 | 30.6 | 14.7 | 82.8 |

*Source:* Simpson, R., Figueroa, I., Teixeira, A., *Food Control,* 18, 458, 2007. With permission.

First, the 300 simulations included random deviations to compare both strategies, the commercial practice and the proportional correction proposed in this study. From each simulation, the exact time that corresponds to the proportional correction and the time according to the commercial correction were obtained. To calculate the percentage of overprocessing for each strategy, the following expression together with Equation 17.2 was utilized:

$$\% \text{ Overprocessing } = \frac{\text{time for the specified correction} - \text{exact time}}{\text{exact time}} \times 100 \quad (17.19)$$

According to simulations, for forced convection products, nearly 79% overprocessing occurred for the commercial correction, and only 36% for the proportional correction. In the case of pure conduction, the two levels of overprocessing were 34% and 2%, respectively. From statistical analysis, it is clear that both online control strategies are from different populations. The probability that both methods differ is 99.97% for forced convection, and 98.93% for pure conduction. Table 17.4 shows the complete results and statistical indicators for the 300 simulations.

Second, to assess quality, 300 simulations were executed to compare the quality retention of processes operated at the registered temperature and those operated at 3°C higher (i.e., a lower margin when compared with normal U.S. industrial practice).

**TABLE 17.4**
**Statistical Comparison between Proportional and Commercial Correction for Deviant Processes**

| | Process Type | |
|---|---|---|
| | **Forced Convection** | **Pure Conduction** |
| Number of simulations | 300 | 300 |
| Prescheduled time (min) | 23.47 | 72.87 |
| Loading time (min) | 5 | 5 |
| Unloading time (min) | 5 | 5 |
| **Proportional Control** | | |
| Overprocessing (%) | 36.54 | 2.22 |
| Standard deviation overprocessing (%) | 9.64 | 1.66 |
| Proportional control reduction capacity (%) | 25.62 | 1.95 |
| **Commercial Control** | | |
| Overprocessing (%) | 78.82 | 34.39 |
| Standard deviation overprocessing (%) | 14.30 | 5.81 |
| Proportional control reduction capacity (%) | 55.27 | 30.24 |
| **Difference** | | |
| Overprocessing (%) | 42.28 | 32.17 |
| Difference standard deviation (%) | 12.45 | 13.97 |

*Source:* Simpson, R., Figueroa, I., Teixeira, A., *Food Control*, 18, 458, 2007. With permission.

**TABLE 17.5**
**Statistical Comparison between Processes Operated at TRT and Processes Operated at 3°C Higher to Avoid Deviant Processes**

| | Process Type (Pure Conduction) |
|---|---|
| Number of simulations | 300 |
| TRT | 110°C–135°C |
| $t_{CUT}$ | 5–15 min |
| $T_{ini}$ | 20°C–70°C |
| Average retention at TRT | 68.8% |
| Average retention at TRT + 3°C | 61.6% |
| % Quality reduction | 10.5% |

*Source:* Simpson, R., Figueroa, I., Teixeira, A., *Food Control,* 18, 458, 2007. With permission.

Table 17.5 indicates that processors can avoid deviant processes (operating at higher retort temperatures) but with a considerable impact in quality retention (10.5% lower). In addition, processors operating at 3°C $TRT_H$ than the registered process will have much higher energy consumption.

### 17.3.5 Correction Strategy without Extending Process Time and Preliminary Validation

Using data from the constant-temperature heat penetration tests carried out in this study, an equivalent process lethality curve (for a target lethality of $F_0=6\,\text{min}$) was constructed for the commercial ravioli product and can size used in this study, and is shown in Figure 17.7.

In order to validate the safety assurance of this new online control strategy, a number of heat penetration experiments were carried out in which process deviations were deliberately perpetrated by manual shut-off of the steam supply to the retort, causing the retort temperature and pressure to fall to a lower level for several minutes, after which the steam supply valve was reopened and the deviation quickly recovered. As soon as the complete anatomy of the deviation was known upon recovery, Equation 17.14 was used to calculate the high temperature equivalent process time ($t_{Hi}$), from which to obtain the $TRT_H$ needed to accomplish the correction, using the isolethality curve in Figure 17.7. The retort controller set point was immediately adjusted upward to the correction temperature ($TRT_H$), and brought back down to the originally scheduled retort temperature after an elapsed time equal to the duration of the initial perpetrated deviation. The process was then allowed to proceed normally for the duration of the remaining originally scheduled process time. During each test, retort and internal product cold spot temperatures were continually

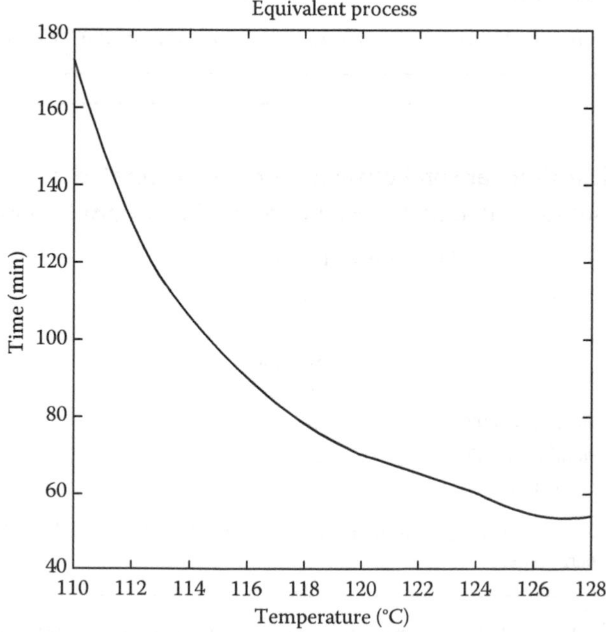


**FIGURE 17.7** Isolethality curve showing equivalent combinations of process time and retort temperature that achieve the same process lethality ($F_0 = 6\,\text{min}$) for ravioli packed in cylindrical cans (0.075 m diameter, 0.113 m height). (From Simpson, R., Figueroa, I., Teixeira, A., *Food Control,* 18, 458, 2007. With permission.)

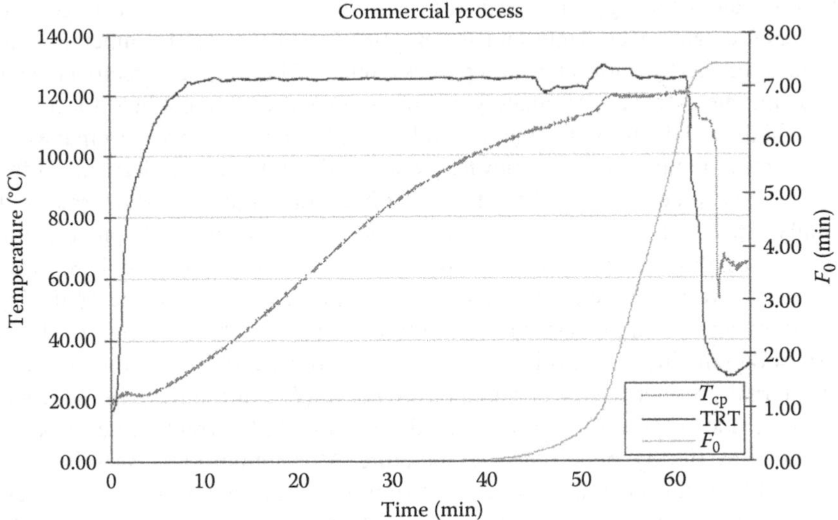


**FIGURE 17.8** Profiles of retort (TRT) and internal product cold spot temperatures ($T_{cp}$) over time (scale on left), along with profile of accumulated lethality over time (scale on right) from heat penetration test with ravioli in cans (0.075 m diameter, 0.113 m height) experiencing perpetrated process deviation immediately followed by temporary high retort temperature correction (calculated online). (From Simpson, R., Figueroa, I., Teixeira, A., *Food Control,* 18, 458, 2007. With permission.)

measured and recorded, and accumulated lethality was calculated as a function of cold spot temperature over time using the General method (assuming a $z$-value of 10°C). Results from a typical test run are presented in Figure 17.8, with the process conditions and parameters used for the test listed in Table 17.6.

Both temperature and lethality are shown as functions of time in Figure 17.8, with the temperature scale shown along the left side vertical axis, and the lethality

**TABLE 17.6**
**Process Conditions and Parameters Used for Heat Penetration Test with Ravioli Packed in Cylindrical Cans (0.075 m Diameter, 0.113 m Height) Producing Results Shown in Figure 17.8**

| Process Parameters (Units) | Value Chosen For Heat Penetration Test |
|---|---|
| Reference temperature (°C) | 121.1 |
| Scheduled process time (min) | 61 |
| Low temperature at deviation (°C) | 122 |
| Initial internal product temperature (°C) | 19 |
| Time duration of deviation, Δ (min) | 5.5 |
| Target process lethality, $F_0$ (min) | 8 |
| Microbial temperature factor, $z$ (°C) | 10 |
| Scheduled retort temperature (°C) | 125 |

*Source:* Simpson, R., Figueroa, I., Teixeira, A., *Food Control,* 18, 458, 2007. With permission.

scale shown along the right side vertical axis. In the case of this test, the target value for process lethality was 8 min and the normal scheduled retort temperature was intended to be 125°C for a scheduled process time of 61 min. The perpetrated deviation was initiated after approximately 45 min into the process, and held for 5.5 min, during which time the retort temperature fell to 122°C. Upon recovery from the deviation, the retort temperature was elevated to approximately 128°C (determined from the calculation procedure described above) for five more minutes, and returned to the originally scheduled 125°C for the remainder of the scheduled 61 min process time.

The measured retort temperature profile can be seen in Figure 17.8, clearly revealing the profile of the perpetrated deviation immediately followed by the high temperature correction process and return to normal, with the cooling cycle beginning right on schedule at the originally appointed process time of 61 min. The measured internal product cold spot temperature curve ($T_{cp}$) in Figure 17.8 reflects the expected erratic response to the combined deviation and correction perturbations experienced by the dynamic retort temperature. Most importantly, in spite of the erratic profile of the internal product cold spot temperature, the final accumulated lethality, calculated as a function of this profile by the General method, still reached the target value of 8 min specified for the process.

## 17.4 CONCLUSIONS

This chapter has described a practical and efficient strategy for online correction of thermal process deviations during retort sterilization of canned foods. This strategy takes into account the duration of the deviation in addition to the magnitude of the temperature drop. It calculates a proportional extended process time at the recovered retort temperature that will deliver the final specified target lethality with very little overprocessing in comparison to current industry practice. In addition, it is described as a strategy that can assure thermal sterilization during unexpected process deviations without extending scheduled process time and with an absolute minimum of overprocessing. They are applicable to all types of foods or containers (solids, liquids, mixtures) or mechanisms of heat transfer (conduction, convection, both combined).

Results from an exhaustive search routine using the complex method support the logic and rationale behind the strategy by showing that the proposed strategy will always result in a corrected process that delivers no less than the final target lethality specified for the originally scheduled process. Economic impact of adopting this strategy over that currently used in industry practice can be a significant increase in production capacity for a typical cannery. In addition, utilizing this novel strategy canned products will attain a much higher quality.

## NOMENCLATURE

| | |
|---|---|
| $A$ | heat transfer area |
| $Cp$ | heat capacity |
| $F_{obj}$ | $F$-value establish for the schedule process |
| $F_p$ | $F$-value for the process with corrected deviations |

| | |
|---|---|
| $F_{proportional}$ | $F$-value with the proportional correction |
| $F_{Tol}$ | $F$-value specified for normal scheduled process |
| $h$ | vertical distance |
| $H$ | operating time of the plant during the season (h) |
| $n$ | number of deviations occurring during the process |
| $N_{Bi}$ | number of batches processed per retort $i$ during the season |
| $r$ | radial distance |
| $t$ | time |
| $t_{ci}$ | time to load retort $i$ with product $j$ (h) |
| $t_{di}$ | time to download retort $i$ with product $j$ (h) |
| $t_D$ | corrected process time |
| $t_{D\ commercial}$ | corrected process time for commercial strategy |
| $t_{Di}$ | process time at the deviation temperature $TRT_i$ |
| $t_{D\ proportional}$ | corrected process time for proportional strategy |
| $t_H$ | equivalent process time at required higher temperature |
| $t_{LDT}$ | equivalent process time at the lower deviation temperature |
| $t_{oj}$ | time to operate retort $i$ (process cycle time) with product $j$ (h) |
| $t_p$ | process time |
| $t_{TRT}$ | preestablished process time at retort temperature TRT |
| $T$ | temperature |
| $T_{cp}$ | temperature in the coldest point |
| $T_{ref}$ | reference temperature for microbial lethality |
| TRT | retort temperature |
| $TRT_H$ | lowest temperature at which must be reestablish process temperature after deviation $i$ |
| $TRT_i$ | lowest temperature during the deviation $i$ |
| $TRT_{max}$ | temperature than cannot be surpass (practical limit) |
| $U$ | universe of feasible process conditions in search routine |
| $U_{ht}$ | global heat transfer coefficient |
| Vol | volume of can |
| $z$ | temperature change necessary to alter the thermal death time (TDT) by one log-cycle |
| $\Delta t$ | duration time of the process deviation |
| $\Delta t_i$ | duration of deviation $i$ |
| $\alpha$ | thermal diffusivity |
| $\rho$ | liquid density |

## REFERENCES


Akterian, S.G. 1996. Studying and controlling thermal sterilization of convection-heated canned foods using functions of sensitivity. *Journal of Food Engineering*, 29: 329–338.

Akterian, S.G. 1999. On-line control strategy for compensating for arbitrary deviations in heating medium temperature during batch thermal sterilization processes. *Journal of Food Engineering*, 39: 1–7.

Ball, C.O. 1928. *Mathematical Solution of Problems on Thermal Processing of Canned Food*. University of California Publications in Public Health 1, N 2, 15–245.

Bigelow, W.D., Bohart, G.S., Richardson, A.C., and Ball, C.O. 1920. *Heat Penetration in Processing Canned Foods*. Bull. No. 16-L. Research Laboratory of the National Canners Association, Washington, DC.

Datta, A.K., Teixeira, A.A., and Manson, J.E. 1986. Computer-based retort control logic for on-line correction of process deviations. *Journal of Food Science*, 51(2): 480–483, 507.

Holdsworth, S.D. 1997. *Thermal Processing of Packaged Foods*. Blackie Academic & Professional, London.

Larkin. 2002. Personal communication. Branch Chief, National Center for Food Safety and Technology, Food and Drug Administration (FDA/NCFST), Chicago, IL.

Manson, J.E., Zahradnik, J.W., and Stumbo, C.R. 1970. Evaluation of lethality and nutrient retentions of conduction-heating food in rectangular containers. *Food Technology*, 24(11): 1297–1301.

Manson, J.E., Zahradnik, J.W., and Stumbo, C.R. 1974. Evaluation of thermal processes for conduction heating foods in pear-shaped containers. *Journal of Food Science*, 39: 276–281.

Pham, Q.T. 1987. Calculation of thermal process lethality for conduction-heated canned foods. *Journal of Food Science*, 52(4): 967–974.

Simpson, R. 2004. Control logic for on-line correction of batch sterilization processes applicable to any kind of canned food. *Symposium of Thermal Processing in the 21st Century: Engineering Modeling and Automation*, at IFT Meeting, Las Vegas, NV.

Simpson, R., Aris, I., and Torres, J.A. 1989. Sterilization of conduction-heated foods in oval-shaped containers. *Journal of Food Science*, 54(5): 1327–1331, 1363.

Stumbo, C.R. 1973. *Thermobacteriology in Food Processing*, 2nd ed. Academic Press, New York.

Teixeira, A., Dixon, J., Zahradnik, J., and Zinsmeiter, G. 1969. Computer optimization of nutrient retention in the thermal processing of conduction-heated foods. *Food Technology*, 23(6): 845–850.

Teixeira, A.A. 1992. Thermal process calculations. In: *Handbook of Food Engineering*, D.R. Heldman and D.B. Lund (Eds.). Marcel Dekker Inc., New York, Chapter 11, pp. 563–619.

Teixeira, A.A. and Manson, J.E. 1982. Computer control of batch retort operations with on-line correction of process deviations. *Food Technology*, 36(4): 85–90.

Teixeira, A.A. and Tucker, G.S. 1997. On-line retort control in thermal sterilization of canned foods. *Food Control*, 8(1): 13–20.

Teixeira, A.A., Balaban, M.O., Germer, S.P.M., Sadahira, M.S., Teixeira-Neto, R.O., and Vitali, A.A. 1999. Heat transfer model performance in simulating process deviations. *Journal of Food Science*, 64 (3): 488–493.

# 18 Plant Automation for Automatic Batch Retort Systems

*Osvaldo Campanella, Clara Rovedo, Jacques Bichier, and Frank Pandelaers*

## CONTENTS



## 18.1 INTRODUCTION

The rapid increase of plant automation in the food processing industry of the early 1980s drove the American Society of Agricultural Engineers to organize the Food and Process Engineering Institute in 1985 "with the objective of improving the information transfer between food processors, food equipment manufacturers, food handlers, and university/government personnel" (Payne, 1990). Manufacturers have the difficult task of growing their businesses in today's competitive market place. For such purpose, they must face challenges of increasing productivity and product quality, while reducing operating costs and safety risks. Plant automation has historically been the main tool to assist the manufacturer in meeting those challenges.

Over the last decade, there have been numerous changes in thermal processing of low-acid foods packaged in hermetically sealed containers. For that reason, retorts and their control systems have been evolved accordingly to fulfill food processors requirements for more flexibility in handling lighter weight containers, better options for process optimization, and with the increasing labor costs and increasingly complex food safety regulations, they are also seeking for fully automated retort operations.

Food engineers have been successfully applying modern computer technology to design and develop process strategies and systems that significantly reduce the amount of defective post-sterile product (Teixeira and Tucker, 1997). Mathematical models can be embedded in the process monitoring and control system so when any unexpected change in the process conditions is detected the model can make automatic adjustments in the process conditions to achieve properly sterilized product.

Given the importance of public health safety upon consuming canned food, processors must operate in strict compliance with the U.S. Food and Drug Administration's (FDA's) low-acid canned food regulations (Simpson et al., 2007a). The FDA has strict rules regarding documentation and record keeping of all critical control points (CCPs) for the sterilization of foods. Therefore, it is essential to clearly identify those CCPs that must be monitored during the process to maintain reliance on the performance of the controller in delivering an adequate product and to provide the mathematical model the information needed to calculate alternate processes when temperature deviations occur.

Automatic control systems must be designed, tested, and validated to assure that they meet the strict process requirements of today's thermal process applications. JBT FoodTech (formerly FMC FoodTech) as one of the world's leading technology and solutions providers to the food industry has many years of experience in designing automatic controlled systems: more than 50% of the world's shelf-stable foods are filled, closed, and sterilized with JBT FoodTech (formerly FMC FoodTech) equipment. Such vast experience and knowledge was combined in the present chapter, which describes the process control and validation considerations required to bring automation to the sterilization process area in a food processing plant. Although discussions are mostly focused on the flexible batch retort systems, the authors have added comments about control strategies for continuous retort systems whenever appropriate.

An overview of the overall system is described, which comprises the loader/unloader system, the basket tracking system (BTS), and the batch retort system. The main CCPs for each system are presented together with a description of the control highlights and the documentation requirements necessary to safely process a product packaged in a hermetic-sealed container. The chapter concludes with theoretical considerations on process optimization using advanced control systems.

## 18.2 AUTOMATED BATCH RETORT SYSTEMS

Unlike retort systems in the 1980s, today's automated retort systems are very sophisticated and fully automated. They consist of multiple subsystems that must work together flawlessly in order to provide exact sterilization processes for each and every retort cycle. The automated batch retort system (ABRS) typically consists of the following subsystems:

- Material handling system, which includes the container loader/unloader and the basket conveyor, all with their own material handling control system
- Retorts with their own process control system
- Basket tracking system (BTS)

### 18.2.1 Material Handling System

Loaders and unloaders are used to automatically place or remove rigid or flexible packages into the retort baskets. Containers can either be loaded into baskets (cans and jars) or onto self-supporting trays assembled as pallets (flexible plastic containers and retortable pouches). Generally, the loading/unloading sequence is controlled by a programmable logic controller (PLC) system.

Once the baskets or pallets are loaded, they need to be transported to an available retort for processing. Typically, three types of transport methods are used: (1) automated guide vehicles (AGVs), (2) X–Y shuttles, and (3) basket conveying systems.

The AGV (Figure 18.1) can be wire guided and follow a wire buried beneath the retort room floor or be laser guided, and use reflectors that are strategically placed around the room, for navigation in delivering the product to the retort. The AGV communicates with a control system to tell the vehicle where to go and what to do.

**FIGURE 18.1** AGV loading a pallet into a batch retort. (Photo courtesy of JBT FoodTech [formerly FMC FoodTech].)

When the AGV arrives at the retort, it communicates with the retort throughout two photocells, both on AGV and retort. When the AGV is in position at the retort docking point, it sends a signal that it is ready for a load transfer. The receiver at the retort senses this signal, and if the retort is also ready to receive baskets, it uses this signal as a permissive to allow the transfer of baskets into the retort, then the AGV activates its conveyor for a load transfer. The retort detects the basket as it enters and starts an internal chain conveyor to index the basket into position in the vessel. Once the retort is full and the AGV has moved away, the retort door automatically closes and the operator is notified to enter the recipe information and begins the process.

The AGV provides the most flexibility of all material handling methods. The system can be used in areas where the building structure prevents the use of other material handling methods. Also, if more retorts are added to the system, the AGV can be programmed to service the new vessels without major changes to the system.

The most commonly used material conveying system is probably the X–Y rail shuttle. The shuttle rides on rails between the retorts and the loader/unloader. As product is filled into the baskets, the full baskets are staged on a buffer conveyor. When a retort is ready to receive product, the shuttle moves in front of the loading buffer and receives a complete load of baskets for one retort. The shuttle is then positioned in front of a waiting retort and the product is transferred to the retort. Similarly, processed product is transferred from the retort to the shuttle and then moved to the unloading station.

A particular type of X–Y shuttle is the dual deck shuttle system (Figure 18.2). In this case, two shuttles are built side by side and move in tandem: one shuttle handles only the unprocessed product and the other handles only the processed product.

**FIGURE 18.2** X–Y dual deck shuttle system loading an JBT [former FMC] SuperAgi.™ (Photo courtesy of JBT FoodTech [formerly FMC FoodTech].)

The simplest basket handling system is the basket conveying system. This system employs a dedicated conveying system that moves individual baskets from the loader to the retort and then after processing from the retort to the unloader. The advantage of the basket conveyor is that it provides a smaller footprint when the plant is tight on space and it is relatively less costly than either the AGV or the shuttle.

#### 18.2.1.1 Preparing the Load for Process

Retorts have a basket detection device, called a pallet sensor. The moment each basket is detected by the pallet sensor, the retort internal conveyor will start its movement and pull the basket inside the retort. The retort conveyor has an absolute encoder connected to the conveyor drive such that the conveyor always knows its position.

The identity of each basket that enters the retort is verified and transmitted to the BTS. This is done by positioning the basket radio frequency (RF) tag above a reader mounted in the retort. When the basket RFID tag reading is successfully completed, the retort receives the OK signal from the BTS. The conveyor will automatically restart the movement of the basket such that the conveyor is positioned correctly to receive and pull the next basket into the retort. When all the baskets from the transportation device (AGV, shuttle, or conveyor) have fully entered the retort, the device knows its loading cycle is finished, so it moves away from the retort docking point to pick up its next load. When all the baskets have been successfully identified, the retort conveyor moves them to their final position in the retort.

### 18.2.2 Retorts and the LOG-TEC™ Momentum Control System

Retorts are automatically controlled by their own process control system. In the late 1990s, JBT (former FMC) developed LOG-TEC™ Momentum process management system, which allows a continuous follow-up of the process with minimum operator interactions required.

The LOG-TEC™ Momentum system (Figure 18.3) has been designed by a unique combination of professionals in the fields of food processing, electrical and software engineering, and computer technology. LOG-TEC™ Momentum was designed as a decentralized system where all controllers for each retort are networked to a host (workstation) computer that serves as the central database. The host behaves as the server in a typical modern client/server computer network and functions as the record keeper for all process parameters and electronic digital copy of logs as well as master of security for the systems.

The host computer is the initiator of all communications. After initial information is downloaded from the host, the controller assumes total independent control of the sterilizer. Cook times, alternate cooks, deviation corrections, process control, and data collection are all calculated or managed by the controllers running a customized application script on Microsoft's Windows CE embedded real-time operating system. Each controller has been designed to effectively interface with the person operating the sterilizer, giving simple prompting, easy input, and correction of information.

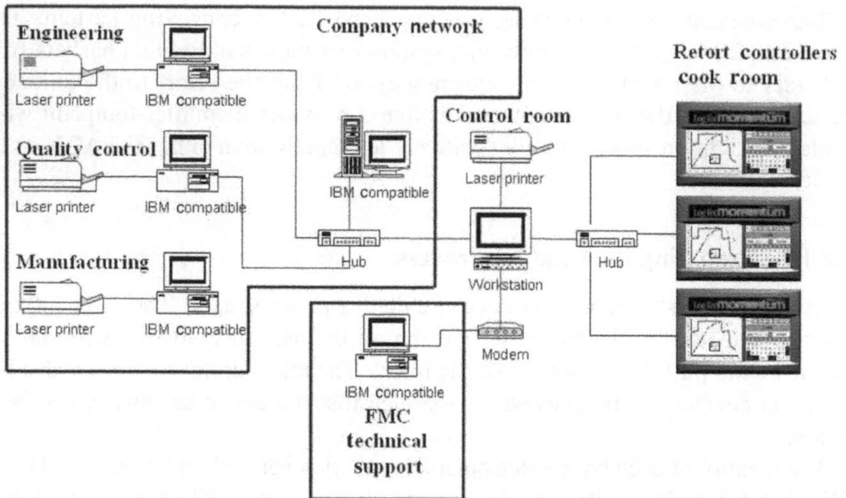


**FIGURE 18.3** LOG-TEC™ Momentum system: all controllers for each retort, the company network, and technical support are networked to a host (workstation) computer that serves as the central database. (Photo courtesy of JBT FoodTech [formerly FMC FoodTech].)

All records, events, logs, and changes are captured and maintained by the controllers. This information is also synchronized and retransmitted back to the host for storage and retrieval. Security is established and managed at the host, transmitted to the controller and maintained as configured by the manager. The areas of engineering, quality control, and manufacturing have direct access to the encrypted data in the secured database at the host computer to retrieve/analyze records through an Ethernet transmission control protocol/Internet protocol (TCP/IP) access protocol. Similarly, if the customer's gives the network permission, JBT (former FMC) has access to the host computer through a direct TCP/IP protocol or through a virtual personal network. This remote and secured access allows technical support specialists and processing experts to provide instantaneous assistance to the costumers irrespective of the plant location in the world.

When a controller boots up, every assigned recipe is downloaded from the host computer's database in the form of an assigned recipe listing. The operator is limited to simply selecting from a displayed list and cannot select a recipe that has not been assigned to the controller by the system administrator (or manager). Therefore, assigning and withdrawing recipes is a very powerful management tool that expedites daily production configuration, maximizes security, and minimizes mistakes. Changes made to any recipe are permanently captured and saved.

Product and container identification, heating factors, sterilization values, process definition, and system information can all be reviewed and updated through the recipe configuration file. The recipe system provides for up to 10,000 recipes, in any mode combination and choice desired. The operator can download a process for any particular product to a controller by selecting the product from the controller's displayed list. All recipes are generated and stored centrally. All process data are recorded as encrypted files, stored centrally on the PC hard disk, printed out on the central printer, and also logged on the local data recorders, meeting the FDA/US Department of Agriculture (USDA) and the latest hazard analysis and critical control point (HACCP) requirements.

Prior to sterilizer operation and following operator entries, the controller prompts for a verification of the selected recipe. Then, the script calls for the recipe model to calculate and determine the process time and temperature profile. While in process, unpredictable temperature process deviation are automatically corrected in real time with alternate processes, which may be electronically reviewed or printed at the host and reviewed on paper.

One key feature of LOG-TEC™ Momentum is that it generates accurate and reliable documentation. In designing the controller as FDA's 21 CFR Part 11 compliant, there are five basic elements that were addressed:

1. *Electronic records*: The electronic records are secured and encrypted to prevent any changes and the system is designed so that records can be reviewed and approved electronically. They can also be printed, but they cannot be changed.
2. *Electronic signatures*: To sign a document the user must enter an electronic signature. The electronic signature consists of two unique means of identification: the users ID and password.
3. *Audit trail*: The system is designed with a complete audit trail. The audit trail is a detailed database that records who made the entry, what was changed, and when it was changed. Any data entry, change in a recipe value, change in a valve tuning parameter, or any other parameter that is critical to system safety or performance is recorded in the audit trail table.
4. *System security*: User and manager IDs are established to assure the security level for anyone entering data at the controller. Typically, IDs and passwords are checked when the user wishes to review the entries made by the operator at the start of the cycle, review the amount of time that will be added to the sterilization phase if a temperature deviation occurs, or display the logs sent to the host computer. The users' privileges of the system can be configured by the system administrator, who determines who can review records, change tuning tables, or change the configuration of the process recipes.
5. *Change control procedure*: It is an important feature in using a control system that is rigorously validated. The base code such as the host system software, controller hardware, and software, as well as the communication software is the same for all installations regardless the type of sterilizer application. The changes that are made for a particular system are the field devices, such as level probes, resistance temperature device (RTD) or analog and digital valves, and the compiled system-specific script. Thus, since only the compiled script program changes from one application to the next, the task of system validation is simplified.

Besides those very important process data management features, the LOG-TEC™ Momentum control system allows an exact repetition of the preprogrammed temperature and pressure profiles, which results in consistent high-level product quality, minimizing risk of product loss.

#### 18.2.2.1 Control Strategies for Critical Control Points

The CCPs that the control system must prompt, monitor, and control in order to attain commercial sterility in this type of equipment are listed below:

- Initial temperature of the product
- Retort temperature
- Retort pressure
- Process time
- Water flow
- Water level
- Rotation speed

During start-up, performance qualification tests are conducted on the equipment. Temperature distribution (TD) tests are conducted to verify that the equipment is performing as designed and that the temperature around the slowest heating zones throughout the vessel has reached the scheduled process temperature when the sterilization phase begins. The results from the TD tests supply the ranges of the CCPs enumerated above. The minimum and or maximum value acceptable for each CCP is configured in the controller so alarms would prompt the operator if any variable is beyond the limits necessary for an appropriate cook.

##### *18.2.2.1.1 Initial Temperature*

A typical food processing plant operation would have a minimum initial temperature (IT) designed for each process cycle. Minimum IT is generally determined according to the value recorded while performing heat penetration (HP) tests to design the process.

Most of today's controllers are programmed to prompt the operator to manually read the IT from a thermometer and enter the value from a keypad. If the operator records an IT lower than the minimum, the controller is programmed to generate an alarm and the lot is set aside as a process deviation; if the IT is higher than the minimum value, it plays a role as safety factor to compensate for any eventual temperature deviation. In this last case, nowadays technology allows to minimize overprocessing by using IT as an input for mathematical models programmed in the controllers that calculate the process at the actual IT in real time.

It is important to mention that there are several types of retorts (batch and continuous) that use water in first step of the process: the recirculating water of a full immersion, a steam water spray, or a cascading batch retort, as well as the infeed leg of an hydrostatic retort, the preheat vessel of a continuous pressure sterilizer, or the cushion water of a crateless retort. In all cases, the temperature of the water must be controlled to remain above the IT of the product to prevent colder water from decreasing the IT, which would result in product under-processed. High technology provides an alternate solution: a control approach that could monitor the water temperature with an RTD and assign the coldest recorded value as the IT for the process. Again in the event that such coldest recorded value is below the minimum, the controller could generate an alarm log and tag the process as a deviation.

#### *18.2.2.1.2 Retort Temperature*

All retort systems with an automatic programmable controller are equipped with at least one RTD to measure the temperature of the heating and cooling media within the processing vessel. The strategy to monitor, to control, and to provide process alternatives vary from the simplest to the most complex algorithms.

The simplest method is to monitor and control the temperature at a well-defined point through a proportional-integral-derivative (PID) control loop. The control set point is typically set at a higher value (+1°F or +2°F) than the scheduled process temperature. This approach is necessary to prevent nuisance process deviations when the retort temperature oscillates about the set point because of insufficient tuning, variation in the steam supply pressure, aging field devices, contamination in the pneumatic lines, or gaps in the production line (for retort systems with continuous container handling such as hydrostatic or continuous agitating retorts). Under the simplest method, the controller alarms the operator that the temperature is below the schedule process temperature. The operator follows manual standard operating procedures to identify, segregate, and document the deviation.

As programmers design systems with higher level of automation, programmable controllers are coded with increasing levels of monitoring, control, and processing decisions. Programmers can add logic to identify RTD failures. For instance, when successive RTD scans vary by an amount that violates the laws of thermodynamic or RTD scans are outside the operating range of the sensor, the controller can alarm the operator, safely park the retort in an idle state, and list a set of control override options to cool the load without damaging the containers. If the control system is equipped with a redundant dual element RTD, the control can automatically swap the input for the control loop from the failed RTD to the backup RTD and resume normal process.

More advanced schemes store a table of precalculated alternate processes. Under this scheme, as the controller detects an RTD reading below the scheduled process temperature, it automatically extends the process time to next alternate value that was precalculated to compensate for the lower retort temperature. Even more, as it is discussed in Section 18.3.1, advanced controllers have the capacity to run finite difference models that can predict the time/temperature history of the coldest temperature within the containers that is being sterilized in the coldest heating zone within the processing vessel. Therefore, in the event that the temperature unpredictably drops below the scheduled process temperature, the process can be recalculated in real time and extend the sterilization time to guarantee a safe process delivery to the target *F*-value.

#### *18.2.2.1.3 Retort Pressure*

Accurate control of pressure within the vessel is essential to maintain the integrity of lightweight cans, glass jars, and flexible containers. A pressure transmitter communicates the actual vessel pressure to the control system. As temperature within the container increases, the internal pressure of the container also increases as a result of thermal expansion of gas and volumetric thermal expansion of the product.

Therefore, the retort pressure needs to be increased to compensate such internal pressure to prevent any deformation of the container or of the sealing cap. Moreover, as the pressure required to maintain the integrity of the container is higher than the saturated steam pressure at a processing temperature, then overpressure is provided during the process by injecting pressurized air and is controlled according to the set points of the recipe. Any excess pressure is vented through an analog vent valve; also a safety valve is added on the top of the retort to prevent pressure built up in the event of any device failure in pressure monitoring hardware, air control valve, vent control valve, or water control valve.

There are two main techniques to develop a pressure profile for a flexible container: the differential pressure method and the linear variable displacement transducer (LVDT) method.

*Differential pressure method*: It is a set of trial-and-error tests conducted with sealed flexible containers or glass jars instrumented with internal pressure sensors, which are placed in a pilot sterilizer equipped with its own sensor that measures the pressure within the vessel. The objective of these tests is to minimize the pressure differential between the internal and the external pressures across the boundaries of the container ($\Delta P$). As time/pressure history is collected from each successive test, data are analyzed and pressure profiles are modified in order to minimize $\Delta P$. This approach is time consuming and often leads to container failures and messy pilot retorts. This method is more appropriate for rigid containers as flexible containers tend to change their shape to minimize pressure differences, thus yielding poor results.

*LVDT*: It is an electromechanical device used for accurate measurement of linear displacements. The electrical output of this device is proportional to the displacement of a sliding stainless steel spring-loaded rod whose end rests on the most flexible surface area of the container. The probe can measure displacements in 1/1000 in. increments. The LVDT probe has a 4–20 mA output signal, which connected to the control system of the retort allows the LVDT signal to be configured as an input for the pressure control loop. As the internal pressure of the container increases and expands, the LVDT detects the expansion causing the pressure control loop to react by adding air pressure to the vessel until the deformation is canceled. Similarly, if the LVDT detects a compression of the flexible container, the pressure control loop compensates by opening the vent valve of the vessel until the deformation is canceled. The control system maintains a record of the time/pressure history necessary to prevent the deformation of the container throughout the entire process cycle. The resulting pressure profile is transferred to the recipe configuration of the product.

#### *18.2.2.1.4 Process Time*

Sterilization times are determined by performing HP tests in every product/container and by calculating the time necessary to process to achieve the proper lethality for that specific product. Modern controllers allow programming process variables as a recipe with heating factors, sterilization values, process definition, and system information, so the script could call for the recipe model to calculate and determine the process time and temperature profiles. For safety purposes, the program should be written in such a way that the set points for temperature, pressure, and time for each segment must be achieved before the next segment begins.

Also for certain type of controllers, alternate processes can be calculated in real time to account for any temperature deviation during the cycle. Different process calculation programs can be used not only to achieve a temperature and overpressure profile that is optimized for the container and the product, but also to correct online process temperature deviations. The industry uses a tabulated generic set of time/temperature combination or mathematical models, such as the Ball method, NumeriCAL™ On-Line (JBT proprietary software based on a finite difference mathematical model, JBT Technologies, Madera, CA, USA), or CTEMP (CCFRA, Campden, United Kingdom), among others. Several numerical methods for online correction of process temperature deviations have been developed in recent years for batch as well as continuous retorts. Those methods are discussed at the end of this chapter.

###### *18.2.2.1.5 Water Level*

The thermal process can be affected by the level of water inside the retort. For immersion processes, if the water level is higher than the target level, the water turbulence in the void space among containers and around the basket decreases, and therefore, the heat transfer to the containers decreases as well. On the contrary, if the water level is too low, certain flexible containers such as plastic bottles and cups may not have enough buoyancy during the process to keep their integrity while under agitation. In particular, for static processes and if steam water spray are used as the heating medium, with high water level any submerged containers could be under-processed, because the overall heat transfer coefficient in water is lower than in an environment exposed to the mixture of steam and high-velocity sprays. In any of those conditions, if the water level reaches values beyond predetermined low and high limits, then the controller alarms, a deviation log is marked on the electronic records, and the product is held and tagged as deviated.

The water level increases as steam condenses, which is continuously monitored. As the level reaches the maximum allowed, the controller drains the excess water throughout a combination of drain valves until the level probe indicates that the proper operating level has been regained.

Some control systems are equipped with an analog level probe within the processing vessel, which gives the engineer an additional programming flexibility.

###### *18.2.2.1.6 Water Flow Rate*

The recirculation water flow forces convection of the heating media throughout the void spaces around the containers, creates high turbulence in the recirculation pipe, distributes uniform water temperature to the spray bars and nozzles, and maintains a high water flow rate through the nozzle. A set of the TD tests are usually designed to challenge the equipment with water flow rates lower than expected during normal production. If the flow rate drops below the minimum designed flow rate, as challenged in TD tests, the controller will alarm and flag a deviation.

Programmers are advised to also include a high flow-rate alarm (5% above the normal production flow rate) as well, to capture the possibility that the maintenance staff have incidentally forgot to restore some nozzles or the end flushing caps of the spray bars after routine cleaning. The missing hardware would provide less resistance to the recirculating water, and thus, the recirculating water flow rate may increase.

The proper functioning of the water recirculation and injection through the nozzles are continuously checked during the total process from come-up to the end of cooling by monitoring the flow rate with the flow meter and the pressure differential between the discharge side of the pump and the process vessel.

*18.2.2.1.7 Rotation Speed*

Rotation improves notoriously the heat transfer from the heating medium to the product. The minimum reel speed programmed as CCP in the controllers is determined according to the minimum value used for the HP tests. During the process cycle, the rotation of the drum will be started when come-up time starts to expose all the containers immediately to the hot process water, and it will stop at the end of cooling segment. The controller will use the rotation speed set point to control the rotational speed and an alarm will mark the process as deviated if the reel speed reaches a value lower than the minimum for the process. In such an event, the product is tagged as a process deviation and is placed on hold until a process authority reviews the corresponding records and determines its disposition.

#### 18.2.2.2 Preparing to Unload

After the sterilization process is finished and the residual pressure inside the retort is zero, the retort door starts the unloading sequence: the conveyor drive is engaged pneumatically, the clamping system is deactivated, and the retort door starts its opening process.

Once the door is opened, the retort controller sends a ready-to-unload signal to the AGV. The retort also sends another signal to the AGV to indicate if the process was successfully completed or if a process deviation occurred during the cycle.

### 18.2.3 Basket Tracking System

A very important component of the automated batch sterilizer system is the BTS. The BTS is a PC-based system that communicates with the retort controller, the loader/unloader, and the basket conveying system. Because baskets are automatically conveyed to and from the retorts, a sophisticated tracking system is employed to track the baskets in real time and assure that the entire product receives the proper process and that there are no retort bypasses.

At any given moment, the BTS knows the location of every basket in the system. Uniquely labeled RF tags are placed on opposing corners of each basket and antenna readers are located at all transfer points, including the retort, loader, unloader, and any holding buffers. The BTS records the time the first layer of containers is placed in the basket, the time the basket loading is complete, and the time the each basket enters and leaves the retort. It will also alarm if a maximum preprocess product hold time is about to be violated. Upon the completion of the retort cycle, the BTS checks with the retort controller to get a "process okay" before allowing the product to be unloaded from the retort. In the event of a process deviation, the deviated baskets are not allowed to go to the unloader but are flagged and moved to an isolated deviation buffer for further action by technical staff with the right privileges.

The whole system is controlled by two computers and a PLC system that works together to move, track, and process each basket loaded with product. The BTS works with the LOG-TEC™ Momentum process management system to ensure that all product-filled baskets received the minimum-scheduled thermal process. The BTS not only tracks all basket movements, but also prevents any unprocessed or under-processed basket from being unloaded at the unloader station.

The BTS has the following functions:

- Keep track of all basket movements and product/basket conditions from the time containers are loaded into baskets until the time-processed containers are unloaded from the baskets. All movements are clearly visualized on the computer screen and pertinent data are recorded.
- Document the process condition of each basket: BTS generates a database with time stamps and product/basket conditions during each product flow period.
- Prevent unprocessed product from being unloaded from the baskets.
- Automatically generate optimized transport commands to the basket conveyor system.

### 18.2.3.1 BTS Host Computer

The BTS host computer is connected with the PLCs, the LOG-TEC™ Momentum host and the transport system through Ethernet networks. The following items are visualized on the BTS host computer:

- Retort status information
- Loader/unloader and transport status information
- System warnings and operator prompts
- Transport command
- Manual operations

In addition to the above data, baskets are shown in real time on the computer monitor of the BTS host with a color code. The basket color code is dependent upon actual status: sterilized, nonsterilized, deviated, rejected, or empty basket. The color code allows the operator to easily visualize basket flow and status in a very clear way throughout the entire system. The data collected and recorded by the BTS for each process cycle are shown in Table 18.1.

### 18.2.3.2 Critical Control Points for Basket Tracking System

The CCPs in the operation, monitoring, and reporting functions of the BTS are listed below:

- Loading time
- Dwell time
- Basket ID
- Retort entry/exit time
- Retort number
- Retort cycle number

**TABLE 18.1**
**Description of Data Collected and Recorded by the BTS**

| Data Field | Description |
|---|---|
| Basket ID | Basket number |
| Batch ID | Batch ID number (identifies the unique batch to which a basket belongs) |
| Product info | Recipe number, product description, and container code |
| Retort | Retort where sterilization took place |
| Cycle | Cycle number from BTS |
| #Container | Number of containers in the basket |
| Status | Empty basket/not sterilized/sterilized—good/ sterilized—deviation |
| Load | Time when containers were loaded into a basket |
| In retort | Time when a basket entered the retort |
| Process | Time for start and end of process indicated by the LOG-TEC™ controller |
| Out retort | Time when a basket leaves the retort |
| Unload | Time when containers were unloaded from a basket |
| Fill time | Time when the first container was loaded into the first basket of a batch |
| | Basket is not sterilized yet but dwell-time violation is not imminent |
| | Basket is sterilized and no dwell-time violation occurred |
| Dwell time | Dwell-time violation in $x$ minutes (dwell-time violation is imminent) |
| | Dwell-time violation (the dwell time is exceeded) |
| Time stamp | Missing time stamp or time stamps in a wrong chronological order cause a basket to need an override (time-stamp violation) |

- Process status
- Recipe number, item number, and lot code
- Disposition of deviated product

#### *18.2.3.2.1 Loading Time*

A basket has a new life cycle as soon as it has been unloaded and it is empty. A basket does not have any time stamp when it arrives at the loader station (stacker). When the first layer is swept on, the life cycle starts. The operator is prompted to enter the time stamp of a container of the first layer of the first basket of each batch. If a sample is taken and the fill time is entered, every basket following that will receive the same fill time. The time stamp is used to detect dwell time violations.

##### *18.2.3.2.2 Dwell Time*

Dwell time is the maximum time that the product may be held since the container is filled with product until the beginning of the retort process. If dwell time is exceeded, it will still receive a full sterilization process, but the thermal process may not be adequate to obtain commercial sterility or the product quality may have decayed beyond acceptable limits. Therefore, if a batch is flagged either with a dwell-time violation or a process deviation, this batch will not be taken out of the retort automatically. Instead, the quality operations manager must designate the product as approved, put it on hold for further analysis, or reject the product and have it destroyed. An early warning alerts the operator that it is time to start the process even if the retort is not full. Dwell-time tracing takes place in the loader buffer, in the shuttle, and in the retort.

##### *18.2.3.2.3 Basket ID*

Each basket has a unique transponder (receiver–transmitter device) that allows the BTS to identify the location of each basket in the system. When a basket is introduced in the system, the transponder of that basket is read and the system builds a unique link to the basket. At a reader station, the BTS reads the basket number and passes the basket information to the PLC of the retort or the loader/unloader system. The PLC keeps track of the position of the baskets at every basket transfer from one location to another.

The BTS can operate in automatic or manual mode. In automatic mode, the system will transport unprocessed baskets from the loader area to a retort, and properly processed baskets from the retort to the unloader area. In manual mode, operators having security clearance can generate basket transportation; however, the system will not allow any transportation that would compromise product safety.

##### *18.2.3.2.4 Retort Entry/Exit Times*

It is the time when each specific basket enters or exits a retort. Baskets are transported from the loader to the retort and from the retort to the unloader or deviation buffer by an AGV, shuttle, or basket conveying system. Every time a basket enters or leaves the retort, the basket is identified using the RF tag reader system. Loading stops when either the retort is full or the operator stops loading due to dwell-time violation, change of product, or even no more product is expected from the filler, that is, end of production. The following steps prepare the retort for the process:

- Baskets are positioned underneath the clamp plates.
- Door is closed and locked.
- Baskets are clamped if there are no empty baskets inside the retort.
- Conveyor is disengaged to allow the drum to rotate.
- Then the retort PLC hands over control to the LOG-TEC™ Momentum controller.

The unloading sequence follows the steps in reverse order.

#### 18.2.3.3 Process Status and Disposition of Deviated Product

The baskets get a batch number upon the start of process and the event is recorded at the start of the come-up phase. The end of process event is recorded at the end of the drain phase. The PLC knows whether or not a deviation has occurred during the process through the data exchange with the LOG-TEC™ Momentum controller.

At the end of the process, the baskets change to different colors depending if any deviation or dwell-time violation occurred or not. Baskets that are sterilized without a deviation are automatically unloaded and are sent to the packaging area. Baskets with dwell-time violation need an "approve" override action from quality assurance (QA) personnel who will either approve the product or reprocess it. When a process deviation occurs and product needs to be held for inspection or incubation, then the "hold" override is used to send the product to the deviation buffer and put it on hold for evaluation. The "send to unloader" override must follow the "approve" or "hold" override.

Baskets that arrive at the unloader station (destacker) will be empty baskets, sterilized basket without deviation or dwell-time violation, or sterilized basket with a deviation or a dwell-time violation that were approved as good product or put on hold by QA staff. When the top layer is swept off, the start of unloading is recorded. In the case of a tray loader, the pallet number is only known at the start of the unload cycle. The bottom frame is taken away after identification. This is the end of the basket life cycle.

#### 18.2.3.4 Safety Measures against Unloading Unprocessed Product

To prevent unprocessed product from being unloaded, BTS verifies the residence time of a basket in the retort, by comparing time stamps of basket entering and leaving the retort. Also BTS obtains the "process OK" verification from the LOG-TEC™ process management system to confirm that the basket has been properly processed. Both conditions must be met or the concerned baskets will not be allowed to enter the unloader buffer. Also, all operator actions that could affect product flow or affect product data are recorded along with the operator ID and time stamp.

As a secondary check, the operator is able to monitor the process status of each basket on the BTS monitor screen. The color code for each basket that receives the proper sterilization process changes to green. Only the green-coded baskets are allowed to move into the unloading buffer. Furthermore, any action that shortens the process either causes a LTM deviation or a BTS time-stamp violation.

In summary, the BTS is responsible for tracking and documenting all basket movements. The BTS is continuously communicating with the LOG-TEC™ process management system and will not let a loaded basket to bypass a retort or to be unloaded if under-processed or if no process was applied. BTS records lot codes that are printed on the package for extra traceability. Each customer may have a different lot code structure that holds date code, plant code, line code, and extra identifier like product of the day.

As the BTS combined with the LOG-TEC™ Momentum track the movement and record the transport and process history of each basket throughout the entire system, the validated event logs constitute a detailed, accurate, and reliable record that helps the processor (or record reviewer at the processing plant) in determining if the basket released to the unloader is safe for consumption. In the event of a process deviation or a user override, the very same records highlight the event that may compromise the safety of the resulting product. The detailed event logs as well as the audit trail generated by the system are a powerful tool that reduce operator errors, and therefore, increases the process safety.

Although the LOG-TEC™ controller downloads the retort system configuration, tuning tables, and the assigned recipe configuration upon reboot, the controller is designed to operate independently in the event of a communication failure. If the communication fails in the middle of the process, the controller generates an alarm to warn the operator, but the normal operation of the controller is not affected. If the communication failure occurs while the controller was turned off, upon reboot, the controller identifies the failure and prompts the user to proceed with the most recent configuration that was successfully downloaded from the host computer.

When communication is restored, all the records that were not successfully downloaded are sent to the host computer. Only when the host computer acknowledges that it has successfully received a complete record group from a given cycle, the controller marks that specific group of records for deletion. The compact flash card inside the controller can keep the event logs for several weeks before running out of memory.

Automation of a sterilization system is complex and there are numerous subsystems of controls that must interact with each other in order to provide an efficient, trouble-free retort system. Some features that are often implemented in ABR systems are line controls interfacing between a loader/unloader and infeed and discharge lines, and interfacing between sterilization systems and utilities controls that control cooling, heating, heat recovery, and water treatment.

## 18.3 PROCESS OPTIMIZATION USING ADVANCED CONTROL SYSTEMS

Canned foods represent a large segment of the processed food industry in the United States. According to the 2005 U.S. Economic Census, over $20 billion of canned foods were produced in the United States. As discussed earlier, during thermal processing of canned foods, process temperature deviations are unavoidable and must be handled properly to ensure food safety. In current industrial practice, food processors stop the process and reprocess the affected containers or discard the product in case of process temperature deviations. These methods severely compromise production efficiency as well as product quality due to overprocessing. In most cases, the damage is so severe that the product must be discarded, which creates unnecessary waste.

Several control algorithms have been developed for automatic adjustment of processing time that guarantees the food safety in case of process temperature deviations. To assure food safety, canned foods must receive sufficient heat-time treatment. This means that when the heating temperature drops, the heating time must be extended to achieve the desired extent of heat treatment.

In retort operations, the food being sterilized must follow prescribed temperature profiles that are determined taking into account food safety and quality considerations. Although thermal processes are defined having microbial inactivation in mind, they produce degradation of valuable components in the food, like heat-labile vitamins, color, and flavor components that may affect their nutritional and organoleptic quality. Thus, the retort temperature profile selected for a process should be one that meets the required degree of sterilization whereas maximizing product final quality and nutrients retention. Calculations of process times providing food safety and maximum nutrition retention can be estimated if the microbial destruction and nutritional degradation kinetics along the temperature history in the product are known. These calculations can be performed by using appropriate models describing these kinetics and the transfer of heat through the product provided the external heating temperature is maintained constant during the process. However, temperature disturbances in the process are often unavoidable and whereas many disturbances are such that they can be handled by the retort temperature control system, there are others, like for example those produced by the interruption of steam supply, in which the temperature controllers may fail to maintain the set retort temperature profile resulting in products whose sterility could be seriously compromised. Currently, food safety is determined in terms of the $F_0$-value, which is defined as the equivalent isothermal, processing time of a hypothetical thermal process at a constant, reference temperature that produces the same effect, in terms of spore destruction as the actual thermal process (Stoforos et al., 1997). An equivalent $F_0$-value is used to assess the degradation of nutrients (Lund, 1975). The $F_0$-value as a way to assess microbial inactivation has been recently challenged (Corradini et al., 2005) as well as the use of first-order kinetics to describe nonlinear survival or nutrient degradation curves. Thus, these new developments should be incorporated in potential algorithms to perform online control of thermal processes.

In the past two decades, effort has focused in developing intelligent online control systems capable of rapid evaluation and online correction of process deviations that may occur in a thermal process. Most of the efforts, mainly concentrated on batch retorts (Teixeira and Tucker, 1997), are based on complex mathematical algorithms that automatically correct process temperature deviations by adjusting the scheduled processing time during the operation of batch retorts in order to get the targeted $F_0$-value (Datta et al., 1986; Simpson et al., 1993; Akterian, 1999; Kumar et al., 2001). These algorithms are able to calculate the temperature history on the coldest point of the product from the external heating conditions and to estimate the momentary process lethality ($F_0$-value). One of the problems faced is that the calculation algorithms should be fast to be implemented in an intelligent control system. Several approaches have been used to develop these models and they are described in the following section.

### 18.3.1 Modeling Heat Transfer Processes during Thermal Processing

As discussed, the basis for thermal process design, optimization, and control is strongly linked to fast algorithms capable of accurately estimating the product temperature profile in different locations of the product (specifically in the coldest point), as a function of time and the external heating/cooling conditions. Considerable efforts have been done for developing algorithms that provide (1) accuracy, (2) flexibility under different external conditions, (3) fast calculation times to allow for real-time process optimization and online control, and (4) generality in the application of the algorithms for different containers' shapes. The many approaches used to predict temperature profiles in different locations of the product are summarized below:

#### 18.3.1.1 Empirical Methods (Formula Methods)

They are based on empirical formulas derived from experimental heating and cooling curves, better known as the Ball's formula (Equation 18.1).

$$\log\frac{T_b - T(t)}{T_b - T_i} = -\frac{t}{f_h} + \log j_h \tag{18.1}$$

where

$T_b$ and $T_i$ are the external heating temperature and the product IT, respectively
$T(t)$ is the momentary product temperature in a specific location, generally the product coldest point

The term affected by the logarithmic function in Equation 18.1 is known as the unaccomplished product temperature, whereas $j_h$ and $f_h$ are defined as the heating lag factor and the heating rate factor, respectively. A similar relationship also applies during the cooling step provided the heating factors $j_h$ and $f_h$ are replaced by the cooling lag factor ($j_c$) and cooling rate factor ($f_c$). In order to use these equations, HP tests are required to obtain the HP parameters ($f_h, f_c, j_h$, and $j_c$), which through Equation 18.1 are used to estimate the product temperature profile under other retort external conditions. It should be noted that the linear relationship between the logarithm of the unaccomplished product temperature and time expressed by Equation 18.1 occurs only after relatively long times, whereas at short times the HP curves exhibit significant curvature. Therefore, the Ball's formula can only describe correctly the heating or cooling curve products for long times. It has been correctly claimed that Ball equations do not take into account the initial heating period due to the negligible accumulated lethality. However, lethality at the initial cooling period could be significant. Consequently, Ball used a hyperbolic function to describe the portion of initial cooling lag (Ball, 1923; Ball and Olson, 1957). Based on the original idea of Ball (1923), several approaches have been developed (Gillespy, 1951; Hayakawa, 1970, 1978; Stumbo, 1973; Pham, 1987, 1990; Larkin and Berry, 1991), which are usually referred to as the "formula" methods. The Ball method is widely used by the food canning industry. Other methods such as the Stumbo (Stumbo, 1973) and the Pham methods (Pham, 1987, 1990) are also used. Stumbo and Longley (1966) published tables that covered a wide range of $z$-values (8°F–200°F) and cooling lag factors $j_c$ (0.4–2.0).

Although the formula methods are used widely due to simplicity and fast calculations, they are not flexible and have several limitations: (1) all of the formula methods require substantial knowledge of their background and development before they can be used correctly and safely (Tucker, 1991), (2) formula methods are only applicable for conditions of constant heating and cooling temperatures, (3) the initial product temperature must be uniform, (4) food products must have homogeneous properties that are independent of temperature, and (5) they do not account for the contribution of overshooting (temperature rise at the geometric center of conduction heating products during cooling) of the sterilization value (Naveh et al., 1983a,b). These methods can be applied to a priori evaluate processing times that provide commercial sterility of the product minimizing degradation of nutrients. In that sense, a proper process control would maximize sterility whereas minimizing nutrient degradation provided corresponding kinetics is known.

#### 18.3.1.2 Numerical Methods

Due to the above mentioned limitations, it is necessary to develop numerical methods to solve heat transfer equations using more realistic conditions. Numerical methods include the finite difference and finite element methods (FEMs). For food containers with regular geometries, the finite differences method is preferred. However, if the container geometry is irregular, the FEM offers some advantages. There exist other approaches, for example, the apparent position numerical solution (APNS) and the artificial neural networks method that given their speed calculations could be implemented in advanced control systems.

##### *18.3.1.2.1 Finite Difference Method*

This is straightforward and simple to implement and was first applied to thermal sterilization of canned food processed in cylindrical containers by Teixeira et al. (1969). In subsequent work, the model was validated experimentally and it showed to predict accurately the temperature history at any product location of cylindrical food containers subjected to a number of different boundary conditions (Teixeira et al., 1975). Finite difference method has been applied for a limited number of irregular containers (Bhowmik and Tandon, 1987; Chau and Snyder, 1988; Pornchaloempong et al., 2003), although longer calculation times could be a serious drawback to implement these algorithms. NumeriCAL™ (JBT FoodTech proprietary software) is a finite difference model designed to optimize the thermal processing of foods in hermetically sealed containers of any shape. NumeriCAL™ is a thermal process calculation method that provides General Method accuracy with the ease of using Ball formula type heating and cooling factors, and it is used by most major food processors of canned food in the United States.

##### *18.3.1.2.2 Finite Element Method*

The FEM is able to handle more complex geometries without sacrificing calculation time at the expenses of more complicated computer programming. Applications of this method to estimate temperature profiles during thermal processes have been reviewed by Puri and Anantheswaran (1993). This method can be applied to several problems of practical importance for the food canning industry, which include (1) variable heating/cooling temperatures, (2) convection boundary conditions, (3) non uniform initial product temperature, (4) non isotropic or temperature-dependent thermal properties of product, and (5) headspace effects. Recently, the FEM has been used for heat transfer calculation of conduction-heated foods prcessed in cylindrical cans (Varga et al., 2000) and retortable pouches (Cristianini and Massaguer, 2002).

##### *18.3.1.2.3 Apparent Position Numerical Solution Method*

Although the APNS method is a semiempirical approach, it can be also considered a numerical method because is based on the numerical solution of the heat conduction equation for a sphere. This method, developed by Noronha et al. (1995), has found application for a number of complex geometries and its validity has been experimentally evaluated by Tucker et al. (1996) and Teixeira et al. (1999). One of the advantages of this method is that computation times decrease significantly due to the fact that the heat transfer problem is reduced to only one dimension (radial). In addition, although the model is applied to one of the pure heat conduction, it may perform well in predicting the temperature evolution at a single location in response to different boundary conditions (Teixeira et al., 1999). A limitation is that the model may not be amenable to process optimization because it can only estimate temperature evolution at a single location in the can.

##### *18.3.1.2.4 Artificial Neural Networks Method*

This method has been applied in several food processing areas (Bochereau et al., 1992; Parmer et al., 1997). Chen and Ramaswamy (2002) implemented the method for evaluation of a thermal process, while a genetic search algorithm was used to find the optimal process conditions for conduction heated foods, that is, to optimize variable retort temperatures whereas minimizing nutrient degradation. The study showed the feasibility of using the method along genetic algorithm searching techniques to model and optimizing variable retort temperature processing. This method, however, relies on commercial software so its implementation on retort control algorithms could not be so straightforward.

##### *18.3.1.2.5 Computational Fluid Dynamics Method*

An alternative method for solving problems in which convection is involved is the use of computational fluid mechanics (CFD). Although CFD has been applied in many different processing industries, its application to food processing has only been done in recent years (Scott and Richardson, 1997).

Advances in computing speed and memory capacity of modern computers make the CFD technique a powerful tool for potential application on process optimization and advanced control systems on retorts. A number of commercial software packages (e.g., FLUENT, FIDAP) could be applied to the processing of foods with temperature-dependent properties and non uniform conditions. However, one of the biggest limitations to implement this method on retort control algorithms is the potential long computation times and the interface between the commercial packages and the data recording systems.

### 18.3.2 Online Correction of Process Deviations in Batch Retorts

During thermal processes of canned foods, the most important aspect is the safety of the food. Safety of a food processed in a retort can be assured by strictly operating the retort under predetermined temperature profiles. When a process temperature deviation occurs, the prescheduled sterilization cannot be achieved unless an online compensation procedure is implemented; otherwise, the product must be discarded or reprocessed. Discarding the product is a costly and wasteful alternative. On the other hand, reprocessing is not a good choice either due to the extra associated cost or potential nutrient loss likely resulting from overprocessing. An alternative method that would eliminate these problems is to apply an appropriate process control algorithm to correct those process deviations.

Given the type of operation in batch retorts, the primary purpose of online retort control is to automatically vary process time to assure that the containers inside the retort receive a treatment that produce the required sterilization value at the cold spot, that is, the slowest heating point of the product. Thus, a control system for batch retorts, acting in the event of a temperature deviation, would consist in determining a new process time required to achieve sterility.

Many methods have been used for online control, but an effective approach is to monitor the cold-spot temperature profile by using a real-time data acquisition system and temperature sensors. The recorded temperature profile allows one to estimate the accumulated sterilization lethality by using standard methods described elsewhere in the literature. Heating process is ended once the target lethality is attained. This type of control system has proved to be effective and reliable. Examples found in the literature are Lappo and Povey (1986), Wojciechowski and Ryniecki (1989), Ryniecki and Jays (1993), and Kumar et al. (2001). However, it has been claimed that this approach may result impractical and cost-prohibitive for large highly automated cook room operations typical of modern food canning industries (Teixeira and Tucker, 1997).

Commercial retort control systems have used the table or correction factor method, which is based on calculation of process times for a range of different constant retort temperatures, that are stored in a table. In the case of a temperature deviation, a new process time is calculated from the data stored in those tables and assuming that the lowest temperature achieved remains for the rest of the project. Although this method does not require long times, the major disadvantage is that once the lower temperature is adopted, the method usually results in overprocessing, particularly because process deviations would quickly recover allowing the retort to resume the normal operating temperature. To minimize this disadvantage, Giannoni-Succar and Hayakawa (1982) determined correction factor that allows the process time to be extended just to the extent needed to compensate for the deviation. The fact that patterns of temperature deviations can be unlimited and there is no way to generate in advance all the correction factors is a serious drawback of this method.

The inconvenience of setting temperature sensors in cans prompted researchers to develop numerical methods able to estimate the temperature of the food in several locations, but particularly at the slowest heating point. These calculations, described in section 18.3.1.2, were accompanied with determinations of lethality using traditional methods. The approach can be easily implemented on a control algorithm by only measuring the retort temperature history. When a process deviation occurs, the numerical method is able to calculate a new process time by assuming that the momentary retort temperature does not change during the rest of the process. The process time is updated, based on the newest reading of the retort temperature. Teixeira and Manson (1982) were the first to use such a model for online correction of process deviations in batch retorts. Their method was improved by Datta et al. (1986). Other methods were also reported by Bown et al. (1986), Kelly and Richardson (1987), and Tucker and Clark (1989). It is important to note that all these control systems are limited to pure conduction-heated foods processed in finite cylindrical containers. This limitation was eliminated in subsequent work of a number of researchers (Akterian and Fikiin, 1994; Bichier et al., 1995; Noronha et al., 1995). A number of recent publications (Simpson et al.,

2006, 2007a,b,c) summarize the work carried out up to date. It also introduces models that include pure convection and purely agitated containers.

NumeriCAL™ (JBT FoodTech proprietary software) has been successfully applied for online correction of process deviations in batch retorts for food canning processes. NumeriCAL™ On-Line embedded into the LOG-TEC™ Momentum control algorithm can continuously and accurately predict the time/temperature history of the slowest heating point within a hermetically sealed product-filled container regardless its shape, while simultaneously monitoring the sterilizer temperature. The temperature and sterilization value at the slowest heating point is calculated in real time; any variation in temperature or pressure that could affect the target lethality is alarmed as a deviation by the controller and an alternate NumeriCAL™ process is calculated and applied. NumeriCAL™ On-Line retains all the benefits of previous proven systems, including automatic record keeping, security, process safety, and FDA and USDA acceptance. The LogTec Momentum with NumeriCAL™ On-Line provides the most reliable, advanced, and efficient control system available for batch retort control.

### 18.3.3 Future Trends on Online Correction of Process Deviations in Continuous Retorts

Continuous retorts, which are not discussed in this chapter, are finding more uses and applications in the food industry; thus, methods to correct temperature deviations in continuous retorts is an area that has not been investigated. The most widely used nonagitating and agitating continuous retorts are hydrostatic (e.g., JBT Hydrostat™), and rotary retorts (e.g., JBT Sterilmatic™) (Gavin and Weddig, 1995). The hydrostatic retort is a versatile food sterilizer that can be operated over a wide range of process temperatures and pressures. The speed of the conveyor moving the can set the process time, which is determined a priori by the processing authority to ensure that each container receives the targeted microbial log-reduction.

A rotary sterilizer is another type of continuous retort. It uses a rotating spiral reel to transport containers through a steam-pressurized processing shell. The rotating reel induces agitation to the food within the containers that improves heat transfer and minimizes process time. The use of a pure conductive heat transfer model for this situation therefore could be problematic if the food in the container is not a solid. The residence time in the processing shell depends on the rotational speed of the spiral reel.

Before discussing potential algorithms to correct temperature deviations in continuous retort processes, it is important to note the main differences between batch and continuous retorts. In batch retorts, all the containers have the same residence time so they are treated equally in the event of temperature deviations. Conversely, for continuous retort operations, at any given time, each container has a different residence time in the retort. Thus, it is important to have a record of the temperature profiles and accumulated lethalities for each container inside the retort at the time that a temperature deviation occurs. In that sense, an online correction algorithm must vary the container conveyor speed to adjust for a new residence time that satisfies sterility of the processed food. However, the development of a correction algorithm for this type of retorts has the additional complexity that each can inside the retort has a different residence time.

The accountability of temperature history and lethality for each can in the retort must continue during the entire period the can is inside the retort. Given the large number of cans processed in commercial retorts, these calculations require prohibitive long computation times, which make unviable their implementation in online correction algorithms to use in this type of retorts. Up to date, there are only two patents (Weng, 2003a,b) dealing with this issue. A recent publication by Chen et al. (2008) describes methods to optimally record container temperature profiles and lethal data so that computation times are minimized and online algorithms aimed to adjust for process temperature deviations can be implemented.

The effects of processing temperature deviations on the product temperature profiles can be determined either by temperature measurements on the product different locations or calculated by suitable heat transfer model or methods as the ones described in the previous section. As discussed, prediction of product temperatures by using accurate heat transfer models offers advantages if these calculations can be integrated to an online correction algorithm. Nevertheless, once a model is assumed, the product temperature profile can be estimated for given external conditions and the accumulated lethality for nonisothermal conditions can be calculated. As discussed, for these calculations, it is important to select a suitable model able to describe the thermal death kinetics of the pertinent microorganisms. Currently, lethality is calculated by assuming first-order kinetics and an integration approach proposed by Stumbo (1965). However, the use of first-order kinetics and the $F_0$-value have been recently challenged (Corradini et al., 2005) and thus it is worth to apply online corrections algorithms for situations that differ from presumed linear relationship, for example, first-order kinetics or Arrhenius-type dependence of microbiological parameters with temperature.

Two algorithms, the fixed point and the worst case, were recently developed by Chen et al. (2008). The algorithms focus on handling retort temperature data and prediction of product temperatures with a suitable heat transfer model; thus, they are independent of the approach used to estimate microbial inactivation during the process. However, estimation of survival rates and lethalities due to changes on product residence time varies with the assumed kinetics. Therefore, two microbial inactivation models, first-order kinetics, and Weibull model along with their respective lethality calculations approaches are considered in the description and evaluation of developed online control algorithms for continuous retorts (Campanella and Chen, 2008).

An aspect to be considered by implementing control algorithm for continuous retorts is that the product temperature profile for each can depends on its relative position within the retort. Thus, recording the temperature–time profile in each individual can of a continuous retort is highly time consuming and may require significant computer space to properly implement this information in an online control system. Suitable algorithms relying on the development of relationships between processing histories of can carriers inside the retort are described by Campanella and Chen (2008).

Regardless being batch or continuous food thermal sterilization, processes must meet required microbiological safety standards that consist in a stipulated reduction in the initial microbial count of target microbial spores at the slowest heating point of the food container. Thus, in order to accurately estimate the efficacy of a thermal process, for online reporting purposes, information of two aspects of the food and the process must be known. The first aspect is related to the microbiology of the system and primarily concerns with the inactivation kinetics of the target microorganisms. Linear and nonlinear models, such as the Weibull model, have been used to describe kinetics of spore inactivation; however, the food industry only uses first-order kinetics as a process validation criterion. New findings and reports (Van Boekel, 2002) show that the presence of nonlinear kinetics is more a rule than an exception. Thus, new methods are necessary to validate processes considering microorganisms whose inactivation kinetics significantly differs from the linear behavior. An attempt to show the differences between both approaches as applied to a proposed control algorithm have been discussed by Campanella and Chen (2008). With the advent of more powerful and faster computer systems and by using algorithms as those described in that work, it would be possible to perform fast calculations to be implemented in online correction methods.

The second characteristic intimately related to the efficacy of the thermal process concerns with the temperature–time history experienced by the product at the slowest heating point. Online correction methods for continuous retorts have not been reported except the two patents mentioned above (Weng, 2003a,b) and recent work by Chen et al. (2008) and Campanella and Chen (2008) so the industry has not had the opportunity of using these methods and yet validating online correction methods for continuous retorts. As discussed, heat transfer models can be used to predict temperature histories in the product. The challenge, however, with continuous retort lies in the difficulty of setting suitable measuring and recording temperature systems of practical feasibility. A suitable system for product temperature measurement and recording in continuous retorts would be a wireless temperature probe (DataTrace™, Ellab™, and ValProbe™). In the past few years, significant progress has been made in improving this technology.

### 18.3.4 Additional Regulatory Considerations

In spite of the advances in automation and modeling, why the food industry in the United States has not been able to file and run an *F*-value based process?

Industry researchers, scientists, and academics have developed the mathematical models to accurately and reliably predict the temperature and the inactivation kinetics of the target microorganisms. Those models have been validated by food scientists and process engineers to safely use them as optimization tools for the production process. Control engineers have developed systems capable of running the mathematical models in real time and regulatory agencies have been working with the industry and academic community in developing plans and publishing guidelines to validate the computer control systems.

Despite the know how listed above, there are more challenges: First, the U.S. FDA inspectors would have a difficult job in determining if a process is adequate if the time/temperature is not filed on the form 2541a and no time/temperature combination is displayed as a schedule process. Second, the computer program that runs the mathematical simulation would have to display the estimated *F*-value in real time and the validated program would become part of the filing process. These challenges add significant cost to the use and implementation of the sophisticated algorithms discussed in this chapter.

## REFERENCES

Akterian, S.G. 1999. On-line control strategy for compensating for arbitrary deviations in heating-medium temperature during batch thermal sterilization processes. *Journal of Food Engineering*, 39: 1–7.

Akterian, S.G. and Fikiin, K.A. 1994. Numerical simulation of unsteady heat conduction in arbitrary shaped canned foods during sterilization processes. *Journal of Food Engineering*, 21(3): 343–354.

Ball, C.O. 1923. *Thermal Process Time for Canned Food*, Bulletin No. 37, Vol. 7, Part 1. Research Council, Washington DC.

Ball, C.O. and Olson, F.C.W. 1957. *Sterilization in Food Technology*. McGraw-Hill, New York.

Bhowmik, S.R. and Tandon, S. 1987. A method for thermal process evaluation of conduction heated foods in retortable pouches. *Journal of Food Science*, 52: 201–209.

Bichier, J.G., Teixeira, A.A., Balaban, M.O., and Heyliger, T.L. 1995. Thermal process simulation of canned foods under mechanical agitation. *Journal of Food Process Engineering*, 18(1): 17–40.

Bochereau, L., Bourgine, P., and Palagos, B. 1992. A method for prediction by combining data analysis and neural networks: Application to prediction of apple quality using near infrared spectra. *Journal of Agricultural Engineering and Research*, 51: 207–216.

Bown, G., Nesaratnam, R., and Peralta-Rodriguez, R.D. 1986. Technical Memorandum No. 442, CFDRA. Chipping Campden Glos, GL55 6LD. *Computer Modeling for the Control of Sterilization Processes*.

Chau, K.N. and Snyder, G.V. 1988. Mathematical model for temperature distribution of thermally processed shrimp. *Transactions of the American Society of Agricultural Engineers*, 31(2): 608–612.

Chen, G., Campanella, O.H., Cowalan, C.M., and Haley, T.A. 2008. On-line correction of process temperature deviations in continuous retorts. *Journal of Food Engineering*, 2(84): 258–269.

Chen, C.R. and Ramaswamy, H.S. 2002. Modeling and optimization of variable retort temperature (VRT) thermal processing using coupled neural networks and genetic algorithms. *Journal of Food Engineering*, 53: 209–220.

Chen, G., Campanella, O.H., Corvalan, C.M., and Haley, T.A. 2008. On-line correction of process temperature deviations in continuous retorts. *Journal of Food Engineering*, 84: 258–269.

Corradini, M.G., Normand, M.D., and Peleg, M. 2005. Calculating the efficacy of heat sterilization processes. *Journal of Food Engineering*, 67: 59–69.

Cristianini, M. and Massaguer, P.R. 2002. Thermal process evaluation of retortable pouches filled with conduction heated foods. *Journal of Food Process Engineering*, 25: 395–405.

Datta, A.K., Teixeira, A.A., and Manson, J.E. 1986. Computer-based retort control logic for on-line correction of process deviations. *Journal of Food Science*, 51: 480–483, 507.

Gavin, A. and Weddig, L. 1995. *Canned Foods Principles of Thermal Process Control, Acidification and Container Closure Evaluation*, 6th edn. The Food Processors Institute, Washington DC.

Giannoni-Succar, E.B. and Hayakawa, K.I. 1982. Correction factor of deviant thermal processes applied to packaged heat conduction food. *Journal of Food Science*, 47(2): 642–646.

Gillespy, T.G. 1951. Estimation of sterilizing values of processes as applied to canned foods. I: Packs heating by conduction. *Journal of the Science of Food and Agriculture*, 2(3): 107–125.

Hayakawa, K.-I. 1970. Experimental formulas for accurate estimation of transient temperature of food and their application to thermal process evaluation. *Food Technology*, 24(12): 89.

Hayakawa, K.-I. 1978. A critical review of mathematical procedures for determining proper heat sterilization processes. *Food Technology*, 32(3): 59.

Kelly, P.T. and Richardson, P.S. 1987. Technical Memorandum No. 459, CFDRA. Chipping Campden Glos, GL55 6LD. *Computer Modeling for the Control of Sterilization Processes.*

Kumar, M.A., Ramesh, M.N., and Rao, S.N. 2001. Retrofitting of a vertical retort for on-line control of the sterilization process. *Journal of Food Engineering*, 47: 89–96.

Lappo, B.P. and Povey, M.J.W. 1986. Microprocessor control system for thermal sterilization operations. *Journal of Food Engineering*, 5: 31–53.

Larkin, J.W. and Berry, R.B. 1991. Estimating cooling process lethality for different cooling j values. *Journal of Food Science*, 56(4): 1063–1067.

Lund, D.B. 1975. Heat processing, *Physical Principles of Food Preservation*, Karel, M., Fennema, O.R., and Lund, D.B. (Eds.). Marcel Dekker, New York, pp. 31–91.

Naveh, D., Kopelman, I.J., Zechman, L., and Pflug, I.J. 1983a. Transient cooling of conduction heating products during sterilization: Temperature histories. *Journal of Food Processing and Preservation*, 7: 259–273.

Naveh, D., Pflug, I.J., and Kopelman, I.J. 1983b. Transient cooling of conduction heating products during sterilization: Sterilization values. *Journal of Food Processing and Preservation*, 7: 275–286.

Noronha, J., Hendrickx, M., Van Loey, A., and Tobback, P. 1995. New semi-empirical approach to handle time-variable boundary conditions during sterilization of non-conductive heating foods. *Journal of Food Engineering*, 24: 249–268.

Parmer, R.S., McClendon, R.W., Hoogenboom, G., Blanlenship, P.D., Cole, R.J., and Doner, J.W. 1997. Estimation of aflatoxin contamination in preharvest peanuts using neural networks. *Transactions of the ASAE*, 40(3): 809–813.

Payne, F.A. 1990. Food processing automation. *Proceedings of the 1990 Conference*, Lexington, KY, p. vi.

Pham, Q.T. 1987. Calculation of thermal process lethality for conduction-heated canned foods. *Journal of Food Science*, 52: 967–974.

Pham, Q.T. 1990. Lethality calculation for thermal processes with different heating and cooling rates. *International Journal of Food Science and Technology*, 25: 148–156.

Pornchaloempong, P., Balaban, M.O., and Chau, K.V., and Teixeira, A.A. 2003. Numerical simulation of conduction heating in conically shaped bodies. *Journal of Food Process Engineering*, 25: 539–555.

Puri, V.M. and Anantheswaran, R.C. 1993. The finite-element method in food processing: A review. *Journal of Food Engineering*, 19: 247–274.

Ryniecki, A. and Jayas, D.S. 1993. Automatic determination of model parameters for computer control of canned food sterilization. *Journal of Food Engineering*, 19(1): 75–94.

Scott, G. and Richardson, P. 1997. The application of computational fluid dynamics in the food industry. *Trends in Food Science and Technology*, 8: 119–124.

Simpson, R., Almonacid-Merino, S.F., and Torres, J.A. 1993. Mathematical models and logic for computer control of batch retorts: Conduction-heated foods. *Journal of Food Engineering*, 20: 283–295.

Simpson, R., Figueroa, I., and Teixeira, A. 2006. Optimum on-line correction of process deviations in batch retorts through simulations. *Food Control*, 117: 665–675.

Simpson, R., Teixeira, A., and Almonacid, S. 2007a. Advances with intelligent on-line retort control and automation in thermal processing of canned foods. *Food Control*, 18: 821–833.

Simpson, R., Figueroa, I., and Teixeira, A. 2007b. Simple, practical and efficient on-line correction of process deviations in batch retorts though simulations. *Food Control*, 18: 458–465.

Simpson, R., Figueroa, I., Llanos, D., and Teixeira, A. 2007c. Preliminary validation of on-line correction of process deviations without extending process time in batch retorting: Any low-acid canned foods. *Food Control*, 18: 983–987.

Stoforos, N.G., Noronha, J., Hendrickx, M., and Tobback, P. 1997. Inverse superposition for calculating food product temperature during in-container thermal processing. *Journal of Food Science*, 62(2): 219–224.

Stumbo, C.R. 1965. *Thermo-Bacteriology in Food Processing*, 1st edn. Academic Press, New York.

Stumbo, C.R. 1973. *Thermobacteriology in Food Processing*, 2nd edn. Academic Press, New York.

Stumbo. C.R. and Longley, R.E. 1966. New parameters for process calculation. *Food Technology*, 20: 341–345.

Teixeira, A.A. and Manson, J.E. 1982. Computer control of batch retort operations with on-line correction of process deviations. *Food Technology*, 36(4): 85–90.

Teixeira, A.A. and Tucker, G.S. 1997. Critical points for on-line computer simulation control of canned food sterilization processes, *Food Engineering 2000*, Fito, P., Ortega-Rodriguez, E., and Barbosa-Cánovas, G.V. (Eds.). Chapman & Hall, International Thomson Publishing, New York, Chapter 16, pp. 291–307.

Teixeira, A.A., Dixon, J.R., Zahradnik, J.W., and Zinsmeister, G.E. 1969. Computer optimization of nutrient retention in the thermal processing of conduction-heated foods. *Food Technology*, 23(6): 137–142.

Teixeira, A.A., Stumbo, C.R., and Zahradnik, J.W. 1975. Experimental evaluation of mathematical and computer models for thermal process evaluation. *Journal of Food Science*, 40: 653–655.

Teixeira, A.A., Balaban, M.O., Germer, S.P.M., Sadahira, M.S., Teixeira-Neto, R.O., and Vitali, A.A., 1999. Heat transfer model performance in simulation of process deviation. *Journal of Food Science*, 64(3): 488–493.

Tucker, G.S. 1991. Development and use of numerical techniques for improved thermal process calculations and control. *Food Control*, 15–19.

Tucker, G.S. and Clark, P. 1989. Technical Memorandum No. 529, CFDRA. Chipping Campden Glos, GL55 6LD. *Computer Modeling for the Control of Sterilization Processes.*

Tucker, G.S., Noronha, J.F., and Heydon, C.J. 1996. Experimental validation of mathematical procedures for the evaluation of thermal processes and process deviations during the sterilization of canned foods. *Transactions of the IChemE*, 74C: 140–147.

Van Boekel, M.A.J.S. 2002. On the use of the Weibull model to describe thermal inactivation of microbial vegetative cells. *International Journal of Food Microbiology*, 74: 139–159.

Varga, S., Oliveira, J.C., and Oliveira, F.A.R. 2000. Influence of variability of processing factors on the *F*-value distribution in batch retorts. *Journal of Food Engineering*, 44: 155–161.

Weng, Z. 2003a. Controller and method for administering and providing on-line handling of deviations in a hydrostatic sterilization process. U.S. Patent 6440361.

Weng, Z. 2003b. Controller and method for administering and providing on-line handling of deviations in a rotary sterilization process. U.S. Patent 6416711.

Wojciechowski, J. and Ryniecki, A. 1989. Computer control of sterilization of canned meat products. *Fleischwirtschaft*, 69(2): 268–270 (in German).

# Index

## C

## D



## E



## F

## I



## L

## M



## N



## O

## P



## R

## U



## W

Milton Keynes UK
Ingram Content Group UK Ltd.
UKHW021919071024
449327UK00022B/1682